THE WAR OF GUNS AND MATHEMATICS

Mathematical Practices and Communities in France and Its Western Allies around World War I

HISTORY OF MATHEMATICS ❖ VOLUME 42

THE WAR OF GUNS AND MATHEMATICS

Mathematical Practices and Communities in France and Its Western Allies around World War I

David Aubin
Catherine Goldstein
Editors

AMS
American Mathematical Society
Providence, Rhode Island

2010 *Mathematics Subject Classification.* Primary 01-02, 01A60, 65-03, 70-03, 97-03.

Cover art used with permission: Severini, Gino: Canon en action, 1915 © Rheinisches Bildarchiv Köln, rba_c013068.

For additional information and updates on this book, visit
www.ams.org/bookpages/hmath-42

Library of Congress Cataloging-in-Publication Data

The war of guns and mathematics : mathematical practices and communities in France and its western allies around World War I / David Aubin, Catherine Goldstein, editors.
pages cm. — (History of mathematics ; 42)
Includes bibliographical references and index.
ISBN 978-1-4704-1469-6 (alk. paper)
1. War and mathematics. 2. World War, 1914–1918—Science. I. Aubin, David, 1967– II. Goldstein, Catherine (Mathematician)

QA10.8.W37 2014
510.944′09041—dc23

2014012563

∞ The paper used in this book is acid-free and falls within the guidelines
established to ensure permanence and durability.
Visit the AMS home page at `http://www.ams.org/`

10 9 8 7 6 5 4 3 2 1   19 18 17 16 15 14

Contents

List of Figures

List of Tables

Contributors

Thomas Archibald is a professor in the Department of Mathematics, Simon Fraser University, Burnaby, Canada. His research interests revolve around mathematical analysis and its applications, mostly in the period 1850–1930. He also has ongoing interests in the relationship between mathematics, mathematicians, and the state.

David Aubin is a professor of the history of science at the Université Pierre et Marie Curie (UPMC) in Paris and a member of the Institut de mathématiques de Jussieu-Paris Rive Gauche. His work concerns interactions of mathematics with science and culture in the nineteenth and twentieth centuries. He is the co-editor with Patrice Bret of *Le Sabre et l'éprouvette: l'invention d'une science de guerre, 1914–1939* (14–18 Aujourd'hui, 2003) and with Charlotte Bigg and Otto Sibum of *The Heavens on Earth: Observatories and Astronomy in Nineteenth-Century Science and Culture* (Duke, 2010).

June Barrow-Green is a senior lecturer in the history of mathematics department at the Open University. She is editor of *Historia Mathematica* and a former president of the British Society for the History of Mathematics. She works on the history of nineteenth- and twentieth-century mathematics with a special interest in dynamical systems. Author of *Poincaré and the Three-Body Problem* (AMS/LMS, 1997), history editor of *The Princeton Companion to Mathematics* (Princeton, 2008), she is currently working on a book on G. D. Birkhoff.

Jean-Luc Chabert is a professor of mathematics at the Université de Picardie, Amiens, France. He has been associate editor of the *Revue d'histoire des mathématiques* from 1995 to 1997. He has co-authored *Integer-valued Polynomials*, Mathematical Surveys and Monographs (AMS, 1996) and edited *A History of Algorithms* (Springer, 1999).

Della Dumbaugh is a professor of mathematics at the University of Richmond in Richmond, Virginia, U.S.A. Her research focuses on the history of mathematics in the late nineteenth and early twentieth centuries, particularly as it relates to algebra and number theory. She has recently collaborated on *Emil Artin (1898–1962): Beiträge zu Leben, Werk und Persönlichkeit* edited by Karin Reich and Alexander Kreuzer (Dr. Erwin Rauner Verlag, 2007) and on the *History of Algebra in the 19th and 20th Centuries*, edited by Jeremy J. Gray and Karen Hunger Parshall (AMS-LMS, 2007).

Christian Gilain is a professor emeritus of mathematics at the Université Pierre et Marie Curie and a member of the Institut de mathématiques de Jussieu-Paris Rive Gauche. Gilain's work principally deals with the history of mathematical analysis in the eighteenth and nineteenth centuries. He has published studies about the theory of differential equations and its applications in the work of Poincaré, Cauchy, Condorcet, and D'Alembert. He is also interested in the debates on the place of mathematics in science and knowledge, especially through the lenses of the history of two major French institutions: the Academy of Sciences and the École polytechnique.

Hélène Gispert is a professor of the history of science at the University Paris-Sud (Orsay, France) and a member of the Groupe d'histoire et de diffusion des sciences d'Orsay (GHDSO, EA 1610). Her work deals with mathematics and mathematicians in France in the first half of the twentieth century. She has in particular studied the activity of the French mathematicians Émile Borel and Henri Lebesgue during WWI. She has published with Juliette Leloup a study of the features of the French mathematical activity during the interwar period with reference to the effects of WWI. She is also interested in the societal, structural, and conceptual changes that occurred in mathematics education over the twentieth century.

Catherine Goldstein is a director of research at the CNRS and a member of the Institut de mathématiques de Jussieu-Paris Rive Gauche. Her work aims at developing a social history of mathematical practices and results, with a special, but not exclusive, focus on the history of number theory. Among her recent publications are "La théorie des nombres en France dans l'entre-deux-guerres : de quelques effets de la première guerre mondiale," *Revue d'histoire des sciences* **62** (2009), pp. 143–176; and with D. Aubin and H. Gispert, "Les mathématiciens français dans la Grande Guerre," in *1914-1918. Identités troublées : les appartenances sociales et nationales à l'épreuve de la guerre*, ed. F. Bouloc, R. Cazals, A. Loez (Privat, 2011), pp. 183–197.

Deborah Kent is an assistant professor in the Department of Mathematics at Hillsdale College in Michigan, U.S.A. Her research activity in the history of mathematics focuses primarily on mathematics in the scientific, institutional, and social context of the nineteenth and twentieth centuries. Her current work addresses questions about the communication and circulation of mathematical knowledge and practices, especially related to the American mathematical research community in the early twentieth century.

Philippe Nabonnand is a professor of the history of mathematics at the University of Lorraine, in Nancy. Among his recent works are *Les uns et les autres... Biographies et prosopographies en histoire des sciences*, ed. with L. Rollet (Presses universitaires de Nancy, 2011) and *Justifier en mathématiques*, ed. with Dominique Flament (Éditions de la Maison des Sciences de l'Homme, 2011).

Pietro Nastasi, now retired, was a professor of the history of mathematics at the University of Palermo. He works on the history of Italian mathematics in the modern and contemporary periods. Among his recent publications are "Towards a Scientific and Personal Biography of Tullio Levi-Civita (1873–1941)," *Historia Mathematica* **32** (2005), pp. 203–236 (with R. Tazzioli); *Italian Mathematics Between the Two World Wars* (Birkhäuser, 2005) (with A. Guerraggio); "Problems of Method in Levi-Civita's Contributions to Hydrodynamics," *Revue d'histoire des mathématiques* **12** (2006), pp. 81–118 (with R. Tazzioli); *Roma 1908: il Congresso internazionale dei matematici* (Boringhieri, 2008) (with A. Guerraggio); and *L'Italia degli scienziati* (Bruno Mondadori, 2010) (with A. Guerraggio).

Laurent Rollet is an associate professor at the Institut National Polytechnique de Lorraine and a researcher at the Laboratoire d'histoire des sciences et de philosophie–Archives Henri Poincaré. He is the editor of Henri Poincaré's private and administrative correspondence and of a forthcoming biographical dictionary of lecturers at the Nancy Faculty of Sciences (1854–1918). He is currently in charge, at the Maison des sciences de l'homme de Lorraine, of a research program devoted to the history of the University of Nancy. He has edited *Henri Poincaré : Scientific Opportunism, An Anthology Compiled by Louis Rougier* (Birkhäuser, 2002) and co-edited with M.-J. Choffel-Mailfert at the Presses universitaires de Nancy, *Aux origines d'un pôle scientifique : faculté des sciences et écoles d'ingénieurs à Nancy du Second Empire aux années 1960*, as well as *Mémoire et culture matérielle de l'université : valorisation, sauvegarde, recherche.*

Rossana Tazzioli is a professor of the history of mathematics at the University of Lille 1 since 2008. She is interested in the history of geometry and mathematical physics between 1850 and 1950 and in the institutional history in Italy from Unification (1861) to the Second World War. Among her recent publications are "I matematici tra scienza, politica e istituzioni," in *Storia d'Italia: Annale 26, Scienza e cultura dell'Italia unita* (Einaudi, 2011); *Riemann: le géomètre de la nature* (Belin, 2010); "Göttingen et Berlin au XIXe siècle," in *La mathématique. I: Les lieux et les temps* (CNRS Editions, 2009), pp. 501–524; and *Mathematicians at War: Volterra and his French Colleagues in World War I* (Springer, 2009) (with L. Mazliak).

Foreword

This book is the outcome of an international research project launched by the team "Histoire des sciences mathématiques" of the Institut de mathématiques de Jussieu-Paris Rive Gauche (CNRS, Sorbonne universities, Université Pierre et Marie Curie, Université Paris, Diderot, Sorbonne Paris Cité) (UMR 7586) in 2005. In several workshops organized in Paris, contributors to this volume as well as many other historians, mathematicians, and military engineers explored various aspects of the life of mathematics and mathematicians around the time of World War I. An international workshop was held in the Centre international de rencontres mathématiques (CIRM), in Marseille–Luminy, in January 2007. The volume that we present now therefore is the result of extensive exchanges among participants of these workshops.

In the following, we have made a few editorial choices that should be explained. For wider understanding, quotations as well as institutional names have almost always been translated into English. We include original texts only when sources are taken from unpublished material that is not readily accessible to scholars. The names of most institutions have been translated, but original acronyms have usually been kept.

This project was continuously supported by our home institutions: the Centre national de la recherche scientifique (CNRS) and the Université Pierre et Marie Curie (UPMC), in particular via the Institut de mathématiques de Jussieu–Paris Rive Gauche and the Atelier Science – Histoire – Cité (ASHiC). The work that led to this book has been supported by a British Academy–CNRS PICS program between the team "Histoire des sciences mathématiques" at the Institut de mathématiques de Jussieu–Paris Rive Gauche and the Center for the History of Mathematical Sciences of the Open University (Milton Keynes), "Constructing modernities: mathematical sciences, mathematicians and the First World War" (2005–2006), and by the young researchers' program "De Humboldt à Gaïa," of the Agence nationale de la Recherche (France).

Many people have helped us along the way. We express special thanks to our colleagues Liliane Beaulieu, Charlotte Bigg, Patrice Bret, Alain Carrière, Moritz Epple, Claudine Fontanon, Sébastien Gauthier, Jean-Marc Ginoux, Jeremy Gray, Pierre Lamandé, Juliette Leloup, Pierre Mounier-Kuhn, Jim Ritter, Norbert Schappacher, Martina Schiavon, Arne Schirrmacher, Florian Schmalz, Reinhard Siegmund-Schultze, Pavel Šišma, Suman Seth, Dominique Tournès, and Annette Vogt.

This book is based on extensive archival research. We are happy to acknowledge the help of several institutions that let us use archival material, manuscripts, and illustrations: we cite in France, the archives of the Académie des Sciences in Paris and especially Florence Greffe, the French National Archives, the Service historique de la défense nationale, the archives of the French Foreign Ministry, the Bibliothèque de la Sorbonne, the Bibliothèque de l'Institut, the Bibliothèque Mathématiques–Informatique–Recherche (CNRS–UPMC–Paris-Diderot) and especially Cédric Dameron, the archives of the École polytechnique, the archives of the University of Nancy and the city of Nancy, the archives of the University of Strasbourg, the archives départementales of Pas–de–Calais, of Meurthe–et–Moselle, and of Bas–Rhin; in the U.K., the Cambridge University Archives and Saint John's College, the archives of Oxford University, the National Archives, the Royal Society Archives, and the archives of the London Mathematical Society; in Italy, the archives of the Istituto per le Applicazioni del Calcolo and the archives of the Accademia nazionale dei Lincei; in the U.S., the archives of Harvard University, of the Carnegie Institution, of the Massachussets Historical Society, of Rice University, and the Library of Congress; and in Sweden, the archives of the Institute Mittag–Leffler.

The cover of this book contains a painting by the Futurist painter Gino Severini, titled *Canon en action*, from 1915. In this painting modern art intermingles mathematical symbolism with military themes. This painting comes from the Museums of Cologne Collections, whom we would like to thank for giving us permission to use this work of art.

Placing World War I in the History of Mathematics

David Aubin and Catherine Goldstein

> *Les vrais adversaires, dans la guerre d'aujourd'hui, ce sont les professeurs de mathématiques à leur table, les physiciens et les chimistes dans leur laboratoire. Guerre à distance, guerre d'industrie.*
> — Jeanne Alexandre, February 6, 1916 [Weis 2005, p. 83].

"Our true adversaries in today's war are mathematics professors at their tables, physicists and chemists in their laboratories. War at a distance, war of industry." Jeanne Alexandre was the sister of the sociologist Maurice Halbwachs who took part at the cabinet level in the early mobilization of the French industry for war. By 1916, she had become a staunch pacifist. In her eyes, the mathematician, and more generally the scientist, stood as the very symbol of what she hated the most about the Great War, the inhuman, destructive nature of fighting at a distance, as well as the tragic alliance of science, industry, and warfare.

One year later, on the Flemish front, British Private Edgar Norman Gladden concurred with Alexandre in attributing a special place to mathematics in this war. At barely twenty-one, he was already a veteran of the Somme and Ypres when in June 1917 he was ordered to take a small ridge previously held by German troops. Reaching a point brutally sapped and shelled, he commented:

> The crater, which I expected to see as an immense jagged hole in the ground, was actually a large flat-bottomed depression like a frying-pan, clear and clean from debris except at the further edge, where vestiges of one of the enemy's trenches showed through its side. The poor devils caught in that terrible cataclysm had no chance.

"Yet what chance was there," Private Gladden went on, "for anyone in that war of guns and mathematics?" [Gladden 1930, p. 121].

David Aubin's address is Université Pierre et Marie Curie, Institut de mathématiques de Jussieu-Paris Rive Gauche, Case 247, 4 place Jussieu, F-75252 Paris, France. His email address is david.aubin@upmc.fr.

Catherine Goldstein's address is Institut de mathématiques de Jussieu-Paris Rive Gauche, Case 247, 4 place Jussieu, F-75252 Paris, France. Her email address is catherine.goldstein@imj-prg.rf.

While heavy artillery posed a set of problems about the location of targets and the computation of trajectories for which mathematical expert knowledge proved crucial, countless technological and military innovations which came to light during WWI also relied on precise mathematical analyses. Keeping an airplane in the air or bringing it down required the solution of several mathematical problems.[1] The development and improvement of technologies such as sonar or wireless telegraphy, as well as the application of cartography or meteorology to daily military usage, required mathematical skills that were perhaps less innovative, but nevertheless rare among the draftees. Soon, mused a reporter tongue-in-cheek, "armies will be commanded by mathematicians whose essential function, in times of war, will precisely be to busy themselves with mathematics" [D 1915, p. 9]. While this prediction was never realized, the French mathematician Paul Painlevé, an aeronautics specialist, did become, in the course of the war, Minister of Inventions, Minister of War, and, briefly, from September to November 1917, head of state. It was in this capacity that he selected Marshall Ferdinand Foch to become chief of the general staff. On November 11, 1918, Foch was admitted to the Academy of Sciences and Painlevé used a string of mathematical metaphors to praise Foch's "logical power." Foch's final assault against the Germans was compared to "a sequence of theorems" and evoked in Painlevé "the old Platonic axiom: 'Divine things are accomplished by geometry'."[2]

Rhetoric aside, the realization that war and mathematics had become entangled was by no means restricted to Allied nations. The first volume in a series of "German booklets for the front and the nation" published by Teubner in Leipzig dealt with *Mathematics in War* (Figure 1). In this booklet, of which a thousand copies were sent free of charge to the front, Dr. Paul Riebesell, a high school teacher in Hamburg, explained:

> Without mathematics, physics would not be in position to state countless laws, it would not be possible for technology to compute and plan its devices. Thus can this war, which has often been called technological, also rightly be called a mathematical war [Riebesell 1916, p. 1].

From high command to the infantrymen, soldiers depended on mathematics often without being aware of it. But make no mistake, Riebesell claimed: "wherever you look in battle, mathematics rules" [Riebesell 1916, p. 2]. Focusing on ballistics and orienteering, this booklet offered a convincing demonstration of the value of basic mathematical thinking in modern warfare.

To speak of the First World War as a "scientific war" [Wells 1915] has become common. But the conflict is mostly associated with chemistry, aerodynamics, and mechanical engineering rather than mathematics. The truth is that adhesion to war aims was so widespread among contemporaries that almost every corner of the scientific, and indeed cultural, landscape was engulfed by what became known as a "total war" [Daudet 1918]. Examples just mentioned bring to fore the feeling that mathematics and mathematicians were also considered major players in what

[1] This is recalled by June Barrow-Green in her contribution to this volume.

[2] See [Painlevé 1918]. On Painlevé's role during WWI, see [Fontanon & Franck 2006, Anizan 2006, Anizan 2012].

FIGURE 1. "Mathematics in War", cover [Riebesell 1916].

George Greenhill also, and before Riebesell, described as a "mathematical war."[3] Still, the historian of mathematics is confronted with some paradoxes.

First, whereas the production of war materials and their use in the field was widely perceived as requiring a certain degree of mathematical aptitude, it was much less obvious whether mathematicians could or should play a special role in the war. The mathematicians' own experience seemed sometimes at odds with the public perception. Eager to contribute to the war effort in the spring of 1917, Oswald

[3]See June Barrow-Green's contribution to this volume.

Veblen expressed the difficulty he had in putting his skills at manipulating abstract notions to the service of his nation. As late as May 19, 1917, the young mathematician from Princeton University confided to George David Birkhoff: "Thus far I have failed to find any place in which mathematical work is of use in connection with the war. If you run across anything of the sort I wish you would let me know."[4]

A second paradox arises from the way in which mathematicians' collective organization shaped their personal responses to their war experiences. By the outbreak of WWI, mathematicians in most countries had acquired a sense of community. National societies had been established in the later part of the nineteenth century and a professional identity structured around definite research agendas and educational missions had emerged. National differences were of course discussed and underscored: looking for the best experts to synthesize every field of the mathematical sciences in his *Encyklopädie der mathematischen Wissenschaften*, Felix Klein specifically sought out Italian writers for chapters on algebraic geometry and British ones for mechanics. The mathematical chronicler in the *Revue du mois* (Émile Borel most probably) wrote that "there was no reason to regret the fact the young French mathematical school leaves aside these number-theoretical studies favored in Germany" [Anonymous 1906, p. 510]. On the basis of a strong national structure of their communities, mathematicians had however started to reinvent the internationalism of their trade. Some mathematicians among the most prominent emphasized the universality of mathematics. Academic voyages had become frequent, at least in Europe; national societies and academies admitted foreign members; several journals attracted an international pool of writers; and International Congresses of Mathematicians had taken place roughly every four years since 1897, in Zurich, Paris, Heidelberg, Rome, and Cambridge. In the prewar years, an International Commission for Mathematical Instruction (ICMI) was especially active in promoting and reforming the teaching of mathematics at all levels of the school and university curricula.[5]

To assess the effect of war on these men and communities therefore is no simple task. When knowledge, as is the case for mathematics, appears highly abstract and remote from mundane reality, even an event as dramatic in its social and political effects as World War I can vanish from sight. For some contemporaries, to "cry for patriotic mathematics" seemed like an all-too-obvious contradiction in terms [Bell 1915, p. 6]. Because of this perception, the First World War had by and large been neglected from the written history of twentieth-century mathematics. In an article about the history of mathematical logic, just to take one example, the period 1915–1930 appears under the heading "The Optimistic Times" [Guillaume 1994, p. 214]. Yet, nearly all mathematicians discussed in the volume from which this article is taken [Pier 1994] had been through life-changing experiences: some had lost sons and students; some had been severely wounded in bloody battles; many had spent several years buried in the mud of the trenches; other had been called to contribute to the war effort in various capacities as truck drivers, translators, computers, mathematical instructors, scientific advisors to general staffs and governments, or members of these governments (Figure 2).

[4][Birkhoff Papers, HUG 4213.2, Box 3]. For some discussion of the war context in Birkhoff's work, see [Aubin 2005]. The American mathematicians' mobilization is discussed in Thomas Archibald, Della Dumbaugh, and Deborah Kent's contribution to this volume.

[5]On these issues the literature is now vast; for international overviews, see [Mehrtens 1990, Goldstein et al. 1996, Lehto 1998, Parshall & Rice 2002, Coray et al. 2003].

FIGURE 2. Paul Painlevé's cabinet as Minister of War, 1917. Gabriel Hanotaux, *Histoire illustrée de la guerre de 1914*, vol. 15 (Bordeaux: Gouhiounou, 1917), p. 201. Repr. from University of Wisconsin Digital Collection.

More inclined to pay attention to the social and institutional context in which science was pursued, some historians of science have clearly identified the Great War as a key episode in the development of powerful scientific communities.[6] With the hindsight of World War II and of the Cold War, and in a context of countercultural denunciations of the role of the military in the sciences, these historians documented the ways in which various forms of scientific mobilization, hastily put together between 1914 and 1918, formed the matrix for a future, deeper reorganization and growth of a scientific complex that was to be placed at the service of the state and of the military. Further studies along those lines have been useful in underscoring the privileged position scientists then carved for themselves as the primary interlocutors of general staffs for all matters regarding inventions and technological innovations.[7] WWI thus appeared as both the root and the imperfect prefiguration of forms of big science to become dominant only in the 1940s and 1950s.

From this historiography, mathematics was generally absent, although mathematicians appeared here and there sometimes playing decisive roles as science diplomats [Schroeder-Gudehus 1978, Dauben 1980]. Only in the 1990s did historians of mathematics start to pay attention to WWI in their studies. In his history of the International Mathematical Union (IMU), Olli Lehto argued for instance that WWI was instrumental in making politics enter international cooperation in science [Lehto 1998].

[6]See, e.g., [Kevles 1995], first published in 1978 but announced by several papers published separately; see also [Forman 1974, MacLeod & Andrews 1971, Hughes 1977, Weart 1979, Varcoe 1970]. For a review of this type of work, see [Hartcup 1988].

[7]From a vast literature, we cite [Pattison 1983, Roussel 1989, Johnson 1990, Hull 1991, Lepick 1998, Aubin & Bret 2003, MacLeod & Johnson 2006, Galvez-Behar 2008].

In emphasizing institutional innovations, historians were echoing assessments widespread among contemporary scientists who saw in the Great War a series of "lessons" to be drawn for future mobilizations, but without much impact on the content of science. The American physicist George K. Burgess was typical in writing in 1919 that, while hundreds if not thousands of new applications of known principles were due to war work, one would be hard pressed to name even two or three new principles developed due to the war [Burgess 1919, p. 98]. Historians of mathematics tended to agree:

> The disciplinary center of mathematics remains autonomous and "pure", untainted by ends to which it would just be a means.... At the same time, however, the web of practices and specialties connecting it to the military ... was made more dense and more stable [Mehrtens 1996a, p. 95].

Dislodged from their usual workplaces and puzzling over problems far afield from their previous occupations, mathematicians often felt that the war had been nothing but a parenthesis as far as normal science was concerned. Yet, contemporaries were divided about the lessons one was to draw from the experiences of WWI [Forman 1973]. While some scientists bracketed their war experience ("the war of course was a big hole," later said the applied mathematician Richard Courant [AIP 1962]), others considered that this was the dawn of a new age, the "scientific and industrial age" [Moureu 1920, p. 376], or a "New World of Science," as an American book proclaimed in its title [Yerkes 1920]. Many of those who had taken part in the formidable technoscientific or industrial mobilization (scientists, engineers, inventors, industrialists, or military personnel) marveled at the number of great realizations made possible during the war. The war confirmed the scientific nature of the modern world and the need to infuse its values and techniques throughout society. These views were not mutually exclusive: Borel, who played major roles in war mobilization and postwar reorganization of science in France, wrote in support of his candidacy to the Academy of Sciences after WWI that, having published no new mathematical paper, he had nothing to add to the report he had written in 1914 [Gispert & Leloup 2009, pp. 47–48]. War then can be both a break in what one considered to be one's properly scientific activity and, at the same time, a means allowing the emergence of new social structures for science policy. An authoritative history of mathematics in the U.S. thus argued that war was a "furlough" away from pure research for most American mathematicians after which they, by and large, had "little inclination" to pursue applications [Parshall & Rowe 1994, p. 444].

Agreeing with the view that the relation between mathematics and the military was necessarily "mediated" by technology and engineering [Mehrtens 1996a, p. 87], a groundbreaking survey of war-related mathematics has attempted to list the "social and epistemic environment for applications of mathematics" in various countries [Siegmund-Schultze 2003]. Emphasizing the need for a long-term history of industrial mathematics, Siegmund-Schultze underscored "the short distance between basic academic research and applied work for the military" and focused on the mathematics explicitly used by the military (mostly during WWII). His study however gives very few clues about larger-scale changes in mathematics that could be ascribed to the war after WWI. Traumas and ruptures caused by WWI are thus restricted, in the literature, to mathematicians' personal lives and to the institutional settings for mathematics. War seemed to leave little imprint, if at all, on the development of mathematics either as discipline or as body of knowledge.

1. Two Theses about the Effect of War on Mathematical Knowledge

There are however two historiographical theses that actually argue that the First World War had a major effect on mathematical knowledge. The "Forman thesis" famously suggested that German mathematicians and physicists accommodated to the irrational mood of Weimar culture and gave up causality and determinism as major demands for scientific explanations, thereby paving the way for the spectacular development of quantum mechanics [Forman 1971]. Although this is not an aspect that is usually put forward, the Forman thesis can be read, as we shall see, as arguing that WWI was instrumental in ushering in modern theoretical physics and mathematics. The second thesis—the "Bourbaki thesis," as we shall dub it—was formulated a few years before Forman's not by historians of science but by mathematicians looking back at their own past. It has since been widely taken up and become a standard trope in the historiography.[8] In this thesis, contrary to Forman's, the ascendency of what was perceived as antimodern trends in postwar French mathematics is also seen as a direct consequence of war.

1.1. The Bourbaki Thesis. The first promoters of the Bourbaki thesis are to be found among the founders of the group set up in 1934–1935 in order to introduce modern elements inspired by David Hilbert's Göttingen school into the French university curriculum.[9] Put briefly, the Bourbakis claimed that the bloodbath of the Great War essentially wiped out an emerging generation of mathematicians in France, later leaving those who were trained in the interwar period alone in front of professors who held outdated views about their discipline. In 1969, Jean Dieudonné explained:

> In the great conflict of 1914–18 the German and French governments did not see things in the same way where science was concerned. The Germans put their scholars to scientific work, to raise the potential of the Army by their discoveries and by the improvement of inventions or processes, which in turn served to augment German fighting power. The French, at least at the beginning of the war and for a year or two, felt that everybody should go to the front; so the young scientists, like the rest of the French, did their duty at the front line. This showed a spirit of democracy and patriotism that we can only respect, but the result was a dreadful hecatomb of young French scientists. When we open the war-time directory of the École Normale, we find enormous gaps which signify that two-thirds of the ranks were mowed down by the war. *This situation had unfortunate repercussions for French mathematics* [Dieudonné 1970, pp. 134–135, our emphasis].

[8]See [Andler 1994], [Houzel 2004, p. 56], [Berger 2005, p. 151], [Audin 2009, pp. 22–27], [Audin 2010, p. 1300].

[9]The Bourbaki literature is growing fast: see, among others, [Beaulieu 1993, Beaulieu 1994, Aubin 1997, Houzel 2004, Mashaal 2006].

FIGURE 3. "To Our Dead." Paul Landowski, war memorial to the alumni of the École normale supérieure, Paris, 1923. For more about the memorial, see [Hottin 2000]. Photo by David Aubin.

Twenty-five years later, André Weil, who had also partaken in the foundation of Bourbaki, concurred in these terms:

> Already when I was at the School [the École normale supérieure, hereafter ENS], I had been deeply struck by the damage wreaked upon mathematics in France by the 1914–18 war. This war had created a vacuum that my own and subsequent generations were hard pressed to fill. In 1914, the Germans had wisely sought to spare the cream of their young scientific elite and, to a large extent, these people had been sheltered. In France, a misguided notion of equality in the face of sacrifice—no doubt praiseworthy in intent—

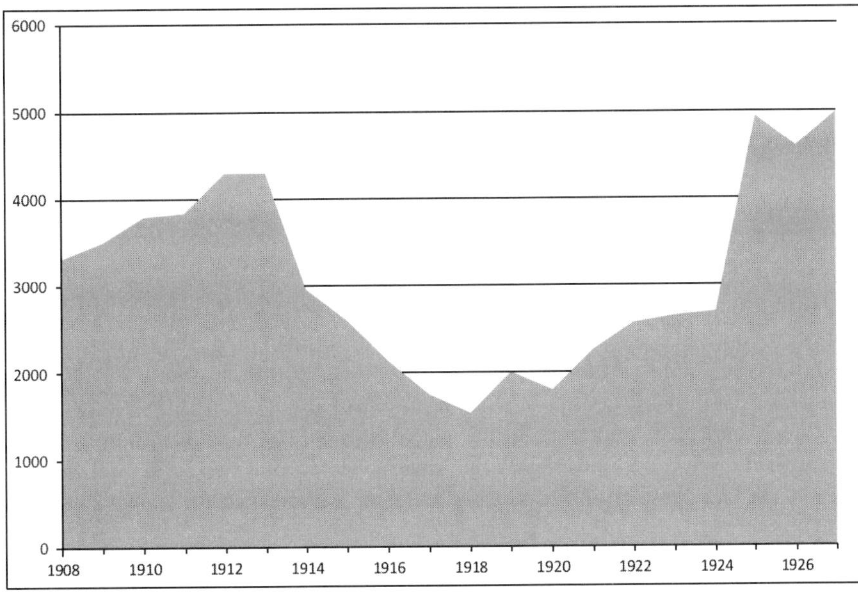

FIGURE 4. Total number of items reviewed in the *Jahrbuch über die Fortschritte der Mathematik* by year of publication, 1908–1927.

had led to the opposite policy, whose disastrous consequences can be read, for example, on the monument to the dead of the École normale [Weil 1991, p. 132] (see Figure 3).

In France, since the end of the nineteenth century, most of those who were able to pursue a professional research career in mathematics had studied at the ENS [Gispert 1991]. The dreadful rate of casualty among its students was naturally interpreted as especially tragic for the future of learning. From the class entering in 1910, more than 6 out of 10 science graduates never came back from the front. Fatal casualties in the following two cohorts were also above 50% of total enrollment in the science section. In speeches made in January 1915, ENS officials noted that out of the 195 mobilized students, at least 34 were confirmed dead, 15 had disappeared, 21 were taken prisoners, 64 wounded: only 54 or 55 remained unharmed![10] It was against this terrible backdrop that the Bourbaki thesis emerged.

That war was a tremendous disruption for mathematical research is obvious. To make this apparent, let us consider how the worldwide mathematical production drastically went down in those years. Figure 4 shows a clear and marked drop in the number of items reviewed in the *Jahrbuch über die Fortschritte der Mathematik* (JFM) already in 1914, down by 31% compared to the prewar peak in 1913, when close to 4,300 papers and books were reviewed. The reviewing activity stayed below the level of 1913 until 1925, reaching a low point in 1918 when the number of items reviewed barely reached 36% of the prewar value. Assuming a constant progression

[10]This issue is further discussed in the contribution on Paris mathematicians by David Aubin, Hélène Gispert, and Catherine Goldstein in this volume. For a comparative study of casualty at the ENS in WWI, see [Aubin 2012].

in the number of items between 1913 and 1925, we could say that more than 25,000 articles or books are, in some sense, "missing."[11]

Quantitative comparisons cannot recapture the contrast felt by young mathematicians between what they saw as the dynamic modernity of Göttingen and the depressing atmosphere of postwar mathematical Paris. Future members of Bourbaki indeed overlapped almost exactly with the students who had the opportunity to travel to Göttingen at the end of the 1920s—often sent with the Rockefeller Foundation's support by the very members of the older generation that was nearly nonexistent according to the Bourbaki thesis [Siegmund-Schultze 2001]. But when they sought to *explain* their impression by the conviction that the slaughter was proportionally greater on the Western side of the trenches, Dieudonné and Weil produced an account whose plausibility will be undermined by much of what is found in this book.[12] They no doubt encountered a community where the memories of WWI took on a wholly different shape than at the ENS, but were they right to assume that German mathematicians were better protected than their French counterparts? As noticed above, differences between researches pursued in France and Germany preexisted the war: the real issue is how or whether the war and the postwar reconstruction changed these research trends. Dieudonné's acknowledgement in 1988 that he had completely ignored the work of topologist Louis Antoine—a veteran of the First World War—even though they both taught at the University of Rennes from 1933 to 1937 is a telling example of the divergence in mathematical interests that seemed to occur between the war generation and the Bourbaki mathematicians. "Retrospectively, I regret not having exchanged with Antoine mathematical ideas which might have been useful to us both; but the obstacle to this exchange was my deep ignorance, at this time, of the topics he had dealt with in his work" [Dugac 1995, p. 11].

Retaining a single aspect of the war period—the tragic loss of a significant number of mathematicians—the Bourbaki thesis has therefore hampered a serious consideration of wartime events and experiences as crucial elements for shaping postwar mathematical cultures. War, as it was lived by mathematicians on the front or in the rear, was put at a distance. To better assess the outlines of the mathematical values held by postwar practitioners, we claim that it may be necessary to reconsider unresolved issues. How was WWI a disruption, if at all, in the mathematical activities of its contemporaries? How did the surviving mathematicians, in France, but also elsewhere, reconnect—if they ever did—with their prewar activities? Did mathematicians who changed occupations due to the war consider the conflict a mere interlude in their professional lives or as a fundamental break?

1.2. The Forman Thesis and the Bracketing of WWI. We believe it is enlightening to consider the way in which the Forman thesis has drawn attention to major postwar disruptions in theoretical physics and mathematics in the context of "Weimar culture." Forman is indeed one of the few historians to suggest that the

[11]About the history of mathematical reviewing and the *JFM*, see [Siegmund-Schultze 1994]. Note that economic hardship in Germany after the war probably reduced the reviewing activity of the *JFM*.

[12]Recent studies about the interwar mathematical community in France have also contributed greatly to the revision of accepted wisdom about postwar French mathematics; see Juliette Leloup's doctoral thesis [Leloup 2009] and a special issue of the *Revue d'histoire des sciences* [Beaulieu 2009], and especially [Goldstein 2009, Gispert & Leloup 2009].

period of WWI might have had an appreciable effect not just on the organization of science, but also on some of its basic principles—and what effect: acausality, quantum mechanics, and so on. In Forman's article war appears as the last glitter of a golden age, now irredeemably lost. This was a time when scientists could share sentiments with "the rest of the German public" [Forman 1971, p. 8]. Indeed, they might have been even more optimistic than most and felt "*self*-confidence and *self*-satisfaction due to their contributions to Germany's military success and to their anticipation of a postwar political environment highly favorable to the prosperity and progress of their disciplines" ([Forman 1971, p. 8]; original italics). In the last months of the war, Felix Klein mapped out a glorious and harmonious future where government money would seal the alliance struck between the university and the military [Forman 1971, p. 9].

The cultural movements discussed by Forman (*Lebensphilosophie*, neoromanticism, antipositivism, belief in acausality) are "intellectual currents, whose sources lay in the prewar period, but which welled up immediately following Germany's defeat and revolution, continued to dominate the intellectual milieu in the mid-1920's as in the first years of the Weimar Republic" [Forman 1971, p. 18]. Only in the "radically rearranged scale of values ascendant in the aftermath of Germany's defeat" [Forman 1971, p. 6], could those undercurrents forcefully emerge. Buried by the discipline, hopes, and requirements of war, a persisting "subterranean acausality current" [Forman 1971, p. 67, footnote] was allowed to spring up and become mainstream in German culture after the Armistice. Forman most clearly expressed the reason for which prewar marginal beliefs could take center stage in the postwar period when he approvingly quoted Arnold Sommerfeld: "The belief in a rational world order was shaken by the way the war ended and the peace dictat" [Forman 1971, p. 13]. The irrationality of defeat, more than the absurdity of trench warfare, induced people, including prominent scientists, to relinquish earlier faith in rationality and determinism and to embrace acausality.[13]

In short, while the Bourbaki thesis is an account of social hindrances standing in the way of scientific progress, the Forman thesis claimed that social anxieties gave rise to new scientific principles. Paradoxically, as opposed to the Forman thesis, the Bourbakis saw the war as a cause for tightening the clutch of outmoded representations of science, blocking the way to modernity. But since we have hinted at the fact that it is not war itself but defeat that actually is central in Forman's argument, the paradox may only be apparent: it seems to make sense that opposite national experiences regarding the end of the war would lead to opposite reactions in scientific communities.

Ultimately, both theses however make the war disappear. In the Bourbaki thesis, as already noticed, nothing more than the death of a few young mathematicians is held as significant for the history of mathematics. In Forman's paper, similarly, the striking war experiences of two of his major protagonists, Hermann Weyl and Richard von Mises, are not discussed. Weyl's whole year of military service in the German Army near Saarbrücken, his release at the request of the Swiss government in May 1916, and his taking refuge to neutral Zurich are all left without mention.[14] Similarly, Forman does not think worthy of discussion the fact that von Mises

[13]Another example where scientific concepts might have been inspired by the war experience might be the emergence of the "polar front" in meteorology [Friedman 1982].

[14]Norbert Schappacher, "Selflimitation & Holism in Mathematics: The example of Hermann Weyl and World War I," talk at the meeting on "Mathematics and Mathematicians through

was intensely involved with aviation throughout the period: as technical advisor to the nationalistically and militaristically informed "Prinz-Heinrich-Flug" race in 1913–1914, as one of the first to be awarded a pilot's licence from the German *Nationalflugspende* in 1914, as an officer of the Austrian Air Force during the war who served actively, who was involved in the conception of the first *Grossflugzeug* (a 600-horsepower airplane), and whose lectures about flight theory to Austrian officers, *Fluglehre*, were published in 1918 and went through several editions (the English translation still being in print) [Siegmund-Schultze 2003]. Fighting, defense work in which mathematicians might have been involved, are nowhere brought up by Forman. In both theses, the effect of war mainly comes from the losses it brought about: the loss of a generation of mathematicians or the loss of trust in rational determinism and prewar values. How experiences *during* the war have brought about these new situations *after* the war still has to be explored. This is the purpose of this book.

Our discussion of the Forman thesis, however, suggests that the result of the conflict (defeat, in the case of Germany) may blur some of the issues that emerged during the war.[15] Faced with an extreme variety of situations, we have decided to compare as far as possible similar agendas of change and thus to restrict ourselves to the Allied Powers. For the same reason, we have not included a chapter on Russia,[16] a case where radical political upheaval produced unmistakable effects on mathematics and mathematicians which would have to be disentangled from the effect of the war itself. Nevertheless, since losses were dramatic for all sides in this conflict, we shall feel free in this introductory chapter to draw our examples from all countries when they illustrate similarities of individual experience in wartime.

2. Varieties of War Experience

> The experience during the war of the man of science has sometimes been confusing [Hale 1919, p. 143].

2.1. Americans in Paris. In the summer of 1914, Lester Randolph Ford was appointed lecturer of mathematics at the University of Edinburgh (Figure 5). A native of Missouri, Ford had studied under Maxime Bôcher at Harvard, graduating with an M.A. in 1913. He had come to Edinburgh with a traveling fellowship to work with Edmund Taylor Whittaker. The student magazine *The Gambolier* stressed how unusual this appointement was: "It is not often we fill up vacancies in, or augment the staff of, our University from across the Atlantic." At a time of war, Ford was introduced as a man who "like the proverbial nation, is happy in the fact that he has no history. So far as we have been able to discover, he has never killed a man."[17] Working on function theory, Ford originally intended, as he wrote in a letter to Birkhoff on January 25, 1915, to go to Cambridge and Paris. Although travel was difficult at the time, Ford managed to reach Paris later

World War I" held at the Centre international de rencontres mathématiques (CIRM) in Luminy, Marseilles, January 21–26, 2007.

[15] The cultural impact of defeat in general is discussed in [Schivelbusch 2001].

[16] On the Russian situation, see [Demidov 1993, Graham & Kantor 2009].

[17] *Lester Ford, M.A., Lecturer in Mathematics, University of Edinburgh*, The Gambolier **7** (Wednesday, January 27, 1915), p. 140; quoted from J. J. O'Connor and E. F. Robertson, "Lester Randolph Ford," MacTutor History of Mathematics archive, `http://www-history.mcs.st-andrews.ac.uk/Biographies/Ford.html` (last accessed November 5, 2012).

FIGURE 5. Lester Randolph Ford in 1925. Courtesy Woodson Research Center, Fondren Library, Rice University.

in 1915 only to find that mathematics was not on anybody's mind. On December 12, Ford shared his disappointment with Birkhoff: "The university has commenced in a half hearted sort of way, but there is nothing that I am interested in beginning until January. I had expected to attend [Gaston] Darboux's course on *Géométrie Supérieure*, but although scheduled to begin Nov. 5 it has not yet started."[18]

Half a year later, the United States was still officially neutral and another mathematics student from Harvard was thinking of traveling to Paris to learn mathematics! After his Harvard Ph.D. thesis on integral tests for the convergence and divergence of infinite series, Raymond W. Brink was awarded the Sheldon Traveling Fellowship. But, again, he found that, as a center for the study of mathematics,

[18]Ford's letters in [Birkhoff Papers, HUG 4213.2, Box 7]. Note that Darboux was to die on February 23, 1917, at age 74. In the chapter by David Aubin, Hélène Gispert, and Catherine Goldstein dealing with Paris mathematicians, we see that this was precisely the time of the largest effort at reorganizing war-related scientific research and that mathematicians were especially involved in this.

Paris was in much disarray. On November 11, 1916, Brink wrote to Birkhoff: "I have been waiting to write you until things at the Sorbonne were sufficiently settled for me to give you some idea of how my work would go here. Until the middle of last week it was impossible to obtain any information as to what courses would be given, or by whom."[19] Brink went on:

> I am much disappointed in affairs at the Sorbonne. There are, to tell the truth, no courses at present that interest me in the least. I go to a course of [Édouard] Goursat's on *Mathématiques générales*, for the sake of what inspiration I can get from that gentleman, and to acquire a mathematical vocabulary in French. I must say that the course itself is not very interesting. Goursat and [Henri] Lebesgue are giving courses such as differential calculus, and no advanced courses at all. In December the Collège de France opens and Hadamard and [Georges] Humbert give courses on partial differential equations and on Abelian integrals, respectively, that from their description in the programme should be interesting.... In the second semester, according to Borel, Picard and Goursat will give advanced courses at the Sorbonne, but nobody knows as yet what the subject will be.

Although war was raging not more than a hundred kilometers away, course offerings seemed relatively diverse in the French capital. But Brink went on to describe "how badly shattered the system at the Sorbonne is by the war." Courses by Borel and others had been announced at the ENS, so Brink went to enquire about it: "The very pretty, very amiable, and very ignorant young lady who is the information bureau said she knew nothing about these courses, and agreed with me that this notice, and the fact that the École was closed on account of the war, seemed to present an amusing paradox." Going to the ENS to try to see Borel, Brink only found the doorman.

> [T]he concierge said M. Borel was not there, but was at the Ministère des Inventions. No, he did not know where that was, but anyway M. Borel was there. A gendarme was of the opinion that the Ministère des Inventions was with the Ministère of Public Instruction, corner of rue de Grenelle and rue de Bellechasse, in which opinion he was confirmed by an important looking bearded person who was just entering the Mairie of the Vme Arrondissement. I walked the several kilometres to the corner of rues Grenelle and Bellechasse. A belle chasse it proved to be[20], for the concierge at this place "forwarded" me to another address 30 minutes away, where at last I really found M. Borel. He received me very pleasantly, apologized for the disordered state of things, refrained from laughing when I invented new French words, said that the notice of courses on the bulletin was merely a meaningless remnant from preceding years, that no courses would be given at the École, that he regretted his inability, on account of the war, to teach me any-

[19] Brink's letters to Birkhoff in [Birkhoff Papers, HUG 4213.2, Box 7].

[20] Literally, a "beautiful hunt."

> thing about series, that my best chance for mathematics was to wait for the courses I have mentioned [by Goursat and others], and said "au revoir."

Paris was not proving to be a very convenient place to learn mathematics.[21] "Libraries that are open only four or five hours a day, and hotel rooms that are too cold to work in, courses that not even professors took seriously, and the general atmosphere of lack of interest and absorption in the war, were very disappointing," Brink complained to Birkhoff on January 21, 1917.

For Brink, luckily, things took a turn for the better. Courses he was finally able to take from Humbert and Hadamard proved worthwhile. Going back to the United States, Brink was immediately offered a position at the University of Minnesota [Olmsted 1974].

But the atmosphere in Humbert's class at the Collège de France must have been peculiar. One cannot help wondering about a possible encounter between Brink and his contemporary Gaston Julia (Figure 6), who was about to defend a doctoral thesis, in December 1917, on topics close to Humbert's interests. A brillant student at the ENS, Julia was now disfigured by gunshots received on the front in 1915. But two years later, "between violent headaches, frequent bandages and operations," soon to be declared "100% invalid," he was attending Humbert's lectures, too.[22]

In 1917, Ford also came back to the U.S. where he defended a Ph.D. thesis on closely related themes, in which he expressed his "indebtedness to Professor Humbert of the Collège de France, who suggested" the subject to him [Ford 1918, p. 2]. Brink and Ford however were caught up by war: when their country joined the fight both became mathematics instructors for army recruits. Before he was recruited at the Rice Institution, Austin, Ford wrote a mathematics textbook for gunners [Ford 1919] to which we shall come back.[23]

2.2. Experiences. Age, country, field of specialization, health, and social status were determining factors in the way in which mathematicians lived through the war. Some were drafted and spent five years in the mud; they got killed or died of illness; they were wounded or taken prisoners. Some lost sons and students. For others, the war provided new opportunities to apply their knowledge and skill in unsuspected directions, to have inspirational encounters with people they might not have met otherwise, to shift drastically their fields of interest, to get involved in science policy at the national and international levels with an effectiveness they might not have dreamt about earlier, or to become influential public figures.

The variety of experiences is matched by the variety of relevant sources that can be used to try and reconstruct them. Scattered in the secondary literature on the history of mathematics of this period are countless hints as to how the war affected the life and work of mathematicians. Journals of the time, mathematical or not, correspondences and military documentation contain many examples of military work and various instances of mathematical techniques being newly applied on the

[21] Another case of major disruption of the university life brought about by WWI is discussed in Laurent Rollet and Philippe Nabonnand's contribution to this volume focusing on the university of Nancy.

[22] See Catherine Goldstein's contribution to [Goldstein & Aubin forthcoming].

[23] For more on the mobilization of American mathematicians, see the contribution by Thomas Archibald, Della Dumbaugh, and Deborah Kent in this volume.

FIGURE 6. Gaston Julia in uniform: in 1914 before he was sent to the front and in 1919 after his release from the Val de Grâce hospital. Source: *Science et monde* **150** (1934), p. 195.

front, in general staff, or in ministries. This dispersion of sources is a major difficulty which confronts the historian wishing to make sense of these wartime experiences.

But also, like Julia's or Ford's, every wartime trajectory is unique. When we look at both of their future work, Humbert's mathematical influence seems greater than their own war-related activities, which could hardly have been more at odds. Reciprocally, common war experiences led to completely opposite reactions. While the loss of his sixteen-year-old son in fighting near Arras may have killed Göttingen differential geometer Paul Stäckel [Renteln 1997], Émile Picard's cruel loss of three sons strengthened his hatred of the Germans and his resolve to act upon it by excluding them from postwar international organizations. As "a refuge from immediate fact," Alfred North Whitehead dedicated his influential *Enquiry concerning the Principles of Natural Knowledge*—"thought out and written amid the sound of guns ... a refuge from immediate fact"—to his son Eric Alfred killed in action over the Forêt de Gobain [Whitehead 1919, p. viii]. The German mathematician Oskar Bolza who had spent more than 20 years in the U.S. also was deeply affected by WWI: the author of a classic text on the calculus of variations practically abandoned all mathematical research for over ten years and devoted himself to the study of religion and Sanskrit, authoring a book titled *Glaubenlose Religion* under the pseudonym of F. H. Marneck [Bliss 1944]. By contrast, the Liverpool professor of mathematics William Henry Young was said to have been driven back to mathematics "as a drug" by his son Frank's death in 1917 [Hardy 1943, p. 310].[24] And if

[24]See also [Grattan-Guinness 1972, pp. 153–156].

both Jacob Klein and Thomas Heath delved into ancient Greek mathematics, the first sought to understand the roots of the symbolic modernity that had made this war possible, while the second took refuge with those Greeks who had "desist[ed] from war and wickedness" [Klein 1934, Heath 1921].

Without engaging in overly theoretical arguments, we would like to claim that a close look at wartime experiences, however diverse they may be in terms of events and reactions to them, is an essential step toward a better understanding of shifting professional identities and disciplinary values in the postwar years. Of course, the concept of *experience* in historiography is notoriously controversial, particularly in the context of WWI [Mosse 1991, Audoin-Rouzeau & Becker 2000, Horne 2005]. Our objective here is not to define a single war experience linked to mathematics, but rather to provide a preliminary survey of mathematicians' multiple, concrete experiences during the war. We concur with the historian Joan Scott in her opinion according to which, counter to common perceptions, "it is not subjects who have experience, but subjects who are constituted through experience. Experience in this definition then becomes not the origin of our explanation, ... but rather that which we seek to explain, that about which this knowledge is produced" [Scott 1991, pp. 779–780]. The effort to reconstitute mathematicians' wartime experiences and confer historical meaning to this scattered amount of evidence will perforce lead us to revise our definitions of what constitutes a mathematician and what properly belongs to mathematics.

2.3. In the Trenches. The number of mathematicians who served as soldiers in WWI will never be known with precision. Just like the great majority of their contemporaries, they carried a gun to the trenches and suffered horribly. "The wounded men," remembered an American hospital worker, "they smell of sweat, camp-fire smoke, leather, and tobacco — all the same, whether the man be a peasant or a professor of mathematics."[25] Evidences we have from their campaigns are many and range from the sadly banal to the heroic and extraordinary. They give flesh to the characteristic features historians of the Great War have already identified: the boredom of months without end spent in muddy trenches, the violence of the attacks, the pain of resulting injuries and death.[26]

The young Willi Böhle, born in 1897, had studied mathematics at the University of Freiburg im Breisgau before the outbreak of WWI. On April 15, 1915, he wrote his family from the trenches:

> Darling Mother, darling little Mother, and you too, my brother and sister, I am ready to endure anything for your sakes, so that you may never see what ruined villages and shell-destroyed fields look like; so that you may never learn what the word *war* really means. Be thankful, my Gretel and Erich—although you are still so young and know nothing of the serious side of life—that our little house is still standing; that you can sleep in beds; that you have a roof over your heads; that your are not tormented by vermin; that you

[25]Testimonial by J. H. G. in [Collective 1916]; available online http://www.gwpda.org/medical/FriendsFrance/ff03.htm (last viewed 3 November, 2009).

[26]The war correspondence between Julia and Borel provides a good illustration of this, as well as concrete details about the military assignments of ENS students [Borel Papers, M143–M154].

> have your meals at the proper times; that you do not know what thirst and hunger are. Be thankful; fulfill all your little duties conscientiously, and never grumble! [Witkop 2002, p. 331, original emphasis].

Conditions experienced on the other side of the no-man's land were the same. No less characteristic, if lighter in tone, was Julia's account in a letter to Borel written on January 21, 1915:

> Life in the trenches is certainly lacking in terms of comfort and mud and water are things most often encountered. We are nose to nose with the Germans, at less than 100 meters. The Germans fire on our positions.... Nevertheless, we keep busy by repairing our shelters,... by cleaning access tunnels and surveillance trenches, by long conversations with colleagues from the battalion, leaving us little leisurely time during which the reading of a book is appreciated. Nonetheless, a more active life would be preferred. But what can you do! [Borel Papers, M151]; [Goldstein 2011].

It was ten days later that, as mentioned above, Julia, who meanwhile was sent to a "more active" sector, sent news he had been hit in the face by a bullet shot at point-blank range [Borel Papers, M152]. Böhle, for his part, was killed on May 26, 1917, near Arras. These are just a few of the countless examples of mathematicians who were not left unscathed from their participation in frontline actions. The German algebraicist Alfred Brauer (the older brother of Richard Brauer) was severely injured, as was the Scottish mathematician Raymond Keiller Butchart, who lost a leg. One remembers Courant, of course, who was badly shot: "it is a miracle I survived" [AIP 1962]. In fact, being wounded was often the only opportunity for many mathematicians to leave the frontline and return to science, either pure or applied to war. Konrad Knopp was wounded as early as September 1914 and before the end of the year, he was teaching in Berlin and working on point-set theory, later partaking in setting up the *Mathematische Zeitschrift* in January 1918 [Kamke & Zeller 1957]. The Austrian statistician Wilhelm Winkler was twice wounded before being invited by his former teacher, now Minister of War, to join the scientific committee he had set in the ministry. But the case of Antoine, who was mentioned earlier, shows that serious wounds hardly guaranteed a safer position: even though he had been wounded twice in 1914, the ENS mathematician was sent back to front duty from which he was finally released after having lost his sight.

On June 22, 1918, the mathematical seminar of the university of Rome honored the memory of four professors of mathematics in Italian universities who had died in battle [Collective 1918]. Luciano Orlando was a *libero docente* in mathematical physics at the University of Rome and Ruggiero Torelli was a *libero docente* in geometry. Both died on the battlefield of the Isonzo in August 1915. The tragic fate of Eugenio Elia Levi, professor of differential calculus at the University of Genoa, struck his contemporaries.[27] Even his specialty as teacher of theoretical geodesy

[27] Levi's brief career is discussed in Pietro Nastasi and Rossana Tazzioli's contribution to this volume.

FIGURE 7. Four Italian university professors of mathematics who died in battle. Upper right: Luciano Orlando (1877–1915); upper left: Ruggiero Torelli (1884–1915); lower right: Eugenio Elia Levi (1883–1917); lower left: Adolfo Viterbi (1896–1917). Photos taken from the commemoration at the Mathematical Seminar in Rome [Collective 1918].

could not save Adolfo Viterbi, of the University of Pavia, from an untimely death while on a delicate topographical mission. British wranglers were also killed.[28]

In Germany and Austria, contrary to Dieudonné's claim, many are the members of the German Mathematicians' Association (*Deutsche Mathematiker-Vereinigung* (DMV)) who were killed on the battlefield. Anton Lackner, *privatdocent* of descriptive geometry at the Technische Hochschule in Vienna, fell before Cracow in November 1914. A former extraordinary professor of applied mathematics and technical physics in Jena, Rudolf Rau, born in 1871, died on December 17, 1914. *Privatdocent* at Göttingen, Dr. Wilhelm Behrens, who was close to Felix Klein, died on June 23, 1917. To this list, one may add the names of Wolfgang Wilhelm Vogt (Heidelberg University), the number theorist Edmund Busche (Hamburg), the

[28]This is documented in June Barrow-Green's contribution to the volume.

director of *Zeitschrift für mathematischen und naturwissenschaftlichen Unterricht* Ernst Grimsehl, a physicist who died at 53 years of age in Langemark near Ypres on October 30, 1914, and the list could go on.[29]

Some mathematicians also fell as indirect casualties and died of illnesses caused by the war. The Polish topologist Zygmunt Janiszewski, who had received his doctorate in Paris, fought for the independence of his country and went on to establish the Polish school of mathematics and the journal *Fundamenta Mathematicae*. He died from the flu epidemic of 1920 [Feferman & Feferman 2004, p. 28]. Better known is the case of Srinivâsa Aiyangâr Ramanujan. When he reached Trinity College, Cambridge, in 1914, working conditions there left much to be desired. He wrote back home:

> I am very slow publishing my results owing to the present war. A lecturer here whom I know well and from whom I received some help to publish my results has gone to war. The other professors here whom I know have lost their interest in Mathematics owing to the present war [Berndt & Rankin 1995, p. 112].[30]

According to his biographers, Ramanujan, who was a strict vegetarian, suffered from a vitamin deficiency during his time in England. He never recovered from his illness and died in India on April 26,1920.

In the mathematical community, the tragic loss of young promising colleagues was felt dearly. Herbert Herkner was a gifted mathematics student Max Born had met in Göttingen. Upon hearing that Herkner was serving on the front, Born tried to bring him to the Artillery Testing Commission (*Artillerie-Prüfungs-Kommission*, hereafter APK) in Berlin. But Herkner was notified of his new assignment just one day before the battle of Cambrai, where he was killed on November 22, 1917. Heartbroken, Born remembered in glowing terms the impression left by Herkner's coming to Göttingen:

> This was an appearance that filled all the colleagues who came into contact with him with reverent wonder and with the hope to see a revival of mathematics arise through him. A tragic destiny has destroyed this hope like so many others.... Just like the gratitude over the services of a busy life deserves an obituary, the mourning over buried hope must find an expression. This is the objective and the sense of these lines [Born 1918, pp. 179–180].

In France, altogether 22 mathematicians who were licensed mathematics teachers (*agrégés*) and graduates from the ENS died in combat, among them Jean Merlin (Lyons Observatory), Alphonse Blondel (Toulouse Observatory), and René Gateaux.[31] Their disappearance left Borel, who "found the school haunted by shadows," unable to resume his scientific directorship of the ENS [Marbo 1967]. After the death of Georges Lery, a specialist on Green functions, Hadamard wrote: "For a certain number of years, there has been in France a remarkable school of young

[29]Most of this information was gathered from the war issues of the *Jahresbericht der Deutschen Mathematiker-Vereinigung*.

[30]The lecturer that had helped Ramanujan and went to war was John E. Littlewood. More detail about his involvement in the British war effort can be found in June Barrow-Green's contribution to this volume.

[31]On French casualties among mathematicians, see [Aubin 2012].

mathematicians, thanks to which our country fear, on that ground, no comparison.... Without possible contestation, Lery has won for himself a position in this phalange" [Hadamard 1916, p. 115]. Julia similarly wrote a touching obituary for his "prodigiously loving friend," the young Paul Lambert, born in 1894 and killed on March 13, 1915, not two months after Julia himself had sustained his devastating injury. Julia vividly recalled Lambert's love of life and his intellectual capabilities. With Borel's nephew and adopted son, Fernand Lebeau, and other now deceased classmates, Julia wrote, they formed a joyous band, cycling around Paris, reciting poetry, and working on mathematics. About his friend's broken future, Julia wrote:

> It would be adventurous to make prognosis, and I cannot here put side by side my friend and such an authentic and indisputable glory as Galois, but I cannot refrain from saying that I often thought of this immortal genius by living close to Lambert: they had many resemblances in their taste, they were both very precocious, they both died very young and tragically [Julia 1919, p. 110].

It is indeed impossible to evaluate what the fallen might have been able to accomplish, even if the discussion was recurrent among surviving contemporaries, and even more difficult perhaps to understand the effect of their loss on the development of mathematics. During the war, however, many wanted to believe that the sacrifice was worthwhile. In a formal speech delivered in December 1915, the president of the Academy of Sciences Edmond Perrier for example praised the way young students had volunteered and fulfilled a useful service to the nation. "They remember that the most educated must lead by example; they will be the first to ... run toward the barbed wires, the first to die [Perrier 1915, p. 804]. But a key issue was survival through mathematics: "the dead ... are for the living a reason to live in order to take their place and fulfill the task that was taken from their hands" [Julia 1919, p. 113]. Besides the writing of detailed obituaries, those mathematicians still alive organized the transmission of the work of their dead colleagues. Gateaux's papers in functional analysis, for instance, were given to Paul Lévy, who carried on his work with great success, although in the framework of probability theory.[32]

Let us consider then the case of Jean Clairin. Before the war, Clairin was a professor of mathematics at the University of Lille; he was killed in battle near Cambrai, as early as August 26, 1914: "his forehead, his noble forehead, broken by a bullet."[33] In his thesis in 1902, Clairin had investigated Bäcklund transformations which generate other solutions to a partial differential equation once at least one solution is known. In the following years, Clairin published several articles (references of which can be found in [Lamb 1976]). Before the war, he had announced some results in the *Comptes rendus hebdomadaires des séances de l'Académie des sciences* (CRAS) without providing complete proofs. In 1920, Émile Gau received some of Clairin's papers, which were in great disorder after the German occupation of Lille,

---

[32]On Gateaux, see Laurent Mazliak, "The ghosts of the Ecole Normale. Life, death and legacy of René Gateaux", `http://fr.arxiv.org/abs/math/0701490v2`.

[33][Ollivier 1917, p. 85]. On Clairin, see also [Pourprix 2009, pp. 42, 55]. There is an obituary by his mentor, Édouard Goursat [Goursat 1916].

and arranged for their publication [Clairin 1920]. Then Clairin's work—and indeed Bäcklund transformations altogether—fell into oblivion until the method developed by Clairin was applied by George L. Lamb in 1974 to the Korteweg-deVries and nonlinear Schrödinger equations and after much delay became a mainstay of soliton theory [Lamb 1974]. Lamb attributed this delay to WWI: "Bäcklund transformations ceased to be an active area of research after World War I. Undoubtedly this was due, at least in part, to the demise of those active in the field (e.g. Clairin)" [Lamb 1976, p. 76]. However, as in Antoine's case, it appears that a bit more than simply the death of promising mathematicians was at play: Édouard Goursat devoted a book to Bäcklund transformations in 1925 [Goursat 1925], and one is then left to ponder the change of mathematical priorities during the interwar period.[34]

In the trenches, mathematics often provided a welcome distraction to fighting men. Recently appointed dozent at the Budapest University, Zoárd Geöcze, who had defended his doctorate in Paris in 1910, found time to write papers on the theory of surfaces which he sent back via military post. First sent to Serbia where he survived a crushing defeat, he became mortally ill in Ukraine. After his death in 1916, his colleague Loránd Eötvös wrote: "no horrors of the trenches or the thunder of guns were able to distract Zoárd Geöcze from concentrating on the solution of his favourite problem and making efforts to widen and deepen our knowledge of the subject" [Szénássy 1992, p. 322].

"At rest,... after my correspondance with my family, mathematics provides me with the best distraction" [Lhermitte 1917, p. 90]. In this letter, Théophile Rousseau, who was a mathematics teacher in Rennes, testified to the fact that his former training offered little more than a welcome escape from the harsh realities of combat. The author of several articles about vector analysis for the teaching of geometry, Rousseau died on the frontline after he was hit by shrapnel on April 11, 1916, without having been asked to put his mathematical skill to better use for defense.

Better known and equally tragic is the frontline experience of Karl Schwarzschild. Volunteering in August 1914, Schwarzschild served in Belgium and France where he was in charge of a weather station and of computing ballistic trajectories. Sent to the Russian front, he found the time and resources to author two papers on Albert Einstein's relativity theory and one on Max Planck's quantum theory. The latter paper explained that the Stark effect, namely the splitting of the spectral lines of hydrogen by an electric field (the amount being proportional to the field strength), could be proved from the postulates of quantum theory. In November 1915, he contributed to the Berlin Academy a paper on "The Effect of Wind and Air-Density on the Path of a Projectile," a theme more directly connected to his present environment. Schwarzschild's relativity papers give the first exact solution of Einstein's general gravitational equations, stating what the geometry of space-time near a point mass should be. As he wrote to Einstein on December 22, 1915, war "was kind to him," because despite "heavy artillery fire in the far," he had been able to enjoy "a stroll through [Einstein's] land of ideas."[35] While in Russia,

---

[34]For France, see [Leloup 2009, Goldstein & Aubin forthcoming].

[35]*Gesammelte Werke*, vol. 1, p. 38. The way in which Schwarzschild kept in touch with recent development in general relativity in discussed in [Rowe 2004, p. 110]. For a short discussion of Schwarzschild's ballistic paper, see [Eddington 1917, p. 315].

however, Schwarzschild fell ill to a rare autoimmune disease of the skin from which he died in 1916.[36]

2.4. Prisoners of War. The publication in May 1916 of an article by a prisoner of war in Hohen-Aspurg with the solution to a mathematical problem posed earlier that year by the professor of geometry at the École polytechnique Maurice d'Ocagne in *L'enseignement mathématique* shows that some prisoners had access to mathematical journals and were able to enjoy a correspondence [Barolet 1916].

In 1916–1918, several Hungarian young men, prisoners of war in the camp of Berezowka in Siberia, decided to organize a mathematical seminar. The most advanced mathematician of the three was Eduard Helly, who had proved the Hahn-Banach theorem in 1912.[37] Voluntarily enrolled in the Austrian Army, he was shot in the lungs in September 1915 and taken prisoner. About the camp, an American relief worker wrote in 1918:

> In every camp we had a school, according to the size and need, enrolling from two hundred to seventeen hundred students, with as many as forty courses, ranging from reading, writing and arithmetic, to advanced modern languages, Latin, Greek, and Hebrew, Physics, Philosophy and Psychology ....
>
> In the great camp at Berezowka, at that time consisting of eighteen thousand prisoners, I started a library of one hundred and five volumes, the last foreign books to be had at Irkutsk. During the succeeding months, as the library grew, although we had a limit of one day for the borrowing of books, there were often thirty or forty men turned away, with not a book in the library. At the time that I left Siberia, a few months ago, we had in this camp more than four thousand volumes, with a central and two branch libraries. As there are several thousand students and professors and men of the university class in this camp, the importance of the school and library can scarcely be overestimated. Such men are much more liable to nervous breakdowns than the uneducated, so that the occupation thus given has saved many a mind agonies which sometimes end in insanity. Not only this, but many of the men have been able to carry on their education and fit themselves better for life and service after the war while confined in these prison camps [Morgan 1918, pp. 9–10].

To pass the time, Helly taught mathematics to his fellow captives. His lessons to Tibor Radó, who had been studying civil engineering before the war, proved so inspirational that when after an adventurous trek through polar regions he was able to resume his course of study at the University of Szeged in 1920, Radó chose mathematics. We may have an inkling of Helly's teaching in the wartime mathematical reflections published by the young Heinrich Elbogen, another Austrian captive who met Helly in Russian prison camps. In his preface to Elbogen's book titled "The Axiomatic Method in Mathematics," Helly wrote that those reflections

[36]Mathematicians also occupied their minds with extra-mathematical activities. Drafted in the Austro-Hungarian Army in 1915, Eduard Čech used the next three years to learn Italian, German, and Russian, skills he could draw on during his career as a world-class topologist.

[37]Note that the Austrian mathematician Hans Hahn was himself severely wounded in 1916.

were written at a time when the author was under "dreadful physical and psychological stress." To Helly, memories of war captivity in Russia were painful, he said, and "one of the rare lightspots was to have met Elbogen there" [Butzer et al. 1980, p. 145, footnote 12].[38]

The case of the French mathematician Henri Eyraud presents some similarities with a happier outcome.[39] Taken prisoner by the Germans, he was finally released in neutral Zurich, where he was able to pursue his scientific training at the Eidgenössische Technische Hochschule (ETH). Before the war, Eyraud had attended the University of Lyons, passing the prestigious national examination for future mathematics teachers (*agrégation*) in 1913. His stay in Zurich where he attended Marcel Grossman's and Hermann Weyl's lectures, on descriptive and projective geometry and on differential geometry and electromagnetic field theory, respectively, had a decisive impact on his professional life. In 1926, he defended the first doctoral thesis on unified field theory in France and later established the Institute of Financial and Insurance Mathematics in Lyons.

Other mathematicians were taken prisoners, including the Polish mathematician Wacław Šierpinsky whom Nikolai Luzin and Dimitri Egorov managed to get released to join them in Moscow where they began the study of analytic sets.[40] In an Italian camp after the end of the war, the Austrian artillery lieutenant Otto Neugebauer (Figure 8), later to become the founding editor of both *Zentralblatt für Mathematik und ihre Grenzgebiete* and *Mathematical Reviews* and one the most important historians of ancient mathematics and astronomy, met his fellow countryman Ludwig Wittgenstein carrying in his rucksack the manuscript of the *Tractatus logico-philosophicus* written during his war service, a time when he took refuge in the life of knowledge to ward off the misery of the world.[41]

2.5. Mobilizing Mathematics. The Russian astronomer Aleksandr Aleksandrovich Friedmann was recently appointed to the Aerological Observatory in Pavlovsk, near Saint Petersburg, when the war broke out.[42] In a letter to Vladimir Steklov, a specialist in the integration of differential equations, from St. Petersburg University, written on February 5, 1915, Friedmann wrote:

> My life is fairly even, except such accidents as a shrapnel explosion twenty feet away, the explosion of an Austrian bomb within half a foot, which turned out almost happily, and falling down on my face and head, which resulted in a ruptured upper lip and headaches. But one gets used to all this, of course, particularly seeing things all around which are a thousand times more awful.

At the time, the Russian Army besieged the town of Przemyśl, occupied by Austrian troops, and Friedmann, who had started flying, took part in bombing

[38]See Heinrich Elbogen, *Die axiomatische Methode in der Mathematik*, edited after his death by his mother Alie Elbogen with the assistance of Prof. Oppenheim and Dozent Dr. Helly, privatly printed, Vienna, 1928, a copy of this text is in the Österreich. Nationalbibl. Wien, Signatur 567444-B. We wish to thank Anne-Sandrine Paumier for providing us a copy of the book. On Helly, Tadó, and Elbogen, see also [Butzer et al. 1980, Butzer et al. 1984, Sigmund 2004, Kreyszig 1992, Nirici & Becknstein 2007] as well as [Young 1981, p. 327]. For more on the Berezowska camp, see [Rachamimov 2002, p. 99].

[39]See Jim Ritter's contribution to [Goldstein & Aubin forthcoming].

[40]About mathematics in Moscow at the time, see [Graham & Kantor 2009].

[41]On Neugebauer's wartime service, see [Swerdlow 1993].

[42]The following is based on [Tropp et al. 1993, pp. 71–77].

FIGURE 8. Otto Neugebauer in Austrian uniform in Val Pampoluzzo on the Italian front, October 6, 1918. Tagebuch 1917–1919, p. 42, Otto Neugebauer papers, Box 13. Courtesy of the Shelby White and Leon Levy Archives Center, Institute for Advanced Study, Princeton, NJ, USA.

expeditions. Thinking of applying his mathematical skill to the study of bomb trajectories, Friedmann asked Steklov's advice regarding the integration of equations he had obtained:

$$\frac{du}{dt} = -au\sqrt{u^2 + v^2}, \qquad \frac{dv}{dt} = -g - av\sqrt{u^2 + v^2},$$

where u, v were the components of the bomb velocity, g the acceleration due to gravity, and a a parameter characterizing the shape and weight of the bomb. Bombs, Friedmann explained, could be divided into two classes: very small a or $a \approx 1$. To solve these two cases, he used approximations, expanding in powers of a in the first case and assuming that $\sqrt{u^2 + v^2} \approx v$ in the second case, which, he noted, was "illegitimate."

On February 28, 1915, Friedmann explained how he used Runge's integration methods to compute results quickly, which although in fairly good agreement with practice made him "uneasy." He added:

> I have recently had a chance to verify my ideas during a flight over Przemyśl; the bombs turned out to be falling almost the way the theory predicts. To have conclusive proof of the theory I'm going to fly again in a few days.

Despite his relative success in applying his mathematical skills to war, Friedmann deplored his poor working conditions and the obligation he was in to drop research topics that were closer to his heart.

> I should say that it is fairly difficult to do investigations in my circumstances, and all kinds of calculations concerning bomb and bomb-sights take a lot of time, therefore I have very little time for the theory of vortices in fluids with changing temperature for which I feel a strong attachment.

Sometimes, such dedication did not go unnoticed and mathematicians were assigned special posts. In 1916, Friedmann was pulled from the front and his mathematical skills were put to use first as an instructor for recruits, then as a researcher in an aeronautical center, until the Revolution broke out in 1917. The Budapest mathematician Geöcze mentioned above was put in charge of a power station and its supply lines. Likewise, other scientists and mathematicians were assigned special duties in sound-ranging units, computing bureaus, and weather stations.

Whether mathematics should be mobilized for war, however, never received a clear-cut answer. For some, mathematics' universalism was drafted in a fight *against* the war. Bertrand Russell, who felt "more allegiance to mathematics than to the State," as he wrote in a wartime letter [Russell 1988–1995, p. 287] took a high-profile position against the war. On December 23, 1916, for example, he published on the front page of the *The New York Times* an open letter to recently reelected U.S. President Woodrow Wilson calling for him to intervene in favor of a negotiated peace between the belligerent nations.[43]

In 1914, the American mathematical educator David Eugene Smith published a problem book for children where he tried to "lay before young people in the elementary schools, at the most impressionable age, the fact of the wastefulness of war" [Smith 1914, p. 5]. Smith asked his readers to evaluate the financial loss of human lives due to the war as well as the lost production due to the 21 million enrolled men. In other sets of problems, the value of battleships was compared to the cost of college tuition, camping trips, baseball tickets, bicycles, etc.

[43]Russell's wartime writings are collected in the vols. 13 and 14 of [Russell 1988–1995]. Other pacifist mathematicians are discussed in June Barrow-Green's contribution to this volume.

> The battleship Alabama cost $4,665,820. In 1913 the total receipts of Alabama for higher education amounted to $533,659. The cost of this single battleship would have kept Alabama in funds for her colleges, universities and schools of technology for how many years? [Smith 1914, p. 12]

But this attitude was rather rare. To express the apparent contradiction between a belief in science's internationalism and heavy involvement in war-related activities, French scientists were quick to recycle one of Louis Pasteur's *bons mots*: "If science has no nation [*patrie*], the man of science must have one" [Pasteur 1888, pp. 29–30]. In the preface to his book on integral equations, the Italian mathematician Giulio Vivanti compared national characters:

> While our sons fight with valor to liberate Europe from the Teutonic yoke it devolves to us, whose age and strength do not permit to offer arms to our country, to work for its scientific emancipation. A *national* science is an absurdity and he would be foolish who would refuse a scientific truth because it arose from beyond the Alps or the sea; but the work of scientific exposition and publication can be and ought to be national. Who does not recognize a German treatise by its minute and sometimes wearisome care of particulars, an English by its good-natured and discursive tone, a French by its form which is sometimes a little vague but always suggestive and elegant?[44]

2.6. Intellectuals and Organizers. Even those who defended the tradition of apoliticism among university professors felt the need to justify afresh their usual activity, that even in wartime scientists could be busy with pure research. In 1915, the applied mathematician Carl Runge (who had lost the youngest of his two sons in battle on October 21, 1914, while the eldest was still prisoner of war) emphasized as prorector of the University of Göttingen what he thought should be university professors' proper conduct during the war:

> The gravity of our time is not favorable to scientific work because all of our thoughts are directed toward a single point, the good and pains of our nation. We are inclined to confer meaning to other reflections only insofar as they play a role for the [national] community. The scientist who was used to consider his research self-justified now ponders upon its validity; he tries to clarify the links it has with and the position it occupies with respect to the intellectual life of our people and consequently to justify, in the eyes of the public and his own, that he [still] devotes his effort to it [Runge 1915, p. 400].

At Christmas 1915, Runge distinguished himself in publishing an article about pure mathematics "Dimensionsbetrachtungen" in a volume prepared by the University of Göttingen with an otherwise strong war flavor. The chemist Otto Wallach, for example, contributed a text entitled "Krieg und Chemie" [Busse 2008, p. 112]. But other German mathematicians were less hostile to a subtle expression of their views, even when addressing topics such as "axiomatic thought." In a speech delivered in September 1917 for the Swiss Mathematical Society, Hilbert opened with

[44] [Vivanti 1916, p. xiii]; translated and quoted in [Miller 1918, p. 117].

the following words: "In the life of nations, the individual nation can prosper only if it gets along with all its neighbors and the interest of the State requires that not only order prevails within each individual state but also among them; the same goes for the sciences" [Hilbert 1917, p. 405].

The most famous—and commented—instance of scientists taking position, this time in favor of their country's war aims, was the manifesto known in German as the "*Aufruf an die Kulturwelt!*" dated October 4, 1914. This manifesto defended German military actions in occupied zones, in particular Belgium. Felix Klein was the only mathematician among its 93 signatories although it is not clear whether he had given much thought to it [Tollmien 1993, Wolff 2003]. The "Declaration of the Professors of the German Reich" (*Erklärung der Hochschullehrer des deutschen Reiches*), dated October 23, 1914, stating that the German Army nurtured scholarship and shared its values, was however signed by over 3,100 professors, including Klein, Hilbert, Moritz Cantor, and Max Noether. The *Aufruf* caused a great uproar abroad, the Paris Academy of Sciences voting, at the secret meetings of March 15, 1915, to exclude foreign correspondants and associates who had signed the manifesto.[45]

But, all over Europe and in the U.S., war was also seen as an "opportunity" for developing research, including pure research [Millikan 1919]. American scientists were most vocal about this.[46] In Italy and France, mathematicians already held positions in the various offices for war-related research put together by academies and governments, and French mathematicians in particular interacted with the military on a number of issues. As soon as scientific institutions were put to the service of war aims, mathematicians naturally assumed leading roles. In 1914, Paul Appell was head of the Commission of Inventions; Painlevé was its secretary. As the perpetual secretary of the Academy of Sciences, Darboux organized the self-mobilization of savants in August 1914. During the war, Picard would take his place at the academy and use it as a platform for rebuilding international research structures. Of a younger generation, Borel was named director of inventions by Painlevé in November 1915. In Italy, Vito Volterra was able to follow this model quite closely as his correspondence with French mathematicians amply shows [Mazliak & Tazzioli 2009]. In the U.K. and the U.S., by contrast, mathematicians did not seem to enjoy the same prewar prominence among scientists as they did in France or Italy. National leaders of the mobilization of science therefore came from other disciplines. In the foundation of the U.S. National Research Council, the physicist Robert Millikan and the astronomer George Ellery Hale were the leaders [Kevles 1995]. In Great Britain, their main interlocutor was the physicist Arthur Schuster. This contrasting situation was mirrored toward the end of the war in the attendees of the Inter-Allied Conferences of Scientific Academies which took place in London in October 1918 and in Paris in November 1918. Mathematicians formed the majority of the French and Italian delegations, with Picard playing a crucial role as the new president of the International Research Council (IRC) [Schroeder-Gudehus 1978]. As a result, the International Mathematical Union was one of the first disciplinary bodies to emerge under the IRC and—controversially,

[45]The implication of this affair for the French mathematicians' involvement in a cultural crusade against the enemy that verged on the irrational is discussed in David Aubin, Hélène Gispert, and Catherine Goldstein's contribution to this volume.

[46]Some examples of such wartime speeches are found in the chapter written by Thomas Archibald, Della Dumbaugh, and Deborah Kent for this volume.

since it was held without the participation of mathematicians coming from defeated and some neutral countries—the first to hold an international meeting in September 1920 in Strasbourg [Lehto 1998].

But the path that led to this situation had never been obvious. One important aspect we bring to the fore here is the sentiment that scientific and mathematical expertise was—at least at the beginning—misused by the military. Had scientists started earlier to play a more prominent role, many felt, precious lives would not only have been saved but also have been put to effective use in bringing out the innovations that were decisive in battle. This for instance was expressed by the president of the Academy of Sciences, after 18 months of fighting, on December 27, 1915:

> The nation is under attack: ... not one [of our young savants] will miss the call. Farewell to the quiet routine of the laboratory; they are mere soldiers now; they do not even think—*and it may be a pity that no one thought of this for them*—to demand special posts because of their knowledge [Perrier 1915, pp. 803–804; our emphasis].

Military assignments were haphazard at best and often remotely connected with prewar activities. While the historian Robert Launay became a meteorologist during the war, the mathematician Maurice Fréchet spent two and a half years as an interpreter with the British. The mathematics professor from Clermont-Ferrand Jules Haag believed that his superiors in the army merely saw him as a "computing machine." Marston Morse, then a young Harvard Ph.D. graduate in mathematics serving as an ambulance driver, wondered whether he might have been "of more use" in some other position.[47]

The experience Richard Courant recounted in an interview may have been typical of the roundabout ways in which mathematicians from all countries got drafted in the technical branches of the army: "For the first two years of the war I was completely involved in strictly military service. And I was in the hospital. I had a very bad shot." Technical work, he added, came later:

> I was transferred to the central research laboratory of the postal office. And I did, I built microphones. I was involved in the first electronics, first amplifiers. That was in Berlin.... But then later I was transferred — I made what the military calls an "invention" — I built some devices to listen — which later had some bearing, some importance, some relation to geophysical questions. I built something to transmit and receive signals. Not radio signals. They couldn't build such things then. But signals through the earth. It was low frequency currents for communication between the frontlines and the artillery and so on. And this then mushroomed into a bigger thing,... then later I got completely involved in mathematics. And I did some theoretical work also, on amplifiers, tubes....

[47]Launay wrote his war memoir under a pseudonym [Bréauté 1919]. Haag's case is discussed in the chapter written by David Aubin. Morse's wartime experiences is mentioned in the contribution to this volume by Thomas Archibald, Della Dumbaugh, and Deborah Kent.

> So when I made this invention, I decided, I have to see that it is being used properly. I had a very interesting time. I got myself a little detachment of people, a few non-commissioned officers and technically trained soldiers, and we experimented in the front lines with this communication. I took part in the Battle of Verdun in some way. I got gassed. But I was quite interested in these matters. I became an organizer [AIP 1962].

Courant's trajectory in World War I may serve as a summary of the variety of experiences undergone by mathematicians. In the next section, we shall focus on aspects of these experiences that can be said to be properly mathematical. This will however lead us to emphasize the way in which this experience helped professional mathematicians forge new forms of collaborative links with a spectrum of people, some of whom they hitherto had little chance to meet in peacetime.

3. Mathematical Collaboration in Wartime

> As far as the mathematician is concerned, he has always been in a difficult position vis-à-vis the public who does not give him credit for his results when it makes them its own and who despises them as being worthless when it does not understand them [Busse 2008, p. 207].

This was how Carl Runge explained, in 1915, the mathematician's predicament, as he saw it, with respect to the applications of his trade to war purposes. Indeed, mathematicians often felt it was impossible to apply the full extent of their special skills to the war effort. After he had been appointed to the Aberdeen Proving Ground to work on ballistics in 1917, the U.S. mathematician Veblen still felt that:

> "there is practically nothing [related to war] that requires real mathematics.... I know of an important application of the problem of Apollonius (circle tangent to 3 given circles) but the real scientific interest in this is physical. I am not expecting to find any considerable math[ematica]l interest in my new department."[48]

As Veblen underscored, war-related research often involved rather elementary mathematics. As a result, many people with some technical training (physicists or astronomers, engineers, artillery officers, etc.) were at least as competent as, if not better suited than, mathematicians for the tasks at hand. In other words, the successful application of mathematical techniques to war purposes required that mathematicians forge new types of collaboration with nonmathematicians in wartime.

The most obvious way in which mathematicians could *as mathematicians* be of immediate service was in teaching. With elementary mathematical skills now required in many combat situations, a massive number of people needed to be trained as quickly as possible, and many mathematicians were employed for this task. "The war presented an exceptional although fleeting opportunity to emphasize the fundamental necessity of mathematics in the technical equipment of the individual.... The student officer was required to learn and use many mathematical ideas that he had never acquired, or had frequently long forgotten" [Bennett 1919]. There

[48]Veblen to Birkhoff, September 10, 1917 [Birkhoff Papers, HUG 4213.2, Box 3].

was a booklet by Alexander Witting, gymnasium professor in Dresden, published in 1916 under the title *Soldaten-Mathematik* [Witting 1916]. It contained introductory remarks about distance measurement and interior and exterior ballistics. Likewise, mathematicians in the Allied countries often served as instructors in artillery schools. Many young French teachers were sent to the Artillery School of Fontainebleau. Lester Ford's 72-page booklet, *Elementary Mathematics for Field Artillery* [Ford 1919], was already mentioned: it contained sections on arithmetic, algebra, geometry, trigonometry, approximate methods, coordinates, aids to calculation, and probability, with an appendix of numerical tables. In the three months prior to the signing of the Armistice, this booklet was taught to about 15,000 students in uniform by a hundred instructors.[49]

When computing became a bit more complicated, interesting types of collaboration were organized between academic mathematicians and officers or military engineers. In his memoirs, Maurice d'Ocagne, the professor of geometry at the École polytechnique who was famous for having invented graphic computing methods he called nomography, explained how he was able to engage the collaboration of practical men:

> when it became imperative to address our shortage of heavy artillery in the shortest delay, the simplicity and the swiftness of the formidable computations required by the autofrettage of the big pieces—computations that until then required the resolution of transcendental equations—was to be maximized. Nomography alone was able to overcome the difficulty, thanks in particular to the nomograms built, following my method and under my supervision, to meet the needs of [the armament complex at] Le Creusot, by the engineer L. Potin.[50]

Besides Louis Potin, d'Ocagne also enlisted the help of navy officer Pierre Goybet, who had served in Verdun: ballistic "computations were long, hard and subject to mistakes due to the fact that they were made under the enemy's fire."[51] Thinking that nomography could help, Goybet was detached to the rear on October 16, 1916, and started to work on his firing nomograms (*nomogrammes de tir*) which reduced computation times from 15 to 3 minutes while eliminating most sources of errors. But d'Ocagne mostly worked with his ex-students of the École polytechique (classes of 1911 and 1913), including Jean Aubert, who published a book about probability theory in wartime firing.[52]

---

[49]Mathematical instruction during the war is addressed in this volume especially in the chapters by Thomas Archibald, Della Dumbaugh, and Deborah Kent and by Jean-Luc Chabert and Christian Gilain.

[50]"*[L]orsque s'imposa la nécessité de remédier dans le plus bref délai possible à notre pénurie d'artillerie lourde, il y eut lieu de porter au maximum la simplicité et la rapidité des formidables calculs requis par l'autofrettage des grosses pièces, calculs qui comportaient jusque là la résolution de certaines équations transcendantes. Ce fut la nomographie qui permit de venir à bout de cette difficulté, notamment grâce aux nomogrammes construits, d'après ma méthode et sous mon contrôle pour les besoins du Creusot, par l'ingénieur L. Potin*" [D'Ocagne Memoirs, Book VI, p. 284].

[51]"*ces calculs étaient longs, pénibles, sujets à erreurs, du fait surtout qu'ils s'opéraient souvent sous le feu de l'ennemi*" [D'Ocagne Memoirs, p. 285].

[52]Jean Aubert, *La probabilité dans les tirs de guerre*, préface de Maurice d'Ocagne, Gauthier-Villars, Paris, 1919; Maurice d'Ocagne, *Principes usuels de nomographie, avec application à divers problèmes concernant l'artillerie et l'aviation, conférences faites à la Section technique*

Like d'Ocagne, mathematicians and scientists seized the opportunity for developing applications of their research field. Often there was no clear distinction between what was a mathematical problem or not, between the types of problems that could be addressed by theoretical or experimental means. While, contrary to WWII, the mathematicians' contribution to cryptology was anecdoctal, since code-breaking was mostly carried out by linguists who marginally used combinatorial and statistical techniques,[53] other mathematical problems encountered, however, proved more challenging for mathematicians and scientists with advanced mathematical training. Here, we shall introduce in broad terms two of them, sound-ranging and exterior ballistics. But aerodynamical research was also lively.[54] In Germany, for instance, Prandtl and coworkers' famous research on the boundary layer was first published during the war, in a top secret journal titled *Zeitschrift der Flugzeugmeisterei* [Prandtl 1918].[55] The historian Detlef Busse has claimed that Ludwig Prandtl indeed instrumentalized war to pursue his research interests in fluid mechanics, succeeding indeed in building a modern wind tunnel in Göttingen and getting army personnel to staff it [Busse 2006, Busse 2008]. Always the statesman of science, Felix Klein was instrumental in brokering the deal with the War Ministry.

3.1. Sound-Ranging. The story of sound-ranging has been told many times, but the chronology presented is often confused and the impact on mathematicians' wartime practices rarely delineated. Most accounts focus either on the technical problem of designing and selecting the proper detector for combat uses and the organizational issues related to the tactical use of sound-ranging in battle.[56] "The chief object of Sound Ranging," explained a participant to Sound-Ranging Sections (SRS) after the war, "is to locate the position of hostile guns; it employs delicate and rather complex instruments and requires a personnel possessing a considerable degree of technical knowledge" [Hinman 1919, p. 14]. As soon as the frontline was stabilized, the practice of artillerymen underwent a tactical change whereby indirect fire became the norm. In other words, the enemy's guns vanished from plain sight and had to be located by indirect methods. In the fall of 1914, therefore, mobilized scientists from various disciplines and military engineers spontaneously started to work on the problem of implementing sound-ranging ideas. Theoretical and experimental studies were carried out without coordination and several systems were developed concurrently. Monopolizing the question, the army's geographical

de l'artillerie (février 1919), par le lieutenant-colonel d'Ocagne, Gauthier-Villars, Paris, 1920, 70 p. D'Ocagne's teaching at the École polytechnique is discussed in Jean-Luc Chabert and Christian Gilain's contribution to this volume.

[53]There is an abundant literature on code-breaking in WWI; see for example [Kahn 1996, Lastours 1998, Collective 2001].

[54]For aeronautics in British context, see June Barrow-Green's chapter; for France, see [Fontanon 1998] and, for the interesting prewar context of the University of Nancy, see the contribution by Philippe Nabonnand and Laurent Rollet in this volume.

[55]On Prandtl, see [Eckert 2006].

[56]The literature on the history of sound-ranging is vast, but there is no systematic study of the various aspects from an international perspective. The story on the French and British sides is summarized in [Chasseaud 1999, pp. 77–78, 96–99]. Innovation and tactical use of sound-ranging in France and Italy are discussed by Martina Schiavon: see [Schiavon 2003a, Schiavon 2003b, Schiavon 2014] and her contribution to [Goldstein & Aubin forthcoming]; see also [Thiberge 1975]; contemporary accounts of surveying in France can be found in [Bourgeois 1920] and [Arthur-Lévy 1923].

service (*Service géographique de l'Armée*), headed by General Robert Bourgeois, a polytechnician having previously specialized in geodesy, succeeded in applying the system to the field by the end of 1915 [Schiavon 2003b, Schiavon 2014].

As already mentioned, many technoscientific problems raised by WWI were rather trivial in mathematical terms. The problem raised by Veblen above had arisen in the context of sound-ranging. Such problems were challenging but nevertheless accessible to anyone with good mathematical training. This question mobilized physicists as well as military and civil engineers, besides mathematicians. Although sound-ranging made it obvious that the solution to mathematical problems was needed in modern warfare, nothing special characterized the mathematicians' approach to a problem that was rather easily solved theoretically by other specialists but that proved difficult to implement for technical, physical, and organizational reasons.

The idea of sound-ranging was simple—basically, locating the opponent's batteries by triangulating the sounds they emitted. It was its practical use on the field that proved rather tricky. Be that as it may, although rather trivial, the mathematical side of the problem highlighted the way mathematical expertise could be brought to bear on military issues. At an early stage of the scientific mobilization, sound-ranging brought mathematicians and scientists to the front where they needed to discuss computing practices, the technical merits of various detectors, the physical characteristics of sound wave propagation, and the tactical use of ranging for the artillery. Occupying a great deal of people with some degree of mathematical skill, sound-ranging therefore shaped the wartime mathematical experience of scientists and combatants, as well as mathematicians' experience of war-related research. While mathematicians no doubt were greatly invested in the problem, some however felt a discrepancy between their theoretical knowledge and the work that was demanded of them.

The mathematical foundation of sound-ranging was rather elementary. Suppose you have three sound detectors A, B, and C on a straight line as in Figure 9. Say the noise made by the discharge was detected first at C, than at B, and finally at A. The time lag between perception at C and at B, $t_c - t_b$, means that the distance between the target P and B is larger than the distance between P and C by a fixed amount corresponding to the time lag multiplied by the speed of sound V, that is $V(t_c - t_b)$. By the definition of a hyperbola, the target P thus belongs to a hyperbola with foci B and C. Applying the same reasoning to A and B led to the conclusion that the battery could be located at the intersection of two hyperbolas, defined by $PB - PA = V(t_b - t_a)$ and $PC - PB = V(t_c - t_b)$.

Alternatively, one could draw three nonintersecting circles centered on A, B, and C, respectively, with radii $V(t_a - T)$, $V(t_b - T)$, and $V(t_c - T)$, where T was a conveniently chosen time. The circle tangent to the three others would be centered at the location of the enemy's gun. Interestingly, mathematicians who dealt with the problem considered the practical aspects of computation. If the three-circle method was used, "the computer may find it useful to have a set of metal disks whose radii differ successively by small amounts, the circle which touches the three circles may then be found very quickly by trial" [Bateman 1918, p. 5].

Another simplification immediately imagined was to replace the hyperbolas by their asymptotes, which worked well when AB was small compared to the distance to the gun. The angle θ of the asymptote to AB could be computed easily as

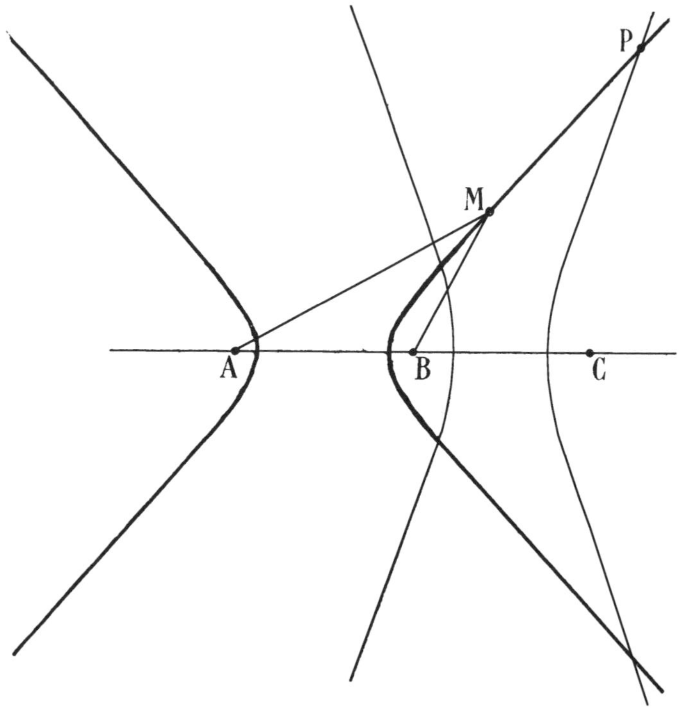

FIGURE 9. The position P of the enemy's battery is determined by the intersection of two hyperbolas with foci A, B and B, C corresponding to the observing stations [Claude 1919, p. 103].

$\cos\theta = (t_b - t_a)/t_{ab}$, where t_{ab} is the time it would take for sound to travel from A to B.[57]

In the course of the year 1915, French scientists altogether developed four systems that took these considerations into account and that indeed seemed to work on the field. Working with Borel and Hadamard, ENS physicists Eugène Cotton and Pierre Weiss built an instrument that measured time intervals electrically. The Cotton-Weiss system was used by the Sound-Ranging Sections (SRS) commanded by Borel on the Champagne front. The second system was developed by the founder of the Liquid Air company, the physicist and industrialist Georges Claude, who called his instrument the "orthophone" (Figure 10) but was frustrated in his attempts to sell it to the military [Claude 1919]. Most widely used by the French Army was a third system known as "TM 1916" (Figure 11). First developed by the hydrographic engineer Ludovic Driencourt, who benefited from the support of Colonel Gustave Ferrié, head of the Military Wireless Service (*Télégraphie militaire*) and Bourgeois' Geographical Service of the army, it was perfected by ENS physicists Henri Abraham and Eugène Bloch [Tilho 1937, pp. 199–200]. The fourth system is due to the astronomer Charles Nordmann; he asked Lucien Bull, a student of the pioneer in chronophotography Étienne-Jules Maray, to design a time

[57]This is the issue debated in Henri Lebesgue's letter to Émile Borel, discussed in David Aubin, Hélène Gispert, and Catherine Goldstein's contribution to this volume.

FIGURE 10. Georges Claude's "ruptor" sending a signal when reached by the sound wave [Claude 1919, p. 99].

recording machine using microphones for which Weiss worked out the mathematical details.[58]

Sound-ranging was quickly adopted by the Allies. Asked by the British to adapt the sound-ranging techniques to their army, the physicist and future Nobel Prize recipient William L. Bragg, assisted by Lieutenant Harold Robinson, who was called "an expert mathematician" even though he was a physicist from Manchester, paid a visit to Ernest Esclangon in Gâvre on March 11, 1916, before they chose Bull's apparatus.[59] Esclangon recounted Bragg's visit to Gâvre. The Bull apparatus was

[58]On Nordmann, see David Aubin's chapter in [Goldstein & Aubin forthcoming].

[59]Quoted from "Report on Survey on the Western Front, 1914–1918," p. 106. This report and other detailed contemporary reports by the British, including international comparisons, are available at http://www.defencesurveyors.org.uk/Images/Historical/

FIGURE 11. Recording unit of French artillery sound-ranging system, model TM 1918. Purchased by Finland in 1927. © Sirkka Ojala/The Artillery Museum of Finland.

later improved upon by Corporal William Sansome Tucker, who had studied physics at Imperial College, London. In 1917, after a visit by French physicists Abraham and Claude Fabry, the Princeton physicist Augustus Trowbridge undertook the study of four different French systems for sound detection and also selected the Bull-Tucker system to equip the U.S. Army.[60]

As for the Germans, they claimed that the idea of sound-ranging had first emerged in Austria before the war. Like the Entente armies, the German military only started to test sound-ranging procedures in the fall of 1914. The following year, the Prussian Artillery Testing Commission (APK) was put in charge of developing a better protocol. The scientific direction of this project was assigned to the

WWI/Report_on_survey_on_western_front___PART_II.pdf (last viewed on July 18, 2014). On British surveying more generally, see [Chasseaud 1999, Palazzo 1999, MacLeod 2000, Van der Kloot 2005]. Esclangon, an astronomer with a thesis in pure mathematics, had written a report on sound-ranging as early as fall 1914; see David Aubin, Hélène Gispert, and Catherine Goldstein's chapter in this volume.

[60] [Trowbridge 1919, Compton 1937]. On American sound-ranging, see [Kevles 1969]; detailed contemporary accounts are to be found in [Hope-Jones 1928].

physicist Rudolf Walter Ladenburg, from Breslau University, assisted by Max Born, the famous physicist then privatdocent at the University of Berlin. At first, Born helped Ladenburg to work out numerical and graphical methods for sound-ranging and to transform them into military regulations. Like his colleagues in the West, he developed ways to take into account wind and altitude, using Hamiltonian dynamics to study the propagation of sound. From the formulas, tables were computed by his staff (composed of students in mathematics and physics) and checked with experiments on the shooting range. "I think it was a good piece of applied mathematics" [Born 1978, p. 171]. Contrary to the Allies, Germany apparently was not able to develop the delicate measuring instruments needed for the task. Only late in the war did Born start to work on the problem posed by head waves, enrolling for this task the help of specialists in ballistics.[61]

However, after an initial phase of excitement, in all countries the mathematical aspects of the problem disappointed mathematicians who, by and large, deserted the study of sound-ranging, while physicists went on to play the major role.[62] But what should be noted is the fact that the adoption of effective sound-ranging methods and the constitution of SRSs in the army furthered the actual use of mathematics near the trenches.

In the dozens of SRSs put together by the French, British, and U.S. Armies, physicists and mathematicians however remained in the minority. Typically, six detectors, placed at known positions, detected the sound of the gun and transmitted a signal to a chronograph (Figure 12). Observers and linesmen had to be positioned on or near the frontline and suffered many casualties. "It must not be imagined that the processes involved are purely mechanical; to insure the proper operation of the apparatus and to obtain information as to the calibre and direction of fire of the enemy's artillery, observers are posted well in advance of the microphones, often in the front-line trenches" [Hinman 1919, p. 14].[63] The sound-ranging central station was usually a few kilometers in the rear: "it is here that the actual data are compiled and submitted to the Chief Intelligence Officer of that section of the front. It is at the central station that the greatest number of men are stationed, linesmen, computers, telephone operators and men who operate the electrical apparatus. Their work is of the utmost importance, but Mr. Censor has forbidden a detailed description of the methods used by them" [Hinman 1919, p. 48].

Time differences recorded by the detectors were reported on a board specially designed for that purpose, where the intersection of strings gave the position of the

[61] Interestingly, however, sound-ranging studies in Berlin meanwhile led to the development of a strong research program in experimental psychology in order to study errors made by human observers [Hoffmann 1994].

[62] We mention the publication of a detailed article about the "mathematical theory of sound-ranging": [Bateman 1918]. A British mathematician who had emigrated to the U.S. in 1910, Harry Bateman had worked for the Weather Bureau in Washington, D.C., before being appointed at Troop College, later to become the California Institute of Technology, together with Hale and Millikan, both major actors in the American scientific mobilization. Surprisingly, this paper was published in January 1918. To be sure, there was an article published earlier by Emmanual Vallier, a member of the French Commission of Inventions, who had also made public the mathematical basis of sound-ranging [Vallier 1915].

[63] Other eyewitness accounts of sound-ranging on the front can be found in [Innes 1935]. There is a French novel for youth that painted the heroic adventures of a sound-ranging section [Spitzmuller 1919].

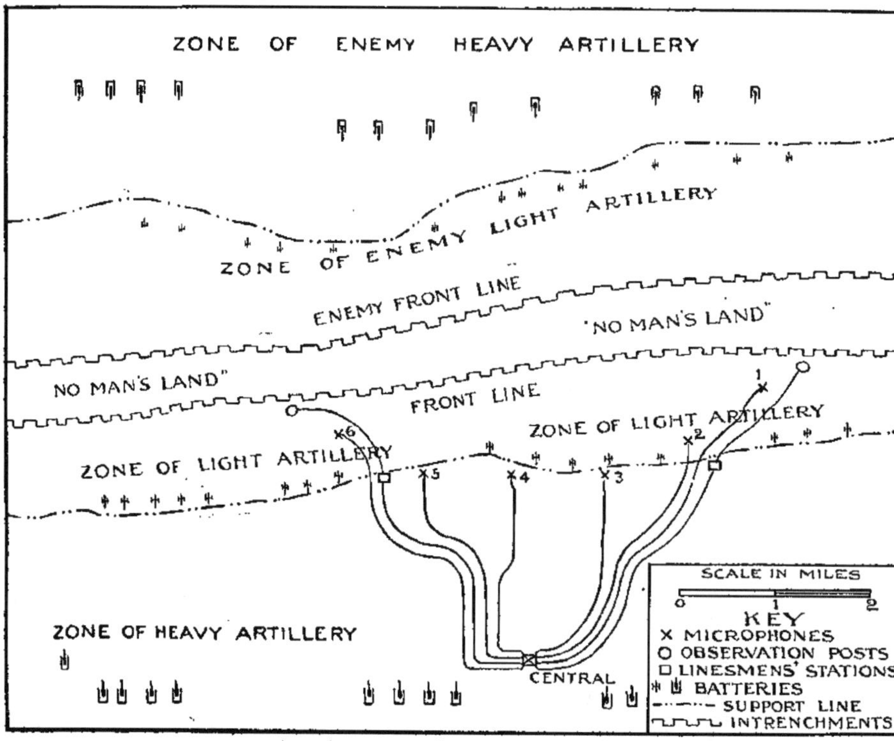

FIGURE 12. Diagram of a sound-ranging base. Taken from [Hinman 1919, p. 28].

piece.[64] Corrections for the direction and velocity of the wind and for temperature were applied. The mathematical and computational aspects of the work were, as we can see, reduced to a minimum. Indeed, if a certain level of mathematical training was recommended for sound-ranging, the British report suggested that the entire training course for officers and computers would take no more than 17 and 14 days, respectively, to cover all practical and theoretical aspects needed to be operational on the field.

> It has often been thought that sound ranging is a peculiarly difficult art, beyond the comprehension of the average man and exacting a knowledge of pure science and mathematics common only to the senior wrangler or a science master. *This is absolutely erroneous*.... Scientific control and experiment are essential but can be done behind and so long as the officers of the sections are able to understand their instruments and the value of the results they obtain it is unnecessary to specify a university degree.[65]

---

[64] A paper representation of this type of board is pictured in David Aubin's contribution to this volume.

[65] "Survey on the Western Front. Preliminary Report," p. 41, our emphasis, `http://www.defencesurveyors.org.uk/Images/Historical/WWI/Winterbottom_Survey_on_Western_Front.pdf`.

Sound-ranging placed mathematics and computing at a central position on the battlefields of WWI, but the mathematics involved was never very advanced. Furthermore, every truly complex mathematical operation could be done once and for all in the rearlines. What was left to do for frontline practitioners was the complex day-to-day maintenance of instruments and telegraphic lines under shelling. For this, "experience and coolness" [Hinman 1919, p. 14] as well as the technical knowledge required to work with delicate electrical apparatuses, were more important than mathematical aptitude.

3.2. Exterior Ballistics. In the field of exterior ballistics, by contrast to sound-ranging, mathematicians played more decisive roles.[66] By 1914, every mouthpiece came with firing tables which were supposed to give the range of its projectiles as a function of the initial angle of shooting and other variables (such as powder load and type or projectile type). Contrary to sound-ranging, ballistics was a well-established science at the start of WWI, although unevenly developed in each belligerent country.[67] The dynamical equations of the ballistic problem were easy to establish, but the need to account for air resistance had plagued theoretical approaches with difficulties. Several suggestions were made for the air drag function, but theoretical ballisticians had to resign themselves to develop methods able to give approximate solutions of differential equations involving a function $f(v)$ whose exact analytical form was unknown:

$$gd(v\cos\theta) = vf(v)d\theta, \tag{3.1}$$

$$gdx = -v^2 d\theta, \tag{3.2}$$

$$gdy = -v^2 \tan\theta d\theta, \tag{3.3}$$

$$gds = -v^2 \frac{d\theta}{\cos\theta}. \tag{3.4}$$

In these equations, x and y represented the Cartesian coordinates of the projectile and s was the intrinsic coordinate along the path of the projectile; one thus has $dx = cos\theta ds$ and the velocity $v = ds/dt$. If integration of (3.1), known as the "hodograph," was possible, all the other quantities would be given by simple integration. But this was possible only for a limited number of special cases, none realistic enough to account for actual ballistic trajectories. By the end of the nineteenth century, the development of practical ballistics had led to the adoption of an empirical law for air drag represented in Figure 13 that clearly was nonintegrable.

At the turn of the century, rational ballistics witnessed spectacular advances, as professional ballisticians developed expedited approximation methods enabling them to compute range tables while relying on a minimum number of costly firing tests. This was mainly due to the work of the military mathematician Francesco Siacci who taught at the School of Applied Artillery and Engineering in Turin as well as at the University of Turin (where he was replaced by Volterra in 1893). Siacci's method was popularized in France by Prosper-Jules Charbonnier, a graduate of the

[66]Ballistics is examined in greatest mathematical details in David Aubin's contribution on Gâvre, but also in the four chapters of this volume dealing with Great Britain, France, Italy, and the U.S. (resp. by June Barrow-Green; David Aubin, Hélène Gispert, and Catherine Goldstein; Pietro Nastasi and Rossana Tazzioli; and Thomas Archibald, Della Dumbaugh, and Deborah Kent).

[67]For a history of ballistics, see [Charbonnier 1928]. As far as we know, there is no equivalent in English. See however [Gluchoff 2011].

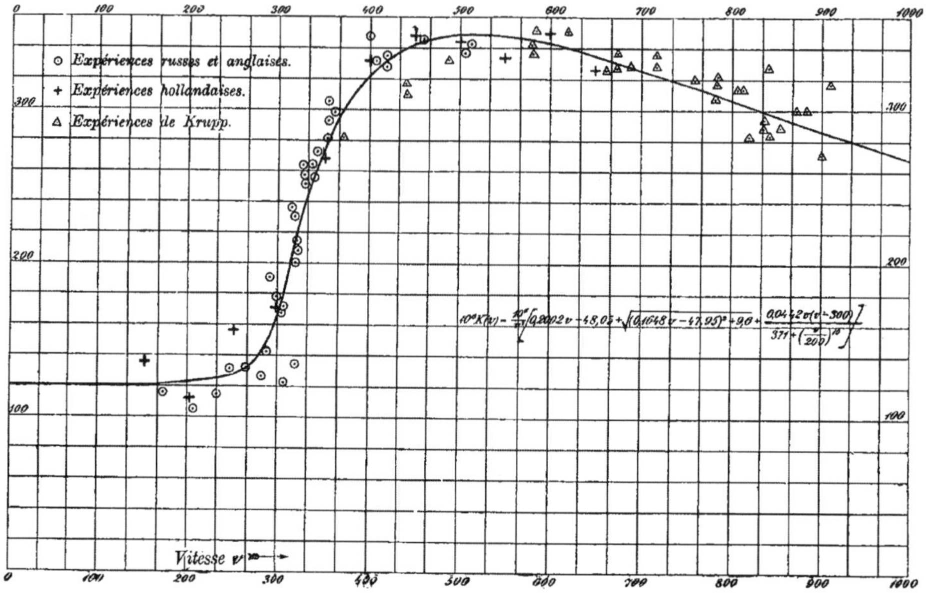

FIGURE 13. Air drag as a function of the velocity. From [Cranz 1903, p. 197]. Note that the experimental data on this curve came from belligerent countries on both sides during the war and one that remained neutral: Russia, Great Britain, Holland, and from the Krupp Company, Germany.

École poytechnique who was named president of the Gâvre Commission in 1906, and in Germany by Carl Cranz, a professor at the Military Technical Academy in Berlin who was known for having introduced precise experimental procedures from physics into ballistics and produced beautiful cinematographic studies of projectiles. Meanwhile, in Great Britain, George Greenhill applied elliptic functions to integrate the hodograph in the case of cubic drag.[68] Despite strong opposition, the mathematical sophistication of exterior ballistics was slowly gaining acceptance from artillery officers, especially in France where many officers were trained at the École polytechnique which had a strong mathematical tradition.

A practical science with its own textbooks and experimental tradition, which was—however badly—taught in military academies, exterior ballistics was in 1914 strongly pursued by military research centers in several European countries and developed into an autonomous branch of military science. But ballistics remained by and large completely foreign to civilian mathematicians and physicists. Reviewing one of Cranz's books, the historian of mathematics and high school teacher Heinrich Wieleitner wrote about ballistics:

[68]On Greenhill, see June Barrow-Green's chapter in this volume. See also Greenhill's *Applications of Elliptic Functions* [Greenhill 1892, p. 244]. Published in 1888, Siacci's treatise on exterior ballistics was quickly translated into French [Siacci 1892]; Charbonnier and Cranz also wrote their own textbooks, while the latter was in charge of articles about ballistics in Klein's encyclopedia of the mathematical sciences [Cranz 1903, Charbonnier 1907]. For more mathematical details concerning Siacci's methods, see David Aubin's chapter in this volume.

> This science, but little mentioned in the courses of mathematicians and physicists, led until recently a lively but retired life in ballistics laboratories, in the lecture halls of military academies, and on the firing ranges of weapon factories. Today, when millions of projectiles from all kinds have been fired for months, all with the purpose of reaching their target, wider circles have become interested in what are the circumstances that determine the trajectory of projectiles [Wieleitner 1915, pp. 52–53].

The wide adoption of Siacci's methods however had a serious drawback, namely that it worked well only for a flat fire, when initial firing angles were less than 10°–15°. But battleground conditions imposed that new tables be computed rather quickly. Before the war, it was usually assumed that artillery would be merely used as a backing for the attacks of rapidly moving infantries and that it would fire mostly at close range. As a consequence, there would be little time for a priori computations of trajectories, which would be of little use since adjusting would easily be made *de visu*. As trench warfare led to the unexpected development of indirect fire, gunners came to rely more than previously on their tables. And they found them wanting. Several factors rendered prewar tables inadequate: the production of new, more powerful materials, especially the heavy artillery on railways, the development of anti-aircraft and mountain gunnery, the necessity to take meteorological conditions into account, etc.[69] Forest R. Moulton, who was in charge of the Ballistics Branch of the Ordnance Department of the U.S. Army in April 1918, described how changes in the practice of artillery impacted theoretical and experimental ballistics:

> The new demands upon artillery and the changes in artillery practice necessitated by barrage fire, counter-battery fire, long-range fire, and anti-aircraft fire, as the war progressed, raised corresponding ballistic problems. Upon entering the army a hasty examination of the classical ballistic methods showed, not only that they were not well suited to the solution of the problem, even under earlier conditions.... Accordingly, the subject was taken up anew by the author as a scientific problem requiring close co-ordination of adequate theory and well-conducted experiments [Moulton 1926].

Soon, exterior ballistics was felt as a pressing need by military authorities who sought out technical assistance wherever it could be found. There were strong contrasts between countries, like France, where parts of the military possessed a strong mathematical culture instilled by polytechnical establishments and those, like the U.S., where since the mathematical skill of army officers was much lower mathematicians played a more active role in introducing Siacci's methods and adapting them to the requirements of the Great War. "In the French field artillery there was a tradition of serious interest in guns, but in the British and German services the focus was on hauling them into action at a gallop, and then firing over open sights. Tables of logarithms were seldom consulted" [Van der Kloot 2005, p. 274].[70] In

[69]This is discussed at some length in several contributions to this volume by June Barrow-Green, by Pietro Nastasi and Rossanna Tazzioli, and, especially, by David Aubin.

[70]The mathematical culture of French army and navy officers is discussed in this volume in the contributions by David Aubin and by Jean-Luc Chabert and Christian Gilain. The ballistic situation in the U.S. is discussed in the contribution to this volume by Thomas Archibald, Della

Britain and Italy, contributions to this volume describe an intermediary situation. As far as Germany is concerned, it is safe to say that while Cranz played a crucial role at the APK in Berlin, many mathematicians contributed detailed studies to ballistic problems.[71] Scientists and mathematicians involved in ballistics clearly produced a great number of studies that contributed to change the field in significant ways, in particular by introducing new computing procedures. Due to military secrecy, however, the publication of this material was tricky: "a former member of the ordnance department has told me that he has in his possession over a hundred copies, mostly unpublished blue-prints, of articles on ballistics" [Kellogg 1921, p. 543].

Like sound-ranging (and flight theory), ballistics involved problems for whose solution mathematical skills proved useful. But practical ballistics also posed problems that required other kinds of expertise. The physicist Gordon Hull from Dartmouth College put it bluntly at the end of the war:

> Enough has been said to show that the fundamental problem of the projectile is not one of mathematics. There are various mathematical methods of handling the problem.... The problem is one of experimental science. We must first determine the complete law of air resistance for every probable form of projectile, then we must determine the variation of force as the axis of the projectile changes in direction.... Mathematicians may then find it necessary, using these known facts, to formulate the differential equations of a twisted trajectory and to evolve methods of integration. But it is quite probable that simple physical methods of integration may be devised [Hull 1920, p. 231].

For the most part, the needs for true mathematical innovation were limited in those fields. What was required mostly consisted in efficient methods to carry out lengthy computations and mathematical analyses of physical problems. Although, as we show in this book, mathematicians sometimes provided powerful insights in some complex problems, for the most part they proved no more useful in that task than mathematically trained physicists and engineers, while lagging behind in most practical matters. By and large, mathematicians were not the best people to address the mathematical problems raised by modern warfare.

Be that as it may, the type of mathematics involved in war-related research covered a rather wide spectrum. In this mechanical war, everything that had to do with rational mechanics was crucial. Mathematicians often had expert knowledge of this domain and its methods, but the war may have helped to distance rational mechanics from mathematical physics and theoretical physics, domains that remained nearer to mathematicians' concerns [Seth 2010]. Closely related of course were issues from mathematical analysis. The exact and approximate solutions of ordinary and partial differential equations were often required: air resistance problems led to research in conformal mapping theory and practical solutions of the hodograph led to nice work in error theory from probabilistic and deterministic viewpoints. All aspects of computational mathematics were of course crucial in a war that required

Dumbaugh, and Deborah Kent. For more mathematical details on ballistics in the U.S. in the context of both military mathematics and numerical integration methods, see [Gluchoff 2011]. About computing in WWI, see [Grier 2005, ch. 9 & ch. 10].

[71] A sample of such studies among many others is [Noether 1919, Wiener 1919].

far more computed data than ever before. Besides the design of algorithms, the analytical treatment of errors, perturbation theory, and applications of probability theory were highly innovative.

4. Mathematicians and Mathematics through the First World War

Besides its obvious human impact, the war allowed many men to experience mathematics differently than they might have otherwise. For some of them, this experience completely changed the course of their professional lives: we will encounter several examples in what follows, including Ralph Fowler and Edward A. Milne in England, or Jules Haag in France. Bracketing the war in their resumes, as several did, was just one among several ways to deal with the return to normalcy—including in some cases the lack of understanding from a younger generation with little or no tolerance for deeply held values sown by traumatic war experiences. A prominent member of those younger cohorts, Weil, after discussing the disastrous losses suffered by ENS students during WWI as quoted above, underlined another consequence of the war. For him, veterans were simply not carrying out very good mathematical research:

> Four years or more of military life—either close to, or far from, death, but in any case far from science—are not a good preparation to come back to it; very few of those who survived recovered an interest they used to have in it [Weil 1991, p. 132].

One of this book's aims is to show that WWI involved more effects on mathematics than the simple loss the Bourbaki thesis is ready to concede. As we have seen, many developed a lifelong, even if in some cases partial, interest in applied mathematics that was absent from their prewar research. Applied mathematics indeed received a boost from the war, and for some, as the Harvard professor Oliver Dimon Kellogg who had worked on submarine detection during the war, this showed the way of the future [Kellogg 1921, p. 543]. Many of those involved in the scientific mobilization retained part-time positions as expert or consultant for the navy, artillery commissions, or industries. Shifting their research interests, mathematicians could also enlarge the type of professional activities they chose to engage in: from research and teaching to institutional reconstruction or political engagement. The Austrian-born mathematician Theodore Vahlen, a number theorist before the war, worked on ballistics during the conflict and turned to applied mathematics afterwards; and he also joined the Nazi Party in 1922, becoming an ardent promoter of so-called *Deutsche Mathematik* [Siegmund-Schultze 1984], which stressed visual intuition and utility against abstraction and formalism. The opposite claims made during and after the war—often much more publicized in the historiography—like Godfrey Hardy's famous dismissal of applied mathematics as trivial mathematics [Hardy 1940, p. 38]—should thus be understood in the context of a much deeper involvement of mathematicians in social matters, on a large variety of scales.

In the chapters that follow, authors have aimed at capturing dynamical features of some of these changes. Focusing on individual experiences, as we have done above, allows us to take the measure of chance meeting, personal relations, and previous disciplinary training. But it makes it more difficult to find the articulation between the various scales at which change occurred. To this end, authors have chosen to analyze transformations that happened in specific *places* through the war.[72] While people moved from or to several places, personal war experiences were partly fossilized in such places that also went through slower modifications. Some chapters of this volume thus illustrate the effects of the war on major centers of research and teaching, like Cambridge University or the École polytechnique. Others emphasize the national level, in the United States, or in Italy, in order to reconstruct chains and shortcuts tying institutional (re)organization to personal experiences of mobilization or to specific war-related technical developments.

But one may wonder if it makes sense to speak of an "Allied" mathematics. During the war, pamphlets, posters, and "days of the Allies" bear testimonies of the efforts made to reinforce the idea that the corresponding, democratic countries were united against a common enemy. But prewar international relations in mathematics did not nicely overlap with political rapprochements. U.S. and Italian mathematicians, for instance, regularly traveled to Germany before the war and were greatly inspired by German mathematicians, and by Klein especially [Parshall & Rowe 1994]. A classical opposition at the time was that between British *civilization* and German *culture*, but the opposition could function as a model to incorporate or as a foil to valorize the achievements of one's country. Some prominent British intellectuals were in the midst of a Celtic revival which made them look with a friendly curiosity towards their German counterparts. The French situation, on the other hand, was very different: the consequences of the 1870 war, and in particular the loss of Alsace and Lorraine, went along with an enduring competition with German centers for mathematical hegemony. This situation partially explains the particularly acerbic attacks in France against German science during and just after the war. However, before the war, this rivalry was accompanied by a keen, if critical, interest in the work of the other: as late as the early summer of 1914, for instance, German scientists visited Gustave Eiffel's aerodynamical laboratory [Fontanon 1998]. This reciprocal awareness would take more than a decade to be reconstructed.

As will appear at several points in this book, wartime alliances led to inter-Allied scientific collaboration: innovations in firing tables were shared; observation trips in the respective Allied proving grounds organized; institutional models copied. After the war, the circulation of mathematics still matched these alliances—at least during the early 1920s. Renewed arguments in favor of the universality of science and mathematics clashed head-on, starting in 1918, with the hope that a new internationalism could grow out of Allied nations by excluding mathematicians from the former Central Powers. In the mid-twenties and thirties, the Rockefeller Foundation would play a decisive role in constructing afresh, if not internationalism *per se*, at least a large circulation of young mathematicians among nations including former Central Powers [Siegmund-Schultze 2001]. But even internationalism restricted to the Allies at the end of the First World War was no guarantee of a unity of objectives.[73] Two nationals of Belgium, a formerly neutral nation that had sided with the Allies after German troops invaded its territory, might serve as an illustration. While the mathematician Charles de la Vallée-Poussin declared in 1924 that "the mathematical sciences ... often borrow from the French mind a definite orientation" [SMF 1924, p. 35], the historian of science Georges Sarton,

[72]Recently, an emphasis on places in the history of science or mathematics has been rather fruitful; this is addressed in [Aubin 2009] where reference to the literature on places can be found.

[73]On this issue, also see [Siegmund-Schultze 2001, ch. 1].

who had taken refuge in the U.S., expressed his "faith in the Anglo-Saxon conception of life, and also [his] love of and [his] hope in the younger civilizations of the world" [Sarton 1919, p. 321].

We have chosen in the first four chapters to draw comparisons by following the chronology of the war. We first discuss two important places for mathematics, Cambridge and Paris, and their very different experience of early mobilization. We then turn to universities in Italy and the United States, as the delayed entrance of their countries into the war, in 1916 and 1917 respectively, allows us to take stock of the debates among mathematicians on the opportunity to join in or not, and of the mathematical links the war fostered. We then move to studies, on a smaller scale, of the fate of two places where collaboration between mathematicians and the military preceded the war, and still was changed by it, the École polytechnique and the Gâvre commission. In the last chapter, we analyze an effect of the redistribution of territory and frontiers in the middle of Europe, through a study of the case of Nancy, which was before 1919 a bridgehead for French mathematics at the limit of the then German Lorraine.[74]

Several common phenomena can be exhibited among the cases surveyed here. In most, the postwar period was for instance a time of intense institutional reconstruction (sometimes forced by the material destruction of older buildings, as in Lille) supported by the wish to draw lessons from the war. Chairs of fluid mechanics, to take but one example, were created almost simultaneously in several countries—sometimes inside new institutes, sometimes in more traditional university settings. This in turn led to an academic disciplinarization (via new textbooks and new curricula) of certain topics which were before the war either a simple collection of results issued from various mathematical fields or left to professionals outside academia. More generally, networks that were sometimes created haphazardly in wartime through chance encounters or personal connections were transformed into officially funded commissions or institutions.

Pure mathematics was also affected in several ways: as seen above, it was for some a refuge and a worthwhile escape from postwar discourses and activities. But the war also left new epistemological marks. For instance, the manner in which computational issues were brought to the fore not only paved the way for the future development of computing machines and of proper technoscientific institutes but also led some mathematicians to reconsider previous theoretical work in this new light. Châtelet's number-theoretical Peccot lectures at the Collège de France delivered just before the outbreak of WWI were for instance later praised for having "revealed ... the use and the effectiveness of these algorithms, the use of which has since become constant in theoretical physics and in mathematics applied to industrial technology."[75]

This leads us to examine afresh the question of modernity and modernization in mathematics. In his pioneering work [Mehrtens 1990, Mehrtens 1996b], the historian Herbert Mehrtens distinguished within the process of modernity a modernist trend (the only one usually associated with the expression "modern mathematics")

---

[74] For the differences and the similarities with the case of Brno in Bohemia, see a paper by Laurent Mazliak and Pavel Šišma, `http://www.proba.jussieu.fr/~mazliak/MazliakSisma2008.pdf`, with Poland and the creation of *Fundamenta Mathematicae* in 1920, see [Duda 1996].

[75] Quoted in Sébastien Gauthier's contribution to [Goldstein & Aubin forthcoming].

from a countermodernist one. For him, important characteristics of modern mathematics were "first, the autonomy of cultural production and, second, the departure from the vision of an immediate representation of the world of experience. [...] Modern axiomatic, conceptual, and structural mathematics becomes largely anti-intuitive. It is the mathematics of autonomous professionals" [Mehrtens 1996b, p. 521]. In this view, professional mathematicians also positioned themselves as internationalists who were opposed to national or cultural biases. Countermodernists, on the contrary, would stress intuition and an externalist basis for the truth and meaning of mathematics, whether in the natural sciences or in society. Moreover, modernists have mostly been identified with an *avant-garde* developing pure mathematics. "Mathematicians of a more conservative type or mathematicians in applications, hybrid fields or in teaching cannot fully identify with the modernist attitude" [Mehrtens 1996b, p. 523]. Severing mathematics from the sciences and society is an important part of the dichotomy [Epple 1999, p. 193]. Confronted with ideals of modernization and rationalization put forward by the Rockefeller philanthropists, Reinhard Siegmund-Schultze underscored the difficulties of equating a modernist position in pure mathematics, as defined above, and simple progress or innovation.

The studies that follow show how the war indeed redistributed these frontiers. Whether they integrated or not some of the internal values of modern mathematics, the point is that many mathematicians of the war generation extended the limits of their profession beyond academic research. Some simply juxtaposed activities of various kinds (most of which being neglected by the historiography), while others, in particular in France, claimed that holding together and knitting all these aspects, from the most theoretical and general theorems in, say, differential equations, to the construction of a center of computations in close contact with industry, was the essential mission of the future, adequately modern, mathematician. These activities of the interwar period certainly required more investigation. But in this respect, the Bourbaki thesis already appears less as a complete, objective, description of the war effect than a consequence of the position of its promotors inside mathematics, and of their choice as to what a mathematical life is made of. Paradoxically enough, their vision widened the gap created by the war experience, while the spokespersons of the war generation (as witnessed above in Châtelet's case) often strove to reconnect beyond the war the threads of their professional lives torn apart by four years of massive destruction.

Abbreviations

- APK: Artillerie-Prüfungs-Kommission.
- CRAS: *Comptes rendus hebdomadaires des séances de l'Académie des sciences.*
- ENS: École normale supérieure, Paris.
- ETH: Eidgenössische Technische Hochschule, Zürich.
- ICMI: International Commission on Mathematical Instruction.
- IRC: International Research Council.

Bibliography

Archival Sources

[Birkhoff Papers] George David Birkhoff Papers, Harvard University Archives.

[Borel Papers] Fonds Émile Borel, Archives of the Academy of Sciences, Institut de France, Paris, 44J.

[D'Ocagne Memoirs] Maurice d'Ocagne, *Souvenirs de Maurice d'Ocagne, membre de l'Académie des Sciences, 1862–1938*. Manuscript at the Bibliothèque de l'Institut de France, Quai Conti, Paris, Ms. 4671–4674, n.d.

Published Works

[AASAE] Association amicale de secours des anciens élèves de l'École normale supérieure, *Réunion générale annuelle*, volume annuel, 1915–2009.

[AIP 1962] Interview with Richard Courant, by Thomas S. Kuhn & Mark Kac, on May 9, 1962, Niels Bohr Library & Archives, American Institute of Physics, College Park, MD, USA, `www.aip.org/history/ohilist/4562.html`.

[Andler 1994] Martin Andler, Les mathématiques à l'École normale supérieure, in *L'École normale supérieure : le livre du bicentenaire*, dir. Jean-François Sirinelli, Presses universitaires de France, Paris, 1994, pp. 351–404.

[Anizan 2006] Anne-Laure Anizan, *Paul Painlevé (1863–1933): un scientifique en politique*, doctoral thesis, Institut d'études politiques, Paris, 2006.

[Anizan 2012] Anne-Laure Anizan, *Paul Painlevé : science et politique de la Belle Époque aux années trente*, Presses universitaires de Rennes, Rennes, 2012.

[Anonymous 1906] Anonymous, *La généralisation d'un théorème de Lagrange*, Revue du mois **2** (1906), pp. 508–510.

[Arthur-Lévy 1923] Arthur-Lévy, *Les coulisses de la guerre : le service géographique de l'armée (1914–1918)*, Revue des deux mondes **17** (1923), pp. 419–452.

[Aubin 1997] David Aubin, *The Withering Immortality of Nicolas Bourbaki: A Cultural Connector at the Confluence of Mathematics, Structuralism and the Oulipo in France*, Science in Context **10** (1997), pp. 297–342.

[Aubin 2005] David Aubin, George David Birkhoff, Dynamical Systems (1927), in *Landmark Writings in Western Mathematics, 1640–1940*, ed. Ivor Grattan-Guinness, Elsevier, Amsterdam, 2005, pp. 871–881.

[Aubin 2009] David Aubin, Observatory Mathematics in the Nineteenth Century, in *Oxford Handbook for the History of Mathematics*, ed. Eleanor Robson & Jacqueline Stedall, Oxford University Press, Oxford, pp. 273–298.

[Aubin 2012] David Aubin, L'élite sous la mitraille : les mathématiciens normaliens "morts pour la France", 1914–1918, in *Aventures de l'analyse de Fermat à Borel. Mélanges en l'honneur de Christian Gilain*, ed. Suzanne Féry, Presses universitaires de Nancy, Nancy, 2012, pp. 681–706.

[Aubin & Bret 2003] David Aubin & Patrice Bret, eds., *Le Sabre et l'éprouvette: l'invention d'une science de guerre, 1914–1939*, 14–18 Aujourd'hui, Agnès Viénot, Paris, 2003.

[Audin 2009] Michèle Audin, *Fatou, Julia, Montel : Le Grand Prix des sciences mathématiques de 1918, et après...*, Springer, Berlin, 2009.

[Audin 2010] Michèle Audin, review of Reinhard Siegmund-Schultze, *Mathematicians Fleeing from Nazi Germany: Individual Fates and Global Impact* (Princeton, 2010), Notices of the American Mathematical Society **57** (2010), pp. 1300–1302.

[Audoin-Rouzeau & Becker 2000] Stéphane Audoin-Rouzeau & Annette Becker, *14–18. Retrouver la guerre*, Gallimard, Paris, 2000.

[Barolet 1916] Henri Barolet, *Question d'analyse indéterminée*, L'Enseignement mathématique **18** (1916), pp. 269–272.

[Bateman 1918] Harry Bateman, *Mathematical Theory of Sound Ranging*, Monthly Weather Review **46** (1918), pp. 4–11.

[Beaulieu 1993] Liliane Beaulieu, *A Parisian Café and Ten Proto-Bourbaki Meetings (1934–1935)*, The Mathematical Intelligencer **15**(1) (1993), pp. 27–35.

[Beaulieu 1994] Liliane Beaulieu, Questions and Answers about Bourbaki's Early Work (1934–1944), in *The Intersection of History and Mathematics*, ed. Sasaki Chikara, Sugiura Mitsuo, & Joseph W. Dauben, Birkhäuser, Basel, 1994, pp. 241–252.

[Beaulieu 2009] Liliane Beaulieu, *Regards sur les mathématiques françaises entre les deux guerres*, Revue d'histoire des sciences **62-1** (2009), pp. 9–37.

[Bell 1915] Clive Bell, *Art and War*, International Journal of Ethics **26** (1915), pp. 1–10.

[Bennett 1919] Albert A. Bennett, Review of Ford's *Elementary Mathematics for Field Artillery*, The American Mathematical Monthly **26** (1919), pp. 353–355.

[Berger 2005] Marcel Berger, *Cinq siècles de mathématiques en France*, Association pour la diffusion de la pensée française & Ministère des affaires étrangères, Paris, 2005.

[Berndt & Rankin 1995] Bruce C. Berndt & Robert A. Rankin, *Ramanujan: Letters and Commentaries*, AMS/LMS, Providence & London, 1995.

[Bliss 1944] Gilbert A. Bliss, *Oskar Bolza—In memoriam*, Bulletin of the American Mathematical Society **50** (1944), pp. 478–489.

[Born 1918] Max Born, *Herbert Herkner*, Die Naturwissenschaften **6** (1918), pp. 179–180.

[Born 1978] Max Born, *My Life: Recollections of a Nobel Laureate*, Taylor & Francios, London, 1978.

[Bourbaki 1960] Nicolas Bourbaki, *Éléments d'histoire des mathématiques*, Hermann, Paris, 1960.

[Bourgeois 1920] Robert Bourgeois, *Le Service géographique de l'Armée pendant la guerre*, Revue scientifique **58** (1920), pp. 673–685.

[Bréauté 1919] Roland Bréauté [Robert Launay], *Un universitaire aux Armées*, Brossard, Paris, 1919.

[Burgess 1919] George K. Burgess, *Science and the After-War Period*, Scientific Monthly **8** (1919), pp. 97–108.

[Busse 2006] Detlef Busse, *Forschung im Krieg – Forschung durch den Krieg: von der Instrumentalisierung militärischer Interessen für die Errichtung der Göttinger Modellversuchtanstalt*, Kollegen - Kommilitonen - Kämpfer: Europäische Universitäten im Ersten Weltkrieg, ed. Trude Maurer, Franz Steiner Verlag, Munich, 2006, pp. 283–296.

[Busse 2008] Detlef Busse, *Engagement oder Rückzug? Göttinger Naturwissenschaften im Ersten Weltkrieg*, Schriften zur Göttinger Universitätsgeschichte **1**, Universitätsverlag, Göttingen, 2008.

[Butzer et al. 1984] Paul L. Butzer, Rolf J. Nessel, & Eberhard L. Stark, *Eduard Helly (1884–1943), in memoriam*, Results in Mathematics **7** (1984), pp. 145–153.

[Butzer et al. 1980] Paul L. Butzer, Stefan Gieseler, Franz Kaufmann, Rolf J. Nessel, & Eberhard L. Stark, *Eduard Helly (1884–1943)*, Jahresbericht der deutschen Mathematiker-Vereinigung **82** (1980), pp. 128–151.

[Charbonnier 1907] Prosper-Jules Charbonnier, *Balistique extérieure rationnelle*, O. Douin, Paris, 1907.

[Charbonnier 1928] Prosper-Jules Charbonnier, *Essais sur l'histoire de la balistique*, Société d'éditions géographiques, maritimes et coloniales, Paris, 1928.

[Chasseaud 1999] Peter Chasseaud, *Artillery's Astrologers: A History of British Survey and Mapping on the Western Front, 1914–18*, Mapbooks, Lewes, 1999.

[Clairin 1920] Jean Clairin, *Sur les invariants des caractéristiques des équations aux dérivées partielles à deux variables indépendantes*, ed. Émile Gau, Annales scientifiques de l'E.N.S. **37** (1920), pp. 107–116.

[Claude 1919] Georges Claude, *Politiciens et polytechniciens: Au creuset de la guerre*, private ed., Boulogne-sur-Seine, 1919.

[Collective 1916] *Friends of France: The Field Service of the American Ambulance Described by its Members*, Houghton-Mifflin Company, Boston & The Riverside Press, Cambridge, 1916.

[Collective 1918] *Onoranze a Luciao Orlando, Rugiero Torelli, Eugenio Elia Levi, Adolfo Viterbi, professori di matematica nelle università italiane, caduti in guerra*, Seminario matematico della facoltà di scienze della Reale università di Roma (Seduta del 22 giugno 1918), Bertero, Roma, 1918.

[Collective 2001] *Le chiffre, le renseignement et la guerre : actes du colloque de l'Historial de la Grande Guerre de Péronne du 21 et 22 mars 2001*, L'Harmattan, Paris, 2001.

[Compton 1937] Karl L. Compton, *Augustus Trowbridge, 1870–1934*, Biographical Memoirs of the National Academy of Sciences **18** (1937), pp. 219-244.

[Coray et al. 2003] Daniel Coray, Fulvia Furinghetti, Hélène Gispert, Brian H. Hodgson, Gert Schubring, eds., *One Hundred Years of* L'Enseignement mathématique. *Moments of Mathematics Education in the Twentieth Century. Proceedings of the EM-ICMI Symposium Geneva, 20-22 October 2000*, L'Enseignement Mathématique, Geneva, 2003.

[Cranz 1903] Carl Kranz, Ballistik, in *Encyklopädie der mathematischen Wissenschaften mit Einschluss ihrer Anwendungen*, IV–18 (1903), pp. 185–279.

[D 1915] D., *La guerre en formules*, Bulletin technique de la Suisse romande **41** (1915), pp. 9–10.

[Dauben 1980] Joseph Dauben, *Mathematicians and World War I: The International Diplomacy of G. H. Hardy and Gosta Mittag-Leffler as Reflected in their Personal Correspondence*, Historia mathematica **7** (1980), pp. 261–288.

[Daudet 1918] Léon Daudet, *La Guerre totale*, Nouvelle librairie nationale, Paris, 1918.

[Delsarte 1944] Jean Delsarte, *Rapport général du Groupe d'études préparatoires en vue d'une meilleure orientation des étudiants vers la recherche scientifique*, Centre de Documentation universitaire Tournier et Constans, Paris, 1944; repr. in *Œuvres*, vol. 2, CNRS, Paris, 1971, pp. 973–1050.

[Deltheil 1917] Robert Deltheil, *F. Lebeau*, Association amicale de secours des anciens élèves de l'École normale supérieure (Paris), Réunion générale annuelle (1917), pp. 130–133.

[Demidov 1993] Sergei Demidov, La revue *Matematicheskii Sbornik* dans les années 1866–1935, in *Messengers of Mathematics*, ed. Elena Ausejo & Mariano Hormigon, Siglo XXI de Espana Editores, Madrid, 1993, pp. 235–256.

[Dieudonné 1969] Jean Dieudonné, *Regards sur Bourbaki*, Analele universitatii Bucaresti, Matematica-mecanica **18** (1969), pp. 13–25.

[Dieudonné 1970] Jean Dieudonné, *The Work of Nicolas Bourbaki*, American Mathematical Monthly **77** (1970), pp. 134–145.

[Duda 1996] Roman Duda, *Fundamenta Mathematicae* and the Warsaw School of Mathematics, in [Goldstein et al. 1996, pp. 479–498].

[Dugac 1995] Pierre Dugac, *Jean Dieudonné : mathématicien complet*, Jacques Gabay, Paris, 1995.

[Eddington 1917] Arthur S. Eddington, *Karl Schwarzschild*, Monthly Notices of the Royal Astronomical Society **77** (1917), pp. 314–319.

[Eckert 2006] Michael Eckert, *The Dawn of Fluid Dynamics: A Discipline between Science and Engineering*, Wiley-VCH, Berlin & Weinheim, 2006.

[Epple 1999] Moritz Epple, *Die Entstehung der Knotentheorie. Kontexte und Konstruktionen einer modernen mathematischen Theorie*, Vieweg, Braunschweig & Wiesbaden, 1999.

[Feferman & Feferman 2004] Anita Burdman Feferman & Solomon Feferman, *Alfred Tarski: Life and Logic*, Cambridge University Press, Cambridge, 2004.

[Fontanon 1998] Claudine Fontanon, ed., *Histoire de la mécanique appliquée : Enseignement, recherche et pratiques mécaniciennes en France après 1880*, Cahiers d'histoire et de philosophie des sciences 46, SFHST-ENS Éditions, Paris, 1998.

[Fontanon & Franck 2006] Claudine Fontanon & Robert Franck, eds., *Paul Painlevé, 1863-1933: un savant en politique*, Presses universitaires de Rennes, Rennes, 2006.

[Ford 1918] Lester R. Ford, *Rational Approximations to an Irrational Complex Number*, Transactions of the American Mathematical Society **19** (1918), pp. 1–42.

[Ford 1919] Lester R. Ford, *Elementary Mathematics for Field Artillery*, prepared and published by direction of the Chief of Field Artillery, Field Artillery Central Officers' Training School, Camp Zachary Taylor, Kentucky (Instruction Memo. No. 20), Courier Journal Press, Louisville, Kentucky, 1919.

[Forman 1971] Paul Forman, *Weimar Culture, Causality, and Quantum Theory, 1918-1927: Adaptation by German Physicists and Mathematicians to a Hostile Intellectual Environment*, Historical Studies in the Physical Sciences **3** (1971), pp. 1–115.

[Forman 1973] Paul Forman, *Scientific Internationalism and the Weimar Physicists: The Ideology and Its Manipulation in Germany after World War I*, Isis **64** (1973), pp. 151–180.

[Forman 1974] Paul Forman, *The Financial Support and Political Alignment of Physicists in Weimar Germany*, Minerva **12**(1974), pp. 39-66.

[Friedman 1982] Robert M. Friedman, *Constituting the Polar Front*, Isis **73** (1982), pp. 343–362.

[Galvez-Behar 2008] Gabriel Galvez-Behar, *La République des inventeurs: propriété et organisation de l'innovation en France (1791–1922)*, Presses universitaires de Rennes, Rennes, 2008.

[Gispert 1991] Hélène Gispert, ed., *La France mathématique: La Société mathématique de France (1870–1914)*, SFHST & SMF, Paris, "Cahiers d'histoire et de philosophie des sciences," **34**, 1991; 2nd ed., 2014.

[Gispert & Leloup 2009] Hélène Gispert & Juliette Leloup, *Des patrons des mathématiques en France dans l'entre-deux-guerres*, Revue d'histoire des sciences **62**, pp. 39–118.

[Gladden 1930] Edgar Norman Gladden, *At Messines Ridge in 1917*, in *Everyman at War*, ed. Charles B. Purdom, J. M. Dent, London, 1930, pp. 114–121.

[Gluchoff 2011] Alan Gluchoff, *Artillerymen and mathematicians: Forest Ray Moulton and changes in American exterior ballistics, 1885–1934*, Historia Mathematica **38** (2011), pp. 506–547.

[Goldstein 2009] Catherine Goldstein, *La théorie des nombres en France dans l'entre-deux-guerres : de quelques effets de la première guerre mondiale*, Revue d'histoire des sciences **62-1** (2009), pp. 143–175.

[Goldstein 2011] Catherine Goldstein, *Gaston Julia, un mathématicien à Hurtebise*, Lettre du Chemin des Dames **23**(2011), pp. 22–25.

[Goldstein et al. 1996] Catherine Goldstein, Jeremy Gray, & Jim Ritter, *Mathematical Europe: history, myth, identity*, Maison des Sciences de l'Homme, Paris, 1996.

[Goldstein & Aubin forthcoming] Catherine Goldstein & David Aubin, eds., *La Grande Guerre des mathématiciens français*, forthcoming.

[Goursat 1916] Edouard Goursat, *Jean Clairin*, Bulletin de la Société mathématique de France **44** (1916), pp. 15–16.

[Goursat 1925] Édouard Goursat, *Le problème de Bäcklund*, Mémorial des sciences mathématiques **6**, Gauthier-Villars, Paris, 1925.

[Graham & Kantor 2009] Loren Graham & Jean-Michel Kantor, *Naming Infinity: A True Story of Religious Mysticism and Mathematical Creativity*, Harvard University Press, Cambridge, 2009.

[Grattan-Guinness 1972] Ivor Grattan-Guinness, *A Mathematical Union: William Henry and Grace Chisholm Young*, Annals of Science **29** (1972), pp. 105–186.

[Greenhill 1892] George Greenhill, *Applications of Elliptic Functions*, Macmillan, New York, 1892.

[Grier 2005] David Alan Grier, *When Computers Were Human*, Princeton University Press, Princeton, 2005.

[Guillaume 1994] Marcel Guillaume, *La logique mathématique en sa jeunesse*, in [Pier 1994, pp. 185–367].

[Hadamard 1916] Jacques Hadamard, *Lery (Georges), né à Limours le 28 avril 1880, tué à l'ennemi le 10 septembre 1914. — Promotion de 1899*, Association amicale de secours des anciens élèves de l'École normale supérieure (Paris) (1916), pp. 112–116.

[Hadamard 1920] Jacques Hadamard, *Rapport sur les travaix examinés et retenus par la Commission de Balistique pendant la durée de la guerre*, Comptes rendus hebdomadaires des séances de l'Académie des sciences **170** (1920), pp. 436–445.

[Hale 1919] George Ellery Hale, *The Responsibilities of the Scientist*, Science **50** (1919), pp. 143–146.

[Hanna 1996] Martha Hanna, *The Mobilization of Intellect: French Scholars and Writers during the Great War*, Harvard, Cambridge, Mass., 1996.

[Hardy 1940] Godfrey H. Hardy, *A Mathematician's Apology*, Cambridge University Press, Cambridge, 1940. Reprinted with Foreword by C. P. Snow, 1967.

[Hardy 1943] Godfrey H. Hardy, *William Henry Young 1863–1942*, Obituary Notices of the Fellows of the Royal Society **4** (1943), pp. 307–323.

[Hartcup 1988] Guy Hartcup, *The War of Invention: Scientific Development, 1914–18*, Brassey's Defence Publ., London, 1988.

[Heath 1921] Thomas Heath, *A History of Greek Mathematics*, Clarendon Press, Oxford, 1921.

[Hilbert 1917] David Hilbert, *Axiomatisches Denken*, Mathematische Annalen **78** (1917), pp. 405–415.

[Hinman 1919] Jesse R. Hinman, *Ranging in France with Flash and Sound*, Press of Dunham Printing Company, Portland, Oregon, 1919.

[Hoffmann 1994] Christoph Hoffmann, *Wissenschaft und Militär: Das Berliner Psychologieche Institut und der I. Weltkrieg*, Psychologie und Geschichte **5**(1994), pp. 261–285.

[Hope-Jones 1928] William Hope-Jones, *Sound-Ranging*, The Mathematical Gazette **14** (1928), pp. 173–186.

[Horne 2005] John Horne, *Entre expérience et mémoire: les soldats français de la Grande Guerre*, Annales: histoire, sciences sociales **60**, pp. 903–919.

[Hottin 2000] Christian Hottin, *80 ans de la vie d'un monument aux morts*, Labyrinthe **5** (2000); online: `http://labyrinthe.revues.org/262`. Last viewed November 20, 2013.

[Houzel 2004] Christian Houzel, *Le rôle de Bourbaki dans les mathématiques du vingtième siècle*, Gazette des Mathématiciens **100** (2004), pp. 52–63.

[Hughes 1977] Thomas P. Hughes, *Elmer Sperry: Inventor and Engineer*, Johns Hopkins Univ. Press, Baltimore, 1977.

[Hull 1920] Gordon F. Hull, *Some Aspects of Physics in War and Peace*, Science **51** (1920), pp. 221–233.

[Hull 1991] Andrew Hull, *War of Words: The Public Science of the British Scientific Community and the Origins of the Department of Scientific and Industrial Research, 1914–16*, British Journal for the History of Science **32** (1991), pp. 461–482.

[Innes 1935] John R. Innes, ed., *Flash Spotters and Sound Rangers: How They Lived, Worked, and Fought in the Great War*, George Allen & Unwin Ltd, London, 1935.

[Jay 2005] Martin Jay, *Songs of Experience: Modern American and European Variations on a Universal Theme*, University of California Press, Berkeley, 2005.

[Johnson 1990] Jeffrey Allan Johnson, *The Kaiser's Chemists: Science and Modernization in Imperial Germany*, University of North Carolina Press, Chapel Hill, 1990.

[Julia 1919] Gaston Julian [sic], *Lambert (Paul-Jean-Étienne), né à Annecy le 27 février 1894, tué à l'ennemi près de Fontenoy (Aisne) le 15 mars 1915. — Promotion de 1911*, Association amicale de secours des anciens élèves de l'École normale supérieure (Paris) **1919**, pp. 109–113.

[Kahn 1996] David Kahn, *The Codebreakers*, rev. ed., Scribner, New York, 1996.

[Kamke & Zeller 1957] Erich Kamke & Karl Longin Zeller, *Konrad Knopp*, Jahresbericht der deutschen Mathematiker-Vereinigung **60** (1957), pp. 43-49.

[Kellogg 1921] Oliver Dimon Kellogg, *A Decade of American Mathematics*, Science **53** (1921), pp. 541–548.

[Kevles 1969] Daniel Kevles, *Flash and Sound in the A[merican] E[xpeditory] F[orce]: The History of a Technical Service*, Military Affairs **33** (1969), pp. 374–384.

[Kevles 1995] Daniel Kevles, *The Physicists: The History of a Scientific Community in Modern America*, 2nd ed., Harvard Univ. Press, Cambridge, Mass., 1995.

[Klein 1934] Jacob Klein, *Die griechische Logistik und die Entstehung der Algebra*, Quellen und Studien zur Geschichte der Mathematik, Astronomie und Physik, Abteilung B: Studien **3**, no. 1 (1934), pp. 18–105; no. 2 (1936), pp. 122–235. English transl.: *Greek Mathematical Thought and the Origin of Algebra*, Cambridge, Mass.: M.I.T. Press, 1969.

[Kreyszig 1992] Erwin Kreyszig, Remarks on the Mathematical Work of Tibor Radó, in [Rassias 1992, pp. 18–32].

[Lamb 1974] George L. Lamb, *Bäcklund transformations for certain nonlinear evolution equations*, Journal of Mathematical Physics **15** (1974), pp. 2157–2165.

[Lamb 1976] George L. Lamb, Bäcklung Transformations at the Turn of the Century, *Bäcklund Transformations, the Inverse Scattering Method, Solitons, and Their Applications*, Lecture Notes in Mathematics **515**, Springer, Berlin, 1976, pp. 69–79.

[Lastours 1998] Sophie de Lastours, *La France gagne la guerre des codes secrets*, Tallandier, Paris, 1998.

[Lebesgue 1991] *Lettres d'Henri Lebesgue à Émile Borel*, Cahiers du séminaire d'histoire des mathématiques **12** (1991), pp. 1–506.

[Lehto 1998] Olli Lehto, *Mathematics Without Borders: A History of the International Mathematical Union*, Springer, New York, 1998.

[Leloup 2009] Juliette Leloup, *L'entre-deux-guerres mathématique, à travers les thèses soutenues en France*, doctoral thesis, Université Pierre et Marie Curie, Paris, 2009, `http://www.math.jussieu.fr/~leloup/These_Leloup_Juliette.pdf`.

[Lepick 1998] Olivier Lepick, *La Grande Guerre chimique*, 1814–1918, Presses universitaires de France, Paris, 1998.

[Levassor-Berrus 1920] André Levassor-Berrus, *Loussert (Maurice-Marie-Joseph), né le 3 mars 1893, mort pour la France, le 25 septembre 1915, devant Saint-Souplet.—Promotion de 1913*, Association amicale de secours des anciens élèves de l'École normale supérieure (Paris) (1920), pp. 129–133.

[Lhermitte 1917] Jules Lhermitte, *Rousseau (Théophile), né à Bligny (Côte-d'Or) le 20 avril 1876, tué à l'ennemi à Avocourt, le 11 avril 1916. — Promotion de 1898*, Association amicale de secours des anciens élèves de l'École normale supérieure (Paris) (1917), pp. 85–91.

[MacLeod 2000] Roy MacLeod, *Sight and Sound on the Western Front: Surveyors, Scientists and the "Battlefield Laboratory," 1915–1918*, War and Society **18** (2000), pp. 23–46.

[MacLeod & Andrews 1971] Roy MacLeod & Elizabeth Kay Andrews, *Scientific Advice in the War at Sea 1915-1917: The Board of Invention and Research*, Journal of Contemporary History **6** (1971), pp. 3–40.

[MacLeod & Johnson 2006] Roy MacLeod & Jeffrey Johnson, eds., *Frontline and Factory: Comparative Perspectives on the Chemical Industry at War, 1914–1924*, Springer, Dordrecht, 2006.

[Marbo 1967] Camille Marbo [Marguerite Borel], *À travers deux siècles, souvenirs et rencontres (1883–1967)*, Grasset, Paris, 1967.

[Mashaal 2006] Maurice Mashaal, *Bourbaki: A Secret Society of Mathematicians*, trans. Anna Pierrehumbert, American Mathematical Society, Providence, 2006.

[Mazliak & Tazzioli 2009] Laurent Mazliak & Rossana Tazzioli, *Mathematicians at War: Volterra and His French Colleagues in World War I*, Springer, Berlin, 2009.

[Mehrtens 1990] Herbert Mehrtens, *Moderne, Sprache, Mathematik: eine Geschichte des Streits um die Grundlagen der Disziplin und des Subjekts formaler Systeme*, Suhrkamp, Frankfurt, 1990.

[Mehrtens 1996a] Herbert Mehrtens, Mathematics and War: Germany, 1900–1945, in *National Military Establishments and the Advancement of Sciences: Studies in the 20th Century History*, ed. Paul Forman & José M. Sánchez, Kluwer, Dordrecht, 1996, pp. 87–134.

[Mehrtens 1996b] Herbert Mehrtens, *Modernism vs. counter-modernism, nationalism vs internationalism: style and politics in mathematics 1900–1950*, in [Goldstein et al. 1996, pp. 517–529].

[Miller 1918] George Armitage Miller, *Scientific Activity and the War*, Science **48** (1918), pp. 117–118.

[Millikan 1919] Robert Millikan, *A New Opportunity in Science*, Science **50** (1919), pp. 285–297.

[Morgan 1918] Jeff H. A. Morgan, *Relief Work among Civil and Military Prisoners in Siberia*, The American Oxonian **5–6** (1918), pp. 6–13.

[Moulton 1926] Forest Ray Moulton, *New Methods in Exterior Ballistics*, University of Chicago Press, Chicago, 1926.

[Moureu 1920] Charles Moureu, *La Chimie et la guerre. Science et avenir*, Masson, Paris, 1920.

[Mosse 1991] George L. Mosse, *Fallen Soldiers: Reshaping the Memory of the World Wars*, Oxford University Press, 1991.

[Nirici & Becknstein 2007] Lawrence Nirici & Edward Becknstein, The Hahn-Banach Theorem and the Sad Life of E. Helly, in *Advanced Courses of Mathematical Analysis III: Proceedings of the Third International School, La Rabida, Spain, 3–7 September 2007*, ed. Juan Delgado Sanchez & Tomas Dominguez Benarides, World Scientific, Singapore, 2008, pp. 97–110.

[Noether 1919] Fritz Noether, *Über analytische Berechnungen der Geschosspendelungen*, Nachrichten von der Gesellschaft der Wissenschaften zu Göttingen, Mathematisch-Physikalische Klasse (1919), pp. 373–391.

[Ollivier 1917] Héloïs Ollivier, *Clairin (Jean), né à Nîmes le 13 novembre 1876, tué à l'ennemi à Thun-l'Évêque (Nord) le 26 août 1914. — Promotion de 1896*, Association amicale de secours des anciens élèves de l'Ecole normale supérieure (Paris), Réunion générale annuelle (1917), pp. 83–85.

[Olmsted 1974] John M. H. Olmsted, *R. W. Brink: An Obituary*, American Mathematical Monthly **81**(8) (1974), pp. 873–875.

[Painlevé 1918] Paul Painlevé, *Élection du Maréchal Foch. Allocution*, Comptes rendus hebdomadaires des séances de l'Académie des sciences **167** (1918), p. 718.

[Palazzo 1999] Albert P. Palazzo, *The British Army's Counter-Battery Staff Office and Control of the Enemy in World War I*, Journal of Military History **63** (1999), pp. 55–74.

[Parshall & Rowe 1994] Karen Hunger Parshall & David Rowe, *The Emergence of the American Mathematical Research Community 1876–1900: J. J. Sylvester, Felix Klein, and E. H. Moore*, AMS/LMS History of Mathematics **8**, Providence/London, 1994.

[Parshall & Rice 2002] Karen Hunger Parshall & Adrian C. Rice, *Mathematics Unbound: The Evolution of an International Mathematical Research Community, 1800–1945*, History of Mathematics **23**, London Mathematical Society, London & American Mathematical Society, Providence, 2002.

[Pasteur 1888] Louis Pasteur, *Discours d'inauguration de l'Institut Pasteur, 14 novembre 1888*, Annales de l'Institut Pasteur **2** suppl. (1888), pp. 26–30.

[Pattison 1983] Michael Pattison, *Scientists, Inventors, and the Military in Britain, 1915-19: The Munitions Inventions Department*, Social Studies of Science **13** (1983), pp. 521–568.

[Perrier 1915] Edmond Perrier, Allocution, *Comptes rendus hebdomadaires des séances de l'Académie des sciences* **161** (1915), pp. 801–819.

[Pier 1994] Jean-Pierre Pier, ed., *Development of Mathematics, 1900–1950*, Birkhäuser, Basel, 1994.

[Pourprix 2009] Marie-Thérèse Pourprix, *Des mathématiciens à la Faculté des sciences de Lille, 1854–1971*, L'Harmattan, Paris, 2009.

[Prandtl 1918] Ludwig Prandtl, *Tragflügeltheorie*, Nachrichten von der Gesellschaft der Wissenschaften zu Göttingen, Mathematisch-Physikalische Klasse (1918), pp. 451-477; (1919), pp. 107–137.

[Prochasson & Rasmussen 1996] Christophe Prochasson & Anne Rasmussen, *Au nom de la patrie: les intellectuels et la première guerre mondiale (1910–1919)*, La Découverte, Paris, 1996.

[Rachamimov 2002] Alon Rachamimov, *POWs and the Great War: Captivity on the Eastern Front*, Berg Publishers, New York, 2002.

[Rassias 1992] Themistocles M. Rassias, ed., *The Problem of Plateau: A Tribute to Jesse Douglas and Tibor Radó*, World Scientific, Singapore, 1992.

[Renteln 1997] Michael von Renteln, *Paul Stäckel (1862–1919): Mathematiker und Mathematikhistoriker*, Überblicke Mathematik (1997), pp. 151–160.

[Riebesell 1916] Paul Riebesell, *Mathematik im Kriege*, Deutsche Feld-und Heimatbücher I.1, Teubner, Leipzig, 1916.

[Roussel 1989] Yves Roussel, *Histoire d'une politique des inventions, 1887–1918*, Cahiers pour l'histoire du CNRS **3** (1989), pp. 19–57.

[Rowe 2004] David Rowe, *Making Mathematics in an Oral Culture: Göttingen in the Era of Klein and Hilbert*, Science in Context **17** (2004), pp. 85–129.

[Runge 1915] Carl Runge, *Mathematik und Bildung*, Jahresbericht der deutschen Mathematiker-Vereinigung **24** (1915), pp. 400–415.

[Russell 1988–1995] Bertrand Russell, *Collected Papers*, Unwin Hyman, London, 1988–1995.

[Sarton 1919] Georges Sarton, *War and Civilization*, Isis **2** (1919), pp. 315–321.

[Saur & Tribout de Morembert 1965–1966] Hubert Saur & Henry Tribout de Morembert, *Ferdinand Daussy*, Mémoires de l'Académie nationale de Metz (1965–1966), pp. 13–21.

[Schiavon 2003a] Martina Schiavon, Des savants-officiers entre science, armée, État et industrie de précision: les géodésiens du Service géographique de l'Armée, 1887–1920, in [Aubin & Bret 2003, pp. 60–73].

[Schiavon 2003b] Martina Schiavon, *Itinéraires de la précision. Géodésiens, savants et fabricants d'instruments en France, 1870-1930 (environ)*, doctoral thesis, EHESS, Paris, 2003.

[Schiavon 2014] Martina Schiavon, *Itinéraires de la précision. Géodésiens, savants et fabricants d'instruments de précision en France, 1870-1930*, Presses universitaires de Nancy, Nancy, 2014.

[Schivelbusch 2001] Wolfgang Schivelbusch, *Die Kultur der Niederlage*, Alexander Fest, Berlin, 2001.

[Schmidt 1990] Marian Schmidt, *Hommes de science*, Hermann, Paris, 1990.

[Schroeder-Gudehus 1978] Brigitte Schroeder-Gudehus, *Les Scientifiques et la paix : la communauté scientifique internationale au cours des années 20*, Presses de l'université de Montréal, Montréal, 1978.

[Scott 1991] Joan Scott, *Evidence of experience*, Critical Inquiry **17** (1991), pp. 773–797.

[Seth 2010] Suman Seth, *Crafting the Quantum: Arnold Sommerfeld and the Practice of Theory, 1890–1926*, MIT Press, Cambridge, Mass., 2010.

[Siacci 1892] Francesco Siacci, *Balistique extérieure*, trans. P. Laurent, Berger-Levrault, Paris & Nancy, 1892.

[Siegmund-Schultze 1984] Reinhard Siegmund-Schultze, *Theodor Vahlen: zum Schuldanteil eines Deutschen Mathematikers am faschistischen Mißbrauch der Wissenschaft*, NTM Schriftenreihe zur Geschichte der Naturwissenschaften, Technik und Medizin **21** (1984), pp. 17–41.

[Siegmund-Schultze 1994] Reinhard Siegmund-Schultze, *"Scientific control" in mathematical reviewing and German-U.S.-American relations between the two World Wars*, Historia Mathematica **21** (1994), pp. 306–329.

[Siegmund-Schultze 2001] Reinhard Siegmund-Schultze, *Rockefeller and the Internationalization of Mathematics Between the Two World Wars*, Birkhäuser, Basel & Boston, 2001.

[Siegmund-Schultze 2003] Reinhard Siegmund-Schultze, Military Work in Mathematics 1914–1945: An Attempt at an International Perspective, in *Mathematics and War*, ed. Bernhelm Booß-Bavnbek & Jens Høyrup, Birkhäuser, Basel, 2003, pp. 23–82.

[Siegmund-Schultze 2004] Reinhard Siegmund-Schultze, *A Non-Conformist Longing for Unity in the Fractures of Modernity: Towards a Scientific Biography of Richard von Mises (1883–1953)*, Science in Context **17** (2004), pp. 333–370.

[Sigmund 2004] Karl Sigmund, *Failing Phoenix: Tauber, Helly, and Viennese War Insurance*, Mathematical Intelligencer **26** (2) (2004), pp. 21–33.

[SMF 1924] Société mathématique de France, *Vie de la société*, Bulletin de la société mathématique de France (1924) pp. 1–67.

[Smith 1914] David Eugene Smith, *Problems about War for Classes in Arithmetic: Suggestions for Makers of Textbooks and for Use in Schools*, Carnagie Endowment for Peace, Washington, 1914.

[Spitzmuller 1919] Georges Spitzmuller, *Le Repérage par le son*, F. Rouff, Paris, 1919.

[Swerdlow 1993] Noel M. Swerdlow, *Otto E. Neugebauer (26 May 1899–19 February 1990)*, Proceedings of the American Philosophical Society **137** (1993), pp. 139–165.

[Szénássy 1992] Barna Szénássy, *History of Mathematics in Hungary until the 20th Century*, Springer, Berlin, 1992.

[Thiberge 1975] Colonel Thiberge, *Le Service des Renseignements d'Artillerie*, Revue historique des Armées **119** (1–2) (1975), pp. 112–131.

[Tilho 1937] Jean Tilho, *Notice sur la vie et les travaux de Gustave Ferrié* (31 mai 1937), Académie des sciences, Notices et discours **2** (1949), pp. 161–237.

[Tollmien 1993] Cordula Tollmien, *Der "Krieg der Geister" in der Provinz: das Beispiel der Universität Göttingen 1914–1918*, Göttinger Jahrbuch **41** (1993), pp. 137–210.

[Tropp et al. 1993] Eduard A. Tropp, Viktor Ya. Frenkel, & Artur D. Chrnin, *Alexander A. Friedmann: The Man Who Made the Universe Expand*, trans. Alexander Dron & Michael Burov, Cambridge University Press, Cambridge, 1993.

[Trowbridge 1919] Augustus Trowbridge, *Sound and Flash Ranging*, Science **49** (1919), pp. 521–523.

[Vallier 1915] Emmanuel Vallier, *Le tir indirect en 1915*, Revue générale des sciences **16** (1916), pp. 297–299.

[Van der Kloot 2005] William Van der Kloot, *Lawrence Bragg's Role in the Developement of Sound-Ranging in World War I*, Notes and Records of the Royal Society **59** (2005), pp. 273–284.

[Varcoe 1970] Ian Varcoe, *Scientists, Government and Organized Research in Great Britain 1914-16. The Early History of the DSIR*, Minerva **8** (1970), pp. 192–216.

[Vivanti 1916] Giulio Vivanti, *Elementi della teoria delle equazioni integrali lineari*, U. Hoepli, Milan, 1916.

[Weart 1979] Spencer Weart, *Scientists in Power*, Harvard University Press, Cambridge, Mass., 1979.

[Weil 1991] André Weil, *Souvenirs d'apprentissage*, Birkhäuser, Boston, 1991. English transl., *The Apprenticeship of a Mathematician*, Springer, New York, 1992.

[Weis 2005] Cédric Weis, *Jeanne Alexandre: une pacifiste intégrale*, Presses de l'université d'Angers, Angers, 2005.

[Wells 1915] H[erbert] G[eorge] Wells, *Scientific War*, Windsor Magazine **41** (1915), p. 237.

[Whitehead 1919] Alfred North Whitehead, *An Inquiry concerning the Principles of Natural Knowledge*, Cambridge University Press, Cambridge, 1919.

[Wieleitner 1915] Heinrich Wieleitner, Review of Carl Cranz's *Lehrbuch der Ballistik*, vols. 1, 3, and 4, Jahresbericht der deutschen Mathematiker-Vereinigung **24** (1915), Sonstiges pp. 52–53.

[Wiener 1919] Otto Wiener, *Die streckenweise Berechnung der Geschossflugbahnen*, Abhandlungen der Sächsischen Akademie der Wissenschaften zu Leipzig, Mathematisch-Naturwissenschaftliche Klasse **36**, Teubner, Leipzig, 1919.

[Witkop 2002] Philipp Witkop, ed., *German Students' War Letters*, transl. Annie F. Wedd, University of Pennsylvania Press, Philadelphia, 2002.

[Witting 1916] Alexander Witting, *Soldaten-Mathematik*, Teubner, Leipzig, 1916.

[Wolff 2003] Stefan L. Wolff, *Physicists in the "Krieg der Geister": Wilhelm Wien's "Proclamation"*, Historical Studies in the Physical and Biological Sciences **33** (2003), pp. 337–368.

[Yerkes 1920] Robert M. Yerkes, *The New World of Science: Its Development During the War*, The Century Co., New York, 1920.

[Young 1981] Laurence Young, *Mathematicians and Their Time*, North Holland Publishing, Amsterdam, 1981.

Part I

Starting Up

Cambridge Mathematicians' Responses to the First World War

June Barrow-Green

On January 11, 1915, *The Times* of London ran a column entitled "A Mathematical War." It was the report of a "remarkable" address given by Sir George Greenhill, formerly professor of mathematics at the Royal Military Academy, Woolwich, on the importance and use of mathematics in warfare. It was only one of many articles about the war published in *The Times* that day but it attracted the attention of readers, letters from two of whom were published in response.[1] But was Greenhill right to describe the war as a mathematical one? In particular, was he right from a British perspective? To what extent was mathematics part of the war effort in Britain? How were British mathematicians deployed during the war? What effect did the war have upon them, and what implications did the war have for their subject?

In seeking answers to such questions, the natural place to begin is Cambridge. Cambridge was the beating heart of British mathematics and the cradle of the nation's mathematical research. Home to more mathematicians than anywhere else in Britain, it graduated more mathematics students, and at a higher level, than any other institution. If war-related mathematics was being done, irrespective of where it was taking place, then Cambridge-trained mathematicians would be in the thick of it. Conversely, if there were mathematicians who chose not to engage in the defense of the realm, then nowhere would it be more evident than in Cambridge.

This chapter examines the responses of Cambridge mathematicians to the war. The scene is set with a description of the situation in Cambridge before the war. The analysis then deals with two distinct groups of mathematicians: those who for reasons of conscience did not use their talents in support of the war effort and those who were actively engaged in war-related mathematics. The discussion of the first group, which is prefaced by a brief account of the political context, centers on mathematicians in Cambridge. The discussion of the second group moves out of Cambridge to various military and scientific establishments, such as the Royal Aircraft Factory, the National Physical Laboratory, and the Anti-Aircraft Experimental Section of the Ministry of Munitions, where the focus is on mathematicians

June Barrow-Green's address is Faculty of Mathematics, Computing and Technology, The Open University, Milton Keynes MK7 6AA, United Kingdom. Her email address is June.Barrow-Green@open.ac.uk.

[1] *The Times*(January 12, 1915; January 14, 1915).

FIGURE 1. Mathematical Tripos candidates and lecturers, St John's College, 1913. The three lecturers are in the front row: T. Bromwich (third left), H. F. Baker (center), E. Cunningham (third right). Among the candidates is H. Jeffreys (front row, far left). By permission of the Master and Fellows of St John's College, Cambridge.

involved in the development of aeronautics and in aspects of ballistics, such as anti-aircraft gunnery. This second group includes both those who were either employed or studying at Cambridge at the outbreak of the war and those who had already left Cambridge by that time but who still had strong Cambridge connections.

1. Cambridge before and during World War I

1.1. Cambridge Prewar. The dominance of Cambridge University over British mathematics and the central role of its undergraduate examination, the Mathematical Tripos, in the nineteenth and early twentieth centuries is well-attested.[2] Indeed, the difference in standard between Cambridge and elsewhere was such that during this period it was not unusual for students to take a mathematics degree at another institution, such as University College London or Edinburgh University, before going up to Cambridge. In 1906 *The Times* summarized the situation:

[2]The Mathematical Tripos began as an unstructured oral examination which became formalized in about 1710; from 1748 honors lists were published and in 1753 the division into wranglers (first class), senior optimes (second class), and junior optimes (third class) was established. The examination underwent several reforms in the nineteenth century. For a discussion of the examination and numerous further references to it and to mathematics at Cambridge in general, see [Warwick 2003].

> The study of Higher Mathematics in the British Empire is now practically concentrated at Cambridge. Thither come graduates from Scotch Universities and all the best men trained in the provincial Universities of England and Wales. Practically all the professorial chairs of mathematics and mathematical physics in London and the provincial universities of England, in the Universities of Scotland, the Colonies and India, are filled by Cambridge men. Many Cambridge men are professors in the United States; Cambridge graduates occupy chairs at Oxford; and even Trinity College, Dublin, that stronghold of local patriotism, has recently taken an Englishman from Cambridge as Professor of Mathematical Astronomy.[3]

However, these other institutions operated on a completely different scale from Cambridge. While their mathematics departments were usually staffed by one, at most two, professors and an assistant, Cambridge had four professors, five university lecturers (from 1892), and a large number of college lecturers, all teaching mathematics (Figure 1). The Cambridge professors' lectures were not part of the Tripos course—they were generally set at a level suitable for postgraduate students or higher—but the professors themselves were often involved with the prestigious Smith's Prize competition, which, since 1885 had been decided on the strength of a postgraduate essay rather than on a second set of examinations immediately after the Tripos.[4]

The only other university that might have provided Cambridge with some competition was Oxford but neither was it any match. It had fewer mathematics staff and students, and its final year undergraduate mathematical examination bore no comparison. In 1912 the prospects for an Oxford mathematics graduate were summed up by an Oxford mathematics lecturer, Arthur Jolliffe, a graduate himself of 1891:

> There is no doubt that Oxford does not offer much of a career to a mathematician. Fellowships are very few, no prize fellowship in mathematics has been offered since 1907, and in the absence of senior scholarships, a man has to proceed to earn his living immediately after taking his degree, usually by teaching in a school, though a few are so fortunate to get into the Civil Service by taking up other subjects [Jolliffe 1912, pp. 263–264].

Thus when the International Congress of Mathematicians (ICM) was held in England in 1912, it was held in Cambridge. It was inconceivable that it could have been held anywhere else in the country.

---

[3] *The Times* (October 20, 1906). The Cambridge graduate appointed to the chair of mathematical astronomy at Trinity College, Dublin, was Edmund Taylor Whittaker.

[4] For a history of the Smith's Prize competition, see [Barrow-Green 1999].

The congress was hosted by the Cambridge Philosophical Society,[5] and the society's president, Sir George Darwin, the Plumian Professor of Astronomy and Experimental Philosophy, delivered the opening address ([Darwin 1912]). The topic, the differences between pure and applied mathematicians, was fitting. In the nineteenth century Cambridge had been a stronghold of applied mathematics and mathematical physics—the names of Airy, Adams, Maxwell, Stokes, Kelvin, and Rayleigh, all students at Cambridge and all of whom, apart from Kelvin, holders of a Cambridge chair, are sufficient to make the point—with pure mathematics being the poor relation, Cayley's long tenure of the Sadleirian chair standing out as a singularity.[6] But in the Cambridge of 1912, pure mathematics was in a much stronger position than it had been in Cayley's day, and, moreover, it was going from strength to strength. This was partly as a result of the 1909 reform of the Mathematical Tripos—the most substantial reform in the examination's history—with its abolition of the order of merit which allowed, for the first time, undergraduate specialization, and which, by making analysis a fundamental constituent, brought Cambridge teaching much closer in touch with its Continental counterparts; and partly due to the twin rise of the geometer Henry Frederick Baker and the analyst Godfrey Harold Hardy. It was no wonder that Darwin felt the need to appeal on behalf of applied mathematicians for mercy and understanding from the international audience of predominantly pure mathematicians: the number of papers in pure mathematics being delivered at the congress outweighing those in applied mathematics by more than two to one.

The consequences of the Tripos reform showed which way the wind was blowing, as Walter William Rouse Ball noted in 1912:

> The reconstitution in 1909 of the Tripos, and the destruction of many of the distinctive features of the former scheme must profoundly modify the future history of mathematics at Cambridge, and perhaps the long-continued efforts to bring students into closer touch with professors and lecturers may be at last crowned with success. The change in the Tripos regulations has been accompanied by a curious alteration in the popular subjects, and to-day but few of the young graduates who desired the change are interesting themselves in those branches of applied mathematics once so generally studied, but rather are turning their attention to subjects like the Theories of Functions and Groups. It is too early to say whether this is only a passing movement [Rouse Ball 1912, p. 322].

The subjects Rouse Ball mentioned were being taught by Baker (both subjects) and by Hardy (theory of functions). Baker, senior wrangler of 1887, had been

[5]The Cambridge Philosophical Society had been established in 1819 with the aim "of promoting scientific inquiry, and of facilitating the communication of facts connected with the advancement of Philosophy and Natural History." Although the society included mathematics among its activities, mathematics did not have a special position. The London Mathematical Society, which by the date of the congress had long since been established as the national (English) mathematical society, had no explicit Cambridge connections (beyond the fact that the majority of its members were Cambridge graduates) and while it officially supported the congress, it played no specific role.

[6]In contrast to Cayley, James Joseph Sylvester was unable to make a career at Cambridge due to the religious restrictions in force at the time—in his case it was because he was Jewish—and he left after sitting for his final undergraduate examinations in 1837.

elected to a university lectureship in 1895. It was a position he held until 1914 when he was elected to the Lowndean Chair of Astronomy and Geometry, the first true geometer to hold the chair, thereby doubling the number of professors in pure mathematics (although still leaving two in astronomy).[7] Elected Fellow of the Royal Society (F.R.S.) in 1898, president of the London Mathematical Society in 1910-1912, and member of the six-man organizing committee of the congress, Baker was one of England's leading pure mathematicians.

Hardy, who came from the generation after Baker, had been elected to a college fellowship in 1900, and awarded a college lectureship in 1906, was one of the most ardent campaigners in support of the Tripos reforms, and his (now classic) *A Course in Pure Mathematics* (1908), a textbook in introductory analysis, had been written to complement the remodeled syllabus.[8] Publishing vigorously from early in his career, he had quickly gained an international reputation. By the outbreak of the war his famous and lifelong collaboration with John Edensor Littlewood, which had begun in 1911, was well under way (they had published several papers together and given a joint paper at the congress) and he had been elected F.R.S. in 1910.

Baker and Hardy were not the only pure mathematicians of substance in prewar Cambridge. There was also the analyst Ernest William Hobson who had succeeded the unfortunate Andrew Russell Forsyth as Sadleirian professor in 1910[9] and, like Baker and Hardy, had been a strong supporter of the Tripos reform. His *Functions of a real variable* (1907) was one of the first books to introduce the modern theory of measure and integration to English readers, although being in the vanguard, its proper recognition was achieved much later with the postwar extended second edition. A less than charismatic figure, Hardy wrote of him that "his influence on Cambridge mathematics was exerted mainly through his books" [Hardy 1934, p. 242].

As far as professors on the applied side were concerned, there was the Lucasian Professor Joseph Larmor who in 1903 had succeeded Sir George Stokes, and the astronomy professors, Arthur Stanley Eddington, who had succeeded to Darwin's chair after the latter's death less than six months after the congress, and Hugh Frank Newall, the first professor of astrophysics and founder of the chair (1909). From the point of view of the impending war, it is pertinent that none of them specialized in ballistics or aeronautics, or indeed in any subject that would be germane to the forthcoming conflict.

But to return to the congress, which, with 574 members from 28 countries, was the biggest ICM yet to be held. The two largest delegations, not counting the British, were from the United States and Germany, with 60 and 53 delegates, respectively. From the end of the nineteenth century, there had been a steady flow of British mathematicians (men and women) making their way to Germany to do postgraduate work. The congress provided an opportunity for German friendships and collaborations to be made and to be renewed. For Baker, who in the 1890s

[7]The astronomers did not lose out completely since Baker, who took the astronomy part of his position seriously, gave courses in dynamical astronomy.

[8]For a study of Hardy's influence on British mathematical analysis, see [Rice & Wilson 2003].

[9]The ostracism that Edwardian Cambridge meted out to Forsyth for allowing Marion Amelia Boys, the estranged and badly treated wife of the physicist Charles Vernon Boys, to move in with him made Forsyth decide to resign his professorship, which he surely would have had to forfeit, and his Trinity fellowship, which was unassailably his for life. He became a professor at Imperial College and married the former Mrs. Boys.

TABLE 1. Students at Cambridge, 1914–1918.

	1914	**1915**	**1916**	**1917**	**1918**
All students	3,181	1,097	712	491	533
Trinity College	503	132	82	52	38
St John's College	229	102	59	34	49

had made two visits to Göttingen to study intensively with Felix Klein, it must have been a sadness that Klein was too ill to attend. While for Hardy, an author in *Mathematische Annalen* (as was Baker) and who "like many of his Edwardian intellectual friends had a strong feeling for Germany" [Snow 1967, p. 38], the congress represented an unparalleled opportunity to meet with German colleagues with whom he had so far only corresponded. It was an opportunity that did not occur again until the ICM in Bologna in 1928 when the exclusion policy against the Central Powers had been broken and British and German mathematicians could once more freely enjoy one another's company at an International Congress.

1.2. Cambridge during the War. At the outbreak of war, the university offered its resources, both material and intellectual, for the benefit of the country.[10] Lecture rooms and laboratories were filled with soldiers, while colleges were converted into schools of instruction for officers, lodgings for nurses, and billets for men. Academics left to take up government positions or commissions as officers, while employees in the laboratories, libraries, museums, and the press left to serve on sea or land. As the war progressed, student numbers declined from more than 3,000 (1914) to less than 550 (1918), with the numbers at Trinity and St John's, the two largest colleges, falling from 503 and 229 to 38 and 49, respectively (Table 1).

As Basil Willey recalled:

> [T]he pulse of the University had almost stopped beating, there were very few undergraduates in residence except boys under military age, Asiatics, and the physically unfit ("infants, Indians and invalids") [Willey 1965, p. 118].[11]

All told, some 16,000 past and present Cambridge men served in the war, of whom over 2,000 were killed and almost 3,000 were wounded.[12]

In mathematics, the number of men students taking Part II of the Mathematical Tripos dropped from 58 in 1914 to 4 in 1918. In 1917, for the first time, more women (13) sat for Part II than men (11). The reduction in undergraduate numbers led to a corresponding reduction in the number of undergraduate courses offered by the university. For the academic year 1913–1914, the Special Board for Mathematics offered 36 different courses to be delivered by 16 different lecturers, while four years

[10]More information about the effect of the war on the university can be found in [Brooke 1993, pp. 330–340] and *The Cambridge University Calendar for the Years 1916–1917*, Cambridge University Press, Cambridge, (1916), p. 376.

[11]Willey, who was professor of English literature at Cambridge after the Second World War, was billeted at Peterhouse, the college he had been due to attend, when he received his call-up in July 1916.

[12]These figures are taken from [Howarth 1978, p. 16]. However, since they are based on information from colleges, they may not be precisely reliable [Brooke 1993, p. 334]. The figures for Oxford are roughly comparable to those for Cambridge; see [Harrison 1994, p. 9].

later the corresponding numbers had reduced to 6 and 4, respectively (Table 2).[13] In contrast, the mathematics professors, all of whom remained in their posts during the war, continued to lecture without interruption (Table 3).

2. Cambridge Mathematicians and Pacifism

In February 1917 Archibald Vivian Hill wrote to Karl Pearson: "I don't know why so many mathematicians are conscientious [objectors]: it is sad though a fact."[14] Although Hill did not explicitly mention Cambridge, it was surely where he had in mind. Cambridge was the center of British mathematical life and it was where Hill and Pearson were each actively seeking recruits to help with their war-related work (discussed below in Section 3). And it is true that a remarkable number of Cambridge mathematicians did indeed display pacifist sympathies.[15]

2.1. Political Background. When war was declared in August 1914 there was no conscription in Britain, and inevitably opposition to the war became tangled up with opposition to its reintroduction. Conscription had been abolished in 1865 a few years after the end of the Crimean War. The many failures of the Boer War at the turn of the twentieth century led to new calls for conscription to be reintroduced. The issue divided largely along party lines, with the Conservatives for it and the Liberals, together with the much smaller Independent Labour Party (ILP), against it. And until 1914 public opinion ran quite strongly against reintroduction and all moves for compulsory military service were defeated.

Opposition to conscription was not the same thing as pacifism; one might after all support the war but not the means by which soldiers were recruited. Outright pacifists were themselves divided into those for which pacifism was first and foremost an expression of their religious faith and their interpretation of Christianity and those for whom pacifism flowed from an identification with the international workers' movement.[16] Among those prominent in the pacifist movement, which naturally opposed conscription, was a Cambridge undergraduate, Reginald Clifford Allen, who had taken up socialism and after graduating in 1911 stayed on to build up the University Socialist Federation. At this time there was talk of making military training compulsory for Cambridge undergraduates, which Allen eloquently opposed in an anonymous article in the short-lived *University Socialist*, the Federation's journal [Allen 1913]. Allen was to be prominently involved in the anticonscription cause throughout the war. Although generally remembered as a most able advocate of total pacifism, Beatrice Webb, in her diary of May 1915,

[13]These numbers are for university lectures and do not take into account the lectures held in the individual colleges.

[14]A. V. Hill to K. Pearson, February 21, 1917 [Pearson Papers, 606].

[15]That is not of course to say that pacifism at the university was confined to mathematicians. There was considerable pacifist activity amongst the student body in general; see, for example, the *Cambridge Magazine* and particularly the writings of its then student editor, Charles Kay Ogden, the English semiotician, during the war period [Howarth 1978, pp. 19–20]. It is also of interest to note that on March 16, 1916, *The Times* published an apparently exaggerated report about the number of conscientious objectors from the university who had recently appeared before a local tribunal. The number announced in the report was 300 whereas, in a response by the Vice-Chancellor published the following day, the actual number was 72. The Cambridge Tribunal is discussed in [Stanley 2007, pp. 134–142].

[16]For a detailed discussion of pacifism during the war, see [Ceadel 1980, Robbins 1976].

TABLE 2. University lectures, 1913 and 1917. The titles of Hardy's courses in 1917 were slightly different from those in 1913: "Theory of functions of a complex variable; theory of functions; elements of mathematics (for nonmathematics students)."

Lecturer	**Course Title**	**1913**	**1917**
R. Herman	Hydrodynamics; planetary & lunar theory; differential geometry	×	
H. W. Richmond	Algebraic geometry; higher solid geometry; synthetic geometry	×	
T. J. Bromwich	Electric waves & electro-optics; vibrations & sound; optics	×	
E. Cunningham	Practical work in electricity; practical work in optics; dynamics; the principle of relativity & the mechanics of electrical systems	×	
J. H. Grace	Theory of numbers; line geometry; elements of Fourier analysis & calculus of variations; Laguerre's writings on geometry	×	
E. W. Barnes	Linear differential equations	×	
A. Berry	Ordinary differential equations; elliptic functions & elementary harmonic analysis	×	
A. Munro	Hydrodynamics & sound	×	×
B. Russell	Fundamental concepts of mathematics; principles of mathematics; empirical basis of physics	×	
J. G. Leathem	Electron theory	×	×
G. H. Hardy	Chapters in the theory of functions; general theory of Dirichlet's series; Fourier's series; elements of mathematics (for nonmathematics students)	×	×
G. Birtwhistle	Hydrodynamics	×	×
F. J. M. Stratton	Stellar physics	×	
L. B. Booth	Geometrical optics; higher integral calculus	×	
J. E. Littlewood	Riemann zeta function & the theory of prime numbers; Taylor's series	×	
G. N. Watson	Difference equations	×	

described him as "a fanatical antiwar, pro-German advocate, who distorts every fact to prove his country wrong" [Cole 1952, p. 33].

When the war began there was a rush of volunteers, and even though this declined a little by Christmas and into 1915, as it became clear that the war might

TABLE 3. Professors' lectures, 1913 and 1917.

Professors	**1913**	**1917**
J. Larmor	Electricity and magnetism; electrodynamic & optical theory	Mathematical physics (introduction); electro-dynamics & optical theory
A. S. Eddington	Spherical astronomy & combination of observations; practical astronomy; celestial mechanics; sidereal observations	Combination of observations; astronomy & determination of orbits; relativity theory of gravitation
H. F. Baker	Theory of functions; geometry of birational transformations; periodic orbits & questions of periodicity & stability of motion of gravitating systems; algebraic functions; plane geometry	Projective geometry; planetary theory; geometry; astronomy
E. W. Hobson	Arithmetic continuum & theory of sets of points; calculus of variations; history of invention of differential & integral calculus	Linear & quadratic forms for infinite number of variables; calculus of variations
J. J. Thomson	Constitution of the atom; electricity & magnetism; discharge of electricity through gases	Properties of matter; electricity & magnetism

go on for some time, nonetheless recruitment ran at a level sufficient to satisfy Lord Kitchener, the Secretary of State for War. However, in September 1915, a Cabinet Committee reported that the voluntary scheme was causing problems for the military. A few days later Kitchener and Lord Asquith, leader of the Liberal Party and current prime minister, persuaded Lord Derby, an active proponent of conscription, to lead a new scheme which he did on the understanding that if it failed there was no alternative but conscription. Unveiled on October 21, 1915, it called for a personal canvass of every man on the Electoral Register, urging them to volunteer. The understanding was that married men would not be called up until every unmarried man had been. The Derby scheme, as it was called, was accepted quite generally as the last chance of showing that voluntary means could provide sufficient manpower for the British war effort. It was against this background that various antiwar organizations were formed, three in particular assuming a special importance.

The *Union for Democratic Control* (UDC),[17] which initially was able to draw on the active support of Bertrand Russell, had been set up on the day after the outbreak of war, August 5, 1914, by Edmund Dene Morel—a notable figure who

[17]For a discussion of the role of the UDC in British politics during the war, see [Swartz 1971].

had led an extremely popular campaign against the iniquities of Belgian rule in the Congo—and Charles Trevelyan, cofounder of the National Peace Council and Liberal member of Parliament. Although not a pacifist organization it was opposed to conscription and it called for a negotiated peace together with parliamentary control over foreign policy. Morel sensed, correctly, that the working men of Britain would not support a totally pacifist stand with sufficient vigor. Moreover, advocates of revolution could hardly plead pacifism with much chance of success; their objection to the First World War was that it was a capitalist's war.

The second group of antiwar activists was the *Fellowship of Reconciliation* (FOR). This was set up at the very end of December 1914, when a group of 130 antiwar Christians, mostly Quakers,[18] met at Trinity Hall Cambridge for a conference on Christianity and the war [Ceadel 1980, pp. 35–37]. They were led by Henry Hodgkin (who was a missionary in China) and Reverend Richard Roberts, although both were often abroad (Roberts fled in 1916). It was largely inconspicuous during the hostilities even though it claimed 8,000 members. However, it was the organization to which Russell would nail his colors in 1916 and it emerged to lead pacifism after the war.

The third group was the *No-Conscription Fellowship* (NCF),[19] and it was to be the most important of all such groups over the next few years. It was a much less intellectual body than the UDC and was formed by Fenner Brockway and Clifford Allen in December 1914. The NCF was an important group opposing conscription and quickly gained members, many of whom were members of the Independent Labour Party. It was not essential to be a pacifist to support it, although many of its members were, and many were Quakers, and practically all became conscientious objectors after conscription.

The Derby scheme ended on December 11, 1915, with the realization that over half a million men had refused to fill in the form. With this, the failure of voluntary recruitment was evident, and on December 15, against a background of popular protest, the Cabinet met to consider conscription. On January 24, 1916, the Military Service Act for Universal Conscription was passed. All single men between 17 and 41 were eligible for call-up. However, the legislation was remarkably liberal. The so-called "conscious clause" exempted men on both moral as well as religious grounds, and a Non-Combatant Corps (NCC) was formed for men who would, for example, serve as ambulance workers but not fight. It was possible, although very rare, to be granted "absolute exemption." The latter was usually given to those medically unfit for service; more usual was exemption from combatant service, and it was these men that were earmarked to do such things as drive an ambulance in the NCC. It was also possible to get exemption on condition of doing work of national importance.

After the initial passing of the Military Service Act in January 1916, the Act underwent four revisions, each time the net for recruits being cast wider, until finally in April 1918, it was revised to include men between the ages of 17 and 51. Between the outbreak of the war and the introduction of the original Military Service Act, there were 3 million volunteers. A further 2.3 million were then conscripted.

[18]The Quakers—or the Religious Society of Friends, as they are more properly called—are a Christian organization which was formed in England in the mid-seventeenth century as a reaction to aspects of the established church.

[19]For a history of the NCF during the war, see [Kennedy 1981].

A network of 2,000 tribunals was set up to administer the system and hear appeals, and there were a further 70 appeal tribunals. There were hundreds of thousands of claims—750,000 applied for exemption between January and July 1916—and inevitably proceedings were of variable quality. On occasions military members of a tribunal were almost impossible to convince, and there were also occasions where bastions of local (county, hunting) community were let off. In total some 16,000 were officially recorded as conscientious objectors during the war.

On the other side were those who felt the whole idea of resisting conscription was distasteful. To these people, refusing conscription smacked of cowardice and of refusing to help the country that provided people with so much, and a discharge looked like a conscientious objector was getting off scot-free while others were risking their lives. Socialism, in these quarters, looked like no excuse at all, and conscientious objectors could find themselves very unpopular indeed.

Tensions within the Liberal party over conscription and tensions within the organizations that opposed either the war or the principle of conscription formed the context within which the young men of Cambridge and their older tutors and professors thought about the First World War and about their individual places within it.

2.2. Pacifists and Conscientious Objectors among Cambridge Mathematicians. On February 5, 1915, Francis Puryer White (Figure 2), a mathematics student at St John's College and a conscientious objector, wrote in his diary:

> After Hall [dinner] half an hour's work at Cunningham's Lecture on Precession. Then with Montagnon to the Blue Boar for the inaugural meeting of the Union of Democratic Control. Bertrand Russell took the chair & Langdon-Davies addressed the meeting, which was quite large, on the objects of the Union. Afterwards came questions and the election of officers: President Russell; Treasurer Lowes Dickinson; Secretaries. G. H. Hardy (Trinity) and Abraham. Then I left, signing a paper on my way out, to express my desire to join.[20]

White kept his diary continuously from January 1, 1915, to July 29, 1916, a time in which he came top in the Tripos[21] and began postgraduate work. The diary provides a wealth of information about his academic, spiritual, and recreational life at Cambridge and at home in London.[22] It includes details about the mathematical texts he was reading,[23] the lectures he attended, and the lecturers he

[20]Diary of F. P. White, February 5, 1915 [White Papers].

[21]White sat for the Tripos in 1915 and although by that time students were no longer ranked, White reported in his diary that he had heard indirectly (the original source being John Hilton Grace, one of the examiners) that he had come first. July 24, 1915 [White Papers].

[22]White's recreational interests included the law, paleography, and novel reading, and he was a keen tennis player and cyclist, regularly cycling the route between London and Cambridge, often in the company of Harold Jeffreys. He was a committed Christian and attended college chapel daily. During vacation he did volunteer work at the St John's Mission in South London.

[23]For example, during the vacation period between March 26 to April 12, 1915, White consulted texts from the following authors at the British Museum: Baker (London Mathematical Society Presidential Address); Poincaré (Relative Stability of Rotating Systems, *Acta* **7**); Stokes (Effect of radiation on the propagation of sound); Whittaker (*Analytical Dynamics*: vibrating & steady motion); Kirchhoff (lecture on mechanics); Chasles (Curves on quadrics); Larmor ("a paper"); Basset (*Hydrodynamics*); Humbert (application of Fuchsian functions to algebraic curves);

FIGURE 2. Francis Puryer White, 1915. By permission of the Master and Fellows of St John's College, Cambridge.

encountered.[24] Mathematical gossip is reported and meetings of the UDC, NCF, FOR and proceedings of tribunals are described. Between April and June 1916, White appeared before five different tribunals, before being given noncombatant exemption and agreeing to work for the Friends Ambulance Unit.[25]

White came from a solid middle-class background—his father was a schoolmaster and the family lived in the quiet area of Islington in North London—and his conscientious objection, which was born out of a deep religious conviction, was, as

Severi (Correspondence); Hurwitz (correspondence, *Math. Ann.* **28**); Baker ("Function," *Encyclopædia Britannica*); Baker (Abel's Theorem) [White Papers].

[24]Joseph Larmor, the Lucasian Professor, came in for particular criticism, e.g. "[Larmor] is really an awful man—made several sneering remarks about the pure mathematicians which I am sure were partly for my benefit. 'In applied mathematics one has to feel one's way, pure mathematicians merely follow their noses.' His 'applied mathematics' is what I prefer to call 'physics.' He would certainly call all Bromwich's work pure mathematics—while I don't know what he would think of Baker's." February 2, 1916 [White Papers].

[25]The Friends Ambulance Unit was a volunteer ambulance service founded by individual members of the Quakers. It was, however, independent of the Quakers' organization, and its staff consisted chiefly of conscientious objectors.

he evocatively described in his diary, a cause of great distress to his family. This was in strong contrast to his situation in Cambridge where he had several friends who went off to fight but who understood and respected his position. The general situation in Cambridge was summarized by Joseph John Thomson:

> In the war there were in the University some conscientious objectors, but not very many; several of these were Quakers, and the greater number objected to war on religious grounds. There were very few whose sincerity could be questioned; indeed it required great moral courage, or exceptional physical cowardice, to face the odium of being a conscientious objector rather than go to the Front [Thomson 1936, p. 234].

A courageous conscientious objector was thus no oxymoron in Cambridge. In London, the situation was different. White's parents, who were painfully aware of the strength of public opinion against conscientious objectors, simply could not understand why their son had adopted such a position.

For several weeks, until he was awarded noncombatant exemption, White lived with the threat of prison hanging over him, much to the consternation of his father. And his father was right to be concerned. Students who had applied for absolute exemption as conscientious objectors but who did not escape incarceration included Lancelot Hogben (who later achieved success on the interface between mathematics and biology and wrote the best-selling *Mathematics for the Million* [1936]), then reading for the Natural Sciences Tripos, and Ralph Homfeld Baker, the son of H. F. Baker. White was well known to Baker, Sr., as he studied with him both pre- and post-Tripos and both were members of St John's College. Baker attended White's tribunals as well as those of his son. Although the father did not subscribe to his son's beliefs, he recognized, as indeed many people did, that, whatever your personal convictions, not all support for the war could be morally justified, as he wrote to Joseph Larmor, the Lucasian Professor:

> I have made a point of telling all no-conscriptionists with whom I have come in contact that I do not approve of the no conscription campaign. My son is well aware that it was on grounds of health only that I have done whatever I have on his behalf....
>
> I have always regarded the Prussian outlook as immoral, and with some personal knowledge, & make no excuses for it. But it is clear to me that a large part of the support of this war is commercialism trying to get by force what it could not get by skill. For this & other reasons I have more respect for many of the dissentients than for the powers that be.
>
> Not only have I refused to have my son's case taken up by politicians, but I have refused to sign petitions — even against what I regard as brutal treatment.[26]

Baker himself did not engage in war-related mathematics but he did offer to help with "the organisation of elementary education," considering himself "in some ways fitted to be a buffer between the working people, and the mad technical-science

[26] H. F. Baker to J. Larmor, January 13, 1917. Papers of Sir Joseph Larmor, St John's College Library.

people on the one hand, and the established order of things on the other—knowing both sides by personal experience."[27] However, it seems that this offer was not taken up.

Many students who were conscientious objectors (mostly on religious grounds) and who applied for absolute exemption ended up, like White, with noncombatant exemption. Several of White's friends appeared before tribunals, although not necessarily for pacifist reasons.[28] They included Robert Stoneley, later a distinguished geophysicist, Herbert William Unthank, John Marshall, and Samuel Pollard (who ended the war on secondment to the Anti-Aircraft Experimental Section), and all of whom were b or b^* (i.e. top class) wranglers between 1914 and 1916.[29]

Of the many mathematicians mentioned in White's diary perhaps the most difficult to place in the pacifist spectrum is G. H. Hardy. Hardy, who was 37 years old when war was declared, signed up under the Derby scheme but was rejected on medical grounds [Hardy 1967, p. 39]. He was, however, a founder member and secretary of the Cambridge branch of the UDC, the branch presided over by Bertrand Russell, and was active in organizing and speaking at their meetings.[30] He was not prepared to do war-related research but he was not a pacifist per se, although he was certainly sympathetic to those with pacifist inclinations, at times interceding on their behalf.[31] He was, for example, a strong supporter of Russell when the authorities of Trinity College turned against him for his pacifist activities. And when an appeal was lodged against Sydney Chapman's exemption which had been awarded "on grounds of national interest" and Chapman had "applied for absolute exemption on the ground that he had a deep religious objection to military service of any kind,"[32] Hardy asked Larmor to write "a strong statement" on Chapman's importance as an applied mathematician and to testify on Chapman's sincerity as a pacifist. For, as Hardy said, "What is the sense of setting a man like him to dig or scrub floors (even if he'd do it)?"[33] But although Hardy went to tribunals to support applicants he knew, he did not like "conscientious objectors as a class."[34]

[27]H. F. Baker to J. Larmor, January 7, 1917. Papers of Sir Joseph Larmor, St John's College Library.

[28]The proceedings at the tribunals attended by White and his friends were reported in the *Cambridge Daily News* and several press cuttings are included in White's diary [White Papers].

[29]The designations b or b^* derive from the structure of the reformed Tripos examination. From 1909, Part II of the examination had two schedules, (A) and (B). Schedule (A) was compulsory for all candidates, while schedule (B) contained alternative subjects in more advanced mathematics.

[30]There are several references in White's diary to meetings of the UDC attended by Hardy, including the first one held at Cambridge at which White himself "expressed the desire to join." Diary of F. P. White, February 5, 1915 [White Papers].

[31]See [Hill 1974, p. 124], quoted in Section 3.3 below.

[32]See undated press cutting from the *Cambridge Daily News* [White Papers]. The previous year Chapman had also featured in White's diary: "A man came in during the hour & was hailed as Chapman—no! Mercer! He turned out to be Chapman after all, the Trinity lecturer, who has not gone to an air-craft factory after all. So he will relieve Wren of the latter's lectures on Hydrostatics, much to his joy." January 26, 1915 [White Papers].

[33]G. H. Hardy to J. Larmor, March 28, 1916 [RS Archives, 680]. Later renowned as a geophysicist, Chapman was given absolute exemption from military service in 1916, Larmor and Frank Dyson testifying as to his sincerity. Since there were so few students at Cambridge, he returned to Greenwich—he had had a position at the observatory prior to his appointment at Cambridge in 1914—and spent the rest of the war teaching at the Royal Naval College [Cowling 1971, p. 55].

[34]G. H. Hardy to F. J. H. Jenkinson, June 15, 1918 [CU Archives, MS Add.6463/8647].

It is certainly the case that Hardy did not believe that the war should be allowed to impede international scientific collaboration. On October 31, 1915, White noted in his diary:

> Then heard an excellent story about G. H. Hardy. He has recently published a *Cambridge Mathematical Tract* in conjunction with Marcel Riesz (and is awfully bucked at having produced something in collaboration with an enemy in war time). Riesz lives at Copenhagen and Hardy had occasion to write to him on some mathematical subject and the letter contained a good many mathematical symbols—also he mentioned the name of a German worker at the end. The censor got hold of the letter, determined from the German name that it must be a code message and spent hours trying to interpret the symbols as a cipher. Failing altogether he refused to let the letter through.[35]

Riesz was a Hungarian subject and therefore presumed to be on the German-Austrian side of the conflict. Hardy did, however, manage to correspond with him during the war, sometimes using Gösta Mittag-Leffler in Stockholm as an intermediary.[36] Nevertheless, as Hardy's correspondence with Mittag-Leffler reveals, the problem of censorship was a very real one for British mathematicians wishing to circulate their work internationally. To send printed matter out of the country required a special application to the Censor's Office, and the Censors, who were "rather alarmed by mathematical formulae," had to be satisfied that the work was innocent.[37]

Hardy never wavered in his support for international collaboration and after the war became one of the strongest British proponents for scientific cooperation and reconciliation.[38]

Unlike Hardy, the case of Arthur Eddington, the Plumian Professor of Astronomy, is clear cut. Eddington was a Quaker and hence unequivocally a pacifist on religious grounds.[39] Although Eddington himself applied for exemption on the grounds of conscience, the university also applied for exemption for him on the basis that his work was of national importance and essential to the running of the observatory. When Eddington's case came up before a tribunal in 1916, White, who was familiar with Eddington through meetings of both the FOR and of the

---

[35]Diary of F. P. White, October 31, 1915 [White Papers]. Riesz was in fact living in Stockholm, not Copenhagen.

[36]On March 30, 1915, Hardy wrote to Mittag-Leffler: "I am as Dr. Riesz suggested to me, addressing my correspondence with him to you. If he is called away on military duties, perhaps you would be so kind as to open anything I send and use your discretion as regards forwarding it. It would be useless to send him in such circumstances, letters which contained only mathematical analysis: but he might be glad to hear from me personally" [Hardy Corr., letter 9].

[37]G. H. Hardy to G. Mittag-Leffler, August 9, 1917 [Hardy Corr., letter 23]. About his application to the Censor' Office, see G. H. Hardy to G. Mittag-Leffler, October 25, 1918 [Hardy Corr., letter 26].

[38]Hardy articulated his position in this regard very clearly in his correspondence with Mittag-Leffler. He placed the highest store on the recognition he had obtained in Germany and declared that it would have been "an irretrievable calamity" should his personal relationships with German mathematicians been lost. Hardy to Mittag-Leffler, January 7, 1919 [Hardy Corr., letter 27]. For the Hardy-Mittag-Leffler correspondence on this subject, see [Dauben 1980].

[39]The significance of religion in Eddington's life is explored in [Stanley 2007].

FIGURE 3. The $\nabla^2 V$ Club, May 1915. Top row: T. L. Wren (center), H. W. Unthank (second right); second row: S. Chapman (second right); front row: A. S. Eddington (far left), H. Jeffreys (second left), H. Glauert (second right). By permission of the University of Cambridge.

$\nabla^2 V$ Club, a small Cambridge club which met regularly to discuss questions in mathematical physics (Figure 3),[40] reported:

> Then to Guildhall for Local Tribunal. Listened to cases of Glauert (who got 1 month to find a munitions job) and Milne (who got off as he is doing anti-aircraft work) and the University appeals for Pigou, Eddington etc. It was most undignified to see the V.C. and the Master of John's etc., pleading to be allowed to keep these men. What is the University coming to?[41]

Eddington's exemption from military service was granted on occupational grounds, his claim for conscientious objection being postponed. In June 1918, with the military's need for able-bodied men growing ever more urgent, his occupational exemption was appealed against and terminated as of August 1, 1918. He then had to face another tribunal at which his case as a conscientious objector was heard. Aided by the intervention of Frank Dyson, the Astronomer Royal, who eloquently argued the case for Eddington to be allowed to continue his work and in particular for him to be available to make the eclipse expedition in 1919 at which Einstein's general theory of relativity would be tested, Eddington was granted a

[40]The $\nabla^2 V$ Club was founded in 1900 and was limited to 25 members [CU Archives, Soc.XXIX.1/1].

[41]Diary of F. P. White, May 10, 1916 [White Papers]. For a report of the tribunal, see the *Cambridge Daily News*, May 10, 1916. For discussions on Glauert and Milne, see Sections 3.3 and 3.1 respectively.

further twelve months exemption, a period which, fortunately, lasted well beyond the date on which the Armistice was signed.[42]

Eddington, like Hardy, was an active promoter of international scientific cooperation, and in May 1916, shortly after he had been granted exemption from military service, entered into a sharp-edged debate with the Oxford astronomer Herbert Hall Turner through the columns of the *Observatory* [Eddington 1916]. Turner held strongly opposing views to Eddington—the infamous Manifesto in praise of German culture and supporting the German invasion of Belgium, which was published in September 1914 and signed by 93 German intellectuals, including 22 scientists, providing grist to Turner's mill [Turner 1916]—and their polemic provoked other mathematicians and scientists, including Joseph Larmor, to add their voices to the debate.[43]

Bertrand Russell's position as a pacifist is well known. He was initially very active in the UDC, having helped to found the Cambridge branch. But before long he began to find the UDC's call for peace to be negotiated irrelevant to the pressing needs of the day and gravitated to the NCF. His ever more vocal advocacy of NCF positions, his protests against the unjust treatment of conscientious objectors, and his criticisms of the war made him more and more unpopular in the country, but he was well connected and difficult to touch. Eventually, however, when in June 1916, after having admitted in a letter to *The Times* to the authorship of an NCF leaflet, he was convicted of "making statements likely to prejudice the recruiting and discipline of His Majesty's forces" and fined, his Cambridge college, Trinity, moved against him. Eleven of the thirteen members of the College Council (made up of fellows) met and agreed unanimously to dismiss him from his lectureship. But opinion in Trinity was deeply divided and the following month twenty-two (of mostly the younger) fellows, sent a protest to the Council against Russell's dismissal. In 1918 Russell was prosecuted again, and this time he was sentenced to six months in prison, a period he used profitably to write his *Introduction to Mathematical Philosophy* (1919). After the war he was reinstated to his lectureship, although he never took it up. Much later, Hardy wrote a careful and dispassionate account of Russell's dismissal which reveals with great clarity the range of attitudes and depths of feelings the war produced [Hardy 1942].

Of those that signed the protest against Russell's dismissal in 1916, several were mathematicians, some of whom were of a pacifist persuasion and some of whom were not. Russell's friend and collaborator Alfred North Whitehead, then a professor at Imperial College, accepted the necessity of the war but was "miserable at differing from [Russell] on so great a question" [Lowe 1990, p. 27]. Both of Whitehead's sons saw active service during the war and it was letters from the elder of the two, North, who was with the British Expeditionary Force in Flanders, which prompted Whitehead to propose a method for streamlining the calculation of the path of a shell in high angle fire [Whitehead 1918]. Whitehead also collaborated with North in producing a report on auditory methods for locating enemy guns [RS Archives, MS/658]. Eric, the younger son, was killed in action in March 1918.

Of the other mathematicians who signed, Hardy, Eddington, and Chapman have already been mentioned, while Ralph Fowler and Joseph Proudman appear

[42]For further details of the statements made at these tribunals, see [Douglas 1956, pp. 92–95] and [Stanley 2007, pp. 145–152].

[43]For a discussion of the *Observatory* debate, see [Stanley 2007, pp. 63–67].

in the following section, which leaves only Ernest Barnes and Eric Neville, both of whom espoused the pacifist cause. Barnes was a friend of Russell's and an active member of the UDC, taking the chair at the inaugural Cambridge meeting in March 1915. He had been second wrangler in 1896 and had a strong reputation as an analyst. But mathematics did not fully sustain his attention and in 1903 he had been ordained as a priest. Outspoken in his views, he was not always at one with other College Fellows. In 1915, in part motivated by his pacifism, he left Cambridge to become Master of the Temple, eventually (in 1924) becoming Bishop of Birmingham, a position from which he gained considerable notoriety for his unorthodox opinions, in particular his "gorilla sermons" supporting evolutionary theory [Bowler 1998, p. 455].

Neville is probably best known as the man who on Hardy's instructions sought out Srinivâsa Aiyangâr Ramanujan in Madras and persuaded him to come to Cambridge in the spring of 1914. Neville made no secret of his hatred of warfare—Russell was "fond of him because of his gentleness" [Langford et al. 1964, p. 132]—and it is generally believed that his pacifism was the reason his College Fellowship was not renewed in 1917. In 1919 he left Cambridge to take up the Chair of Mathematics at University College, Reading (now Reading University).

Another staunch Cambridge pacifist was Ebenezer Cunningham, senior wrangler of 1902 and a lecturer and Fellow at St John's, who in 1914 had been the first in England to publish a book on relativity theory. Cunningham's pacifism, like that of Eddington and White, was based on a religious conviction, and it had at its roots the horrors he had felt at the events of the Boer War. To Cunningham, as he recorded in his autobiography, "the whole idea of war seemed irrational, impossible of achieving good, and, in particular, entirely foreign to the meaning and practice of Christianity" [Thompson 1996, p. 551].

A member of the FOR and young enough to be called up, he was initially granted exemption from military service to work as a schoolteacher. Later the exemption was repealed because, due to his beliefs, he was considered unfit to work with schoolchildren, and he spent the latter part of the war years working on a farm, before returning to Cambridge, his best mathematics behind him. Cunningham, who was not only in the same college as White but had been to the same school as him in Islington, features regularly in White's diary. White took Cunningham's lecture courses in dynamics, and they often met outside lectures at meetings of the FOR:

> ...I went to Cunningham's rooms for a meeting of the FOR.... Professor Eddington, Thompson of Queens' were there besides others whom I did not know even by sight.... It is generally recommended that all pacifists should appear before the Tribunal to claim exemption from combatant services and allied work.[44]

Unlike Cunningham, Max Newman, who was later to play an important role in code-breaking during World War II, was not a through and through pacifist: he described himself as conscientious objector "more or less" (see letter reproduced below). Newman, who had a German father, went up to St John's College in 1915. White, who invited Newman as a freshman to tea, noted in his diary that he was "the senior Maths Fresher at John's (though only a £60 scholar)" and that he "did not get money from the L.C.C. [London County Council] solely on account of his

[44]Diary of F. P. White, February 5, 1916 [White Papers].

name [then Neumann], though he was born in England and is a Christian."[45] In 1916 Newman changed his name by deed poll. After a year he left Cambridge to become a paymaster in the army before becoming a schoolteacher. In February 1918, while working at Archbishop Holgate's School in York, he wrote to one of his Cambridge friends, Harold Jeffreys, to ask him to support his application for exemption from military service. Jeffreys, who had graduated as a b^* wrangler in 1913, was at that time employed in the Meteorological Office as "a research assistant working on hydrodynamical problems coming from the Front in France" having spent the early part of the war working part-time in the Cavendish Laboratory on war-related problems [Cook 1990, p. 307].[46] The two were clearly good friends and Newman freely admitted his "Conchy" (i.e., conscientious objector) position:

> Having received on Feb. the 9th a letter asking me, with a singular lack of courtesy, to present myself at Cambridge R.O. [Recruiting Office] on the 22nd inst., after various shufflings I have become a Conchy, more or less, skillfully welding any objections, as I call them, on to my unfortunate father's place of birth. I am seeking exemption from military service, but am not refusing Work of National Importance.
>
> Among the twenty odd soul searching questions I am at present wrestling with is this:
>
> 5(b): What evidence can you produce in support of your statement. Please forward written evidence (from persons of standing if possible) which should be quite definite as to the nature & sincerity of your conscientious ob.
>
> Casting around vainly for anything remotely resembling a person of standing among my friends, relations & acquaintances, I was suddenly illumined by the fact that you are a Fellow of an Illustrious College [St John's], & moreover may write Dr before your name or DSc after it according to the degree of sophistication of the person with whom you are dealing. Surely a person who could stand with anybody. The rest is simple. All you have to do is to grasp a pen, address a sheet of paper to the Chairman of the York Tribunal & tell him you know I have a rooted objection to taking part in the war. You'd better not say much more on this point or you may put your foot in it, as the system of beliefs I've evolved for this Tribunal is Queer; it only allows its master to work it.
>
> An impression of immense age would be good. (Don't forget to splash a bit round your signature.)
>
> ...I am learning Esperanto inbetween paroxysms of Pacifism. It is a rotten language; it sounds worse than G—.[47]

Since there is no evidence of Newman having done military service, his application was presumably successful.

[45] Diary of F. P. White, October 26, 1915 [White Papers].

[46] Jeffreys later gained distinction as geophysicist, was a good friend and colleague of White's at St John's, and features prominently in White's diary. For a discussion of wartime activities at the Cavendish Laboratory, see [Kim 2002, ch. 6].

[47] Newman to Jeffreys, February 26, 1918. St John's College Library, Papers of Max Newman. "G—" stands of course for "German."

Although Cambridge in wartime did not have a monopoly on pacifist mathematicians, traces of others have been hard to find. One well-known exception is the Quaker Lewis Fry Richardson, a Cambridge graduate of the 1903 Natural Sciences Tripos and who, because his pacifist principles underlie one of his wartime contributions to mathematics, is noted briefly here. (Another is Arthur Doodson, who is considered in Section 3 below.)

In 1913 Richardson had been appointed Superintendent of the Meteorological and Magnetic Observatory at Eskdalemuir in the Scottish Lowlands.[48] Since his work as a meteorologist was considered of national importance, he could probably have remained at Eskdalemuir for the duration of the war. But in 1916 he resigned to join the Friends Ambulance Unit and spent the remainder of the war driving ambulances in France. The experience proved to be seminal, for, as Ashford has described, "His war experiences and Quaker beliefs led him at the same time to turn his thoughts to the causes of war and how to prevent them. He felt that it might be useful to tackle this problem by an objective scientific approach" [Ashford 2004, p. 833]. The result was his pioneering booklet *The Mathematical Psychology of War*, in which he developed a mathematical model for describing the change in animosity between two opposing sides during a war. His aim was to ascertain the relation between what he called the "Vigor-to-War" and the "Warlike Activity" of a side and determine how these characteristics evolve over the duration of a conflict [Richardson 1919, pp. 11–13]. At the time of Richardson's booklet, applying mathematics to a social problem of this type was very uncommon. That Richardson was able to even think of adopting a mathematical approach to the problem may, ironically, be due to the fact that he had had little by way of advanced mathematical training and was thus unconstrained by the mathematical conventions of the time.[49] What is certain, however, is that Bertrand Russell was an important influence on him—Richardson sent Russell an early draft of the paper—and it was Russell's writings that got Richardson thinking about the problem in the first place [Ashford 1985, pp. 63–64]. Richardson was reappointed by the Meteorological Office after the war, although to a different observatory, but in 1920 he resigned his position on pacifist grounds when the Office was taken over by the Air Ministry.

3. Cambridge Mathematicians Employed in the War Effort

Although there were a number of fields in which mathematicians made use of their knowledge and skills during the First World War—ballistics, aeronautics, sound-ranging, statistics, etc.—the greatest concentration of talent ended up working in something to do with aeronautics, whether it was helping to keep airplanes up in the air or developing methods for shooting airplanes down, and it is these mathematicians on which this section is focussed.

At the outbreak of the war, Britain had 113 aircraft in naval and military service [Edgerton 1991, p. 10]. Aeronautical theory was still in its infancy and these aircraft were essentially deployed in reconnaissance and artillery observation—it had barely been imagined that they would be used for anything else. They had a maximum speed of 80 miles per hour and could climb to 5,000 feet. By October 1918 the newly established Royal Air Force possessed more than 22,000 aircraft and bombing

[48]For a biography of Richardson, see [Ashford 1985].

[49]Richardson's mathematical experience only consisted in a few months spent working with Karl Pearson at University College in 1907 learning about statistical proof [Ashford 1985, p. 24].

FIGURE 4. The B.E.2c airplane. From *Applied Aerodynamics*, by George P. Thomson, 1919.

was a matter of course. The maximum speed had increased to 140 mph, while the maximum height attainable was now 25,000 feet.[50] Numerous engineers and physicists contributed to this rapid advance but mathematicians too played their part. The two main centers of aeronautical research were the Royal Aircraft Factory at Farnborough and the National Physical Laboratory (NPL) at Teddington, both of which were within easy reach of London. The two establishments, although separated by a distance of some 25 miles and concentrating on different aspects of aeronautical theory—Farnborough was the center of aircraft design while the NPL focussed on aerodynamics—had close contact with one another, and their research was published in the *Reports and Memoranda of the Advisory Committee for Aeronautics.* A third center of activity was at Whale Island, near Portsmouth, on the south coast, where a group of mathematicians concerned with anti-aircraft gunnery, designated the Anti-Aircraft Experimental Section (AAES), were located. As it turned out, the latter evolved into the largest and most coherent of all the groups of mathematicians working on war-related problems.

3.1. The Royal Aircraft Factory. The Royal Aircraft Factory (formerly the Royal Balloon Factory) acquired its name in 1912. It retained this name until April 1918 when, due to the formation of the Royal Air Force, it was renamed again (as the Royal Aircraft Establishment). The Factory's main responsibility was the generation of aircraft designs. These designs were mostly for research aircraft but

[50]For a summary of the advances in engineering and science during the war, which includes these numbers, see [Petavel 1920]. Broadly similar numbers are given in [Bairstow 1920a, p. 266], except for the maximum climbing height for airplanes at the beginning of the war which is given as 10,000 feet. Both [Petavel 1920] and [Bairstow 1920a] were presented by their authors at the 1919 annual meeting of the British Association for the Advancement of Science.

a few, such as the *B.E.2c.* (Figure 4), went into mass production, especially during the war. The Factory itself built some aircraft but the majority were built by private companies. The head of the Factory was Mervyn O'Gorman, who although not a Cambridge man, gathered Cambridge graduates in mathematics and engineering around him. Advice was also sought from mathematicians away from Farnborough, such as the fluid dynamicist Horace Lamb, professor of mathematics and natural philosophy at Manchester, who worked on problems connected with the strength of aircraft [Glazebrook & Love 1935, p. 391]. Although the Factory was in a mode of expansion when war was declared, it naturally grew substantially during the war until it was employing some 3,500 men and 1,500 women.[51]

Among the older mathematicians at Farnborough was Arthur Berry, senior wrangler in 1885 and mathematical lecturer from King's College, Cambridge. Berry was a pure mathematician who worked primarily on the theory of elliptic functions and differential equations. Like several of the mathematicians at Farnborough he arrived there more by accident than design. In 1915, while on active service, he had come to the attention of the Admiralty Air Department (AAD) when he had recast the solution to the problem of the behavior of a continuous beam over a number of supports with different axial loads in the separate bays into a more elegant and accurate mathematical form [Sutton-Pippard 1966, p. 70]. To do this he modified the standard form of the three-moment equation to include the effects of axial load, introducing and tabulating what are now known as "Berry functions," and his method is today recognized as an important contribution to the theory of structures ([Berry 1924, Day 1966]). He was then seconded to the AAD in London where he was working on problems connected with aircraft structures and bomb trajectories.

While in London, Berry had a team of women mathematicians working with him on stress calculations. Their leader was the gifted geometer Hilda Hudson who had been equal to the seventh wrangler in the Tripos of 1903. Described as "a terrific worker" [Chitty 1966], Hudson stepped outside her area of research (Cremona transformations) with her war work.[52] She contributed to the contents of the *Handbook of Strength Calculations*, the confidential publication in which Berry's above-mentioned result was first published in 1918; and in June 1918 she was made an Officer of the Order of the British Empire (OBE) for her work in "Aero-Dynamics Technical Research."[53]

One night in late 1917, the London office in which Berry had been working was destroyed by a zeppelin bomb, and he was immediately moved to Farnborough [Garner 1966, p. 95]. At Farnborough, where he was assisted by Harry Mason Garner, "a very competent young Cambridge mathematician,"[54] he continued to apply his mathematical skill to engineering problems, producing reports on such diverse subjects as the torsion of a fuselage and the whirling of an airscrew (propeller) shaft.

---

[51] A description of the work of the Royal Aircraft Factory in the years around the war is given in [Anonymous 1920].

[52] See, for example, her paper on "Incidence Wires" [Hudson 1920].

[53] The Order of the British Empire is an order of chivalry which was established by King George V in 1917 largely, although not exclusively, to honor the many people who had served in a noncombatant role during the First World War.

[54] Garner was a b^* wrangler in 1914. A. Berry to K. Pearson, May 29, 1918 [Pearson Papers, 213].

The employment at Farnborough of the Cambridge mathematical physicist Geoffrey Ingram Taylor was also somewhat accidental. When war was declared, Taylor was a young Fellow of Trinity College and midway through a three-year appointment as a reader in dynamical meteorology. He had immediately volunteered to set up a weather forecasting service for the army but had been turned down because "Soldiers don't go into battle under umbrellas, they go whether it is raining or not."[55] And it was only due to a chance meeting between the officer who interviewed him and the officer recruiting scientists for the Royal Aircraft Factory that Taylor was offered the opportunity to work at Farnborough. It suited him perfectly. The developing theory of aeronautics drew on his expertise in fluid dynamics—his work there included problems of pressure distribution, turbulence, and airscrews—and he found flying exciting. Taylor's interest in the theory of flight had developed well before his arrival at Farnborough, as is evident in his unpublished Adams Prize essay *Turbulent Motion in Fluids*, which he completed at the end of 1914.[56]

Hermann Glauert (Figure 5) is another Cambridge man whose engagement at Farnborough was described as "accidental" [Farren & Tizard 1935, p. 607]. Glauert, embarrassed by the fact that his father was German, had initially decided to spend the war years in Cambridge where, as a research student in astronomy and optics, he could study untroubled. But as time progressed, he felt increasingly uncomfortable with his position, and in 1916 when by chance he was offered a position at the Royal Aircraft Factory, he accepted it. Of all the mathematicians at Farnborough, he made the greatest impression with his mathematical ability (Figure 6). Unlike many of the theoreticians, such as Taylor, he had no desire to get into the air himself. Described as "a mathematician with quite unusual powers of verbal exposition" [Stephens 1966, p. 77], he could deal with theoretical and practical problems alike, although most of his work involved the analysis of experimental results, and he had a particular talent for making his work accessible to his less theoretically inclined colleagues:

> In almost every paper [Glauert] pursued his objective to a numerical solution and left the matter in a form in which less gifted—or as he would have put it *differently gifted*—workers could make full use of it. All these publications show his sense of judgement of the requirements of the engineer. He delighted and excelled in the solution of a difficult problem in a neat and elegant way; but his main objective was to put an essential piece of fundamental theory effectively at the disposal of those who wished to make practical use of it [Farren & Tizard 1935, p. 609].

Among the problems tackled by Glauert was the spin of an airplane, then a major cause of accidents and fatalities but which, due to a lack of experimental data, had been little researched. In this work Glauert teamed up with Frederick Lindemann (later Lord Cherwell), who fearlessly carried out some of the first experiments on spinning in 1917 (which showed that a spin is essentially a spiral glide in a very steep path with the wing incidence above the stalling point), and their first account

[55] Taylor gave several lectures recounting his early adventures in aeronautics. The text of one of them, originally entitled "Aeronautical Experience before 1919" is published in [Batchelor 1996, pp. 63–78].

[56] For a discussion of Taylor's thesis, see [Bloor 2011, pp. 145–148].

FIGURE 5. Scientific staff of the Royal Aircraft Factory outside the Chudleigh Mess in 1917. Left wall: H. Glauert, G. P. Thompson, F. Lindemann; top step: H. Grinsted, W. S. Farren; lower step: D. H. Pinsent, R. McKinnon Wood; right wall: F. W. Aston, H. Renwick. Courtesy of Farnborough Air Sciences Trust.

of the theory of spin was published in March 1918.[57] While Lindemann provided the observations, Glauert determined the equations of motion (of an airplane in spin) and used them to analyse the data. The following year Glauert produced a more detailed report of their investigations which also contained a summary of the history of spinning and recovery to date [Glauert 1919].[58] Glauert was one of the most popular men at Chudleigh, the large house run as a private mess where practically everyone of note in aeronautics at the Royal Aircraft Factory stayed at some time during the war, including Lindemann and Taylor.

[57]See [Farren & Thomson 1958, pp. 50–55] for a description of Lindemann's contribution to the theory of spinning.

[58]The significance of the research on the theory of spin is summarized in [ACA 1922, p. 6].

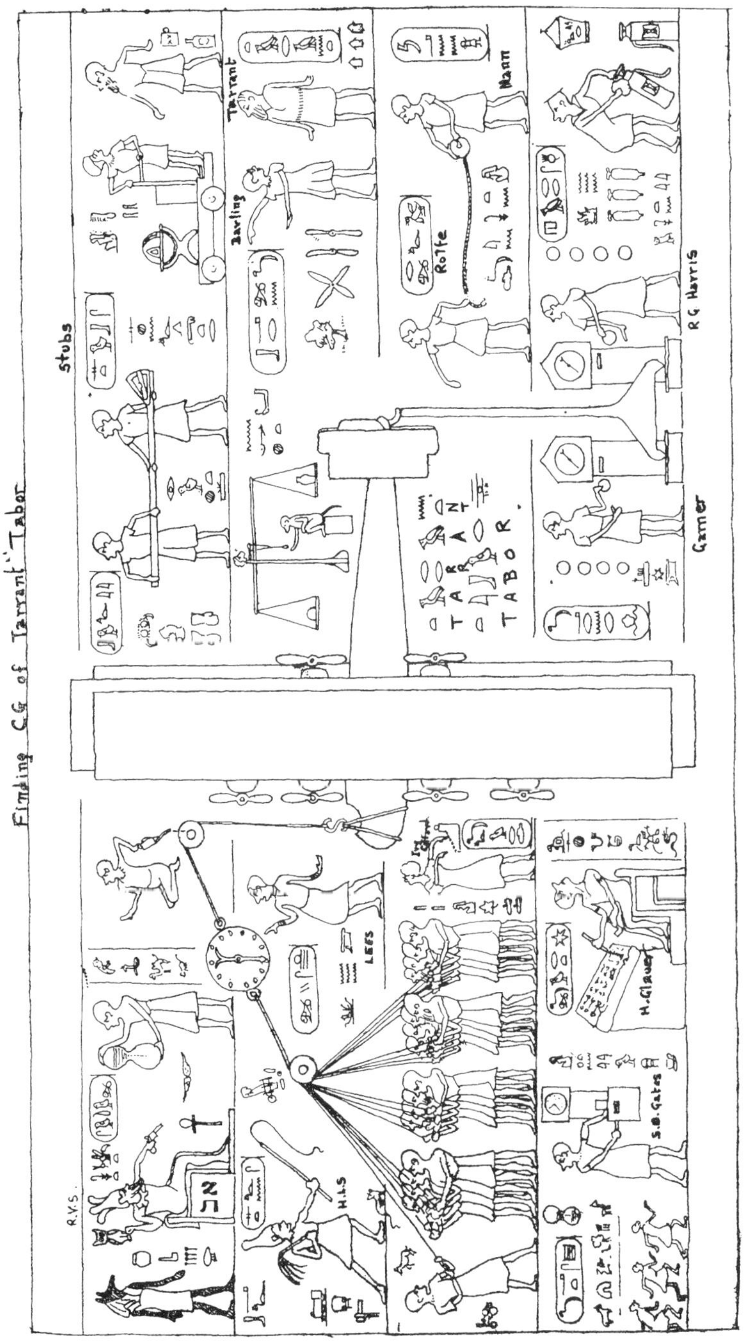

FIGURE 6. A pastiche of an Egyptian hieroglyphic text in which workers at Farnborough are depicted busily determining the center of gravity of the short-lived Tarrant Tabor, a giant British triplane; Glauert is portrayed sitting at a desk calculating while the others are engaged in more practical tasks such as measuring (from *Book of Aeron*, in [Southwell et al. 1966, p. 87]). Courtesy of the *Journal of the Royal Aeronautical Society.*

As well as Glauert, Taylor, and Lindemann, other occupants of Chudleigh included David Pinsent and George Thomson (son of the physicist J. J. Thomson), great friends and contemporaries also of Glauert's at Cambridge—Thomson was a b^* wrangler in 1912, while Glauert and Pinsent were b^* wranglers in 1913. Pinsent, a brilliant mathematician and close friend of Ludwig Wittgenstein,[59] initially joined Farnborough in 1916 as a mechanic (he had been rejected for active service due to his slight physique) but was persuaded by Thomson and others that he would be more useful to the war effort as a mathematician.[60] He was killed accidentally while flying in May 1918. Thomson, who was attached to Farnborough in 1915 after a period serving in France, worked on problems of stability and of spin. After the war he published a substantial textbook on applied aerodynamics that collected together results found at Farnborough and which had been previously published in the Advisory Committee for Aeronautics reports [Thomson 1919].

But not all the mathematics done at the Royal Aircraft Factory was done by trained mathematicians. Keith Lucas, who had studied for the Natural Sciences Tripos, was well settled into a career as a physiologist at Cambridge when he was assigned to Farnborough in September 1914. At that time pilots were having navigational problems, particularly when turning on a northerly course or when flying through clouds, due to the erratic behavior of their compasses. The pilots thought that the behavior was due to variations in the magnetic fields in the clouds. Lucas, having established that the same behavior occurred in clear air, attacked the problem mathematically. He discovered that the problem was caused by the fact that the vertical component of the earth's magnetic field varies across the globe and designed a new compass which took this variation into account, thus allowing pilots to navigate safely in clouds over enemy territory while flying in any direction.[61] He was killed in a midair collision in 1916.

A late joiner at Farnborough was the applied mathematician and engineer Richard Southwell. Southwell, a graduate of the Mechanical Sciences Tripos of 1910, who had been due to start as a college lecturer in October 1914, joined up on the outbreak of war and was in France by the end of the year. He was recalled the following year to work on the development of nonrigid airships but by 1917 he had been promoted to an office job in the Admiralty. It was only through his Cambridge connections, notably the intervention of Bertram Hopkinson, the professor of mechanism and applied mechanics, that he was transferred to Farnborough, and from 1918, he was in charge of the aerodynamics and structural departments [Christopherson 1972, p. 550].

Among the women employed at Farnborough, one of the most outstanding was Lorna Swain, a don from Newnham College, Cambridge. A wrangler of 1913 and exact contemporary of Glauert, as a postgraduate she had studied in Göttingen and at Manchester (with Lamb) where she had begun research in fluid dynamics. While

---

[59] Wittgenstein studied aeronautics at the University of Manchester from 1908 to 1911 before going to Cambridge in 1912 to study philosophy with Bertrand Russell. For a discussion of Wittgenstein's work in aeronautics, see [Lemco 2007]. Pinsent's friendship with Wittgenstein is chronicled in Pinsent's diary and is the subject of [Von Wright 1990].

[60] Thomson, who in 1937 won a Nobel Prize for physics, later reminisced that Pinsent was "the most brilliant man of my year [at Cambridge], among the most brilliant I have ever met" [Von Wright 1990, p. xv].

[61] For further details about Lucas's compass, see [Melvill Jones 1966] and [Davis 2009, pp. 86–90].

at Farnborough she worked on problems of propeller vibration and was one of the few women to have her name attached to an *Advisory Committee for Aeronautics Report* [Kennedy 1934, p. 156].

3.2. The National Physical Laboratory. At the end of the nineteenth century British scientists, aware of the widening gap between the universities and industry, pressed the government to provide them with a state laboratory which could be used for both experimental work and training purposes, and they looked to Germany, with its network of scientific research institutes, to provide the model. In 1900 the National Physical Laboratory (NPL), modeled on the Physikalisch-Technische Reichsanstalt in Berlin and administered by the Royal Society, was founded.[62] The Department of Aerodynamics was established in 1909 as part of an initiative by Lord Haldane, Secretary of State for War, which included the formation in May of that year of the *Advisory Committee for Aeronautics* (ACA) and which came into being after Haldane had consulted with George Darwin [Bloor 2011, p. 13].[63]

The ACA, whose remit was to give advice on scientific problems arising in connection with the Admiralty and War Office in aerial construction and navigation, provided the link between the NPL, the Royal Aircraft Factory, the universities, and industry. It planned research and experiments at the NPL and at Farnborough, and it brought together a group of eminent scientists interested in flying.[64] The president was the distinguished mathematical and experimental physicist, Lord Rayleigh, and the chairman was the director of the NPL, Richard Glazebrook. Other members included Mervyn O'Gorman, superintendent of the Royal Aircraft Factory; Sir George Greenhill (discussed in Section 4 below); Napier Shaw, director of the Meteorological Office; Horace Darwin, founder of the Cambridge Scientific Instrument Company[65] and brother of George; Joseph Petavel, Professor of Engineering at Manchester[66]; Arnulph Mallock, an engineer and physicist; Frederick Lanchester, the aeronautical engineer; and representatives from the army and navy. Of those on the committee, several had strong Cambridge connections. Glazebrook and Shaw, wranglers in 1876, had worked together as demonstrators in the Cavendish Laboratory under the leadership of Rayleigh. Mallock had also worked with Rayleigh in the Cavendish Laboratory, while Darwin, who had been a Senior Optime (i.e. second class) in the Mathematical Tripos of 1874, supplied instruments to the Cavendish Laboratory.

Of those working on aerodynamics at the NPL during the war, the leading figure was Leonard Bairstow, who ran the aerodynamics department from its inception in 1909 until 1917, when, due to differences with Glazebrook [Nayler 1966b, p. 83], he left to work as a scientific researcher and consultant for the newly formed Air Board

---

[62]For the early history of the NPL see [Moseley 1978].

[63]For a history of the ACA and its successor, the Aeronautical Research Council, see [Nayler 1966a].

[64]The selection of those chosen to serve on the ACA was not without controversy. In particular, the aeronauts thought that there were too many theorists and not enough practical men on the committee. For a report of the aeronauts' point of view, see *The Observer* (May 9, 1909).

[65]For further information about the Cambridge Scientific Instrument Company, see [Cattermole 1987].

[66]In 1917 Petavel became the first chairman of the ACA aerodynamics subcommittee. In 1919 he succeeded Glazebrook as director of the NPL.

FIGURE 7. Wind channel at the National Physical Laboratory, 1913. From Leonard Bairstow, *Applied Aerodynamics*, 1920.

(later the Air Ministry).[67] Bairstow, like O'Gorman, was not a Cambridge man. He had been educated at the Royal College of Science (one of the three colleges that in 1907 merged to form Imperial College) and several of the permanent staff at the NPL had Imperial College connections, e.g. Ernest Relf and Arthur Fage, the former having won an Imperial College scholarship tenable at the NPL. One exception was the Cambridge graduate Joseph Lawrence Nayler, a b^* wrangler of 1912. But the war brought in men from across the country, including Robert Jones and Dan Williams, students of George Hartley Bryan in Bangor, and several from Cambridge, the latter including Robert Frazer, a b^* wrangler and exact contemporary of Nayler. They were joined later by two more Cambridge men, both from Gonville and Caius College, Edward Gunner Gallop, a mathematical lecturer, the second wrangler (from Trinity) of 1883, and Christopher Noel Hunter Lock, the only b^* wrangler of 1917, both of whom were members of the AAES and who came to the NPL to work on the dynamics of shells.

Until moving to the NPL it seems that Frazer had been destined for a career in pure mathematics. Tutored by Baker and Grace, in 1914 he won a Rayleigh prize for an essay on the theory of point involution. On arrival at the NPL he was appointed to the Aeronautics Section of the engineering department and there began work on viscous flow and, together with Nayler, produced a report for the ACA. But it was not long before circumstances dictated to divert him away from fluid mechanics and into full-scale experiments on the performance of airships, leaving purely mathematical research behind him.

[67]For further details about Bairstow's career at the NPL, see [Hashimoto 2000].

Nayler coauthored many reports for the ACA during the war, the most important of which were on the theory of aeronautical stability. In 1912, the NPL, under Bairstow's direction, had embarked on a program of stability experiments in which models were tested in wind tunnels (Figure 7) in order to derive the data that was needed in order to use the stability theory developed in [Bryan 1911].[68] The success of Bairstow's group in determining the "derivatives" to put into Bryan's equations, combined with the experimental work of Edward Busk, resulted in early 1914 in the production of the first truly stable airplane. In the early stages of the war, Nayler, together with Bairstow and Jones, extended the theory from steady rectilinear motion to steady curvilinear motion (either in a horizontal plane or in a spiral). This important development required the solution of a complicated octic equation and their *ACA Report* "Investigation of the Stability of an Aeroplane when Circling in Flight"[69] gives the complete mathematical theory leading to the general conditions for stability. It also includes an appendix by Bairstow in which he shows, by extending methods developed by Gräffe and Runge, how to solve octic (and higher degree) equations with several pairs of complex roots. As Hashimoto has described, an interesting aspect of this report is how Bairstow came to acquire the knowledge to solve the complicated algebraic equations required by the theory [Hashimoto 1990, pp. 37–41]. Unable to find a sufficiently high-level English book on the subject, Bairstow asked Jones, who had been studying in Göttingen with Ludwig Prandtl when war was declared, to bring back a new German textbook, Horst von Sanden's *Praktische Analysis* (1914), with him on his return. But even this book turned out to be inadequate for his purposes, and he was led to consult an encyclopædia article by Runge which in turn led him back to the 1837 computing method of Gräffe. Of all the aeronautical work carried out at the NPL during the war, the research into stability theory was considered "especially valuable" because it gave the "constructors and pilots a very material advantage" [ACA 1922, p. 3]. Since it was known that the Germans were particularly anxious to get their hands on this research, none of the stability theory reports were published during hostilities [Nayler 1966b, p. 83].

3.3. Anti-aircraft gunnery. When the zeppelin raids began in 1915, the need for effective anti-aircraft gunnery (Figure 8) became abundantly clear. On September 9, 1915, Adelaide Davin, a member of Karl Pearson's computing staff at University College London (UCL), wrote to Pearson to describe the zeppelin raid over central London that she had witnessed the night before. The result of the raid was devastating: 20 people were reported killed and 86 injured, and it was estimated that over half a million pounds worth of property damage had been inflicted on the area.[70] And, as Miss Davin noted at the end of her letter, the anti-aircraft guns "all went very wide of the mark." However, this was not surprising; indeed it would have been more surprising had they hit their target. Combat in the air was a new phenomenon and the theory and practice of anti-aircraft gunnery was in its infancy. But as the numbers of zeppelin raids increased with, from May 1917, Gotha biplane bombers as well as zeppelins being used, the need for

[68]Bryan and his work are discussed below in Section 4.2. Two of the experimental methods used in these experiments are described in [Hashimoto 1990, pp. 31–35].

[69]The report, which is dated 1914, is No. 154 of the ACA's *Reports and Memoranda*.

[70]A. Davin to K. Pearson, September 9, 1915 [Pearson Papers, 774/9]. For casualty figures, see *The Times*, September 11, 1915.

FIGURE 8. Anti-aircraft gun, Shoeburyness, July 1917. Courtesy of the Royal Artillery Historical Trust.

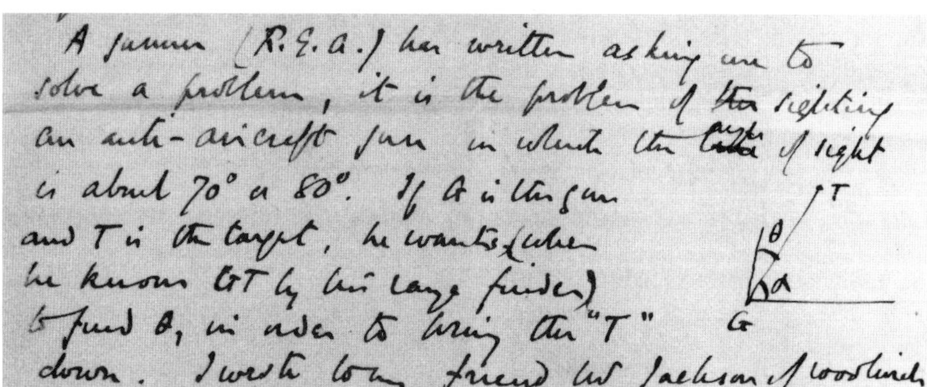

A gunner (R.G.A.) has written asking me to solve a problem, it is the problem of the sighting an anti-aircraft gun in which the angle of sight is about 70° or 80°. If G is the gun and T is the target, he wants (when he knows GT by his range finder) to find θ, in order to bring the "T" down. I wrote to my friend Mr Jackson of Woolwich

FIGURE 9. "Find θ in order to bring the T down." Extract from a letter from Frank Carey to Joseph Larmor, February 1916. © The Royal Society.

defense against air reconnaissance by the enemy became ever more important, and the improvement of anti-aircraft systems became a matter of urgency. However, making such improvements, which involved creating range tables, was no trivial task and involved some difficult mathematics, as the following letter to Joseph Larmor from Frank Carey, the professor of mathematics at Liverpool University, of February 1916, indicates.

> I am writing to bother you on a matter which is troubling me because I feel sure that there is no-one else in the House of Commons[71] who can appreciate the nature of the problem which has been referred to me.
>
> [See Figure 9.]
>
> A gunner (R.G.A.) has written asking me to solve a problem, it is the problem of sighting an anti-aircraft gun in which the angle of sight is about 70° or 80°. If G is the gun and T is the target, he wants (when he knows GT by his range-finder) to find θ, in order to bring the "T" down. I went to my friend C.S. Jackson of Woolwich[72] and he has sent me the manual of gunnery, the last copy issued to the public — 1902 — but he says there is nothing in the later copies, and of course if there were, the gunner who sent me the problem ought to have it in his possession. Now as far as I can judge, the problem is one of considerable difficulty: but a first-rate applied mathematician, like Proudman, could give tabular solutions in a month or six weeks.[73] The problem is of really first-rate importance in view of the necessity of reconnaissance work by airplanes at the Front.
>
> It would be easy enough to create a hub-bub over the matter in view of the sensitiveness of people to zeppelins. But I don't want that—I do however want the job done well—honestly I believe that it requires better brains than the present editors of the manual possess. If it is left to them, it will mean delay because second-rate men when faced with a new problem rarely succeed, if they succeed at all, at the first attempt.
>
> I hope very much that you will be able to press this important matter on the government and get a satisfactory set of tables for our keen to learn gunners, — after all we can't expect them to solve differential equations at the Front.[74]

That Carey, a mathematics professor, should have been contacted directly by a gunner not only highlights the urgency of the problem but also shows that soldiers on the ground had at least some idea about the mathematical difficulties involved. How Larmor replied to Carey is unknown but, as we shall now see, plans to produce high-angled range tables for anti-aircraft guns were already getting underway.

3.3.1. *Formation of the AAES.* The development of air defenses was part of the brief of the Ministry of Munitions, the ministry that had been created in the

---

[71]From 1603 until 1950 the University of Cambridge was a university constituency with the power to elect two members to the House of Commons. Joseph Larmor served as one of the members from 1911 to 1922.

[72]Charles Samuel Jackson, known colloquially as "Slide Rule," was eighth wrangler in 1889. A man "with great ability and wide scientific knowledge," he was an instructor in mathematics at Woolwich from 1891 until his sudden death in October 1916; see [Maurice-Jones 1954, p. 36].

[73]Joseph Proudman, a b^* wrangler of 1912 who since 1913 had been a mathematics lecturer at Liverpool University.

[74]F. S. Carey to J. Larmor, February 10, 1916. MS/603, Larmor Papers, The Royal Society, London.

aftermath of the artillery shell crisis of 1915 that had helped to bring down the liberal government under Herbert Asquith.[75] The ministry, which was initially headed by Lloyd George, had various divisions, one of which was the Munitions Inventions Department (MID). When the MID was formed in August 1915, Lloyd George appointed Ernest Moir, a civil engineer with military experience and previously in charge of machine gun production, as the comptroller. Moir then assembled an Advisory Panel. This was a group of 20 scientists on whose shoulders the success or failure of the department would rest; the members included Horace Darwin and the electrical engineer Sir Alexander Kennedy who was responsible for advising on gun sights. Among the principal functions of the MID were to assess the viability of inventions "likely to be of service in carrying on land warfare" and "to investigate and experiment upon means for dealing with aircraft attack" [Pattison 1983, pp. 529–530]. Since the latter was an entirely new area of research, a special group, the Anti-Aircraft Experimental Section (AAES), was formed early in 1916 to take care of it.

The architects of the AAES were Darwin and Kennedy who, recognizing that dealing with aircraft attack needed not only anti-aircraft weapons but personnel trained to operate them, had persuaded the MID to form a new section with a brief to deal with height and range-finding [Hartcup 1988, p. 161]. Charged with the task of finding someone to run the AAES, Darwin and Kennedy sought out Archibald Vivian Hill, a physiologist from Cambridge who had been third wrangler in the Mathematical Tripos in 1907, then serving as Brigade Musketry Instructor in the infantry, and he was appointed as the director.[76]

Hill's priority was to find staff and so he went to Cambridge to consult with Hardy, then a Fellow of Trinity and with whom he had studied mathematics as an undergraduate, hoping that Hardy might be willing to come himself. As Hill later recalled:

> [Hardy] replied at once, with indignation, that he was not going to prostitute his brains on such a job. He was not a pacifist he protested, and was ready if called to serve in the army—but prostitute his brains. No! [Hill 1974, p. 124].

However, Hardy did not seem to have any qualms about prostituting other people's brains. He recommended two students. One, Edward Arthur Milne, then a second year undergraduate who was unfit for military service due to shortsightedness and whom Hardy regarded as "one of his very best pupils in pure mathematics."[77] The other, Ralph Howard Fowler, had been a research student in pure mathematics when war broke out and was convalescing after having been badly wounded in the shoulder at Gallipoli while a gunner officer in the marines. Shortly after Fowler and Milne had been engaged by Hill, Herbert William Richmond, a geometry don of some 30 years standing from King's College, volunteered his services; and it was

[75]Within three years of its formation, the Ministry of Munitions had become the "largest government department the country had ever seen with as many as 1,600,000 men and 800,000 women employed on protected munitions work" [Hazlehurst 1919–1922, introduction].

[76]After the war, Hill went on to have a very distinguished career as a physiologist, winning the 1922 Nobel Prize for medicine. He also continued to devote a large part of his life to public service and in 1935 was part of the committee responsible for setting up radar in Britain.

[77]Hill to McCrea, November 2, 1950 [Milne Papers, 38]. In the same letter Hill described Milne as "an essential member in our party" and one who "did some of the finest work they achieved."

these three, Fowler, Milne, and Richmond, who became the mathematical mainstay of the AAES.

As the war progressed other mathematicians and electrical engineers of all ages were recruited or volunteered—the main requirement being an ability to calculate trajectories and fuze curves. Cambridge supplied several members of the group, including Thomas Lancaster Wren, a geometer "addicted to the purest of pure geometry" [Hill 1960, p. 307], and two former senior wranglers: Robert Alfred Herman, a lecturer in applied mathematics, and Geoffrey Thomas Bennett, a mathematician renowned for his versatility[78] and, in particular, for solving geometrical problems by mechanical means. William Hartree, who had retired from the teaching staff of the Engineering Laboratory at Cambridge in 1913, was an early member,[79] and his son, Douglas, a student at Cambridge, joined soon after him.[80] Another undergraduate, Arthur Charles Hawkins, was also recruited early on. Later Cambridge recruits included Gallop and Lock, who were assigned to the NPL, as well as Samuel Pollard, a b^* wrangler of 1916. Also in the group was the pure mathematician William Edward Hodgson Berwick, an assistant lecturer at University College of North Wales, who had been fourth wrangler in 1909, the last year in which wranglers were ranked.

In assembling his staff Hill met with considerable resistance from the recruiting authorities who had other ideas for their employment, and without the ongoing support of Horace Darwin the whole enterprise might have collapsed.[81]

As well as being a sound mathematician, Hill was an inspirational and good humored leader and collectively the group became known as Hill's Brigands. Hill was also blessed with natural tact and diplomacy, attributes which would stand him in good stead as the war developed.

At the end of the Lent term 1916, Hill, Fowler, and Milne went down from Cambridge to Northolt Aerodrome in Middlesex to test the instrument later known as the Darwin-Hill Mirror Position Finder.[82] However, finding the attitude of the authorities there somewhat chilly and encountering a general apathy from the people at the aerodrome, Hill ended up moving his staff to the National Physical Laboratory (NPL) for four months. Eventually the group settled on board *HMS Excellent*,

---

[78]When Bennett was elected to the Royal Society in 1914, the citation on his election certificate (which, incidentally, included the signatures of Richmond and Hardy) stated that he was "distinguished for his knowledge of, and contributions to, several distinct branches of mathematics." Royal Society, EC/1914/03.

[79]For a description of William Hartree's work for the AAES, see [Hill 1960, pp. 173–178].

[80]Douglas Hartree's main work for the AAES was the integration of differential equations for trajectories. He introduced the practice of using the time rather than, as had been traditionally used, the angle of elevation as the independent variable. For a short description of the numerical methods he developed for ballistic calculations, see [Hartree 1920].

[81]When Hill was having difficulty in getting Fowler formally attached to the MID—the Admiralty wanted to appoint Fowler as an inspector of steel, a subject about which the latter knew nothing—it was Darwin who came to the rescue by writing to the First Lord of the Admiralty. Hill also had trouble with engaging Milne—the ministry hesitated about signing him on as, aged 20, Milne was liable for military service although due to poor eyesight this would have been for noncombatant duty only. See [Weston Smith 1990, pp. 242–243].

[82]An instrument originally designed to record the flight of airplanes, consisting of two plane horizontal mirrors at the two ends of a long base, inscribed with coordinate parallels and provided with eye-holes at a fixed height above the mirrors and by means of which it was possible to determine the position of a shell burst in the sky with a reasonable degree of accuracy [Hogg 1978, p. 46].

at Whale Island, Portsmouth, where they were located for the rest of the war. It was here using the Royal Navy's proving ground that experiments were carried out, observations made, and the theory developed. Although Darwin continued to be consulted for advice,[83] "Hill was the inspirer of most of the researches that were conducted whilst Fowler [who had been appointed assistant director] executed them in the field" [Milne 1945, p. 66]. In addition to Whale Island, the group used the NPL for experiments and consulted members of staff at the Royal Aircraft Factory for advice. Later on the group also had a "travelling circus," a mobile party which toured military positions in Britain and France, recording the performance of guns and instruments [Anonymous 1920].

While based on board *HMS Excellent*, Milne kept up a regular correspondence with his younger brother Geoffrey, who was still at school.[84] His letters paint a vivid picture of both life in the AAES and the public attitude to anti-aircraft defense. On October 19, 1917, there had been a particularly bad zeppelin raid, in which 11 airships had attacked, bombing London, and not only had none been shot down, no shots were even fired. This caused an outrage in the popular press and questions were asked in the House of Commons.[85] What made matters worse was the fact that after the attack the zeppelins had then flown over France where most of them had been shot down. But as Milne explained to his brother, the public outrage was based on complete ignorance of the situation.[86] The raid took place on a cloudy night, which meant that it would have been impossible for the anti-aircraft gunners to see the zeppelins and consequently there would have been no point in their shooting. Moreover, not only were the zeppelins over France in daylight but also, due to the strong wind, the pilots would not have realized where they were until it was too late. Thus, Milne continued, the French should not be congratulated for shooting down 4 or 5 of the zeppelins (which were easy targets), but they should be rebuked for not shooting them all down.

In the same letter Milne went on to describe the role of the AAES: "designing new guns, new fuzes, new methods of fire" and "pointing out faults" in the current weaponry, while making the point that whatever improvements or devices they invented, there was an inevitable time lag, sometimes almost as long as a year, before they could be implemented. He also drew attention to the barrage system being employed in London at night:

> Because Londoners panicked G.H.Q. had to adopt a barrage system, which consists in simply blazing away an indefinite no. of rounds into an assigned spot. Previously each gun tried to fire only at a definite target, and consequently only got off a limited number of rounds. Now each gun gets off hundreds in a night, and the result is that *nearly all 3″ A.A. guns in London have had to be condemned as worn out.* Our section will in the future save the

[83]In November 1916 Horace Darwin produced a report for the ACA in which he indicated the direction in which he thought the research at Whale Island into anti-aircraft fire against both airplanes and zeppelins should be going and in particular advocating the formation of a school of instruction in anti-aircraft fire. [Nat. Arch., DSIR 23/866].

[84]The letters are preserved in [Milne Papers].

[85]For a report of the House of Commons debate, see *The Times*, October 23, 1917, p. 9.

[86]E. A. Milne to G. Milne, October 30, 1917 [Milne Papers, A8].

> War Office thousands of pounds for we have demonstrated that the high velocity 3 inch gun is useless for anti-aircraft purposes and that you want a much lower muzzle velocity which will consequently last much longer.

In other words, the anti-aircraft guns were wearing out without having served the real purpose for which they were designed. Not only was this policy costly in financial terms but replacing the guns was not easy either. It was an expensive way to reassure the public.

Away from Whale Island, the computational work of the AAES was being done in the Biometric Laboratory at UCL under the leadership of Karl Pearson. From the outbreak of the war, Pearson—a Cambridge graduate who had begun his career at UCL as professor of applied mathematics and in 1910 had changed his position to become professor of applied statistics—had been involved in war-related work. He had worked for the Board of Trade producing unemployment charts and import tonnage charts, for the Royal Aircraft Factory calculating the torsional strain in the blades of airplane propellers, for the Admiralty Air Department calculating elastic constants of wood, and for the Advisory Committee on Aeronautics preparing numerical tables for the calculation of bomb trajectories [ACA 1922, p. 18].

At the end of 1916 Pearson offered to help Hill with his calculations and Hill jumped at the offer:

> Ballistics is of course more or less specified, but the work we are doing on fuzes is quite new and has been developed by us: it is therefore unlikely that your people will know much about it at present. But if you could take on large pieces of work for us I could send one of my people, e.g. HW Richmond of King's, whom you probably know, or EA Milne, ... to show your people what wants doing and how to get at the required data out of ballistics tables etc. and how to use our experimental results contained there. It seems to me that you might be a very great help to us, to the Admiralty and to the Ordnance Committee. Your people being trained to work together could probably get results out very much more quickly than we being amateurs at computation possibly can: and we have a large amount of work we could take on if only there was a possibility of getting it finished.[87]

Shortly afterwards Hill wrote to Pearson again to ask him if he would be able to undertake work for the Ordnance Committee at Woolwich as well as for the AAES:

> [Milne] is coming to you tomorrow afternoon. I have arranged for him to see and discuss things with JE Littlewood in the morning.... Littlewood you probably know is a fellow of Trinity—now a Lieut. RGA and working for the Ordnance Committee at Woolwich. The latter are also very anxious to avail themselves of your help, and if you can find that you are prepared to put up with even more labour than I can inflict upon you, Littlewood will come and bring you some things too ....
>
> It will be a great help to us getting rid of a lot of the computation, and will give us more time for what is our proper function

[87]A. V. Hill to K. Pearson, December 15, 1915 [Pearson Papers, 606].

FIGURE 10. John E. Littlewood. By permission of the Master and Fellows of Trinity College, Cambridge.

> here—the collection of accurate data, and the investigation of the problems arising. From time to time also I expect you will help us with the problems as well as the arithmetic.[88]

Milne did indeed visit Pearson but at that time, with the work of the AAES still in the early stages of its development and the paucity of knowledge about the ballistics of anti-aircraft gunnery, he was rather limited in what information he could convey. Pearson later recorded that while Milne had transmitted "all the knowledge they [i.e. Milne, Hill, and Fowler] possessed" it seemed "very narrow" in the light of what he and his staff afterwards taught themselves.[89]

Littlewood (Figure 10) would have known Hill from their undergraduate days: they were fellow mathematics students at Trinity College, with Littlewood being one year ahead. Littlewood, who served as a second lieutenant in the Royal Garrison Artillery throughout the war, had been seconded for ballistics work at Woolwich in early 1916. Hill regularly consulted him on theoretical matters of ballistics, and Littlewood collaborated closely with other members of Hill's team, especially Fowler and Milne. Pearson and Littlewood may also have had a link through one of the female mathematicians at Cambridge, Frances Cave-Browne-Cave, to

[88]A. V. Hill to K. Pearson, December 31, 1916 [Pearson Papers, 720/30].

[89]K. Pearson. Journal of the Galton Laboratory, May 25, 1918 [Pearson Papers, 246].

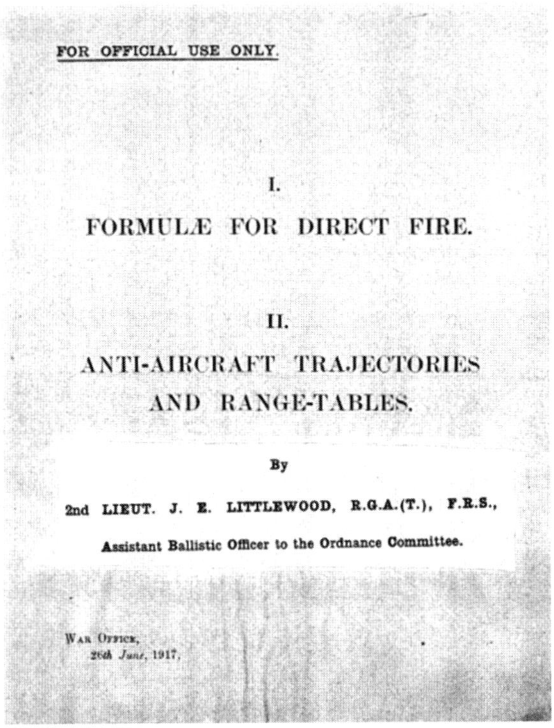

FOR OFFICIAL USE ONLY.

I.

FORMULÆ FOR DIRECT FIRE.

II.

ANTI-AIRCRAFT TRAJECTORIES AND RANGE-TABLES.

By

2nd LIEUT. J. E. LITTLEWOOD, R.G.A.(T.), F.R.S.,

Assistant Ballistic Officer to the Ordnance Committee.

WAR OFFICE,
26th June, 1917.

FIGURE 11. Title page of Littlewood's 1917 article on anti-aircraft gunnery. Pearson Papers, UCL Library Services, Special Collections.

whom Littlewood turned for help with finicky ballistics calculations.[90] Frances Cave-Browne-Cave was a lecturer at Girton College who had worked with Pearson before the war, and her sister Beatrice was one of Pearson's computers—Pearson employed several female computers—during the early years of the war, although she had left in the autumn of 1916 to work (for more money) at the Ministry of Munitions [Grier 2005, pp. 130–131]. Both sisters had studied at Cambridge.[91]

In his ballistics work Littlewood consulted with Augustus Love, Sedleian Professor of Natural Philosophy at Oxford. Love had devised a method for the calculation of high-angle trajectories by "small arcs" which was in use at Woolwich. Littlewood himself

> devised his own rapid and powerful method of calculating high-angle trajectories avoiding the painfully long process of 'small arc' integration.... But no-one knew whether projectiles really followed Littlewood's trajectories or not. [However, at the range-table firings] to the astonishment and joy of all concerned, the observed

[90] For further details about the assistance given by Pearson's laboratory to the authorities at Woolwich, see [Barrow-Green forthcoming].

[91] Frances Cave-Brown-Cave was ranked equal to the fifth wrangler in 1898 while Beatrice was ranked between the thirty-third and thrity-fourth wranglers in the same year. Beatrice also helped Leonard Bairstow with the calculation of tables for his textbook *Applied Aerodynamics* [Bairstow 1920b].

> positions of the shell-bursts fell exactly on Littlewood's trajectories, at the correct time-markings, within the very small errors of observation [Milne 1945, p. 66].

Littlewood gave an account of his method in a classified document (Figure 11) [Littlewood 1917] and much later published a description [Littlewood 1971–1972].

While the military at Woolwich certainly valued having a mathematician of Littlewood's caliber among them, not everyone had the same appreciation for Littlewood's work in ballistics. Hardy, Littlewood's mathematical collaborator, wrote: "even Littlewood could not make ballistics respectable, and if he could not who can?" [Hardy 1967, p. 140]. Although the remark is typical of Hardy, famous as a proponent of pure mathematics, it is conceivable that there was an additional reason for it. Time spent by Littlewood on ballistics was time not spent in their joint collaboration, a collaboration which had begun not long before war was declared—the first joint publication appeared in 1913—and which had quickly gathered momentum. The collaboration was not entirely interrupted by the war—at least one joint paper appeared during each year of the conflict—but it was surely affected.[92]

Among those employed on AAES work in Pearson's laboratory were two Cambridge research students: Lindsey Ince, who had originally graduated from Edinburgh and whom Pearson described as "a good mathematician, but not a very strong computer,"[93] and Adolph Zaiman, a b^* wrangler of 1916. Zaiman's route to Pearson's laboratory was rather convoluted and following it gives a glimpse of the rather disorganized nature of the recruitment process.[94] Both Pearson and Hill, who had each become aware of Zaiman's existence through the University of Cambridge Appointments Board, simultaneously offered him a position: Pearson at his laboratory and Hill at Whale Island. But shortly after Zaiman had accepted the offer from Hill, Hill discovered that he was not after all allowed to increase his staff. Pearson, however, was still able to take on new recruits, so in May 1917 Zaiman went to work for him. But less than three months later Zaiman had been called up and Pearson and his superiors in the Ministry of Munitions were having to fight for his exemption from military service.[95]

But not all Cambridge mathematicians who made themselves available to work on war-related calculations were prepared to toil in Pearson's laboratory. The analyst George Neville Watson, who had been an exact contemporary of Hill's at Trinity and senior wrangler, evidently considered the computing work beneath him, as a letter from Hill to Pearson reveals:

---

[92]For a bibliography of Littlewood's publications, see [Burkhill 1979].

[93]K. Pearson, *Journal of the Galton Laboratory*, May 25, 1918 [Pearson Papers, 246]. This entry includes the names of several students from London colleges and elsewhere, female as well as male, who were employed in his laboratory during the war. Ince later became well known as the author of textbooks on differential equations.

[94]H. A. Roberts to K. Pearson, April 3, 1917; A. Zaiman to K. Pearson, April 17, 1917; A. V. Hill to K. Pearson, April 24, 1917 [Pearson Papers, 606]. H. A. Roberts was secretary to the University of Cambridge Appointments Board.

[95]It is not clear where Zaiman was based during the rest of 1917 but by July 1918 he was working for the Meteorological Section in Shoeburyness. He had been due to go to France to work on sound-ranging but due to a 'slump' in activity in that department he had transferred. It turned out that much of the work he was doing there was similar to the work he had been doing with Pearson and he admitted that he did "sometimes yearn for a Brunsviga!" A. Zaiman to K. Pearson, July 13, 1918 [Pearson Papers, 906/1].

> I have known Watson these last 12 years, and am interested (and distressed) to hear he is exactly like he used to be. I didn't know that the war, and coming to London, had not made him more humble and human, but from what you say it obviously has not. When one sees people here like Richmond and Bennett giving up their whole time with the utmost goodwill to what is often sheer drudgery and your people, and countless other people doing the same, one can only marvel at a man who thinks himself above such things. Since Kennedy & Coker[96] asked me about him I thought there was a possibility that he might be willing to forget himself for a bit & try to be useful: & I therefore didn't tell you that I thought, from my previous knowledge of him, that he might not be much use. I have simply told Kennedy that you think he would not be the sort of man to get much out of. [97]

3.3.2. *Work of the AAES.* As director of the AAES, Hill was called upon to produce reports on the work and organization of the section. In a report produced towards the end of the war,[98] he divided the functions of the section into three main categories:

(1) The general investigation of the theory of anti-aircraft (AA) gunnery and connected experimental work.
   (a) Investigation, experimental and theoretical, of the shell's motion in air.
   (b) Construction of AA range tables and graphic range tables for the Ordnance Committee and the Director of Naval Ordnance.
   (c) Investigation, in gun trials and in the laboratory, of fuse-burning, causes of blinds, effects of temperature, pressure, and spin; theoretical analysis of results.
   (d) Investigation, experimental and theoretical, of height finders, range finders, predictors, and other AA instruments.
   (e) Investigation of AA gun-sights and central post instruments.
   (f) Construction of means of exhibiting AA deflections.
   (g) Testing new guns, shells, fuses, and apparatus of any kind, for the purpose of AA.
(2) The investigations of some branches of ordinary gunnery.
   (a) Construction of graphic range tables for "Air Burst Ranging."
   (b) Calculations of the effects of winds, and changes of air-density, etc., for flat fire.
   (c) Trials of guns on the flat, by observation.

---

[96]Ernest George Coker was Kennedy Professor of Civil and Mechanical Engineering, University College, London.

[97]A. V. Hill to K. Pearson, February 20, 1917 [Pearson Papers, 606].

[98]The report from which this information is taken is undated but it must have been written some time after June 3, 1918 since it acknowledges the OBE which was awarded to Fowler on that date. Hill Papers, Churchill College Library, AVHL 1/1.

(d) Ballistic work for the Ballistic Office of the Ordnance Committee.
(e) Investigations of the dynamics of shell flight, initial stability of forces on shells, rotation of the earth, drift, etc.

(3) The investigations of problems connected with the location of aircraft by sound.

Different guns and different shells each required their own tests and observations, from which the individual range tables had to be constructed, so there was much work to be done, and the results were urgently required by those operating in the field. The novelty of the tasks and the difficulty of retaining personnel put great pressure on all involved.

An important consideration in preparing the tables was the degree of accuracy to which they should be made. There was no point taking the calculations beyond the point at which they could be of practical use. Or so Pearson thought. But Hill had to inform him otherwise:

> The reason why Fowler ... asked you for times and fuzes to 1/100 sec or setting, was that various people desire to plot results, and unless the results are quite smooth two different people plotting the same thing may obtain very slightly different results. This would not matter only some people are anxious to strain out gnats though quite prepared to swallow camels: and it was thought to be perhaps a good thing to prevent them from wasting their own—and other people's time on the gnats. The camels—in the shape of the inevitable uncertainty of the experimental data—are a provision of nature, and nobody seems to mind them.... Everybody realises quite well that it is absolutely impossible to obtain results accurate to 1/100: do not take any further trouble to do this; but if without any further trouble you can appease the gnat-strainers it may be a good thing.[99]

The letter also exemplifies an ongoing problem for those working in the AAES: the lack of understanding by the military of the theory underlying the construction of the range tables. Unlike for flat fire, for anti-aircraft gunnery every point of the arc of the trajectory of the shell should be known and the flight time is critical (since the time fuze has to be set). Since the differential equations which describe the motion of the shell cannot be solved exactly, either numerical integration ("small arcs" method) or interpolation has to be used. The problem is further complicated by the need to take into account varying wind speed, varying air density, speed of aircraft, rapidity of maneuver of aircraft, height of aircraft, lack of visibility of aircraft, magnitude of unknown errors affecting gunfire at great heights, and the speed with which it is necessary to carry out the firing.[100] Hill was regularly annoyed by staff officers who needed to be convinced not only that it was possible to find the height of an airplane from the ground, but also that it was actually useful to know the height when it was found. His letter continued in a similar vein:

---

[99] A. V. Hill to K. Pearson, April 16, 1917 [Pearson Papers, 606].

[100] See [Barrow-Green forthcoming] for details of the work involved in preparing the range tables.

> Our work is largely educational and not merely scientific—otherwise we should get on 5 times as fast! The ignorant people at the head of things who continually fail to appreciate the urgency and the point of a new idea and the new instruments are a far worse difficulty than nature. Nature is easy to investigate being merely inanimate and reluctant: but to persuade military officers of high standing to unlearn their prejudices and to give one—especially if one is a junior officer—the credit for knowing one's job is infinitely harder, and requires tact and patience, beyond understanding. It has taken a long time to persuade them to adopt *height* [emphasis in original] instead of range: and the urgency of it has made one perfectly sick sometimes of the whole show. But things are improving very fast now.

So not only were there members of the military who did not understand the construction of the tables, there were those who could not see the point of constructing them at all: those who thought that the tables for flat fire would do just as well! Hill had very definite views on the (lack of) scientific knowledge of the army and he did not hesitate to express them: he had no doubt that "a General (or even a Major) in a large red hat is capable of misunderstanding anything."[101] His discontent in this respect did not, however, extend to the navy of whom he had a very positive opinion and with whom he had extensive contact during his work for the AAES.[102]

Nevertheless, despite these difficulties, the diagrams produced by Pearson's laboratory were a success. When Hill showed the first ones to the Ordnance Committee at Woolwich they immediately asked Hill to get 200 copies made and printed on linen for use in the field and requested similar ones to be made for other guns.[103] So far no account has come to light of the diagrams being used in a conflict situation but they were certainly used in the field. In May 1917 Hill was with the British Expeditionary Force in France from where he reported to Pearson:

> I think you and your people would be really gratified if you could be out here for a few hours and see how your G.L. [Galton Laboratory] diagrams are regarded. I have seen several of the Army Group Commanders and the Officers ... at G.H.Q. They affirm without hesitation that these diagrams are invaluable and fulfill a long felt want: that they are beautifully done and printed ... and that people here are really grateful for these charts. They seem to spend hours every day poring over them, measuring them and reading things off them. And they want to have a dozen each so they can draw on them and use them generally. Please tell your people of this.[104]

---

[101] A. V. Hill to K. Pearson, May 14, 1917 [Pearson Papers, 606].

[102] The superior scientific knowledge of the navy was widely acknowledged at the time. See, for example, the memorandum drawn up in early 1916 by a committee concerned by the general neglect of science in public institutions and which was signed by Karl Pearson. *The Times*, February 2, 1916.

[103] A. V. Hill to K. Pearson, February 12, [1917] [Pearson Papers, 720/3].

[104] A. V. Hill to K. Pearson, May 22, 1917 [Pearson Papers, 606].

Fowler too told Pearson that the anti-aircraft group commanders in France were very pleased with the diagrams and wanted more of them.[105]

As the war continued, the intensity of German bombing raids increased. But the anti-aircraft defenses were not effective. Pearson obviously felt the strain and by the end of September began to question the value of the work he was doing. By this time he and his staff had been working solidly for over nine months without a break. The hours were long, the calculations tedious and repetitive,[106] and he was battling to keep his staff. So when a letter came from Richmond on October 4, 1917, querying some of his results, it seemed like the last straw. He was ready to throw in the towel and Richmond had given him the wherewithal to do so.

Richmond had discovered what he believed to be a mistake in some calculations from one of Pearson's tables and wrote in a matter of fact manner to Pearson asking him whether it was possible that his computers had made a mistake in the sign of some correcting terms.[107] But Pearson took immediate umbrage. He considered a slur had been cast on the ability of his computers and declared his intention to resign. While Richmond might have been more unctuous in his tone, Pearson's reaction was all out of proportion to the event. And it meant that Hill had to bring all his diplomatic skills in to play to keep Pearson on board, even going right to the top of the administrative tree in order to do so.[108]

The "Richmond incident" reverberated around the staff of the AAES.[109] In a letter to his brother Milne commented:

> Richmond here has had a violent quarrel with K.P., so that it is doubtful whether, in spite of Hill's mediation, K.P. will ever do any more calculations for us again. Thus my *contretemps* [emphasis in original] with K.P. last January is in a way justified.[110]

So Richmond was not the first to have a disagreement with Pearson. In Milne's case Pearson had rather tetchily questioned Milne about some discrepancies which had arisen from using two types of formulae for plotting trajectories. But Milne provided a measured response and the situation remained under control.

Pearson and his staff worked tirelessly on ballistics problems for the whole of 1917 almost without a break. Apart from one week at Christmas the Biometric Laboratory was never closed from January 1, 1917, to January 1, 1918. Given the nature of the work and the long hours he spent on it, it was perhaps not surprising that Pearson somewhat overreacted to Richmond's remarks. But his efforts and those of his staff were certainly valued both by those who were using their results as well as by the officials in charge of the administration of the enterprise.[111]

But as the new year wore on, Pearson became increasingly ground down by the work and the lack of time he had for doing his biometric and eugenics work.

---

[105] R. H. Fowler to K. Pearson, May 23, 1917 [Pearson Papers, 606].

[106] Since each type of gun required several range tables (involving different combinations of tangent elevations, angles of sight, heights, and fuze settings) for each type of fuze, the calculations had to be repeated many times; see [Barrow-Green forthcoming].

[107] H. W. Richmond to K. Pearson, October 4, 1917 [Pearson Papers, 606].

[108] Colonel Goold Adams, comptroller of the MID, to A.V. Hill, October 18, 1917 [Pearson Papers, 606].

[109] The "Richmond incident" is described in greater detail in [Barrow-Green forthcoming].

[110] E. A. Milne to G. Milne, October 21, 1917 [Milne Papers, A8].

[111] See the letter from R. H. S. Bacon to K. Pearson, February 14, 1918 [Pearson Papers, 606], reproduced in [Barrow-Green forthcoming].

Finally towards the end of March 1918, he decided to give up the ballistics work.[112] He handed over the reins to Arthur Doodson, a Liverpool graduate and conscientious objector who had joined his staff at UCL in 1916, prior to the formation of the AAES.[113] Without Pearson, there was little reason for the work to be done at UCL, and Pearson made arrangements for the team to be transferred, almost in its entirety, to Woolwich, which was in many ways a more practical location than UCL since there ballisticians and computers could work at the same place. Littlewood was already there, as was Henry Hassé, a lecturer in mathematics from Manchester University and Cambridge graduate of 1908, who was working on the mathematics of anti-aircraft gunnery [Vint 1956, p. 253]. They were joined by Joseph Proudman, a high achieving Cambridge wrangler of 1912, who arrived during 1918 and worked on interior ballistics [Proudman 1921]. Proudman, a mathematical lecturer at Liverpool University—he had not been called up for medical reasons—was presumably recruited by Doodson, who had been his student before the war [Cartwright & Ursell 1976, p. 320].

Not all members of the AAES remained with the group until the end of the war. Bennett, for example, who had made his mark by inventing a device for determining the height of an attacking plane, left in 1917 to work at the gyro-compass department in the Admiralty, a position in which "his grasp of dynamical principles would have had full scope" [Baker 1944, p. 109]. And such was the pressure of his work for the Admiralty that he had to resign from his position as secretary of the London Mathematical Society.[114] On the other hand, there were members who continued taking part in experiments at the firing ground of *HMS Excellent* for several months after the war was over, a notable result of which was a long and important paper on the aerodynamics of spinning shells ([Fowler et al. 1920, Fowler et al. 1922]), which was influential on the development of ballistics for the Second World War. This was followed by further joint work of Fowler and Lock on the problem of establishing the cause of the disturbance in the initial motion of a shell [Fowler & Lock 1921]. Meanwhile Fowler also worked with Bairstow and Douglas Hartree on the pressure distribution on the head of a shell moving at high velocities [Bairstow et al. 1920] and with Milne on the problem of location of aircraft by sound [Milne & Fowler 1921].

After the war was over, Richmond took on the enormous task of compiling a record of the theoretical and experimental advances in which the AAES had been involved. He was given an office in Cambridge designated for this purpose, and the result, the confidential *Textbook of Anti-Aircraft Gunnery* [Richmond 1924–1925],

---

[112] A. V. Hill to K. Pearson, March 23, 1918 [Pearson Papers, 606].

[113] Doodson had written to Pearson in June 1916: "I am glad to have further details of the nature of the Government work undertaken by your staff. I have noted particularly the statement that your staff is not recruited for the prime purpose of doing Government work, and, as I understand it, any work of this nature is more or less incidental to your ordinary work. This is the principle on which I have been acting—to acquiesce in any war work incidental to my ordinary occupation and which I was doing because of that reason, not because it was war work. I have recently had to refuse a position which was in fact created for me but which was concerned purely and simply with military matters. I take up this line because I recognise that it is impossible for me to take up any occupation which does not in some degree contribute to the progress of the war.... Certainly I should prefer to do as little war work as possible and I certainly should refuse to take a position for the sake of doing war work... My objection to military service is based on conscientious grounds." A. Doodson to K. Pearson, June 6, 1916 [Pearson Papers, 676/5].

[114] Council Minutes of the London Mathematical Society, October 11, 1917.

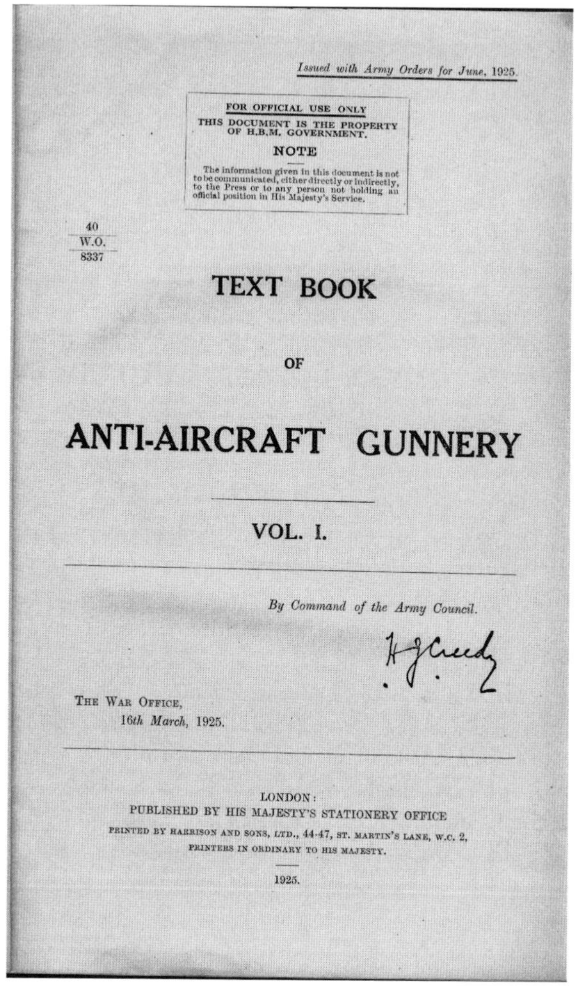

Issued with Army Orders for June, 1925.

FOR OFFICIAL USE ONLY
THIS DOCUMENT IS THE PROPERTY OF H.B.M. GOVERNMENT.

NOTE

The information given in this document is not to be communicated, either directly or indirectly, to the Press or to any person not holding an official position in His Majesty's Service.

40
W.O.
8337

TEXT BOOK

OF

ANTI-AIRCRAFT GUNNERY

VOL. I.

By Command of the Army Council.

THE WAR OFFICE,
16th March, 1925.

LONDON:
PUBLISHED BY HIS MAJESTY'S STATIONERY OFFICE
PRINTED BY HARRISON AND SONS, LTD., 44-47, ST. MARTIN'S LANE, W.C. 2,
PRINTERS IN ORDINARY TO HIS MAJESTY.

1925.

FIGURE 12. Title page of *Textbook of Anti-Aircraft Gunnery*, 1925. Copy belonging to the author.

was a two-volume work of 1,300 pages to which many members of the AAES contributed, including those at UCL and the NPL (Figure 12). The information it contained was comprehensive and it was used extensively in the Second World War.[115]

4. Cambridge-Trained Mathematicians

Of the mathematicians too old to be called up but who actively sought to use their mathematical skills in support of the war effort, there were a number who, although not based in Cambridge when war broke out, had been Cambridge trained and had kept up their Cambridge connections. Among the most notable were Sir George Greenhill, retired professor of mathematics at the Royal Military Academy, Woolwich; George Hartley Bryan, professor of pure and applied mathematics at the

[115] [Milne & White 1948, p. 227].

University College of North Wales; and William Burnside, professor of mathematics at the Royal Naval College, Greenwich. Each worked in a different branch of mathematics and each was keen to make the best use of his abilities in the service of his country. But prevailing attitudes towards mathematics meant that they did not always find it easy to do so.

4.1. George Greenhill: The Military Mathematician. George Greenhill, who was professor of mathematics at the Royal Military Academy at Woolwich for over 30 years, spent most of his career teaching mathematics to young soldiers. The author of several textbooks, he is best known for his research in hydrodynamics and in elliptic functions. Although primarily interested in the mathematics, he was always on the lookout for practical applications and was renowned for using physical demonstrations to illustrate his lectures. After his retirement from Woolwich in 1908, he continued to play a very active role in the mathematical and scientific communities. He was, for example, a member of the three-man International Commission for Mathematical Instruction,[116] as well as a member of the ACA. A somewhat eccentric figure, he was never shy about stating his opinions. Well before 1914 and on more than one occasion, he publicly expressed his concern about the state of British military science and in particular its deficiencies in comparison with that in Germany.[117]

For Greenhill in January 1915, there was no doubt about the nature of the war. "This is a Mathematical War," he declared in his widely circulated presidential address to the Mathematical Association [Greenhill 1915, p. 25].[118] The address was reproduced in full in both *Nature* and *The Times*, as well as in the Mathematical Gazette, the periodical of the Association.[119] To demonstrate his point, Greenhill used elementary examples, e.g. applying Euclidean geometry to the fragment of a shell in order to determine its caliber. Then, in an echo of earlier prewar remarks, he launched into a fierce criticism of the "shabby and undisciplined" situation at Woolwich ([Greenhill 1915, p. 28]), comparing it extremely unfavorably with the Military Technical Academy in Berlin, which he had visited, under the guidance of the ballistician Carl Cranz, some five years before. Pointing the finger of blame firmly at "the rule of Thumb, the official genius" ([Greenhill 1915, p. 29]), he vented his ire on the fact that several years ago officials had deemed calculations connected with long-range fire no longer useful.

[116]The Commission was formed at the International Congress of Mathematicians in Rome in 1908; the other two members were Felix Klein and Henri Fehr.

[117]See, for example, Greenhill's review of H. A. Bethell's *The Science of Modern Artillery* in Nature **87**, October 12, 1911, pp. 478–479.

[118]Greenhill was neither the first nor the last to make such an explicit connection between war and mathematics. On August 9, 1914, Herbert Kaufman, an American writer and journalist who was visiting England contributed a lengthy column on the war to *The Times* in which he wrote, "The range of weapons and modern tactics have set the struggle upon a basis of mathematics." Nurse May Sinclair, in her memoir written in 1915, portrayed No. 1 Belgian Field Hospital as "a world apart, a world of insufferable space and agonizing time, ruled over by some inhuman mathematics and given over to pre-transcendent pain," quoted in [Mayhew 2013, p. 239, Note 5]. Then two years later, Private Gladden, a 20-year-old soldier who took part in the assault on Messines Ridge in 1917, hauntingly described his feelings on the crossing of the shell-torn hillside: "The poor devils caught in that terrible cataclysm had no chance. Yet what chance was there for anyone in that war of guns and mathematics?" [Gladden 1930, p. 121].

[119]*Nature* **94** (1915), pp. 553–554; *The Times* (January 11, 1915). A résumé of Greenhill's address was also published in *The Observer* (January 10, 1915).

The subject of long-range fire was one to which Greenhill returned in March 1918 when the French capital came under fire from the German so-called "Paris Guns." These guns had a range of 75 miles, a distance far in excess of any gun known previously. Again Greenhill commanded several column inches in *Nature* and *The Times*.[120] But, this time, in contrast to his earlier invective, he wrote with an optimistic air. The intervening period had seen such a ratcheting up of scientific work at Woolwich, he now felt sure "that the scientific artillery officer is on his mettle to rival and exceed these recent [German] performances which have taken us by surprise."[121]

Nevertheless, however optimistic Greenhill might have been feeling in March 1918 about the future outcome of the war, the bombardment of Paris had underlined the enemy's superiority in ballistics. It therefore seems likely that it was this incident that prompted him to embark on the translation of Volumes 1, 3, and 4 of Cranz's *Lehrbuch der Ballistik*.[122] Cranz's text was known to contain much new material on both exterior and interior ballistics and Greenhill, having met Cranz and having written extensively on ballistics, was well suited to the task.[123]

Greenhill's early interest in hydrodynamics progressed naturally into an interest in aeronautics. As noted earlier, he was a founding member of the ACA, and he was also the author of two important reports for the Committee. The first, published in 1910, on the theory of streamline motion past a plane barrier, with particular application to the airplane, is a highly theoretical work which extended the theory by providing solutions to problems which require elliptic functions. However, the report's almost exclusively mathematical content prompted criticism from those of a more practical persuasion, and it was therefore seen as typifying the general lack of cooperation which then existed between the theoretical and practical constituencies.[124] But Greenhill did at least endeavor to redress the balance. In 1910 and in 1911, he gave a course of lectures at Imperial College, the object of which was to provide a guide to the report and in which he made frequent recourse to physical examples.[125] A supplement to the report, bringing the research up to date, appeared in 1916. Greenhill's second report for the ACA, a *Report on Gyroscopic Theory* (1914), also contained elliptic functions but, in contrast to the earlier report, it included plenty of practical applications, was well received, and was acknowledged to be of practical value during the war [Smith 1915, p. 794].

[120] *Nature* **101** (March 28, 1918), pp. 65–66; *The Times* (March 26, 1918).

[121] *The Times* (March 26, 1918).

[122] In the translation, which was published by H.M.S.O. in 1921, the name of the translator is not given. However, Greenhill's name is given as the translator in the Licence Application made by the Stationery Office to the Patent Courts for the translation. See "German Book of Gunnery," *The Times* (October 11, 1918).

[123] Greenhill was, for example, the author of the article on "Ballistics" in the acclaimed 11th edition of the *Encyclopædia Britannica* (1911).

[124] One reviewer, who considered the ACA to be out of touch with the "average man," suggested that the (lack of) cooperation problem could be resolved if the Committee would "produce a volume giving a quantitative comparison between solutions of problems as calculated mathematically and as obtained by actual experiment," [B. G. C. 1911].

[125] The publication of the lectures [Greenhill 1912] was warmly received by G. H. Bryan who, in a review, characteristically used the opportunity to bemoan the general lack of interest of mathematicians in such questions [Bryan 1912].

4.2. George Hartley Bryan and the Importance of Mathematics for Aeronautics. G. H. Bryan was a pioneer in the study of aeronautics whose own work centered on questions of stability.[126] Fifth wrangler in 1886, Bryan obtained a Smith's Prize the following year with an essay entitled "The waves on a rotating liquid spheroid of finite ellipticity" and was elected a fellow of his college (Peterhouse) in 1889. In 1896 he was appointed professor of pure and applied mathematics at the University College of North Wales, Bangor, where among his best students were Robert Jones and Dan Williams, both of whom went on to have distinguished careers as aerodynamicists at the NPL, having begun their work there during the war. Bryan maintained the connections with his Cambridge college and in 1915 was made an honorary fellow.

For Bryan the importance of mathematics for aeronautics research was indisputable; and during the war it was a theme to which he returned repeatedly. His zeal in this respect, which was evident well before the war, clearly made an impression on those who heard him speak. As one member of the audience at his Royal Institution lectures in 1909 reported:[127]

> One thing which must have been noticed was the lecturer's repeated references to the necessity of mathematical knowledge on the part of those who undertook research work connected with aerial navigation. It was a remarkable fact, he said, that with very few exceptions those who had experimented with aeroplanes and airships did not possess a large amount of mathematical knowledge or knowledge of experimental physics. It was a pity that the practical and theoretical men had not worked side by side. Few of the experimenters knew what a biquadratic equation was, and the sooner they did the better it would be for them.[128]

The following month the announcement of the formation of the ACA gave Bryan another opportunity to articulate his concerns. Writing in *Nature* on "The Government and Aeronautical Research" he was unequivocal:

> It would be more correct to describe the present position of aeronautics in England as "too much theorising and too little theory." Many papers have found their way into aeronautical and other periodicals, some of them full of algebraic symbols and formulae, but an investigation is not necessarily mathematical because it contains equations, and the author is not necessarily a mathematician because he employs them. Indeed, in many cases, it is the "practical man" who revels in the excessive use and abuse of formulae, and the mathematician and physicist who would like to bring themselves in touch with practical problems are consequently deterred from reading such literature. Moreover, there is a want of suitable journals for the publication of mathematical and physical investigations bearing on aeronautics [Bryan 1909, p. 314].

The fact that actually building airplanes was for most people a rather more attractive proposition than understanding the fundamental principles underlying

---

[126]For Bryan's biography see [Bairstow 1933].

[127]Bryan gave two lectures at the Royal Institution on "Aerial Flight in Theory and Practice" which were reported in *The Times* (April 2, 1909) under the subtitle "The Dangers of Aeroplanes."

[128]*The Engineer* **107** (April 2, 1909), p. 353.

their construction should not, he argued, diminish the importance of maintaining support for mathematical investigations.[129]

In 1911 Bryan published his best known work, *Stability in Aviation*.[130] Based on the study of small oscillations and using the theory of dynamic stability for rigid bodies first formulated by his Tripos coach Edward Routh, it contained discussions of both longitudinal and lateral stability and convincingly made the case for the mathematical treatment of stability problems. Up until this time there had been little by way of practical investigation into the stability of airplanes, a point not lost on Bryan, who in the introduction to his book complained about the apparent disinterest of the NPL in this respect [Bryan 1911, p. 8]. But with the appearance of Bryan's book the situation changed rapidly. In 1912 Bairstow and his team at the NPL set up wind tunnel experiments to study stability using Bryan's formulation of the theory, and before long Bryan's stability theory was incorporated into all airplane design.[131] The NPL, working together with the Royal Aircraft Factory at Farnborough, applied Bryan's theory to the construction of airplanes, and in 1914 the B.E.2c (Figure 4) was built to demonstrate the theory's effectiveness. It was the first plane to intercept and destroy German zeppelins using on-board gunnery. As Bairstow later remarked:

> Scientifically and technically the B.E.2c is one of the most interesting aeroplanes ever built. It showed conclusively that the line of argument developed by Bryan could be used to calculate the degree of stability of an aeroplane and so paved the way for the design of aeroplanes with the desired characteristics of indifference or stability as required for fighting or bombing [Bairstow 1919, p. 2].

Bairstow's successful application of Bryan's theory was noted across the Atlantic. In December 1916, in an address to the Mathematical Association of America, Edwin Bidwell Wilson, professor of physics at the Massachusetts Institute of Technology, drew attention to the work of Bryan and Bairstow, ascribing the rapid British advance in aeronautics during the war to the strength and longevity of Cambridge training in mathematical physics.

> I have been told that the theory of Bryan, recast by Bairstow, and supplemented by data of wind tunnel experiments, has enabled a machine to be designed and constructed which actually has been flown "hands off." If this be true, it marks a great triumph for mathematical physics; whether it is true or not, we can confidently assert that to the English training in mathematical physics is in no small degree due the great and sudden advance in airplane design and the great success in aircraft warfare which have been realized in England in relatively few months. [...]
>
> I have shown how the theoretical work has come in that stage of the development of aeronautical engineering where it has been of real help in rapidly advancing the art. There are always with

[129]The mathematics department at Bangor did in fact build and fly a "flying machine" in 1911 [Sanderson 1972, pp. 133–134].

[130][Bryan 1911]. For a more detailed discussion of Bryan's work, see [Hashimoto 1990, pp. 41–50].

[131]For a description of two of these experiments, see [Hashimoto 1990, pp. 31–35].

> us branches of engineering and physics in which the right kind of mathematics is of great value for the rapid advance of those branches. This right kind of mathematics is the good old traditional Cambridge, England, type, the mathematics of Newton, of Green, of Maxwell, of Kelvin, of Rayleigh, a type of mathematics which in this country, owing perhaps to the preponderance of German-trained mathematicians, has all too little prestige [Wilson 1918, p. 297].

Moreover, Wilson was not wrong in his judgement. Bryan had both undergone the Tripos, with its rigorous program of courses (many in subjects of mathematical physics) and done postgraduate work in an environment led by professors such as Stokes and J. J. Thomson.

Meanwhile Bryan had been working with his student Robert Jones on the problem of discontinuous stream motion past a bent plane. Their investigation [Bryan & Jones 1915], which was based on the Kirchhoff-Helmholtz theory of discontinuous streamline motion, followed on from [Greenhill 1910]. Although the paper exemplified Bryan's mathematical mastery in dealing with airplane problems, the calculation methods he used were old-fashioned and, as a historian has pointed out [Hashimoto 1990, p. 150], more efficient and elegant techniques for dealing with this type of problem had already been developed elsewhere.[132]

During the war Bryan emphasized ever more strongly the importance of mathematics for aeronautics. In his Wilbur Wright Memorial Lecture on *The Rigid Dynamics of Circling Flight* [Bryan 1915], the message was clear. The lecture was chaired by Frederick Lanchester, and Bryan used the former's work on stability (avowedly unmathematical) to make his point, as the following extract from his introduction, which was published as an indirect quotation, shows:

> We had all heard lately the insistent cry, "Wanted—more shells." Those who would read his book, "Stability in Aviation," and in fact everything that he has published on the subject, as far back as 1897, would find that throughout his work the insistent cry was to be found, "Wanted—more algebra."
>
> He [Bryan] hated algebra and no doubt Mr Lanchester also hated it, for in the investigations he had published into airplane stability he had tried to do without algebra, and he, Professor Bryan, would like to say that Mr Lanchester had obtained remarkable results and had carried the subject as far as any man could hope to do without using algebra. Professor Bryan had seen only too clearly from the commencement that unfortunately algebra could not be done without, and therefore had to face the distasteful task as best he could [Bryan 1915, p. 49].

As the war continued, so too did the lack of mathematicians working on aeronautical problems, as was evident from the columns of *Nature*.[133] The subject was even ignored by Section A (Mathematical and Physical Sciences) of the British

---

[132]For an example of the more modern approach (which was based on work of Tullio Levi-Civita) to a similar problem, see [Levy 1916], and for a discussion of the related work of Henri Villat, see [Aubin forthcoming].

[133]See, for example, *Nature* **94** (January 1915), p. 541.

Association for the Advancement of Science. For Bryan, who regarded every airplane "as a collection of unsolved mathematical problems" [Bryan 1916, p. 510], the situation had to be remedied and in August 1915 he advertised for research students to work on such problems.[134] His call was answered by the young Selig Brodetsky.[135] Brodetsky, who had been bracketed (joint) senior wrangler in 1908 and awarded a PhD from Leipzig in 1913, was no longer a student but a lecturer in applied mathematics at Bristol University. He began working with Bryan at the end of 1915 on the effects of air gusts on airplane motion ([Bryan & Brodetsky 1916]). It was a productive collaboration and in 1917 Bryan received a three-year grant from the newly established Department of Scientific and Industrial Research[136] so that he could work with Brodetsky in Bristol.[137] Based at the British and Colonial Aeroplane Company [Sanderson 1972, p. 229], together they worked on "the theory of canonical forms for dealing with the general motion of an aeroplane" [Brodetsky 1928, p. 850], and after the war Bryan produced a number of reports for the Advisory Committee for Aeronautics detailing the results of this research and other work on the effect of air compressibility on streamline motion.

Bryan's influence on Brodetsky is evident in the latter's undergraduate text on the mathematical problems of flight which appeared in 1921—a large proportion of which was too far removed from reality to be of any practical value—in which he mentioned his debt to Bryan no less than three times in the short preface [Brodetsky 1921]. However, the book did not take into account the more recent developments, a point which was noted by Glauert, who, in a review published in the same year, rather crisply observed that it "would have been a good book six years ago" [Glauert 1921, p. 378].[138] Nine months later Bryan, who no doubt felt stung by Glauert's remark to his protégée, responded in characteristically robust fashion. He used his review not only to exaggerate the merits of the text—"Dr Brodetsky is performing for aviation what Clerk Maxwell accomplished in anticipation of modern electrical engineering" [Bryan 1922, p. 297]—but also to respond to the "mathematicians of repute" who had recently criticized him for encouraging greater mathematization of the subject (ibid., p. 295), remarking that Brodetsky's text was "thoroughly imbued with the ideals and spirit of the best mathematical school of Cambridge" (ibid., p. 297).

4.3. Code-breaking and William Burnside. The vital contribution of mathematicians to code-breaking, during the Second World War, in particular that of Alan Turing and Max Newman in breaking the Enigma cipher, is well known and rightly celebrated. In the First World War, the situation with respect to mathematicians and code-breaking could hardly have been more different, as a memorandum

---

[134] University and Educational Intelligence, *Nature* **95** (August 1915), p. 664.

[135] University and Educational Intelligence, *Nature* **95** (August 1915), p. 664.

[136] For a discussion of the origins of the Department of Scientific and Industrial Research, see [Hull 1999]. An announcement of the grant appeared in *Nature* **100** (October 1916), p. 112.

[137] When conscription was introduced in 1916, Brodetsky was given conditional exemption. When his case came up for reconsideration in July 1917, the Dean of the Science Faculty arranged for a representative from the Board of Education to attend the tribunal. The latter successfully argued for the importance of science in war and, rather remarkably, got total exemptions for the lecturers in mathematics, physics and chemistry [Brodetsky 1960, p. 85].

[138] Glauert, who had visited Prandtl in Göttingen soon after the war, was primarily responsible for initiating the acceptance of the Lanchester–Prandtl circulatory theory of lift in Britain.

of March 1918 from an official in the British Embassy to the Cipher Bureau in Washington, D.C., makes clear:

> I should like to emphasize in this connection the importance of selecting the right kind of brain to do this research [code-breaking]. For research of this kind, requires an active well-trained, and scholarly mind; not mathematical but classical [Barker 1979, p. 19].

However, this is not as surprising as it might at first appear. In the First World War code-making and code-breaking relied heavily on text (in the form of codebooks). This was in sharp contrast to the situation in the Second World War when both activities were driven by machine. Nevertheless, while it is certainly the case that an education in the classics, linguistics, or paleography could provide a perfectly sound background for cryptology in an essentially text-based discipline, to deny categorically the usefulness of a mathematically trained mind in this context does, at least to modern eyes, appear rather bizarre. Yet despite such strictures, one or two mathematicians were (or tried to get) involved in code-breaking.

The task of devising means of code-breaking and of setting up an organization to do so fell to Sir Alfred Ewing, the director of Naval Education.[139] Ewing's group became known as "Room 40," being named after the number of the room at the Admiralty where the work took place. For his recruits Ewing looked first to the schoolmasters at the Royal Naval Colleges of Osborne and Dartmouth: they could be relied on for discretion and some were German linguists. Only three members of Room 40 had any mathematical background, of which the most distinguished was James Henderson, professor of applied mechanics at Greenwich Naval College. Henderson, who had studied under Kelvin and Helmholtz, specialized in the study of gyroscopes and applied his knowledge throughout the war to development of various gyroscopic and optical instruments, despite finding it "very difficult to maintain sufficient enthusiasm to carry on inventing without apparent benefit and without official thanks or encouragement" [Hartcup 1988, p. 121]. However, it was not Henderson's knowledge of mathematics that was put to use in Room 40 but his knowledge of German.

But as well as not seeking out mathematicians, Ewing also ignored them. In June 1915, the group theorist William Burnside, professor of mathematics at Greenwich, wrote to Larmor:

> I am writing to ask for your advice. In a fortnight's time I shall be free for 3 months.... This year I feel that if it were possible to put in the 3 months in some useful way I would gladly give up my holiday altogether. Soon after the war started I wrote to Ewing offering to make myself useful in any way at the Admiralty if that were possible and at the beginning of October I renewed the offer as regards my spare time from Greenwich, but he took no notice of it. One recognizes that in such times as the present mere mathematics are pretty useless, and I am afraid that I am not much good with my hands; but there ought to be some way of making such abilities & such leisure as one has available. I wonder whether you could make any suggestion. There must surely be a

[139]Ewing, an engineering graduate of Edinburgh University, had spent thirteen years as professor of mechanism and applied mechanics in Cambridge before being appointed in 1903 as director of Naval Education. For details of Ewing's life and his work in Room 40, see [Ewing 1939].

> lot of calculating work to be done in many directions, and though as you know my line in mathematics is not very practical, I am not afraid of calculations.[140]

Burnside's letter apparently galvanized Larmor into action on his behalf because a few days later Burnside wrote to him again to say that Ewing had forwarded his letter to the director of naval construction (the naval architect and engineer Sir Eustace Tennyson-d'Eyncourt) and that as a result he was going to work on the problem of screw propulsion,[141] which indeed he did [Burnside 1918]. It was also at this time that Burnside first started thinking about the theory of probability and it is possible that Burnside's work in this area originated from the mathematics of war problems, although Burnside himself never made this explicit.[142]

Burnside's difficulty in finding war work was far from unique. J. J. Thomson, who was on the governing body of the Board of Invention and Research (BIR), the body set up to aid the Admiralty in organizing and promoting scientific effort for the navy, reported:

> My experience at the B.I.R. brought home to me how intense and widespread was the eagerness of men of science to do something to help to win the war. Many problems came before us on which it was important to get expert opinion from physicists, chemists, engineers and mathematicians. These frequently involved special investigations, experimental or mathematical, and those who undertook them had to put their own work aside. So far from grudging this, they welcomed the opportunity of doing the war work. We had many appeals for work of this kind from those who had not had such work assigned to them [Thomson 1936, p. 222].

Conclusion

Conspicuous in the above discussion is the lack of any organized mobilization of mathematicians. There were no governement initiatives that recognized that mathematicians per se had something to contribute to the war effort. Of the mathematicians that were employed in war work, the majority were engaged through their own volition, and many of them only found positions by chance; others had to fight to keep the recruiting sergeants at bay. But it was not only the governement war machine that overlooked mathematicians' potential: nothing was organized from within the profession itself. But this is neither surprising nor a cause for condemnation. In 1914 mathematics as a profession in Britain had barely come of age. The London Mathematical Society (LMS), which had been founded in 1865, had been

[140]Burnside to Larmor, June 13, 1915. St John's College Library, Papers of Sir Joseph Larmor. Burnside was not a complete stranger to applied mathematics. From 1875, the year he graduated as second wrangler, until 1885, when he took up the position in Greenwich, he taught applied topics at Cambridge. He also published a number of papers on applied topics. For a discussion on Burnside's work in applied mathematics, see [Barrow-Green 2004].

[141]Burnside to Larmor, June 26, 1915. St John's College Library, Papers of Sir Joseph Larmor.

[142]In the introduction to Burnside's text on probability, Forsyth wrote that Burnside's first paper on the subject, which was written in 1918 [Burnside 1919], "deals with what is manifestly a military question, reduced (for purposes of calculation) to a purely mathematical form" [Burnside 1928, p. v]. For a discussion of Burnside's work on probability and statistics, see [Aldrich 2009].

established as a vehicle for promoting mathematical research and in its 50 years of existence had not changed its basic remit. It had no aspiration to get involved in or influence issues of national policy. The LMS Council Minutes for the war period make almost no reference to the conflict, apart from noting the agreement to stop copies of the society's *Proceedings* going to Germany and Austria.[143] One body that might have acted to mobilize mathematicians was the Royal Society (RS), which counted many mathematicians among its membership. But when, in 1914, the RS set up its system of war committees, mathematics was conspicuously absent. It would seem that this was a conscious decision since a war committee for "Mathematics and Physics" (to deal with problems connected with aerostatics, artillery, transport, submarines, and mines) was initially proposed, but when the committee was actually formed, the word "mathematics" had been dropped from the title [RS Archives, MS/625].

As far as Cambridge itself was concerned, the university decided it was better to take matters of mobilization into its own hands rather than wait for the government to intervene, as the University Correspondent of *The Times* reported in June 1915:

> Many of the senior mathematical and classical tutors have given up their vacations to teach in schools and so to free the younger schoolmasters for duty at the front, or to give a greatly-needed holiday to the over-worked staff of the Government offices in Whitehall. Others have taken positions as naval instructors. All the members of our very large Science School have placed their resources at the service of the Government.
>
> On the whole, Cambridge has found it more advantageous to mobilize itself than to await Government mobilization. A complete list of members of the University in residence has recently been compiled, stating the service which each resident-member feels he can most usefully offer, and the numerous laboratories in Cambridge have been placed at the disposal of the Government, with a clear and succinct statement of the work each laboratory can best do to promote the progress of our armies in the field.[144]

While Cambridge may have put their mathematicians at the disposal of the government, there is little evidence to show that the government recognized either the value of mathematicians or of the use of mathematics as tools in the development of effective systems of defense and attack.[145] As the case of William Burnside attests, there were mathematicians who wanted to use their abilities to support their country during the war but the country did little to encourage them. Others such as Glauert and Taylor did get involved in the war effort, but more by good luck than good management. But mathematics was not alone in this respect. At the beginning of the war the organization of science suffered from a general lack of

---

[143] Council Minutes of the London Mathematical Society, November 12, 1914 [LMS Archives].

[144] *The Times* (June 15, 1915).

[145] Further evidence of this is given by the resolution adopted in December 1915 by the Conference of Headmasters: "That in the opinion of this conference very grave loss to the country is caused by the employment of young students of exceptional mathematical and scientific ability as subalterns in Line battalions." It was, a member of the Conference argued, "eminently a war of science, and the mathematical and scientific ability of the country should be husbanded jealously." *The Times* (December 24, 1915).

coherence and poor communication.[146] In Britain scientists did not have the same reputation as their Continental counterparts and the fighting forces often resented the offering of scientific advice. Very few civil servants, let alone the military, had any scientific knowledge. The startling lack of comprehension by some senior officers of the nature of the work being done resulted in the government's failure to support, at least initially, some of the most urgent and essential research.[147] As far as mathematics was concerned, there was ignorance at the most elementary level, as Greenhill so forcefully announced.

This ignorance also resulted in the tragic loss of many talented scientists, the most notable being that of the gifted young physicist Henry Moseley who had done remarkable work on X-ray spectra confirming Niels Bohr's quantum theory of atomic radiation and whose death at Gallipoli in 1915 prompted Ernest Rutherford to begin lobbying behind the scenes for selective exemption from conscription.[148] Hardy supported him, and in the event fewer British mathematicians and scientists were killed in action than their French counterparts. Even so, after the war, Ernest Barnes said, "Of my pupils at Cambridge at least one-half, and practically all the best, have been killed or maimed for life" [Kanigel 1991, p. 261]. Edward Kingsley Wakeford, a leading wrangler of 1914 who early in his career had been recognized as a geometer of uncommon knowledge and power, was killed on the Western front in July 1916 when only 22, an unfinished manuscript in his kitbag.[149] He wrote a fellowship dissertation on the theory of canonical forms, and when he died, although not a member, the London Mathematical Society exceptionally published an obituary. It closed with the haunting words: "He only needed a chance, and he never got it" [Grace 1917, p. lvii]. Another loss was Samuel McLaren, third wrangler of 1899, who died of wounds in Abbeville in 1916. Professor of mathematics at University College, Reading, McLaren was "a learned and philosophical investigator in the difficult domain of statistical molecular dynamics and the relations of the aether to material systems" [Larmor 1917, p. 6], who in 1913 had been awarded an Adams Prize—Cambridge's most prestigious prize in mathematics—for his work

[146]The situation was aptly summarized in 1919 by Joseph Petavel, the newly appointed director of the NPL, in his presidential address to the Engineering Section of the British Association for the Advancement of Science: "Before the war the British nation was anti-militarist, non-scientific, and strongly individualistic. To achieve victory the nation accepted universal conscription, and submitted to the mixture of Socialism and tyranny which necessity dictated. Under extreme pressure, scientific knowledge, technical skill, industrial ability, military and naval experience welded into a homogenous organisation" [Petavel 1920, p. 260].

[147]On February 2, 1916, *The Times* published a memorandum drawn up by a committee of leading scientists, including Lord Rayleigh and Karl Pearson, concerned about the neglect of science. It opened: "It is admitted on all sides that we have suffered checks since the war began, due directly as well as indirectly to a lack of knowledge on the part of our legislators and administrative officials of what is called 'science,' or 'physical science.' By these terms we mean the ascertained facts and principles of mechanics, chemistry, physics, biology, geography, and geology. Not only are our highest Ministers of State ignorant of science, but the same defect runs throughout almost all the public departments of the Civil Service. It is nearly universal in the House of Commons." For further discussion on this topic, see [Turner 1980, pp. 603–608].

[148]As Rutherford wrote in 1916, "It is a matter of great regret that the services of Moseley could not have been utilized for the war in some scientific capacity rather than as a combatant in the firing line" [Rutherford 1916, p. xxviii].

[149]After the war, the London Mathematical Society had the manuscript, which was on chords of twisted cubics, refereed and published [Wakeford 1923]. Council Minutes of the London Mathematical Society, February 13, 1919 [LMS Archives].

on the theory of radiation. All told, thirty-five fellows of the Royal Society and fifty-five members of the Royal Institution were killed [Pattison 1983, p. 562].

Undoubtedly the most coherent group of mathematicians involved in the war effort consisted of those connected with the AAES and who played a fundamental role in the development of anti-aircraft gunnery. But the evidence shows that their success had rather more to do with the efforts of their leader, Captain A. V. Hill, than the efforts of the official bodies. Hill, although a physiologist by inclination and profession, was not only a competent mathematician who understood what mathematics was required and recognized the urgency of the task, but also, and importantly, he provided the social glue which kept the group together. He was also an expert when it came to dealing with officialdom and in obtaining the resources necessary in order to get the job done.

The scientific exigencies of the war did contribute towards a change in attitude towards the public funding of science and "In late 1916 the British Government finally bowed to pressure from scientists and sympathetic elements of the public to organise and fund science centrally and established the Department of Scientific and Industrial Research (DSIR)" [Hull 1999, p. 461].[150] However, as Sir Richard Glazebrook, chairman of the Advisory Committee for Aeronautics, clearly articulated in his postwar report of 1919, it was important that the research momentum generated by the war should not be lost:

> The need of progress in aeronautical science and technology, in education and research is now universally recognized as much more pressing, the demand for the greatly increased expenditure required is now much more readily granted, in this as in other fields of investigation, than it was in 1913. It has taken a European war to give us a national Department of Scientific and Industrial Research. More complete organisation and co-ordination of research in aeronautics for civil and military purposes will be needed for future work. The Committee feel that it is appropriate to conclude this review of their activities during the war by urging and insisting once again on the essential and immediate importance of greatly increased activity in the investigation of the new problem of flight, which is undoubtedly destined to exercise an ever-increasing influence on the future history of the world [ACA 1922, pp. 20–21].

Sir Richard's words did not fall on deaf ears. As a result of increased funding and changed funding models, scientific institutions such as the Royal Aircraft Establishment and the NPL were indeed able to retain a number of their wartime staff, several of whom were mathematicians [Anonymous 1920–1924, p. 3]. Thus theoreticians and practitioners who were brought together because of the war worked together afterwards. Sometimes, as in the case of Berry and Bairstow who published together after the war, the end of hostilities meant only the end of working in the same environment (Berry returned to Cambridge); it did not mean the end of their collaboration.

---

[150] As Hull has described, the story of the formation of the DSIR is a complicated one, with origins stretching back to before the war [Hull 1999], and it is not one for retelling here. Further information can be found in [Turner 1980, pp. 603–608] and [Gummett 1980, pp. 22–25].

TABLE 4. Mathematical Tripos results, 1912–1922. Number of wranglers are given in parentheses in Part II columns.

	MEN			WOMEN		
Year	Part I	Part II	Total	Part I	Part II	Total
1912	120	54 (27)	174	15	16 (2)	31
1913	124	77 (30)	201	16	8 (1)	24
1914	92	58 (25)	150	20	11 (2)	31
1915	58	28 (14)	86	25	8 (2)	33
1916	23	20 (11)	43	16	12 (2)	28
1917	14	11 (5)	25	17	13 (2)	30
1918	12	4 (3)	16	17	7(0)	24
1919	119	9 (7)	128	17	13 (3)	30
1920	184	18 (6)	202	14	7 (0)	21
1921	108	56 (32)	164	19	8 (2)	27
1922	110	58 (35)	168	24	8 (1)	32

Meanwhile in Cambridge:

> We have the authority of the Cambridge University Handbook to assure us that by the end of the year 1920 all the normal activities of the University had been resumed.... The number of undergraduates in residence had fallen from 1178 in 1914 to 235 in 1916. By January 1919 the number was 2635; by May 3844, including 190 United States servicemen on a short course and 400 naval officers whose education [at Dartmouth] had been interrupted [Howarth 1978, p. 25].

The number of male undergraduates sitting for the Mathematical Tripos reflected the overall pattern and by 1920 it had surpassed its prewar level (see Table 4). On the other hand, the number of women sitting for the Mathematical Tripos remained fairly constant throughout the whole period.

Mathematicians like Richmond, Littlewood, and Taylor, who had left to join the war effort returned to their former positions, while those who had been students when the conflict began, like Fowler and Milne, were appointed to positions commensurate with their age and wartime experience. Mathematics lecture courses resumed with little apparent difference in content, apart from the addition of one or two more physics-based courses (Table 5).

In 1920, however, there was a striking innovation: the inclusion of a woman's name on the list of lecturers. The Girtonian Dorothy Wrinch, who had been ranked as a wrangler in 1918 and had subsequently lectured at University College London for three years (having been awarded an M.Sc. and D.Sc. from the University of London), gave a course on asymptotic expansions.[151] While it is hard to make a direct connection between the war and Wrinch's appointment, it is nonetheless the case that the war did provide new and different opportunities for women to become involved in mathematical and computational activities and thereby provide

[151] For further information on Dorothy Wrinch, see the recent biography [Senechal 2012].

TABLE 5. University lectures, 1919 and 1921.

Lecturer	1919	1921
A. Berry	Ordinary differential equations; elliptic functions (theory of transformation)	Ordinary differential equations; elliptic functions (theory of transformation)
G. Birtwhistle	Hydrodynamics	Hydrodynamics
T. J. Bromwich	Electric waves or the differential equations of mathematical physics	Special problems in electrostatics & magnetism; electromagnetic (& optical) waves (mathematics of wireless telegraphy); differential equations of mathematical physics
E. Cunningham	Electron theory; general dynamics; elements of vector analysis;	Electron theory (2 terms); elements of vector analysis
C. G. Darwin	Thermodynamics & dynamical theory of gases (twice)	General dynamics; quantum theory
R. H. Fowler	Gamma & Bessel's functions	Dynamical theory of gases (elementary) (2 terms)
J. H. Grace	Theory of invariants; classical theory of numbers; line geometry	Theory of invariants; theory of numbers (classical theory); line geometry
W. J. Harris	Hydrodynamics (viscous liquids); sound & vibrations	Hydrodynamics (viscous liquids); sound & vibrations; computational & statistical methods
R. Herman	Lunar & planetary theory; differential geometry (2 terms); geometrical optics	Lunar & planetary theory differential geometry (2 terms); geometrical optics
H. Jeffreys		Physics of the earth's interior
H. Lamb		Elastic waves & vibrations with applications to seismology
J. G. Leathem	Conformal representation applied to physical problems	
J. E. Littlewood	Functions of a real variable; theory of functions	Functions of a real variable (foundations); advanced analysis (double limit problems); theory of functions
J. Mercer	Functions of a complex variable; algebraic functions & their integrals; theory of functions, special problems	
S. Pollard	Functions of a real variable	Functions of a real variable (introduction); trigonometric series & integrals; mathematical logic
H. W. Richmond		Algebraic geometry: surfaces and twisted curves
G. I. Taylor		Elasticity
D. Wrinch		Asymptotic expansions

evidence of women's capacity for such work.[152] Although Pearson had employed women in his laboratories before 1914, the war provided many additional openings. Hilda Hudson, who worked for the Admiralty Aircraft Production Department, and Lorna Swain, who worked at the Royal Aircraft Factory, each successfully applied their mathematical abilities to wartime problems. Nevertheless, however willingly Cambridge may have sanctioned the presence of a female lecturer in 1920, the university was still not ready to admit women to full membership. Another World War would have to be fought and won before, in 1948, women achieved equality in that respect.[153]

As far as new positions in Cambridge were concerned, while there were none in mathematics directly, there was the foundation in 1919 of the Francis Mond chair of aeronautical engineering, which was held from its inception until 1952 by Melvill Jones.[154] The next year Jones gained the assistance of William Farren, a graduate of the Mechanical Sciences Tripos of 1914 who had spent the war at the Royal Aircraft Factory in charge of the aerodynamics department, who began to lecture in engineering and aeronautics.

In 1919, the Zaharoff chair of aviation was founded at Imperial College.[155] This was held first by Glazebrook (after his retirement as director of the NPL), and from 1923 by Bairstow. The study of aeronautics at Imperial College had a further boost when, in 1920, the college was awarded a substantial grant from public funds towards the cost of its Department of Aeronautics.[156]

On an administrative level, the war had the palpable effect of speeding up the introduction of the PhD to Cambridge. There had been two attempts, one in the early 1870s and one in 1895, to introduce such a degree but both had failed. However, with the exigencies of wartime and the lack of foreign students, the degree's supporters felt the time was right to try again. As Kim has described, in 1916 J. J. Thomson, a strong proponent of the degree, made a plea in the senate for the introduction of new degrees for research students, asking:

> whether it might not be possible to give to students who after six terms' residence have completed a satisfactory piece of research the titular degree of Doctor. I think it is probable that after the war many students from neutral countries will be unwilling to go to Germany for their post-graduate studies, and would much prefer

[152] To take two examples: on August 31, 1917, it was reported in *The Times* that there was "a big demand for women with degrees in economics, mathematics and science, for work in various Government Departments;" while on March 8, 1918, it was reported in the same newspaper that women employed in the drawing offices of the Royal Naval Air Service "must reach the London matriculation standard in mathematics and geometry; a mere talent for drawing is no recommendation—mathematicians, not artists, are required."

[153] Women at Oxford fared rather better: they have been entitled to take degrees since 1920.

[154] Jones, a graduate of the Cambridge Mechanical Sciences Tripos in 1909, worked on aeronautics at the NPL before the war and, after his call-up, was sent by the army initially to the Royal Aircraft Factory and then to the Air Armament Experimental Station at Orford Ness, the latter being run by Bertram Hopkinson, the professor of mechanism and applied mechanics at Cambridge.

[155] The chair was endowed by Sir Basil Zaharoff, a notorious arms dealer who also funded chairs in aviation at the Sorbonne and Petrograd (now St. Petersburg) State University.

[156] A summary of the report issued in February 1920 by the Committee on Education and Research in Aeronautics (appointed October 1918) in which the establishment of a Department of Aeronautics at Imperial College is recommended is given in *The Times* (February 19, 1920). Further details of the report are given in *Nature* **105** (March 1920), pp. 14–17.

> to come to England. I have reason to believe that such students attach considerable importance to the attainment of the degree of Doctor (a titular degree is all they require) and unless some such degrees were obtainable they would not be likely to come in any considerable number.[157]

It was a persuasive argument but the fact that the PhD was perceived as a German institution militated against its immediate introduction and it was not until 1920 that the degree was finally adopted, with the first doctorates in mathematics being awarded in 1924.[158]

But if at an institutional level the war did not create the decisive break that might have been expected, at an individual level the situation was different. While there were many, mostly established mathematicians, such as Richmond, Littlewood, and Pearson, who returned to their original prewar work, their careers delayed but essentially unchanged, and some, like Cunningham, who never recovered their former mathematical powers, there were several (mostly younger) mathematicians who had their careers set on new paths, and in this way the conflict did have an influence on the future of the nation's mathematics.

At the outbreak of war both Fowler and Milne had been set for careers in pure mathematics but as a result of their experience with Hill, by the end of the war their interests had completely changed. As Hill later wrote: "Kaiser William and I, jointly, did good service to science in diverting both R. H. Fowler and E. A. Milne from pure mathematics to other fields" [Hill 1974, p. 245]. For Milne, the years spent with the AAES under the leadership of Hill and Fowler had a profound effect:

> At directing research, both [Hill and Fowler] were superb; to behold the way they set about a new problem, often in a new terrain with new material, to see the way they made inferences and came to conclusions and sound judgments and to take part in it all, was far better training than most universities can offer to aspirants in research. The present writer [Milne] desires to put on record his realization that this three years' period under Hill and Fowler, at the most formative period of his life (age twenty to twenty-three) was a most vivid, enjoyable and valuable experience; to both of them he can never be sufficiently grateful for that schooling in research [Milne 1945, p. 66].

Both Fowler and Milne had impressive careers in areas of applied mathematics: Fowler in statistical mechanics and Milne in astrophysics. Fowler, who was awarded an OBE for his wartime work, became Plummer Professor of Mathematical Physics at Cambridge in 1932 and director of the NPL in 1938. Milne became Beyer Professor of Applied Mathematics at Manchester in 1924 and Rouse Ball Professor of Applied Mathematics at Oxford four years later. For both men, the years working with the AAES had equipped them for the future in ways far beyond anything that a purely academic life at Cambridge would have had to offer.

[157] *Cambridge University Reporter* (November 14, 1916), p. 219, quoted in [Kim 2002, p. 179].

[158] The first Cambridge students to be awarded PhDs in mathematics were Thomas MacFarland Cherry and Leslie Comrie, both of whom had come from the Antipodes: Cherry from the University of Melbourne and Comrie from the University College of Auckland.

Then, there were those whose careers were specifically defined either by their work during the war or by their attitude to the conflict. Glauert, for example, did not return to Cambridge but remained at the Royal Aircraft Establishment where he wrote numerous influential reports, eventually becoming head of the aerodynamics department in 1934, although he was tragically killed in a freak accident on Fleet Common shortly afterwards. Barry Gates also remained at the Royal Aircraft Establishment where he had a distinguished career as an aerodynamicist, renowned for his work on stability and control and eventually rising to become a Chief Scientific Officer. Likewise, Robert Frazer remained at the NPL for the rest of his career where he did important work on wing-flutter problems and suspension bridge design. Although propelled by circumstances into a career in engineering, Frazer remained a mathematician at heart, always seeking mathematical solutions first and turning to engineering approaches second. The NPL also became the permanent academic home for Christopher Lock and Joseph Nayler, both of whom continued to work on aeronautics. Douglas Hartree's experience in calculating trajectories for the AAES marked the beginning of his work in computation, a subject in which he later came to be regarded as "a world leader" [Darwin 1958, p. 105]. Bryan's students Robert Jones and Dan Williams went on to have distinguished careers as aerodynamicists at the NPL. On the pacifist side there was Neville, who was sidelined from Cambridge; Barnes, who left mathematics altogether; and Richardson, whose career and mathematical interests were both changed by the war.

The war had a very particular effect on the career of Hardy. Trinity College's wartime treatment of Russell had left him disillusioned with Cambridge and was one of the reasons he moved in 1920 to Oxford to take up the Savilian Chair in geometry. He stayed at Oxford for 11 years until, in 1931, he returned to Cambridge as Sadleirian Professor of Pure Mathematics. Hardy had an enormous influence on every aspect of Oxford's mathematical life, not only with his immense mathematical ability but also by actively promoting mathematics in many quarters, arguing for an increase in the number of college fellows and making the case for a dedicated mathematics institute. With Hardy's move to Oxford, no longer could it be said that "Oxford does not offer much of a career to a mathematician" [Jolliffe 1912, p. 263].

Abbreviations

- ACA: Advisory Committee for Aeronautics.
- AA: Anti-Aircraft.
- AAD: Admiralty Air Department.
- AAES: Anti-Aircraft Experimental Section.
- BIR: Board of Invention and Research.
- DSIR: Department of Scientific and Industrial Research.
- FOR: Fellowship of Reconciliation.
- FRS: Fellow of the Royal Society.
- GHQ: General Headquarters.
- ICM: International Congress of Mathematicians.
- ILP: Independent Labour Party.
- LMS: London Mathematical Society.
- MID: Munitions Inventions Department.

- NCC: Non-Combatant Corps.
- NCF: No-Conscription Fellowship.
- NPL: National Physical Laboratory.
- OBE: Order of the British Empire.
- RS: Royal Society.
- TUC: Trades Union Congress.
- UDC: Union for Democratic Control.
- UCL: University College London.

Bibliography

Archival Sources

[CU Archives] Cambridge University Archives, Cambridge.

[Hardy Corr.] Hardy Correspondence, Institut Mittag-Leffler, Djursholm, Sweden.

[LMS Archives] London Mathematical Society Archives, London.

[Milne Papers] Papers of E. A. Milne, Oxford University.

[Nat. Arch.] National Archives, Kew, Richmond, Surrey.

[Pearson Papers] Papers and correspondence of Karl Pearson, University College, London.

[RS Archives] Royal Society Archives, London.

[White Papers] Papers of Francis Puryer White, St John's College Library, Cambridge.

Published Works

[ACA 1922] *Technical Report of the Advisory Committee for Aeronautics, for the year 1918–19*, vol. 1, H.M.S.O., London, 1922.

[Aldrich 2009] John Aldrich, *Burnside's Engagement with the "Modern Theory of Statistics"*, Archive for History of Exact Sciences **63** (2009), pp. 1–79.

[Allen 1913] [Richard Clifford Allen], *Cambridge and Conscription*, The University Socialist, no. 2 (1913), pp. 104–107. The author of the article was given as "a B.A. of that University."

[Anonymous 1920] Anonymous, *Some Applications of Physics to War Problems*, Nature **105** (1920), pp. 237–238.

[Anonymous 1920–1924] Anonymous, *The Royal Aircraft Establishment* The Scientific Worker **1** (1920–1924), pp. 3–5.

[Ashford 1985] Oliver Ashford, *Prophet or Professor? The Life and Work of Lewis Fry Richardson*, Adam Hilger, Bristol, 1985.

[Ashford 2004] Oliver Ashford, Lewis Fry Richardson, s.v. in *Oxford Dictionary of National Biography*, vol. 46, University Press, Oxford, 2004, pp. 832–834.

[Aubin forthcoming] David Aubin, *"Audacity or Precision": The Paradoxes of Henri Villat's Fluid Mechanics in Interwar France*, paper presented at a "Workshop on the History of Fluid Mechanics", Rauischholzhausen, October 15–18, 2006, forthcoming. `oai:hal.upmc.fr:hal-00639877`.

[Bairstow 1919] Leonard Bairstow, *Progress of Aviation in the War Period: Some Items of Scientific and Technical Interest*, The Aeronautical Journal **23** (1919), pp. 1–24.

[Bairstow 1920a] Leonard Bairstow, *The Scientific Progress of Aviation during the War*, Report of the British Association for the Advancement of Science: Annual Meeting 1919, BAAS, London, 1920, pp. 266–267.

[Bairstow 1920b] Leonard Bairstow, *Applied Aerodynamics*, Longmans, London, 1920.

[Bairstow 1933] Leonard Bairstow, *George Hartley Bryan, 1864–1928*, Obituary Notices of the Royal Society **1** (1933), pp. 139–142.

[Bairstow et al. 1920] Leonard Bairstow, Ralph H. Fowler, & Douglas Hartree, *The Pressure Distribution on the Head of a Shell Moving at High Velocities*, Proceedings of the Royal Society of London **97** (1920), pp. 202–218.

[Baker 1944] Henry F. Baker, *Geoffrey Thomas Bennett*, Journal of the London Mathematical Society **19** (1944), pp. 107–128.

[Barker 1979] Wayne G. Barker, ed., *The History of Codes and Ciphers in the United States during World War I*, Aegean Park Press, Laguna Hills, 1979.

[Barrow-Green 1999] June Barrow-Green, *"A Corrective to the Spirit of too Exclusively Pure Mathematics": Robert Smith and his Prizes at Cambridge University*, Annals of Science **56** (1999), pp. 271–316.

[Barrow-Green 2004] June Barrow-Green, *Burnside's Applied Mathematics* in *The Collected Papers of William Burnside*, ed. Peter M. Neumann, Anthony J. S. Mann, & Julia C. Tompson, Oxford University Press, Oxford, 2004, vol. 1, pp. 63–70.

[Barrow-Green forthcoming] June Barrow-Green, *"Anti-aircraft guns all day long": K.F. Pearson and computing for the Ministry of Munitions*, forthcoming.

[Batchelor 1996] George Batchelor, *The Life and Legacy of G. I. Taylor*, Cambridge University Press, Cambridge, 1996.

[Berry 1924] Arthur Berry, Theorem of Three Moments with End Loads, in *Handbook of Strength Calculations*, HMSO, London, 1924, pp. 67–77. First edition (classified) 1918.

[B. G. C. 1911] B. G. C., Review of *Report on the Theory of a Stream Line past a Plane Barrier, and of the Discontinuity arising at the Edge with an Application of the Theory to an Aeroplane*, by Sir George Greenhill, The Aeronautical Journal **15** (1911), p. 94.

[Bloor 2011] David Bloor, *The Enigma of the Aerofoil. Rival Theories in Aerodynamics, 1909–1930*, University of Chicago Press, Chicago, 2011.

[Bowler 1998] Peter J. Bowler, *Evolution and the Eucharist: Bishop E.W. Barnes on Science and Religion in the 1920s and 1930s*, British Journal for the History of Science **31** (1998), pp. 453–467.

[Brodetsky 1921] Selig Brodetsky, *The Mechanical Principles of the Aeroplane*, J. & A. Churchill, London, 1921.

[Brodetsky 1928] Selig Brodetsky, *Prof. G. H. Bryan, F.R.S.*, Nature **122** (1928), pp. 849–850.

[Brodetsky 1960] Selig Brodetsky, *Memoirs: From Ghetto to Israel*, Weidenfeld & Nicholson, London, 1960.

[Brooke 1993] Christopher N. L. Brooke, *A History of the University of Cambridge*, vol. 4, Cambridge University Press, Cambridge, 1993.

[Bryan 1909] George H. Bryan, *The Government and Aeronautical Research*, Nature **80** (1909), pp. 313–314.

[Bryan 1911] George H. Bryan, *Stability in Aviation: An Introduction to Dynamical Stability as applied to the Motions of Aeroplanes*, Macmillan, London, 1911.

[Bryan 1912] George H. Bryan, Review of G. Greenhill's *The Dynamics of Mechanical Flight*, Aeronautical Journal **16** (1912), pp. 264–267.

[Bryan 1915] George H. Bryan, *The Rigid Dynamics of Circling Flight: Wilbur Wright Memorial Lecture*, Aeronautical Journal **19** (1915), pp. 49–74.

[Bryan 1916] George H. Bryan, *Researches in Aeronautical Mathematics*, Nature **96** (1916), pp. 509–510.

[Bryan 1922] George H. Bryan, *Principles and Problems of Aeronautics*, Nature **109** (1922), pp. 296–298.

[Bryan & Brodetsky 1916] George H. Bryan & Selig Brodetsky, *The Longitudinal Initial Motion and Forced Oscillations of a Disturbed Aeroplane*, Aeronautical Journal **20** (1916), pp. 139–156.

[Bryan & Jones 1915] George H. Bryan & Robert Jones, *Discontinuous Fluid Motion past a Bent Plane, with Special References to Aeroplane Problems*, Proceedings of the Royal Society of London, Series A **91** (1915), pp. 354–370.

[Burkhill 1979] John C. Burkhill, *John Edensor Littlewood*, Bulletin of the London Mathematical Society **11** (1979), pp. 59–103.

[Burnside 1918] William Burnside, *On the Efficiency of a Surface of Pressure Discontinuity Regarded as a Propeller*, Proceedings of the London Mathematical Society (2) **17** (1918), pp. 42–53.

[Burnside 1919] William Burnside, *On the Probable Regularity of a Random Distribution of Points*, Messenger of Mathematics **48** (1919), pp. 47–48.

[Burnside 1928] William Burnside, *Theory of Probability*, edited with a memoir by Andrew R. Forsyth, Cambridge University Press, Cambridge, 1928.

[Cartwright & Ursell 1976] David E. Cartwright & Fritz Ursell, *Joseph Proudman, 30 December, 1888 – 26 June, 1975*, Biographical Memoirs of Fellows of the Royal Society **22** (1976), pp. 319–333.

[Cattermole 1987] Michael J. G. Cattermole, *Horace Darwin's Shop: A History of the Cambridge Scientific Instrument Company, 1878–1968*, Adam Hilger, Bristol, 1987.

[Ceadel 1980] Martin Ceadel, *Pacifism in Britain 1914-1945: the Defining of a Faith*, Clarendon Press, Oxford, 1980.

[Chitty 1966] Lorna Chitty, [untitled], Journal of the Royal Aeronautical Society **70** (1966), pp. 67–68.

[Christopherson 1972] Derman G. Christopherson, *Richard Vynne Southwell. 1888-1970*, Biographical Memoirs of Fellows of the Royal Society **18** (1972), pp. 549–565.

[Cole 1952] Margaret I. Cole, ed., *Beatrice Webb's Diaries 1912-1924*, Longman's, London, 1952.

[Cook 1990] Alan Cook, *Sir Harold Jeffreys, April 2, 1891 – March 18, 1989*, Biographical Memoirs of Fellows of the Royal Society **36** (1990), pp. 303–333.

[Cowling 1971] Thomas G. Cowling, *Sydney Chapman, 1888–1970*, Biographical Memoirs of Fellows of the Royal Society **17** (1971), pp. 53–89.

[Darwin 1958] Charles G. Darwin, *Douglas Rayner Hartree, 1897–1958*, Biographical Memoirs of Fellows of the Royal Society **4** (1958), pp. 102–116.

[Darwin 1912] George H. Darwin, Opening address, in *Proceedings of the Fifth International Congress of Mathematicians*, ed. Ernest W. Hobson & Augustus E. H. Love, Cambridge University Press, Cambridge, 1912, pp. 33–36.

[Dauben 1980] Joseph Dauben, *Mathematicians and World War I: The International Diplomacy of G. H. Hardy and Gösta Mittag–Leffler as Reflected in their Personal Correspondence*, Historia Mathematica **7** (1980), pp. 261–288.

[Davis 2009] Sophia Davis, *Raising the Aerocompass in Early Twentieth-Century Britain*, British Journal for the History of Science **42** (2009), pp. 73–93.

[Day 1966] Albert Day, *Design Requirements for Military Aircraft*, Journal of the Royal Aeronautical Society **70** (1966), pp. 250–252.

[Douglas 1956] Allie Vibert Douglas, *Arthur Stanley Eddington*, Thomas Nelson & Sons Ltd., London, 1956.

[Eddington 1916] Arthur S. Eddington, *The Future of International Science*, The Observatory **39** (1916), pp. 270–272.

[Edgerton 1991] David Edgerton, *England and the Aeroplane. An Essay on a Militant and Technological Nation*, Macmillan, Basingstoke, 1991.

[Ewing 1939] Alfred Ewing, *The Man of Room 40: The Life of Sir Alfred Ewing*, Hutchinson & Co., London, 1939.

[Farren & Tizard 1935] William S. Farren & Henry T. Tizard, *Hermann Glauert, 1892–1934*, Obituary Notices of Fellows of the Royal Society **1** (1935), pp. 607–610.

[Farren & Thomson 1958] William S. Farren & George P. Thomson, *Frederick Alexander Lindemann, Viscount Cherwell, 1886–1957*, Biographical Memoirs of Fellows of the Royal Society **4** (1958), pp. 45–71.

[Fowler et al. 1920] Ralph H. Fowler, Edward G. Gallop, Christopher N. H. Lock, & Herbert W. Richmond, *The Aerodynamics of a Spinning Shell*, Philosophical Transactions of the Royal Society **A221** (1920), pp. 295–387.

[Fowler & Lock 1921] Ralph H. Fowler & Christopher N. H. Lock, *The Origin of the Disturbances in the Initial Motion of a Shell*, Proceedings of the Cambridge Philosophical Society **20** (1921), pp. 311–319.

[Fowler et al. 1922] Ralph H. Fowler, Edward G. Gallop, & Christopher N. H. Lock, *The Aerodynamics of a Spinning Shell. Part II*, Philosophical Transactions of the Royal Society **A222** (1922), pp. 227–247.

[Garner 1966] Harry Mason Garner, *Early Research at Farnborough and Felixstowe*, Journal of the Royal Aeronautical Society **70** (1966), pp. 95–97.

[Gladden 1930] Edgar Norman Gladden, *At Messines Ridge in 1917*, in *Everyman at War*, ed. Charles B. Purdom, J. M. Dent, London, 1930, pp. 114–121.

[Glauert 1919] Hermann Glauert, *The Investigation of the Spin of an Aeroplane*, Reports and Memoranda of the Advisory Committee for Aeronautics **618** (1919), 32 pp.

[Glauert 1921] Hermann Glauert, *The Mechanical Principles of the Aeroplane: S. Brodetsky*, The Aeronautical Journal **25** (1921), p. 378.

[Glazebrook & Love 1935] Richard T. Glazebrook & Augustus E. H. Love, *Sir Horace Lamb, 1849-1934*, Obituary Notices of Fellows of the Royal Society **1** (1935), pp. 374–392.

[Grace 1917] John H. Grace, *Edward Kingsley Wakeford*, Proceedings of the London Mathematical Society **16** (1917), pp. liv–lvii.

[Greenhill 1910] George Greenhill, *Report on the Theory of a Stream Line Past a Plane Barrier and of the Discontinuity Arising at the Edge with an Application to the Theory to an Aeroplane*, Advisory Committee for Aeronautics Report **19**, H.M.S.O., London, 96 pp.

[Greenhill 1912] George Greenhill, *The Dynamics of Mechanical Flight: Lectures delivered at Imperial College of Science and Technology, March, 1910 and 1911*, Constable & Company Ltd., London, 1912.

[Greenhill 1915] George Greenhill, *Mathematics in Artillery Science*, The Mathematical Gazette **8** (1915), pp. 25–29.

[Grier 2005] David Grier, *When Computers were Human*, Princeton University Press, Princeton, 2005.

[Gummett 1980] Philip Gummett, *Scientists in Whitehall*, Manchester University Press, Manchester, 1980.

[Hardy 1934] Godfrey H. Hardy, *Ernest William Hobson*, Obituary Notices of Fellows of the Royal Society **1** (1934), pp. 237–249.

[Hardy 1942] Godfrey H. Hardy, *Bertrand Russell and Trinity*, Cambridge University Press, Cambridge, 1942.

[Hardy 1967] Godfrey H. Hardy, *A Mathematician's Apology*, Cambridge University Press, Cambridge, 1940; 2nd ed., 1967.

[Harrison 1994] Brian Harrison, ed., *The History of the University of Oxford. The Twentieth Century*, Clarendon Press, Oxford, 1994.

[Hartcup 1988] Guy Hartcup, *The War of Invention: Scientific Developments, 1914–18*, Brassey's, London, 1988.

[Hartree 1920] Douglas Hartree, *Ballistic Calculations*, Nature **106** (1920), pp. 152–154.

[Hashimoto 1990] Takehiko Hashimoto, *Theory, Experiment and Design Practice: The Formation of Aeronautical Research, 1909-1930*, Ph.D. thesis, John Hopkins University, 1990.

[Hashimoto 2000] Takehiko Hashimoto, The Wind Tunnel and the Emergence of Aeronautical Research in Britain, in *Atmospheric Flight in the Twentieth Century*, ed. Peter Galison & Alex Roland, Kluwer Academic Publishers, Dordrecht, 2000, pp. 223–240.

[Hazlehurst 1919–1922] Cameron Hazlehurst, *History of the Ministry of Munitions*, 12 vols., British Library Microfiche, London, 1919–1922.

[Hill 1960] Archibald V. Hill, *The Ethical Dilemma of Science*, The Rockefeller Institute Press, New York, 1960.

[Hill 1974] Archibald V. Hill, *Memories and Reflections*, typescript, *Churchill College Archives*, Cambridge, 1974.

[Hogg 1978] Ian V. Hogg, *Anti-Aircraft Attack: A History of Air Defence*, Macdonald & Jane's, London, 1978.

[Howarth 1978] Thomas E. B. Howarth, *Cambridge Between Two World Wars*, Collins, London, 1978.

[Hudson 1920] Hilda Hudson, *Incidence Wires*, Aeronautical Journal **24** (1920), pp. 505–516.

[Hull 1999] Andrew Hull, *War of Words: The Public Science of the British Scientific Community and the Origins of the Department of Scientific and Industrial Research, 1914–16*, British Journal for the History of Science **32** (1999), pp. 461–481.

[Jolliffe 1912] Arthur E. Jolliffe, Examinations for Mathematical Scholarships at Oxford, *Special Reports on Educational Subjects* **27**: "The Teaching of Mathematics in the United Kingdom. Part II," H.M.S.O., London, 1912.

[Kanigel 1991] Robert Kanigel, *The Man Who Knew Infinity: A Life of the Genius Ramanujan*, Scribner, New York, 1991.

[Kennedy 1934] Margaret Delina Kennedy, *Lorna Mary Swain*, Journal of the London Mathematical Society **9** (1934), pp. 155–157.

[Kennedy 1981] Thomas C. Kennedy, *The Hound of Conscience: A History of the No-Conscription Fellowship, 1914–1919*, University of Arkansas Press, Fayetteville, 1981.

[Kim 2002] Dong-Won Kim, *Leadership and Creativity: A History of the Cavendish Laboratory, 1871-1919*, Kluwer, Dordrecht, 2002.

[Langford et al. 1964] Walter James Langford, Thomas Arthur Alan Broadbent, & Reuben Louis Goodstein, *Professor Eric Harold Neville*, The Mathematical Gazette **47** (1964), pp. 131–145.

[Larmor 1917] Joseph Larmor, *Address by the retiring President*, Proceedings of the London Mathematical Society **16** (1917), pp. 1–7.

[Lemco 2007] Ian Lemco, *Wittgenstein's aeronautical investigation*, Notes and Records of the Royal Society **61** (2007), pp. 39–51.

[Levy 1916] Hyman Levy, *Discontinuous Fluid Motion Past a Curved Boundary*, Proceedings of the Royal Society of London **92** (1916), pp. 285–304.

[Littlewood 1917] John E. Littlewood, I. Formulae for Direct Fire; II. Anti-aircraft trajectories and range-tables, in [Pearson Papers, 603], 1917.

[Littlewood 1971–1972] John E. Littlewood, *Adventures in Ballistics, 1915–1918*, Mathematical Spectrum **4**(1) (1971), pp. 31–38; **4**(2) (1972), pp. 80–86.

[Lowe 1990] Victor Lowe, *Alfred North Whitehead: The Man and His Work*, 2 vols., The Johns Hopkins University Press, Baltimore, 1990.

[Maurice-Jones 1954] Kenneth Wyn Maurice-Jones, *The Shop Story 1900–1939*, The Royal Artillery Institution, Woolwich, 1954.

[Mayhew 2013] Emily Mayhew, *Wounded: From Battlefield to Blighty. 1914–1918*, Bodley Head, London, 2013.

[Melvill Jones 1966] Bennett Melvill Jones, *Cloud Flying in the First World War*, Journal of the Royal Aeronautical Society **70** (1966), p. 207.

[Milne 1945] Edward A. Milne, *Ralph Howard Fowler*, Obituary Notices of Fellows of the Royal Society **5** (1945), pp. 61–78.

[Milne & Fowler 1921] Edward A. Milne & Ralph H. Fowler, *Siren Harmonics and a Pure Tone Siren*, Proceedings of the Royal Society **A98** (1921), pp. 414–427.

[Milne & White 1948] Edward A. Milne & Francis P. White, *H. W. Richmond*, Obituary Notices of Fellows of the Royal Society **6** (1948), pp. 219–230.

[Moseley 1978] Russell Moseley, *The Origins and Early Years of the National Physical Laboratory: A Chapter in the Pre-history of British Science Policy*, Minerva **16** (1978), pp. 222–250.

[Nayler 1966a] Joseph Nayler, *The Aeronautical Research Council*, Journal of the Royal Aeronautical Society **70** (1966), pp. 79–82.

[Nayler 1966b] Joseph Nayler, *Aeronautical Research at the NPL*, Journal of the Royal Aeronautical Society **70** (1966), pp. 82–84.

[Pattison 1983] Michael Pattison, *Scientists, Inventors and the Military in Britain, 1915–19: The Munitions Inventions Department*, Social Studies of Science **13** (1983), pp. 521–568.

[Petavel 1920] Joseph Petavel, Presidential Address to Section G (Engineering) of the British Association, *Report of the British Association for the Advancement of Science: Annual Meeting 1919*, BAAS, London, 1920, pp. 256–263.

[Proudman 1921] Joseph Proudman, *The Principles of Internal Ballistics*, Proceedings of the Royal Society of London Series **A100** (1921), pp. 289–305.

[Richardson 1919] Lewis Fry Richardson, *The Mathematical Psychology of War*, Wm. Hunt, Oxford, 1919.

[Rice & Wilson 2003] Adrian Rice & Robin Wilson, *The Rise of British Analysis in the Early 20th Century: The Role of G.H. Hardy and the London Mathematical Society*, Historia Mathematica **30** (2003), 173–194.

[Richmond 1924–1925] Herbert W. Richmond, ed., *Textbook of Anti-Aircraft Gunnery*, 2 vols., H.M.S.O., London, 1924–1925.

[Robbins 1976] Keith Robbins, *The Abolition of War: The Peace Movement in Britain 1914–1919*, University of Wales Press, Cardiff, 1976.

[Rouse Ball 1912] Walter W. Rouse Ball, *The Cambridge School of Mathematics*, The Mathematical Gazette **6** (1912), pp. 311–323.

[Rutherford 1916] Ernest Rutherford, *H. G. J. Moseley, 1887–1915*, Proceedings of the Royal Society of London, **A93**, xxii–xxxviii.

[Sanderson 1972] Michael Sanderson, *The Universities and British Industry 1850–1970*, Routledge & Kegan Paul, London, 1972.

[Senechal 2012] Marjorie Senechal, *I Died for Beauty: Dorothy Wrinch and the Cultures of Science*, Oxford University Press, Oxford, 2012.

[Smith 1915] David E. Smith, Review of George Greenhill's *Report on Gyroscopic Theory*, Science **41** (1915), pp. 793–794.

[Snow 1967] Charles P. Snow, Foreword, in [Hardy 1967, pp. 9–58]

[Southwell et al. 1966] Richard V. Southwell, Hermann Glauert, William D. Douglas, & Herbert L. Stevens, *The Book of Aeron, Commonly Called the Revelations of Abah the Experimenter*, Journal of the Royal Aeronautical Society **70** (1966), pp. 85–87.

[Stanley 2007] Matthew Stanley, *Practical Mystic: Religion, Science, and A. S. Eddington*, University of Chicago Press, Chicago, 2007.

[Stanley 2003] Matthew Stanley, *"An Expedition to Heal the Wounds of War": The 1919 Eclipse and Eddington as Quaker Adventurer*, Isis **94** (2003), pp. 57–89.

[Stephens 1966] Arthur Veryan Stephens, *Some British Contributions to Aerodynamics*, Journal of the Royal Aeronautical Society **70** (1966), pp. 71–78.

[Sutton-Pippard 1966] Alfred J. Sutton-Pippard, *Admiralty Air Department 1915*, Journal of the Royal Aeronautical Society **70** (1966), pp. 69–70.

[Swartz 1971] Marvin Swartz, *The Union of Democratic Control in British Politics during the First World War*, Clarendon Press, Oxford, 1971.

[Thompson 1996] Andrew C. Thompson, *Logical Nonconformity? Conscientious Objection in the Cambridge Free Churches after 1914*, Journal of the United Reform Church History Society **5** (1996), pp. 548–561.

[Thomson 1919] George P. Thomson, *Applied Aerodynamics*, Hodder & Stoughton, London, 1919.

[Thomson 1936] Joseph J. Thomson, *Recollections and Reflections*, G. Bell & Sons, London, 1936.

[Turner 1980] Frank M. Turner, *Public Science in Britain, 1880-1919*, Isis **71** (1990), pp. 589–608.

[Turner 1916] Herbert H. Turner, *From an Oxford Notebook*, The Observatory **39** (1916), pp. 240–242.

[Vint 1956] James Vint, *Henry Ronald Hassé*, Journal of the London Mathematical Society **31** (1956), pp. 252–255.

[Von Wright 1990] Georg H. Von Wright, *A Portrait of Wittgenstein as a Young Man: From the Diary of David Hume Pinsent 1912-1914*, Blackwell, Oxford, 1990.

[Wakeford 1923] Edward Kingsley Wakeford, *Chords of Twisted Cubics*, Proceedings of the London Mathematical Society **21** (1923), pp. 98–113.

[Warwick 2003] Andrew Warwick, *Masters of Theory: Cambridge and the Rise of Mathematical Physics*, The University of Chicago Press, Chicago, 2003.

[Weston Smith 1990] Meg Weston Smith, *E. A. Milne and the Creation of Air Defence: Some Letters From an Unprincipled Brigand, 1916–1919*, Notes and Records of the Royal Society **44** (1990), pp. 241–255.

[Whitehead 1918] Alfred North Whitehead *Graphical Solution for High-Angle Fire*, Proceedings of the Royal Society **A94** (1918), pp. 301–307.

[Wilson 1918] Edwin B. Wilson, *The Mathematics of Aerodynamics*, The American Mathematical Monthly **25** (1918), pp. 292–297.

[Willey 1965] Basil Willey, *Spots of Time: A Retrospective of the Years 1897-1920*, Chatto & Windus, London, 1965.

The Total War of Paris Mathematicians

David Aubin, Hélène Gispert, and Catherine Goldstein

From the very day war broke out in August 1914, Paris mathematicians were prominent actors in the scientific mobilization consented to by France. In the decades preceding the First World War, they had enjoyed a rather comfortable situation. France had a great tradition in mathematics and Paris remained one of the most attractive places for mathematical research—perhaps *the* place to be in the domains of real and complex analysis where significant advances had occurred since 1900, when the International Congress of Mathematicians (hereinafter ICM) had taken place in the French capital. As often in the past, Paris mathematicians held many center stage positions in scientific institutions, universities, observatories, as well as on an international level. From 1900 to 1917, for instance, the Perpetual Secretary of the Academy of Sciences was a mathematician, Gaston Darboux (1842–1917), and, after his death, another mathematician, the analyst Émile Picard (1856–1941), was elected to his seat. Mathematicians had also become influential cultural and political figures in the capital, and some of them were increasingly involved in public affairs, as the analyst and mathematical physicist Paul Painlevé (1863–1933; Figures 7 and 9), elected member of Parliament for Paris in 1906.[1]

Because of their high stature in the Parisian circles of academic and university power, mathematicians' participation in the wartime scientific mobilization was therefore decisive. Paul Appell (1855–1930; Figure 1), for instance, the president of the Academy of Sciences for 1914 and the dean of the Sorbonne Faculty of Sciences (*Faculté des sciences de l'université de Paris*) was the president of the Commission of Inventions; he also organized the *Secours national* to help civilians who were

David Aubin's address is Université Pierre et Marie Curie, Institut de mathématiques de Jussieu-Paris Rive Gauche, Case 247, 4 place Jussieu, F-75252 Paris, France. His email address is david.aubin@upmc.fr.

Hélène Gispert's address is Université Paris–Sud, Groupe d'histoire et de diffusion des sciences d'Orsay, Bâtiment 407, Centre Scientifique, 91405, Orsay Cedex, France. Her email address is helene.gispert@u-psud.fr.

Catherine Goldstein's address is CNRS, Institut de mathématiques de Jussieu-Paris Rive Gauche, Sorbonne Universités, UPMC Univ Paris 06, Univ Paris Diderot, Sorbonne Paris Cité, F-75005, Paris, France. Her email address is cgolds@math.jussieu.fr.

[1] The main reference for the mathematical life in 1870–1914 remains [Gispert 1991]. Complements can be found on the French Association for the Advancement of Science in [Gispert 2002] and on the history of mathematical education in [Belhoste et al. 1996, Gispert et al. 2007]. The notion of cultural capital and the particular example of Paris has been discussed in [Roche & Charle 2002, Dierig et al. 2003].

FIGURE 1. Paul Appell (1855–1930).

victims of the war.[2] A professor at the Collège de France and a member of the Academy, Jacques Hadamard (1865–1963) would sit on several academic commissions for national defense. Already vice director of the École normale supérieure (hereinafter ENS) and the editor of an influential intellectual magazine entitled *La Revue du mois* before the start of the war, Émile Borel (1871–1956) would occupy various important cabinet positions in the Ministry of Inventions and, as such, was instrumental in first organizing and coordinating the scientific war effort. And

[2]See "L'Œuvre du secours national" in [Appell 1922, pp. 266–281]. In the absence of a recent biographical study devoted to Appell, his autobiography [Appell 1923] may be consulted on his life and work.

Painlevé himself became Minister of Education, Arts, and Inventions, then Minister of War from March 1917, and even the prime minister for a short period, from September 12 to November 13, 1917.[3]

In the 1920s, this enviable situation was followed by a national mathematical scene that some have described as being in marked decline. Prominent before the war and, as we shall see in more detail, especially active in the mobilization for promoting modern scientific warfare, Paris mathematical circles would paradoxically emerge from the war scathed and verging on archaism. Roots of this picture can be traced to what we call the founding members of Bourbaki's "ego-history."[4] It is said that the mathematicians who should have trained the new elite had been decimated on the front due to a concern for Republican egalitarianism and that the next generation beginning their studies after the war was, as a consequence, cut from novel tendencies in the field. More than a decade, it was claimed, would thus be necessary for a new *avant-garde* to rise.[5]

However, recent studies have led to a significant reexamination of French mathematics during the interwar period [Leloup 2009, Beaulieu 2009]. An abundant secondary literature also deals with various aspects of the scientific and industrial mobilization of France around the period of WWI.[6] To assess more precisely the value of such descriptions and the role of WWI in the transformations of mathematics and of its dominant images, we consider the fate of mathematicians especially in so far as they intervened in public debates as technical experts, as intellectuals, or as the architects of science policies.[7] We follow rather closely and chronologically the evolution of mathematicians' roles at the academy, in the military, and in government during the war. We argue that, for them, the Great War was total. War touched most aspects of their personal, professional, but also mathematical, lives. Inversely, through their mobilization, mathematicians contributed to make it

---

[3]Technically, the head of the French government was then called *Président du Conseil*, which is translated here as prime minister. Contrary to the present presidential regime in France, the prime minister, not the President of the Republic, was the most powerful elected official at the time. The biography of Painlevé has recently been examined in great detail [Anizan 2006, Fontanon & Franck 2006, Anizan 2012]. On Hadamard, see [Maz'ya & Shaposhnikova 1998]. On Borel, besides the biographical studies [Guiraldenq 1999, Gispert 2012], one may consult the memoirs of his wife, Camille Marbo [Marbo 1967].

[4]This view has indeed been publicized by some mathematicians of Bourbaki's founding generation; see [Leloup 2004, Leloup 2009, Goldstein 2009, Aubin 2012]. The well-known Bourbaki group was then formed, in 1934, with the ambition of renovating French mathematics by assimilating and developing the abstract, axiomatic currents promoted by German mathematicians in Göttingen. On the history of Bourbaki, see [Beaulieu 1989, Beaulieu 1993, Beaulieu 1994, Aubin 1997, Mashaal 2002].

[5]There is no single reference summarizing this common view of interwar French mathematics. For a recent reference in English, see [Siegmund-Schultze 2001].

[6]French research policies have been studied by several historians, although none of them has emphasized the impact of mathematicians' important involvement [Picard & Pradoura 1988, Roussel 1989, Picard 1991, Pinault 2006, Galvez-Béhar 2008]. Two general studies about the cultural history of WWI pay significant attention to savants, without focusing on mathematicians [Prochasson & Rasmussen 1996, Audouin-Rouzeau & Becker 2000]. Longer-term studies about scientists and universities in France include [Charle 1994, Fox 1995]. For a special focus on mathematicians and WWI, see [Schiavon 2003, Mazliak & Tazzioli 2009, Durand et al. 2013, Schiavon 2014]. Among useful recent biographical studies besides those already quoted (footnote 3), we mention [Letté 2004, Moissinac & Roussel 2010, Auvinet 2013].

[7]For studies on images of mathematics, see [Dahan Dalmedico & Bottazzini 2001], in particular, Leo Corry's contribution on Bourbaki in this volume; see also [Corry 2004].

a total war for French society in general.[8] From this study, we draw the conclusion that mathematicians emerged from the war having modified their forms of social involvement in significant ways.

1. A Mathematical Capital, Paris 1900–1914

"Would mathematics be in fashion?" ironically asked the literary chronicler of the newspaper *Le Figaro* on January 29, 1909. That day, Henri Poincaré was introduced among the members of the prestigious French Academy (*Académie française*), the academy in charge of matters pertaining to the French language, and a crowd turned out at the Palais de l'Institut to witness the show (Figure 2). The mathematician's image that commentators painted on that occasion certainly is evocative of a certain *état d'esprit* about mathematics at the end of the first decade of the twentieth century: "Such minds are located outside time and space. They live and work under the auspices of eternity. Because of that they are not easily intelligible" [Beaunier 1909]. In the first decade of the twentieth century, the image of the absent-minded, slightly autistic savant detached from mundane reality might have enjoyed renewed popularity.[9]

This discourse about mathematicians being "in fashion" but out of touch with the outside world was a response to the double reality of a discipline both increasingly unintelligible even to the most educated public and ever more indispensable due to its numerous applications to science and technology.[10] As Appell recalled in a speech in January 1914 when he was elected president of the Academy of Sciences, scientific progress was quickly changing the materialities of life at that time, and the scientist's role in society was likewise changing [Appell 1914a, p. 18]. Always an astute observer, the novelist Anatole France, like his famous British contemporary H. G. Wells, [Wells 1901, ch. 6], took notice that the next war would be terrible and that mathematics would, as never before, have a role to play in the tragedy. "[T]o massacre each other in beauty," one of France's characters declared, "by using artillery and the art of fortification, they have introduced chemistry and mathematics in the necessary destruction," adding ironically: "This is a sublime invention" [France 1903, pp. 166–167].

1.1. Transformations in the Mathematical Professional Community. Historians have emphasized the traumatic impact that the defeat against Germany in 1870 had on scientists. Both the French Association for the Advancement of Science (hereinafter AFAS) and the French Mathematical Society (*Société mathématique de France*) (hereinafter SMF) were created as a direct consequence

[8]The expression "total war" (which emphasized the fact that in modern conflicts most resources of a society are affected by the war, outside the battlefield *stricto sensu*) was already coined in France during WWI; see [Daudet 1918]. For the concept of "total war" itself, see for instance [Guiomar 2004, Chickering & Förster 2006].

[9]About Poincaré's presence in the French media, see [Rollet 1999–2000, Rollet 2000, Ginoux & Gérini 2012]. One may also think of Anatole France's *L'île des pingouins* (1909), a novel that featured a certain Professor Obnubile—"*qui menait depuis soixante ans une vie solitaire et recluse, dans son laboratoire où ne pénétraient point les bruits du dehors*" [France 1909, p. 175]. Still earlier, the "savant Cosinus," a character that appeared in one of the first comic books ever published [Christophe 1900], in its fifth printing in 1912, was said to have been inspired by Jacques Hadamard [Maz'ya & Shaposhnikova 1998, pp. 91–94].

[10]For a discussion of this apparent contradiction, see [Aubin 2009].

FIGURE 2. J. André Castaigne, "Members of the *Académie française* crossing the bridge after a session". Note the dome of the Palais de l'Institut in the background, where the Academy of Sciences also met. Originally published in *Artistic Paris* by Richard Whiteing, Century **60** (July 1900), p. 406. Library of Congress, Prints & Photographs Division, DLC/PP-1935:0022.

of the Franco-Prussian War [Gispert 1991, Gispert 2002]. The former French Minister of Education noted: "we have been vanquished by von Moltke's mathematicians and geographers."[11] Universities were reformed according to the Humboldtian model and the "new Sorbonne" was built for the University of Paris [Charle 1994, Hottin 2001]. An Alsatian by birth, dean of the Faculty of Sciences from 1903 onward, Appell expressed the priorities of the government at the turn of the century:

> After the 1870–71 disasters, the Republican Government understood that a great democracy's intelligence and conscience stem from institutions of higher education as they foster free research and innovativeness, the foundation of industrial and practical progress; the government thus generously sponsored the organization of the fundamental function of the faculties: high scientific culture.[12]

[11] "*[N]ous avons été* [vaincus] *par les mathématiciens et les géographes de de Moltke.*" Victor Duruy to Urbain Le Verrier, July 18, 1871 [Le Verrier Papers, ms. 3710, 281]. Helmuth von Moltke headed the Prussian armies' advance during that war.

[12] Discours d'Appell au banquet des anciens élèves de l'Institut de chimie appliquée, December 27, 1906; repr. [Appell 1922, pp. 67–71; quote on pp. 67–68].

As a result, the University of Paris and, in particular, its Faculty of Sciences were greatly expanded in the period. Under Darboux's tenure as dean (1889–1903), its budget increased from 775,000 to 1,267,000 francs. In 1894–1895, there were 860 students in the Faculty of Sciences (including 27 women); in 1908–1909, this number had risen to 2,151 students (including 323 women). In the same period, the personnel doubled, from 80 to 163. Thirteen new chairs were created, reaching a total of 34 chairs in 1908–1909. The number of associate professors (*maîtres de conférences*) went from 23 to 34 and laboratory personnel (*chefs de travaux* and *maîtres préparateurs*) skyrocketed from 36 to 95.[13]

At the end of 1903, the ENS was attached to the University of Paris, but—as Appell once explained to a British audience [Appell 1922, p. 58]—the program of the competitive admission exams required to enter the science section remained highly mathematical, just like those opening the doors to the École polytechnique. After general courses in calculus and rational mechanics, students specialized in (research-oriented) higher mathematics or in applied mathematics, like astronomy or experimental mechanics. In applied domains, students had access to an observatory and a small laboratory with machines, although Appell longed for closer interactions with industrial machines. Just before the war, higher mathematics curricula included higher geometry, celestial mechanics, higher analysis, the theory of functions, mathematical physics, and probability theory. This reflected, as we can see, a deep commitment to the unity of the mathematical sciences construed broadly.

The chairs in the mathematical sciences at the Sorbonne corresponded rather closely to these courses. Three new chairs appeared during the 1903–1904 reform of the curriculum and, in 1909, a chair in the theory of functions was created for Borel.[14] This institutional evolution reflected noticeable changes in the mathematical production, as analysis now exerted a strong dominance at the forefront of the Parisian mathematical research. In two domains more specifically, the French production was recognized internationally as excellent: the modern theory of functions (with set theory and the Lebesgue integral) and functional analysis [Gispert 1991]. As a symbol of this new trend, one can consider Borel's scientific work and his series of *Monographies sur la théorie des fonctions* (where 18 volumes were published between 1898 and 1913 authored by Henri Baire, Henri Lebesgue, Borel himself, and several others). This even led to the creation of something akin to a doctoral school, due to Borel's presence at the ENS [Gispert 1991].[15] Hadamard also took a leadership role at the time in functional analysis.

As far as applications were concerned, a new chair in aviation was created at the Sorbonne with funds coming from the arms dealer Basil Zaharoff. The chair was attached to the Aerotechnical Institute in Saint-Cyr, which was funded by the

[13]These figures come from Paul Appell, "La Faculté des Sciences de l'université de Paris (1895–1910)," *Revue de Paris* (1910); repr. [Appell 1922, pp. 128–161; numbers on pp. 132–133]; for the budget of the Faculty of Sciences, see ibid., p. 178.

[14]The three chairs corresponded to a chair in "General Mathematics" to cover the beginners' special needs (held by Painlevé until 1912) and two chairs resulting from the integration of the ENS within the Faculty of Sciences: a second chair in calculus and a chair in the applications of analysis to geometry, which was held by Louis Raffy until 1910 and then left unoccupied until 1918; see [Maurain & Pacaud 1940, pp. 23–28].

[15]Jeremy Gray coined the expression " French modernists" to describe this trend, see [Gray 2008, pp. 216–225]. However, their characterization as modernists in the sense proposed by Gray has been contested by [Gispert 2012].

petroleum magnate Henry Deutsch de la Meurthe.[16] Like the Applied Mechanics Laboratory created for Gabriel Kœnigs, this establishment fostered close collaborations between experimenters and mathematicians, some of whom authored influential books on aviation [Painlevé & Borel 1910]. Painlevé, who was a professor both at the École polytechnique and at the Sorbonne (and a member of the Academy of Sciences from 1900), flew on the Wright biplane in 1908 and studied aeronautics both from scientific and strategic viewpoints. When a member of Parliament in 1910, he actively promoted aviation for the military. In a new domain like aviation, Painlevé could thus transform his mathematical expertise into important military considerations that were in return highly instrumental in launching his brilliant political career [Fontanon 2006].

At the same time, mathematics' wider professional community in Paris went through dramatic transformations. Engineers trained at the École polytechnique and amateurs were increasingly replaced by graduates from the ENS teaching at the university or in the *classes préparatoires* [Gispert 1991].[17] The physiognomy of the French Mathematical Society reflected this change: among its French members (186 in 1900 and 178 in 1914), there was a sharp drop in the number of polytechnicians (from 46% to 29%) and a parallel increase in the number of *normaliens* (from 24% to 35%). The number of members teaching in a university similarly increased from 38 (20%) to 60 (33%).

At the Academy of Sciences, mathematicians occupied a prominent position: in addition to the six members of the Geometry Section, half of the members of the Mechanics Section and a third of those of the Astronomy Section had defended doctoral theses in the mathematical sciences.[18] Here, renewal seemed less obvious since only three new members had been elected to any of these sections since 1910: Hadamard in the Geometry Section; Léon Lecornu, professor of mechanics at the École polytechnique, in the Mechanics Section; and Pierre Puiseux, from the Paris Observatory, in the Astronomy Section. Still, as we shall see, academicians will be very active during WWI. Darboux, who turned 72 in 1914 and died before the end of the war, initiated for instance the discussions that would lead to the Inter-Allied Conferences of Academies and the international reorganization of scientific institutions in 1918.

Closely knit family networks put into close and frequent contact those scientists who otherwise might have been at odds over their political and religious views. Picard and Appell had married cousins, both nieces of a mathematician who had been Perpetual Secretary of the Academy until his death, Joseph Bertrand (1822–1900). Picard's wife, Marie, was also the daughter of Bertrand's brother-in-law Charles Hermite (1822–1901), another prominent mathematician of the previous century who, like Picard, was a Catholic Conservative. Borel in turn married Appell's daughter, Marguerite, a novelist known as Camille Marbo. Borel had met her

[16]See Appell's inauguration speech at the institute in July 1911 [Appell 1922, pp. 167–172].

[17]The *classes préparatoires aux grandes écoles* are part of a typically French system of higher education explained and discussed in the contribution by Jean-Luc Chabert and Christian Gilain to this volume.

[18]In the academic context,"geometry" meant "pure mathematics." The mathematicians of the Geometry Section in 1914 were Paul Appell (1855–1930; 59 years old in 1914), Gaston Darboux (1842–1917; 72), Jacques Hadamard (1865–1963; 49), Camille Jordan (1838–1922; 78), Paul Painlevé (1863–1933; 51), Émile Picard (1856–1941; 58). For statistical analyses of the composition of the academy, see [Gispert & Leloup 2009, appendices].

when he was invited by her father to join informal gatherings that met regularly at Appell's house with Painlevé and Darboux [Marbo 1967, p. 18]. Close to the "*clan Borel*"—he was Borel's best man at his wedding [Anizan 2006, p. 109]—Painlevé was also in regular contact with Lucien Poincaré (a physicist who was Henri's cousin and the brother of the president of the republic, Raymond Poincaré). A Left-leaning politican, Painlevé is nonetheless mentioned in the memoirs of the geometry professor of the École polytechnique, Maurice d'Ocagne, a "man of deep faith and tradition" [de Broglie 1938], who remembered fondly the vacations Painlevé spent in his house at Villiers-sur-Mer in 1898 (on which occasion Painlevé met his wife) [Ocagne, n.d., book 5, p. 9].[19]

These mathematical circles remained highly attractive viewed from the outside and Paris was a mainstay to many foreigners. Out of 298 members of the SMF in 1914, 110 had countries of residence other than France. At the Faculty of Sciences in Paris, 237 students out of 1,546 were foreigners (15%) in 1904 [Urin 1904, p. 29]. In 1911–1912, the proportion had skyrocketed to 27%: there was 520 foreigners (the great majority Russian) for about 1,400 French students [Collective 1918, p. 67].[20] For more advanced research students, a trip to the French capital often remained a necessity.[21]

1.2. Mathematicians as Technical Experts. In the social life of *Belle-Époque* Paris, mathematicians now played a multifaceted role as scientific and technical experts in science policy, in the domain of education at various levels, as well as in military affairs. To the traditional mission assumed by the academy of distributing prizes to reward important scientific achievements were slowly added instruments intended to sponsor and guide the development of scientific research in new promising directions. Although the Scientific Research Fund (*Caisse des recherches scientifiques*, CRS) established in 1901 barely involved mathematicians, private endowments of the Academy of Sciences, such as the large Loutreuil foundation (3,500,000 francs donated in 1912), allowed it to play a prominent part in the development of a research policy that broke away from its previous prize-as-reward policy [Pestre 1984, Paul 1985].[22] "The importance of our Academy's role," Appell reminded his colleagues when he started his tenure as president of that body in January 1914, "increases year after year" [Appell 1914a, p. 18]. It functioned as a clearing house where official communications and private conversations fostered the continuous exchange of ideas between scientists of various disciplines and, via the reporters attending its weekly *séances*, towards the general public. Mathematicians also played a significant role in the Bureau des longitudes where they were involved in scientific enterprises that had clear political undertones, such as meridian measurements in Ecuador [Schiavon 2006].

[19]On the Poincaré network, see the memoirs of Henri's sister, Aline Boutroux [Boutroux 2012]. On family networks among Paris mathematicians, see [Zerner 1991].

[20]Slighty different global figures, but with the same order of magnitude, are given in [Appell 1922, pp. 132–133].

[21]A striking example of this can be found in the story of Harvard students who decided to visit Paris during the war (see the introduction to this volume).

[22]The decree establishing the CRS can be found in *Journal officiel* (July 23, 1901); for secondary literature on the topic, see [Picard & Pradoura 1988, Duclert 1999, Pinault 2006, Galvez-Béhar 2008].

Recent curricular reforms in secondary education—to which mathematicians contributed abundantly, Darboux being president of the Science Curriculum Reform Committee—considerably increased the place of mathematics and science in the *lycée*. Between 1902 and 1905, mathematicians such as Borel and Poincaré publicized a view of high school that prepared pupils for the realities of modern industrial society and left considerable room for the teaching of mathematical and scientific disciplines. In tune with the dominant image of mathematics we will describe below, many insisted on the necessity of adopting a concrete approach to the study of mathematics and on its tight connections with the sciences.[23] At the same time, as mentioned earlier, the training of the elite of the nation, especially engineers at the École polytechnique, still hinged on a curriculum characterized by high-level mathematical training.[24]

Military research structures, where polytechnicians always occupied the majority of influential positions, were reformed to adapt to the increasing industrialization of the conduct of the war, which had been made clear by the defeat of 1870–1871. On October 23, 1887, a permanent commission was put in place by the army to assess the military worth of inventions submitted from all parts of the country. This Commission of Inventions (*Commission des inventions intéressant la défense nationale*) was the root of all invention boards put together by various countries at the start of WWI. At the Gâvre Commission, established in 1829 by the navy to deal with everything concerning ballistics, important doctrinal changes also took place that very year and the "brisk wind of theory" (*l'air vivifiant de la théorie*) began to blow, thus clearing the way for the adoption of highly mathematical approaches to ballistic problems.[25]

These military institutions at first functioned only with military personnel, but in 1894 scientists from the academy were asked to take part in the meetings of the Commission of Inventions. For twenty years, Appell and Joseph Boussinesq (1842–1929), from the Mechanics Section, took part in bimonthly meetings with colleagues from other disciplines and military officers. Around 1907–1908, Poincaré was put in charge of a scientific commission that investigated the circumstances surrounding the explosion of the battleship *Iéna* due to unstable gunpowder. This commission recommended that the respective attributions of the various military laboratories dealing with gunpowder be made clearer [Bret 2008]. Although fragmented in various independent, or even concurrent, institutions, relations between military men and scientists, and between industrialists and scientists many of whom came from the mathematical sciences, had become a matter of course [Pestre 1997]. "[T]he organization of national defense itself," Appell made sure to emphasize in his inaugural speech as president of the academy, "rests on the use of scientific methods" [Appell 1914a, p. 17].

[23]See, e.g., Borel's talk on practical training in mathematics [Borel 1904]. On French reforms in the *lycée* curriculum, there is an abundant literature. We mention [Bkouche 1991, Belhoste et al. 1996, Gispert et al. 2007]. This movement of course had an international dimension where Felix Klein played an important role.

[24]This is discussed by Jean-Luc Chabert and Christian Gilain in this volume; see also [Gispert 1994].

[25]On the history of the Commission of Inventions, see [Roussel 1989]. On Gâvre, where mathematicians were able to make significant contributions to the war effort during WWI, see David Aubin's contribution to this volume. Quote from [Charbonnier 1906, p. 415].

1.3. Mathematicians as Intellectuals. Paris mathematicians had become quite present in various institutions at the beginning of the twentieth century, as well as in the public eye, as witnessed by the popularity of the caricatures mentioned above. The role taken by scientists as experts in charge of innovation in industry and the military also led them more visibly to take positions in the intellectual and social scenes. When Poincaré died unexpectedly after a banal surgical intervention on July 17, 1912, the portrait of the mathematician painted in the press again emphasized both the extreme abstraction of his daily concerns and the profound ways in which it mattered for the common people. Like that of most of his colleagues, Poincaré's work struck the profane by its inaccessibility: the joke was made that his writings could barely be understood by "three or four of [his] followers—should we say four?"[26] But beyond this first impression, the type of mathematician Poincaré embodied was much lauded by his contemporaries. The Minister of Education Gabriel Guist'hau, for example, declared at his funeral: "while the mathematician's work is accessible to the very few, everyone knew that Henri Poincaré stood for the purest, the most disinterested, and the best" type of scientist in France [Darboux 1914, p. 110]. Mathematicians were portrayed as detached from mundane concerns and in touch with their artistic and poetic inclinations, an image that stood in positive contrast with the mechanical thinking of the modern age. Mathematicians were the true muses of modernity.[27]

But contrary to ancient muses, mathematicians had stepped into the public arena. "To better understand Poincaré the popularizer," wrote a contemporary journalist, "*Poincaré in the street*, I insist—it suffices to mention ... Émile Borel who does active sociology in the *Revue du mois* [and] Painlevé, future Minister of the Navy" [Seylor 1912, p. 8; added emphasis].[28] The Italian mathematician Vito Volterra explained why he thought that mathematicians had caught the public's attention:

> The scientific movement, the relationship between science and [daily] life, between the greater public and the scientists have changed deeply lately. The causes of this are easy to understand, its effects striking. Brilliant discoveries have encroached in all manifestations of life. This is why science has become popular and, from the mathematical and physical sciences especially, new and useful results are expected without ends.... The scientist who, only a few years back, used to stay hidden in his study or his laboratory is now mingling with other scientists and with the public [Volterra 1913, pp. 130–131].

In Paris, Poincaré and, with him, a new generation of mathematicians had become public figures on the occasion of the Dreyfus affair, when a Jewish officer

[26][Lebon 1912, p. 3]. One may recognize here a widespread anecdote applied to many of his contemporaries, most famously Arthur S. Eddington in 1919, or, of course, Albert Einstein.

[27]On the links between practical concerns and theoretical investigations in physics at the time, see [Galison 2003, Walter et al. 2010]. For Poincaré's public activities, see [Mawhin 2004] and for his image in the press, see [Ginoux & Gérini 2012]. On biographical writing about Poincaré more generally, see [Rollet & Nabonnand 2012a].

[28]Three books by Poincaré were published in that period: *La Science et l'hypothèse* (1902) [16,000 copies sold in 1912], *La Valeur de la science* (1905), *Science et méthode* (1908), and, posthumously, *Dernières pensées*. The first two greatly contributed to his election to the *Académie française*. About Poincaré's public engagement, see [Rollet 2000, ch. 5].

and graduate of the École polytechnique was in 1894 falsely accused of spying and deported to French Guyana. This was indeed a defining moment for the emergence of the figure of the public intellectual whose engagement as concerned citizen was subservient to his rational analysis of political and social circumstances. As the historian Vincent Duclert has argued, the Dreyfus affair played a part in sparking the parallel processes by which the nineteenth-century *savant* became a "scientist"—that is, the professional, technical expert serving state and society as described above—and an "intellectual" [Duclert 1999, p. 93]. Many mathematicians were indeed directly involved as technical experts in Alfred Dreyfus's second and third trials in 1899 and 1904. A member of Dreyfus's family, Hadamard was called to the witness stand in 1899, an occasion which prompted a life-long involvement with human rights.[29] His confident Painlevé was also questioned at length [Anonymous 1900, vol. 3, pp. 325–353]. In Dreyfus's trials, the accusation in part rested on a graphological analysis by Alphonse Bertillon that made use of faulty probabilistic reasoning. In a report from 1904, Poincaré, Darboux, and Appell used their stature as mathematicians—and the micrometers of the Paris Observatory—to denounce Bertillon's "theory whose ridiculous absurdity is made glaringly obvious by our material observations."[30] Borel's contribution to the debate is quite interesting: his *Revue du mois*, which was launched in 1905 with the help of his friends of the ENS, was conceived as a vehicle for scientists to intervene in public intellectual debates. Borel used statistical data from a survey he conducted among the *Revue du mois*'s readership to show that graphology should have no claim to scientific status and that it should be used with utmost caution in judicial trials [Borel 1906]. The scientific method and mathematical procedures here served to delineate claims that counted as scientifically proven in a court of law.

For the mathematicians involved, their intellectual mobilization in the Dreyfus affair thus had great significance. These men took the position of scientific professionals and based their conclusions on a "scientific method" that, they claimed, was common to every discipline and critical of naive scientism [Duclert 1999]. This understanding of the unified foundation of science allowing scientists to intervene not only in scientific and technical debates, but also in various political, cultural, and social issues, would also crucially inform mathematicians' public involvement in the First World War.

1.4. The Scientific Ethos. The specific and important role played by Paris mathematicians in the French scientific mobilization during WWI was therefore rooted in a particular ideological context that needs to be taken into account when one wants to understand this specificity. As we shall see, prominent themes involved in prewar discourses about mathematics would prove to be powerful tools in wartime. It is remarkable that these discourses cut across political positions, generations, and even mathematical priorities.

[29] From this time onward, Hadamard, Borel, Painlevé, and Jules Tannery belonged to, and played important roles in, the Human Right League (*Ligue des droits de l'homme*) founded in 1898 in response to Dreyfus's first trial.

[30] Une "*théorie dont l'absurdité ridicule est rendue éclatante par les constatations matérielles que nous avons pu faire*" [Poincaré et al. 1908–1909, p. 579], quoted in [Rollet 1999]. See excerpts of Poincaré's report in [Leblois 1929, pp. 172–185] and the papers [Rollet 1999, Mawhin 2004, Mansuy & Mazliak 2005, Mansuy & Mazliak 2011]. For more mathematicians' writings about the Dreyfus affair, see [Duclert 2006] and also [Anizan 2006, pp. 81–100]. On the affair more generally, see [Bredin 1986].

First, the operational dichotomy in the *Belle Époque* was not so much the opposition between pure and applied mathematics, or between fundamental science and technological applications, as it was between disinterested science and for-profit invention. The French meritocratic system now allowed some men who were without independent means to derive a stable income from their talent in science (usually as professors) but the true savant was the one who had no product to sell. One should not conclude, however, that disinterested science was seen as useless for society. On the contrary, as we have seen in discourses about Poincaré, the moral and economic value of disinterested science was constantly emphasized. "Theoretical ideas have often been the fruitful seed from which emerged important progress in industry, agriculture, [and] medicine. Scientific dreamers who seem to be lost in their speculations are in their manner practical men" [Picard 1909, pp. 8–9]. While presenting the struggle between "science and its fruitful applications" as unequal, Picard foresaw that research institutions solely devoted to the pursuit of disinterested science would soon have to be established [Picard 1912, p. 581]. According to a historian of invention, it was precisely because science and industry seemed to draw closer and closer together that the condition of *savant* was redefined by sublimating his material conditions [Galvez-Béhar 2008, p. 237].

Second, at the same time as they praised disinterest in research, mathematicians emphasized the unity of science, as well as its connections with industry. "Modern science appears to us as a work of reflection, but also of action, of disinterested research, but also of useful applications" [Appell 1922, p. 129]. From the most abstract mathematical concepts to the most mundane applications, science was a continuum, unified by a single scientific method that applied to all types of reasoning. "In the diverse branches of science, matter and instruments differ; the way to invention is the same," wrote the mathematician Jules Tannery in the two-volume collective work *On Method in the Sciences* [Tannery 1909, p. 66].[31] In that respect, mathematics was a science like any other.

While there was a wide consensus on the unity of the sciences including mathematics, what was debated in Parisian mathematical and philosophical circles was the relation of mathematics to logic and especially logic's role as a foundation of mathematics.[32] Many working mathematicians shared Borel's view, according to which "mathematics was a natural science in which logic did not play a larger role than in other natural sciences" [Borel 1907, p. 276]. To study the scientific method, Borel argued, nothing was more enlightening than to look at "the 'practice' of men belonging to the elite who in the laboratories, observatories, and libraries, devote themselves to disinterested work" [Borel 1911, pp. ii–iii]. In the early part of 1914, he explicitly expressed his disagreement with Hilbert's point of view:

> that is exactly why I always endeavored to set the parts of set theory that in effect contributed to the progress of the theory of functions apart from purely verbal logical constructs in which

[31]Tannery and several other French mathematicians were clearly inspired by the naturalistic view of mathematics expounded by Hermite, according to whom mathematics was a science of observation in the strong sense of the term; see [Goldstein 2011].

[32]On debates about logic and the philosophy of mathematics in early twentieth-century France, see [Grattan-Guinness 2000, passim].

> one juggles with symbols corresponding to no intuition at all [Borel 1914, p. 84].[33]

As we have seen, such conviction had likewise structured then recent curricular reforms in high schools and universities. To introduce practical training in the secondary teaching of mathematics was a strong emphasis of Borel's for curricular reforms, as he wrote in 1904: "a mathematical education that is both theoretical and practical ... can exert the most favorable influence on the formation of the mind" [Borel 1904, p. 440]. It may be useful to add that this conviction that mathematics was anchored in physical reality was in no way an automatic contradiction of modern developments. An interesting example comes from Baire's innovative *Leçons sur les fonctions discontinues* in which he introduced his theory of classes. In his preface, Baire explained that since modern physics was more and more interested in modeling discontinuous phenomena, "the duty of the mathematician was ... to start by studying *in abstracto* the relation between the notions of the continuous and the discontinuous" [Baire 1905, p. vi, original emphasis].

Third, it was fitting that the concrete and the practical were so much valued since mathematics as a body of scientific knowledge was believed above all to be based on *expérience*, where the notorious polysemy of the French substantive (which stands for both *experience* and *experiment*) was conspicuously not disentangled. For example, to reconcile Poincaré's conventionalism and a more realist position according to which the axioms of geometry were objective and derived from experimentation, Appell suggested that "the axioms of geometry are definitions afforded by experience."[34] Picard similarly argued that Euclidean geometry was to be preferred because: "Geometry started as a physical science, the diverse sensations triggered in us by the outside world having led to the notion of sensitive space, and this one led us to the construction of geometrical concepts" [Picard 1909, p. 21].[35] Not only geometry but also arithmetic and analysis should likewise be rooted in experience. According to Painlevé, for example,

> It is today an undeniable truth that there exists no rational science in the strict sense of the word: there is no branch of science whose foundation can do without borrowing some undefinable notions from our perception of the world. Abstract science par excellence, arithmetic and analysis rely on the notions of number and of the sum of two numbers. However intuitive they may appear to us, these notions nonetheless have an experimental origin [Painlevé 1909, p. 367].

Finally, while doubts about the foundation of mathematical certainty were often voiced, the end of a strict belief in positivism (that was sometimes dramatically forecast) did not question the validity of knowledge acquired through experience. Warning about the risk of elevating science to the status of a religious belief, Painlevé wrote that the doctrine of positivism "remains excellent provided

---

[33] This was a view Borel held at least since he had published his first volume on the theory of functions in 1898; see [Gispert 1995].

[34] Appell's preface to Freycinet's *De l'expérience en géométrie*, Gauthier-Villars, Paris, 1903; repr. [Appell 1922, pp. 28–36, quote on p. 32].

[35] This view was widely shared: it was for example expressed by Poincaré and Hadamard in 1904 in the *Conférences du musée pédagogique*.

it is taken to mean that to every period corresponds a *provisional* domain of scientific research, a domain that is every day wider, but beyond which it is reckless to venture" [Painlevé 1906, pp. 661–662; original emphasis]. Insistence on the uncertainties of knowledge even in the domain of pure mathematics perhaps accounted for the popular success of Poincaré's philosophical writings. If anything, discussions about the limits of knowledge reinforced the special position of mathematics which seemed to be based on a more fundamental form of experience. Mathematics' ancillary role to the mechanical and physical sciences was proof of its great usefulness, and it was expected that in due time the life sciences and the social sciences would follow the same path [Volterra 1905].

In this ideological context, one is not surprised to find mathematicians at the top of the academic ladder. It is from that position that after 1914 they became a decisive force of the scientific mobilization which would consequently reflect some of their values. The belief in a mathematical knowledge that was close to application, based on experience, and acquired through practice as well as in the fundamental unity of the scientific method had well prepared French mathematicians to make truly useful contributions to the war effort. Although they conceived of science and technology as a unified whole, mathematicians in positions of power often emphasized the value of fundamental research, an attitude for which they drew incomprehension on the part of some military officers, engineers, and industrialists. More fundamentally, the view that the mathematical sciences appeared to be as morally good as they were useful—"a fortifying thought for those who devote their life to [them]," as Picard wrote in 1909 [Picard 1909, p. 10]—would be challenged.

2. Mathematicians' Wartime Experiences

For an overwhelming majority of Paris mathematicians, war was a time of hardship, pain, and great personal investment. *Engagement* was the rule. Contrary to Cambridge, Paris (or France for that matter) seemed to harbor no pacifist mathematician: our extensive investigations have unearthed no such cases.[36] Even an anarchist like the mathematician Charles-Ange Laisant acknowledged, as one of the signatories of a manifesto published on February 28, 1916, that

> we, anarchists, we antimilitarists, we, enemies of war, we, passionate partisans of peace and of fraternity of the people, we sided with resistance and ... we think that unless the German population ... refuses to be the instrument of projects of Pan-Germanistic political domination, there can be no question of peace.[37]

The nature of French mathematicians' engagement greatly hinged on the generation to which mathematicians belonged. The youngest who had not completed their training were—like most of the cohort—sent to the front. Mathematicians

[36] June Barrow-Green discusses the case of Cambridge pacifist mathematicians in her contribution to this volume. A case of apparent indifference, that of Joseph Pérès, is studied in [Mazliak & Tazzioli 2009, ch. 7, pp. 155–164]: spending the war as a teacher in the south of France, Pérès however made computations for the Ministry of Inventions.

[37] Known as the Manifesto of the Sixteen, this text, inspired by Peter Kropotkine, was published in *La Bataille* in March 1916. Laisant's position is discussed in Lamandé's contribution to [Goldstein & Aubin forthcoming]. On Laisant, see [Auvinet 2013].

slightly more advanced in their careers—typically starting as professors in a university after having spent a few years in various research or teaching positions—mostly intervened in war as scientists in general terms and not as mathematicians in a narrow sense: their contribution was noticed at the end of the war and hinged upon their computing and analytic skills and sometimes on their technical knowledge. Often, members of this age group also took part in the training of military officers. Established men were in charge of organizing the scientific mobilization. Some of them laid the foundations of the national bodies for scientific research that would be formally established later in the interwar period. Others took part in the war effort as intellectuals: they devoted themselves to producing inspirational discourses as well as downright propaganda, but also to setting up charity organizations and to reorganizing international research instances. Of course, these diverse types of experience were not contradictory, and the same university professor preserved from the most exposed positions and involved in the scientific mobilization could, for example, later in the conflict, volunteer for frontline duty: this was Borel's experience.[38]

Of these types of mobilization, the last have been discussed more extensively. Ideological and organizational engagements have moreover fared differently in the historiography. While the intellectual crusade was often dismissed as a high point in irrationality which was seen, at best, as of little importance with respect to research directions taken by French mathematicians during and after the war and, at worst, as a likely explanation for its decline in the 1920s, the efforts in organizing scientific research, which was clearly influenced by such discourses, have been interpreted as the root of national and international research councils set up after the Armistice. With respect to mathematicians' direct involvement with military affairs, either as soldiers, instructors, or technical experts, the historiography has been scant until recently.[39]

2.1. Mourning the Dead. Viewed from Paris, however, all types of experience were not equally striking.[40] A policy according to which all able-bodied men of a certain age were sent to the front caused massive casualty. Scientists and mathematicians were not spared from this. As noted in the introduction of this volume, more than half the students attending the ENS at the outbreak of the war were killed (50.7%); some scientific section even had a mortality rate of 60%.[41] Recent graduates of the ENS were killed at a rate from 5% to 30%. Once we realize that almost all mathematicians who pursued an academic career in France had attended the ENS, the loss becomes even more significant. The idea that an elite had been sacrificed, whether or not this sacrifice was approved, already took shape during the war.[42]

[38]French mathematicians' generational wartime experiences are discussed more extensively in [Aubin et al. 2011]. For discussions on successive generations of patrons of French mathematics, see [Gispert & Leloup 2009]. Note that the issue of generations is commonly raised with respect to WWI: see in particular [Cohen 1984, Sirinelli 1988].

[39]For a pioneering investigation of some of these issues, see [Siegmund-Schultze 2003].

[40]This section follows closely [Aubin et al. 2011, pp. 185–186].

[41]This particularly high mortality rate, about twice that of the Polytechnicians, is analyzed in [Mariot 2012]. For comparative data on this issue, see [Aubin 2012].

[42]This question is studied in more detail in [Aubin 2012, Mariot 2012, Mariot 2013]. See also, on the École polytechnique, [Villermet 1993].

Consider the case of Jean Piglowski, born on August 4, 1889. He entered the ENS in 1910 and passed the examination for becoming a high school teacher (*agrégation*) in 1914. Before he could reach his post in Albi, war broke out and he joined a fighting unit on the front, where he died heroically defending his position on February 18, 1915. His campaign and sacrifice were reported on the rearlines and struck the Parisian public. Painlevé praised his name "that would later become legendary" [Painlevé 1916, pp. 194–195]. In its realities as much as in its published representations, this *normaliens*' war, however, seemed remote from the "mathematical war" or the "scientific war" reported by the press. Contrary to polytechnicians who served in the artillery or on the general staff, *normaliens* were most often of a middle rank in the infantry (which explains their higher mortality rate) and their scientific skills were of no importance. If many of the accounts of their wartime service insist on the correspondence between their training and the tasks they were called to accomplish, it was above all to underscore how this training helps them to fulfill their duty as "leaders" (*chefs*), guaranteeing the good conduct of their troops and standing up to the enemy.

The feeling of a terrible waste of talent and unfulfilled promise was reinforced by the family character of the Parisian mathematical community we have already mentioned. Many a mathematician too old to fight lost a child or a close relative. In November 1914, Picard wrote: "I have two sons and two sons-in-law in the battle; both of the latter are now wounded. These are hours of anguish."[43] In the end, Picard lost three of his five children because of the war. Borel's nephew and adoptive son, Fernand Lebeau, a physicist who had studied at the ENS and served in sound-ranging sections (SRS), was mowed down by machine-gun fire on the Champagne front on September 26, 1915. Hadamard lost two sons, one of whom was a Polytechnician.[44] Jordan lost three sons and his oldest grandson. The list could go on....

The trauma of such losses for Paris mathematicians was expressed in various ways. Élie Cartan devoted himself to the organization of a hospital for war wounded in the ENS compound. Sorrow determined some of their later choices. In his jubilee in 1936, Hadamard acknowledged that after his sons' deaths "no joy was allowed to be truly pure to me" [Hadamard 1937, p. 51]. After the war, Borel was unable to resume his scientific directorship of the ENS, finding "the school haunted by shadows" [Marbo 1967]. In October 1916, Picard wrote:

> My health has not been brilliant for the last few months, my nervous system was unable to get its balance back. I force myself however to work in view of the postwar on various committees and commissions. No matter how distant the end of the war may now appear, it will nonetheless come one day, and the fight will then erupt again under a new form against a perfidious enemy who, however beaten he may be, will try to rise again from his defeat. Men of my age will never know quiet days again.[45]

[43]Picard to Villat, November 13, 1914 [Villat Papers]. See also [Mazliak & Tazzioli 2009, pp. 50, 61–62].

[44]Relevant letters have been published in [Mazliak & Tazzioli 2009, pp. 89–92].

[45]"*Ma santé n'a pas été brillante depuis quelques mois, mon système nerveux ne pouvant retrouver son équilibre. Je m'efforce cependant, dans la faible mesure de mes moyens, de travailler en vue de l'après-guerre dans des comités ou commissions diverses. Si lointaine qu'apparaisse encore la fin de la guerre, elle viendra cependant un jour, et la lutte reprendra alors sous une autre*

To give meaning to the loss, much activity was geared toward the perpetuation of fallen *normaliens*' mathematical name. In 1915, Hadamard published parts of the work on the Goldbach conjecture in number theory done by the astronomer Jean Merlin, who died on August 27, 1914. Hadamard also asked Paul Lévy to prepare for publication chapters from a doctoral thesis on functional analysis begun by René Gateaux, who fell on October 2, 1914. Lévy published Gateaux's papers and developed his work into a book containing many new developments and applications.[46] In December 1915, Gateaux, as well as 15 of his unfortunate colleagues, received posthumous prizes from the Academy of Sciences.

Of course, Paris savants likewise devoted themselves—perhaps more than ever before—to the mentoring of some of the survivors. They became greatly involved in the doctoral studies of *normaliens* who came back badly wounded to the ENS hospital. Gaston Julia had received a gunshot in the face in his first days on the front in 1915; Louis Antoine was blinded when he was wounded for the third time. After a convalescence in Paris where they were helped by their teachers to renew their mathematical research, both went on to become university professors in Paris and in Rennes, respectively.[47]

2.2. Academic Self-Mobilization: Committee and Ideological Warfare. Concerted efforts to enroll scientists in war-related research have often been seen as the prefiguration of research organizations like the *Centre national de la recherche scientifique* (CNRS), which was established in 1939 as the successor of research bodies set up in the immediate postwar period. Because of this, organization is among the scientists' activities during World War I that have been studied the most.[48] What has attracted much less attention is the special position occupied by Paris mathematicians in shaping military innovation.

Scientists' first reactions to the war declaration came from the Academy of Sciences and its president, Paul Appell. On August 3, 1914, he officially stated, on his colleagues' behalf, that all nonmobilized academicians were at the French government's disposal to help national defense.[49] Immediately, the academy formed six commissions for war-related research: (1) mechanics, including aviation; (2) wireless telegraphy; (3) radiography; (4) chemisty, including explosives; (5) medicine, surgery, and hygiene; and (6) food. Mathematicians who volunteered to take part in the effort included Darboux, Picard, Appell, and Hadamard. Like their predecessors in 1791 and 1870, and despite the intensity of then recent internationalist discourse, they were prepared to play their part in the coming war [CDN Archives].[50]

Grandiloquent, the government's response to the Academy's offer showed that the leaders of the French state saw the role of the venerable assembly in terms

---

forme avec un ennemi perfide qui, si abattu soit-il, cherchera d'une manière ou d'une autre à se relever de sa défaite. Les hommes de mon âge ne connaîtront plus jamais des jours tranquilles" [Villat Papers].

[46]For a study of Gateaux's work and legacy, see Laurent Mazliak's contribution to [Goldstein & Aubin forthcoming].

[47]On Julia, see Catherine Goldstein's contribution to [Goldstein & Aubin forthcoming]; on Antoine, see [IRMAR 1988, Lefort 2007].

[48]On this, see [Paul 1985, Roussel 1989, Pinault 2006, Galvez-Béhar 2008]. On the history of the CNRS, see [Picard & Pradoura 1988, Guthleben 2011]. Anne Rasmussen has characterized the scientists' first period of engagement as "self-mobilization" [Rasmussen 2003].

[49]*Comptes rendus hebdomadaires des séances de l'Académie des sciences* **159** (1914), p. 349.

[50]On internationalism before WWI, see [Schroeder-Gudehus 1978].

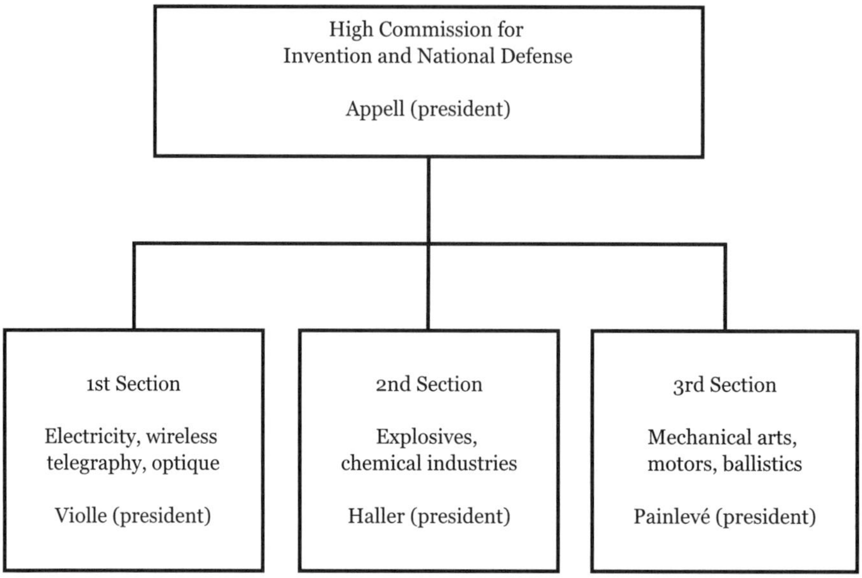

FIGURE 3. High Commission for Inventions and National Defense, August 11, 1914. Redrawn from [Anizan 2006].

that were more moral than material. For the prime minister, the Academy of Sciences seemed only to be useful as a reminder of the deep values of progress and civilization France stood for in the face of aggression.[51] But, on September 2, 1914, the government was retreating to Bordeaux, and in the trains that departed, special seats were booked for academicians. According to his memoirs, Appell refused to leave and took up residence at the Borels' only to preside over an almost deserted academy [Appell 1923, pp. 260ff].[52] After the battle of the Marne and the French victorious resistance to the German offensive, the front stabilized over the next two months. Academic affairs could be resumed in Paris at a more or less normal pace.

One had not sensed in the government's reaction to the Academy's offer any need to mobilize quickly and efficiently the scientific resources of the country—simply because this need was nowhere to be felt [Rasmussen 2003]. To address some technical aspects of the present war, however, the army had by then already restructured its old Commission of Inventions, where scientists from the Academy sat together with military officers. On August 11, 1911, 46 personalities, including 20 members of the Academy, were nominated to sit on the High Commission for Inventions and National Defense (*Commission supérieure des inventions intéressant la défense nationale*) [Roussel 1989, p. 36]. With many officers on the front and

[51]"*En s'associant ainsi à l'élan patriotique qui unit en ce moment tous les Français dans un même idéal de dévouement et de sacrifice, l'Académie des sciences n'apporte pas seulement au Gouvernement de la République le concours infiniment précieux des illustres savants qui honorent notre pays. Elle atteste, dans ce qu'il y a de plus élevé, le sentiment de la solidarité nationale, et affirme devant l'Europe le calme et l'énergie de la France, qui lutte non seulement pour son existence, mais pour le progrès et la civilisation du monde entier.*" Letter from Prime Minister René Viviani to the Perpetual Secretary of the Academy of Sciences, August 14, 1914 [CDN Archives].

[52]Appell recalled the "*terrible session*" of September 7, 1914: "*une inquiétude poignante, l'attente du désastre, planait sur nous*" [Appell 1923, p. 265].

unable to attend, they were joined by some engineers including Gustave Eiffel and Georges Claude, the founder of the Liquid Air Company, and politicians, including two senators and five members of Parliament (among whom was Jules-Louis Breton, who ultimately played an important role in invention policies). In the High Commission, a third of the academicians were mathematicians; Appell was president and Painlevé headed the third section in charge of the mechanical arts and ballistics—and effectively ran the High Commission instead of Appell (Figure 3) [Anizan 2006, pp. 240–242]. Over the duration of the war, the High Commission received 44,976 inventions to examine and retained 1,958 of them [Marin 1919, p. 6].

Meanwhile, several academicians felt compelled to engage in cultural warfare. One of the most infamous instances in which scientists took a public position in favor of their country's war aims was the "Manifesto of the Ninety-Three," known in German as the "*Aufruf an die Kulturwelt!*" and dated October 4, 1914. Among the 93 intellectuals who signed the manifesto, there was only one mathematician, Felix Klein [Tollmien 1993]. The *Aufruf* caused a great uproar abroad.[53] The reaction of the French Academy of Sciences perhaps best encapsulates the escalation in this war of words. After a long series of debates, at the secret meeting of March 15, 1915, the academy voted to exclude foreign correspondents and associates who had signed the manifesto [Rasmussen 2004]. Mathematicans were not unanimous. While Hadamard thought that the impression given to the neutrals might be unfavorable, a majority including Appell, Darboux, and Picard voted for Klein's exclusion together with three of his compatriots, two chemists, and one anatomist [Acad. Sci. Arch., comité secret].[54] The French Mathematical Society went much further; its council decided in 1915 to stop "relations with those of its members who belong to enemy nations" and erased their names from the membership list [SMF 1915].

Interestingly, Picard emphasized at the Academy that the German scientists' exclusion was to be argued, not on scientific, but on moral grounds. On October 15, 1914, the historian Émile Boutroux, who had married Poincaré's sister, had put science under accusation, claiming that it had been enrolled by the Germans in their quest for power: "brutality is here calculated and systematized:... this is scientific barbarity."[55] Scientists felt compelled to disculpate science while still inculpating the Germans. At the beginning of 1915, the mathematical physicist Pierre Duhem had amended his earlier description of the French "deep and narrow" mind and of the British "broad and shallow" mind to oppose the German "geometrical" mind to the French "subtle" mind (*esprit de finesse*). His demonstration was in great part based on trends he perceived in German mathematics: German scientists being especially gifted in deductive reasoning, they had a dangerous tendency to catch the various branches of science in the "net woven by Algebra" [Duhem 1915, p. 117]. This type of rhetorical exercise proved appealing to several academicians. The mathematical analyst Picard also wrote copiously to denounce what he presented

[53]The manifesto was translated into French and published in the *Revue scientifique* (August 8–November 14, 1914, pp. 170–171).

[54]For a more detailed discussion of Picard's wartime trajectory, see Pierre Lamandé's contribution to [Goldstein & Aubin forthcoming].

[55]"*La brutalité est ici calculée et systématisée : ... il y a une barbarie savante*" [Boutroux 1914, p. 388].

as the wrong-headedness of German science [Picard 1916].[56] Opposed to what he called "Kantian subjectivism," Picard argued that the goal of science—including mathematics—was the discovery, not the invention, of the truth. The Germans followed, he insisted, abstract reasonings to their logical conclusion without paying attention to reality, relishing even in eschewing common sense:

> How many studies [by German mathematicians] about the most bizarre geometries and to the strangest symbolisms could be cited; these are exercices of formal logic where no concern whatsoever is given to what could be useful for the future development of mathematics [Picard 1916, p. 36].

While we may be tempted to dismiss such propaganda as the kind of aberrant thought that only war tension allowed to be voiced, the ideas expressed in those texts are consistant with the deeply held values about science we have met earlier. Many of Picard's opinions, without the rabid anti-German sentiment, already infused his prewar reflexive writing about mathematics which were reprinted without significant change in 1921 [Picard 1914]. Mathematics (as a natural science) was all the more fruitful when developed in close link to its applications and allied domains, and not when prioritizing logical foundation. Expressed by influential figures like Picard, such views lastingly shaped the landscape of French mathematics. We shall come back to this issue.

2.3. Mathematics for the Trenches: The Case of Sound-Ranging. Scientists' self-mobilization quickly produced a striking impact on the field. Very often, however, this was not as a result of the committees put in place by the Academy of Sciences nor even of those of the army, but rather of the complex interaction of uncoordinated initiatives taken by younger scientists, engineers, and military officers outside the main scientific centers. As far as mathematicians were concerned, the most emblematic example of this fact was the case of "sound-ranging" where interactions of mathematicians with other scientists and with the military indeed led to methods that quickly were implemented with success on the front. To understand later changes in national research policies decided in the Paris centers of power, it is therefore necessary to discuss the lessons drawn from the successes and failures of scientists' experience with sound-ranging.

The aim of sound-ranging is to locate the opponent's batteries by triangulating the sounds they emit. If one posits two sound detectors at, say, A and B, the difference of the distances between the battery (located at P) and the two sound detectors is the speed of sound V multiplied by the time lag, $t_b - t_a$, with t_a and t_b the times when the discharge of the battery at P is perceived at A and B, respectively. In other words, since the measured time difference is a fixed parameter of the situation, point P is located on a given hyperbola of foci A and B. By using a third sound detector, one finds P (the battery) as the intersection of two hyperbolas.[57]

---

[56] Duhem's earlier contrast between the French and British minds is to be found in his famous book on *The Aim and Structure of Physical Theory*, originally published in 1906 [Duhem 1991]. Other intellectuals also contributed to this florishing literature, for instance [Durkheim 1915]. On wartime intellectual propaganda, see [Prochasson & Rasmussen 1996, Hanna 1996, Aubin forthcoming].

[57] For a more detailed account of these methods, see David Aubin and Catherine Goldstein's introduction to this volume.

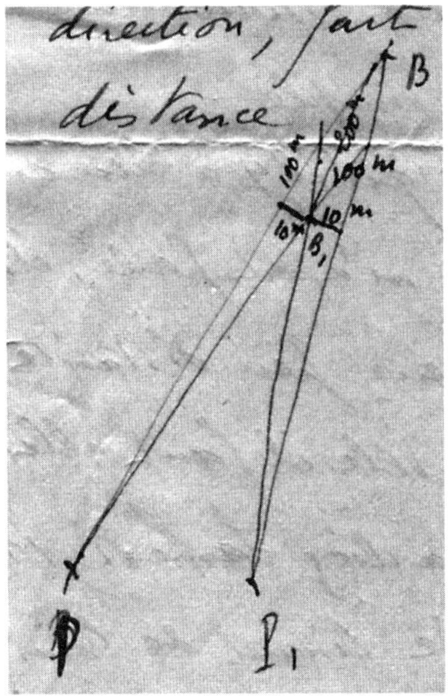

FIGURE 4. Figure from Henri Lebesgue's letter to Émile Borel, ca. February 1915, letter 217. With permission of the Académie des sciences–Institut de France. See [Lebesgue 1991, p. 321].

Various simplifications of the procedure were of course suggested, in particular an approximation by which hyperbolas were replaced by their asymptotes.

The accuracy of approximations in sound-ranging was discussed by Borel and Lebesgue.[58] In an undated letter, probably written in February 1915, Lebesgue highlighted the way in which purely mathematical considerations mingled with military tactics, computing procedures, and physics. He was in particular concerned with the value of approximations and the choice between graphical and computational methods to find P:

> Regarding the approximation studies that you mentioned, I am hopeless at that [*ça n'est pas dans mes cordes*]; I never had the sense for those things. Moreover, it is a question of making approximations in a situation where I—who have never seen the apparatus nor know what it provides—am ignorant of the order of magnitude of anything.
>
> On this topic. At first, ... I was told: to reach an approximation of 200 m would be for us ideal. I based my reflection on this fact and on what I had been told: 2 stations of 300 m distant [from one another] by one kilometer positioned between the French battery and the other at B. And this concerned field batteries whose normal combat distance is from 3 to 4 km. We are far from the 10 km you mentioned last Saturday.

[58] On the interesting correspondence between Lebesgue and Borel, see Helène Gispert's contribution to [Goldstein & Aubin forthcoming].

> To be sure, with these 10 km the graphic would be a source of errors, but if the angle at B is acute for the graphic [Figure 4] it will be [acute] in all respects. I was convinced of that and, last Saturday, I was wrongly looking for the effect of the acuteness in the effect of an error on the middle point of one of the stations. I should have looked for it in an error on the directions of a station. An error of $\frac{1}{1000}$ on this direction, in each direction, makes an error of 200 m on the distance [of B]. [This is a] natural thing for that matter: PBP_1 is known to $\frac{2}{1000}$, thus with a relative error of $\frac{1}{50}$,[59] and we find a relative error of $\frac{200}{1000} = \frac{1}{50}$ on the distances.
>
> Now, to account for $\frac{1}{1000}$ in the direction PB seems to me, a layman, to account for $\frac{1}{1000}$ of a second as far as time is concerned, and of many other things. Consequently, in the case of the 10 km, the graphic becomes more imprecise, but in sum, just like the data itself.
>
> True, computing has this advantage, that it allows us mathematicians to wash our hands from any error. It suffices to let the physicists make all determinations and to make exact computations with a great deal of decimal places if need be. But this is not the practical, simple method. I also think that one should distinguish [several cases]. The case which I was told was that of field batteries at a small distance but very mobile, for which one needs to be quick. The case you mentioned last Saturday, 10 km, is that of a heavy battery firing from afar, hardly mobile, allowing enough time for long computations and worthwhile of long computations. Then, *en avant!* with theodolites and Army geographers.
>
> I would all the more use computation in this case because I well believe that we shall be obliged here to do probability theory, something which will not displease you [Lebesgue 1991, pp. 321–323].

Lebesgue further considered that for central observation stations that remained in the same spot for a long time, one could advantageously compute double-entry tables giving the positions of the enemy's gun as a function of the angles measured from observations. He also pointed out that, because of systematic errors, replacing the hyperbolas by their asymptotes does not always provide a better approximation, even when the enemy batteries are far away. Revealingly, Lebesgue expressed the disappointment he felt as a mathematician regarding the problems posed by sound-ranging: "At bottom, I have lost interest in this:

> 1° because I realized that I only see in the telescopes objects that are brightly lit;

[59] An error of $\frac{1}{1000}$ in the direction P_1B gives a variation of 10 m (pointing to the B_1 in the figure). The relative error is the error (of $\frac{2}{1000}$) on the angle PBP_1, divided approximatively by the ratio of the distance between the stations to the distance of the enemy battery B, or here $\frac{1}{10}$; that is, $\frac{2}{1000} \times 10 = \frac{1}{50}$.

> 2° because the problem is one for topographers and because we will not (luckily perhaps) prevent topographers from doing as they wish since this is a thing they know;
> 3° because, if there was by chance a problem to be solved in this practical matter, Esclangon will have all the insights (and skills) that I am lacking totally to do this [Lebesgue 1991, p. 323].

Ernest Esclangon, whose name appeared in Lebesgue's letter, indeed devoted a great part of his wartime activities to the problem of sound-ranging. Trained as a mathematician at the ENS, he defended his doctoral thesis on quasi-periodic functions in Paris in 1904, before he was hired as an astronomer by the Bordeaux observatory. The father of a large family, he was spared from active duty in 1914. According to his own account, he met General Bourgeois (1857–1945), a polytechnician heading the army's Geographical Service, when the government retreated to Bordeaux in September 1914, and wrote him a summary account of the principles of sound-ranging.[60] Bourgeois would comment a few years later: "Once the problem was posed, its mathematical study, even when envisioned in all its practical computing details, presents no difficulty. ... For this reason, I would be personally inclined to assign but mediocre importance to the very idea and sole principle of sound-ranging" [Esclangon 1925, p. 3].

Having failed at the time to draw the army's attention to sound-ranging, the great quality of his report notwithstanding, Esclangon was indeed embittered by the fact that the Paris Observatory astronomer Charles Nordmann had gained credit for being the first to imagine and to implement sound-ranging procedures. Mobilized in an artillery unit, Nordmann convinced his superior, Colonel Robert Nivelle (later named general and head of the French Army in Verdun), who strongly supported the initiative, to develop sound-ranging into a tactically useful method for locating hidden guns. Relying on human observers equipped with stopwatches and telegraphic tapping keys, Nordmann's first experiments were, although successful, rather crude. The first German battery was located on December 8, 1914, near Soissons. This battery was 6.5 kilometers away from the observation basis consisting of a central observation position and one station on each side (Figure 5).

To show how widespread the idea of sound-ranging was, consider the mining engineer Ferdinand Daussy (1878–1964), mobilized as a soldier in the 45th Regiment of the Territorial Infantry, who also claimed to have invented sound-ranging. Denied permission to leave Verdun at the start of 1915, he had to build his recording instrument from scrap material available to him on the front: the motor of a phonograph, wrapping paper treated with hyposulfite as recorder, and a tuning fork given to him by a school teacher from Verdun. That he seems to have been able to work out the triangulation problem and apply the method on the front is also revealing of the diffusion of such technical skills at the time.[61]

For French mathematicians and scientists who, beginning in August 1914, were looking for ways to put their special skills to the nation's service, sound-ranging suddenly appeared, in the fall of 1914, as an opportunity not to be missed. Problems hindering the large-scale application of the method to the battlefield involved the design of an appropriate detection device and the careful study of the acoustics

[60] Titled "Sur le repérage à distance de l'emplacement des batteries," this note is reproduced in [Esclangon 1925, pp. 377–384].

[61] [Saur & Tribout de Morembert 1965–1966, pp. 19–20].

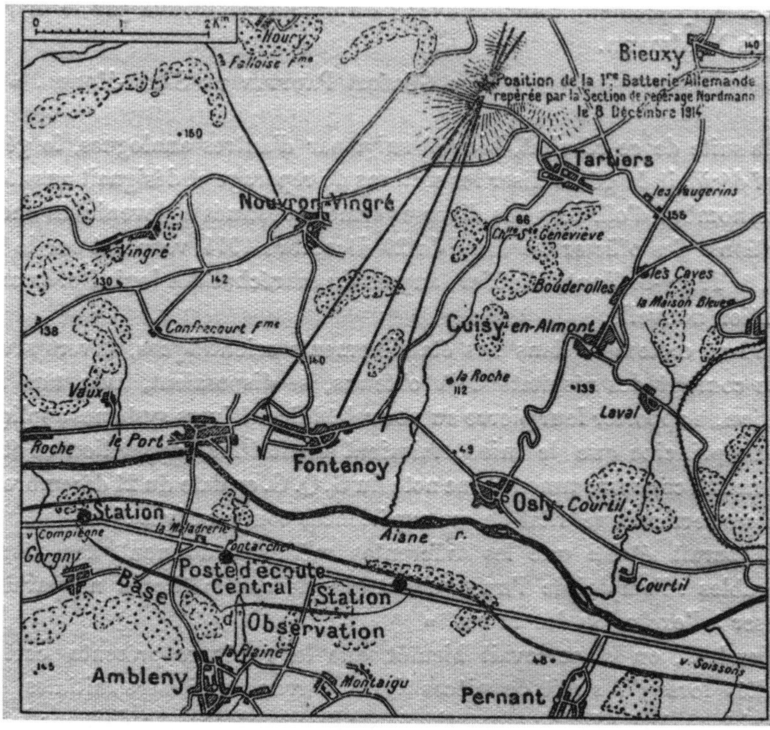

FIGURE 5. The first sound-ranging of a German battery, near Soissons, on December 8, 1914. One may note that the hyperbolas here pictured are nearly straight lines corresponding to the asymptotes [Nordmann 1928, p. 29]. With permission of the Académie des sciences–Institut de France.

of projectiles, neither of which was purely mathematical. Right after Nordmann met Painlevé in October 1914 to discuss his ideas, many scientists devised new instruments. Three schemes were investigated simultaneously: (1) Nordmann's telegraphic tappers; (2) microphones with or without relays; and (3) galvanometers. Sound waves were either recorded on smoked cylinders or signaled by needle variations. But the main problem was to detect the right sound in the midst of many spurious noises. The report emitted at the time of a gun's discharge was often fainter and detected later than the conical head wave, or shock wave, emitted by the projectile as it passed the speed of sound. The two sounds could "be heard very distinctly, the sharp crack of the shell being followed by the duller sound of the report" [Hinman 1919, p. 45]. As Lebesgue inferred, the study of this delicate phenomenon required a subtle mixture of experimentation with real projectiles and complex mathematical analyses of the physical situation, which was carried out by Esclangon after he reached the Gâvre ballistic experimental station in January 1915.[62] By the end of that year, Esclangon sent several unpublished reports to Paris, where he established that the best way to distinguish between waves was to measure the disturbances they caused to atmospheric pressure.

[62]The history of the Gâvre Commission before and during WWI is discussed in David Aubin's contribution to this volume.

The number of young scientists employed in the SRS in France was indeed impressive. The physicist Jean Cabannes together with Borel implemented the Cotton-Weiss method in Champagne, while the geophysicist Charles Maurain and the astronomer Jules Baillaud—the son of the director of the Paris Observatory, Benjamin Baillaud—commanded frontline sections in the north of Paris, under General Pétain [Cotton 1949]. Borel's adoptive son, the physicist Fernand Lebeau, and his brother-in-law, the chemist Jacques Duclaux, were also mobilized in the SRS, like the physicists Marcel Pichot and Georges Bruhat, the metrologist Albert Pérard, but also the surrealist poet Benjamin Péret. The phonetician cleric Jean Rousselot worked on sound-ranging while at the Fontainebleau artillery school. Among the few mathematicians who took part in the work of the SRS may be counted the celestial mechanician Jean Chazy and Jacques Chapelon.[63] The success of the SRS in locating the *Pariser Kanone* (often confused with the "Big Berthas") firing on the capital in 1918 has been greatly lauded [Huyon 2008].

2.4. Reclaiming Power for Government. For mathematicians in influential positions in Paris, the anarchic way effective methods for using sound-ranging on the battlefield were developed highlighted the need for more systematic and coordinated research policies. In the summer of 1915, the respective role of the academy's defense commissions and the military High Commission were defined more clearly. With academicians heavily involved in both, it was suggested that the Academy of Sciences be put in charge of "coordinat[ing] the efforts of the [scientific] workers in the country."[64] On June 23, 1915, a plenary meeting of all academic defense commissions was organized with representatives of the army. After some debate, where mathematicians like Appell, Darboux, Picard, and Hadamard, as well as specialists in experimental and rational mechanics such as Emmanuel Vallier, Jules Violle, and Léon Lecornu were most active, it was agreed that the academy's role would thereafter be restricted to two functions: to suggest new research directions (eventually to be examined by military commissions) and to respond to inquiries specifically formulated by the government. The military inventions boards, in turn, would submit to the academy's investigation all ideas it received which "were interesting in themselves but not ready to be implemented."[65]

To this suggestion, War Minister Alexandre Millerand replied that the offer was welcome, but he made sure to specify that the academy should "exert its fruitful action" by encouraging scientific research without "mingling with the work" of the High Commission of Inventions and the army's technical bodies. One should avoid, he added, theoretical and practical investigations susceptible of becoming "critical inquiries concerning the merits of already adopted solutions or giving rise to discussions for which secrecy . . . could only be preserved with difficulty."[66] Clearly

[63]Chapelon defended a thesis in number theory in May 1914. However, after the war, Chapelon turned to more applied topics, in particular he taught one of the first courses on probability theory at the École polytechnique. See [Goldstein 2009] and Jean-Luc Chabert and Christian Gilain's contribution to this volume.

[64]*Comité secret de l'Académie des sciences* (June 14, 1915), p. 198 [Acad. Sci. Arch.]. Discussions of the Secret Committee between May 7 and August 23 concern these issues.

[65]"*des idées intéressantes en elles-mêmes, mais qui ne sont pas au point*" [Acad. Sci. Arch., Comité secret, p. 200].

[66]The full quote reads: "*l'Académie exercera son action féconde, de haute impulsion scientifique sans s'immiscer dans les travaux de la Commission des Inventions ou des Comités Techniques. Vous estimerez en effet certainement avec moi, que, dans les circonstances présentes, il*

debates about various sound-ranging instruments and the publicity that had been given to such methods in the press had not pleased the Minister of War.

To face up to this new challenge, the Academy's defense commissions were opened to scientists who did not belong to the Academy of Sciences, many of whom were already working on their own on war-related issues. In an effort to organize all war-related scientific research around the Academy's leadership, defense commissions were reorganized in August 1915. There were now only four commissions: (1) mechanics, which included among others Darboux, Boussinesq, Picard, Appell, Painlevé, Lecornu, and Hadamard; (2) physics, again with Darboux, Appell, Painlevé, and Hadamard among others; (3) chemistry; and (4) health. Many subcommittees were moreover set up in order to deal with specific topics: artillery and sound-ranging (with Darboux, Hadamard, Gabriel Lippmann, and others); optics and the spectroscopic study of flares (for which, yet again, Hadamard volunteered); the study of the speed of gases in steel; submarine and aerial warfare; construction processes for barbed wire networks; ballistics; chemical poisons; pathogenous and poisonous microbes, etc.[67] The minister's delegate suggested that the subcommittees meet every week until the problems they were asked to tackle were solved. One subcommittee decided to organize a meeting on the spot, another in Henry Le Chatelier's laboratory at the Sorbonne, still another at the office of the army's Aeronautical Service in the Hôtel des Invalides.

As we can see, in 1914 and 1915, the Academy of Sciences was clearly taking the lead in the effort to gear up the warring nation scientifically, and, with it, the many mathematicians it counted among its members committed to a wide range of different tasks. But one may wonder whether the Academy truly had the means to realize its ambition. Soon, academic commissions wished they had laboratories at their disposal, but this was opposed by the army [Roussel 1989, p. 41]. The archives of the academy [CDN Archives] reveal a story that is at odds with some triumphant narratives. If the Chemistry and Physics Commissions indeed held regular meetings from July to November 1915, the register of the Health Commission remains completely empty while the labels intended for the Mechanics Commission register were never even glued on the cover! The significant ballistics file that may be found in the archives [CDN Archives] was assembled after the war by Hadamard for the purpose of determining the scientific merit of unpublished work rather than directly from the committee's work [Hadamard 1920]. Clearly, this organization was far from being as effective as one might have wished.

On November 13, 1915, a report was addressed to the president of the republic Raymond Poincaré by three ministers (education, war, and navy). A new government had just been sworn into office on October 29, in which the mathematician Painlevé (Figure 7) was Minister of Education and Inventions (in full, *Ministre de l'instruction publique et des inventions intéressant la défense nationale*). Voicing some of Painlevé's concerns, the report stressed the changing character of warfare and the urgent need to reorganize war-related scientific research. Every day, war increasingly appeared as "a fight of science and machines." Overwhelmed by

---

est essentiel de laisser fonctionner régulièrement les rouages normaux et d'éviter que les études théoriques et pratiques à effectuer, pour satisfaire aux besoins de la guerre actuelle, prennent le caractère d'enquêtes critiques sur le mérite des solutions précédemment adoptées ou donnent lieu à des discussion sur lesquelles le seecret pourtant indispensable aux fabricants de la défense nationale ne pourrait être que difficilement conservé" [Acad. Sci. Arch., Comité secret, p. 207].

[67] On chemical warfare in France, see [Lepick 1998, MacLeod & Johnson 2006].

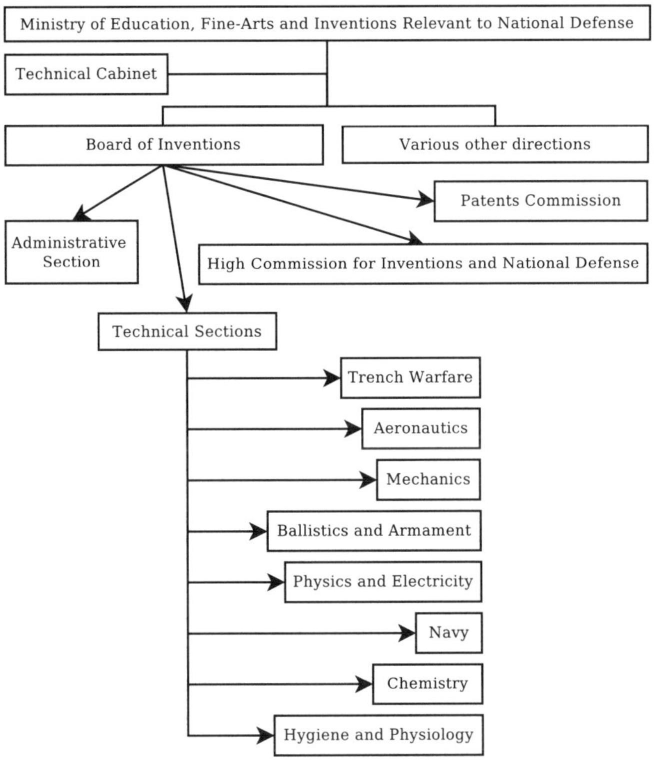

FIGURE 6. Inventions at the Ministry of Education, Fine Arts, and Inventions, November 1915. From [Anizan 2006, p. 337].

the need to keep up the production of war material, the army and the navy were hard-pressed to stand up to the challenge. The scientific resources of the nation, the report emphasized, had not been used to their fullest: "researchers isolated in their laboratories or their workshops are wasting their effort due to a lack of sufficient connections with the frontline."[68] Yves Roussel has noted that the word "researcher" (*chercheurs*), which was systematically used in the decree, letters, and the press at the time as a shortcut for both scientists and inventors, was new in this context [Roussel 1989, p. 48, footnote]. The science chronicler Nordmann commented on the tremendous change this represented in the relationship between science and the state.

> To my knowledge it is the first time that in a document produced by government, it is officially acknowledged that science is called to play its part in the business of the State.... The institution of a national organism whose exclusive aim is to help science fulfill its part in the needs of our time is nothing less than a kind of revolution [Nordmann 1915, p. 698].

[68] *Journal officiel de la République française* (November 14, 1915), quoted in [Lebesgue 1991, pp. 443–444]. See also the "Rapport secret sur la direction des inventions" addressed to Painlevé (December 1916) [Arch. nat., 313/AP/62]. On Painlevé's political role in WWI, see [Anizan 2006, vol. 2, pp. 231–542] and [Fontanon & Franck 2006, pp. 57–82].

FIGURE 7. War Minister Paul Painlevé and French President Raymond Poincaré inspecting a test site in Mailly in June 1917 [Claude 1919, p. 192].

The Academy of Sciences went almost unmentioned in the report. Instead, a Board of Inventions (*Direction des inventions intéressant la défense nationale*) was created at the Ministry of Education to which the High Commission was now attached (Figure 6). The board had two principal objectives that had been among the Academy's: to examine *quickly* proposals submitted by inventors and to develop in agreement with them the ones that seemed most interesting for defense purposes.[69] After that, the organization of national scientific and industrial research remained an issue for some members of the Academy of Sciences, especially the chemist Henry Le Chatelier. In May 1916, the academy discussed the creation of a national laboratory and a new section for industrial science. In February 1917, Painlevé argued for establishing an *Office des sciences appliquées*. But the Academy never regained the leadership of war-related research it had once had [Galvez-Béhar 2008, Anizan 2006].

[69] "*Orienter vers des buts précis les tentatives des inventeurs et coordonner leurs recherches, démêler dans la multitude des propositions celles qui sont susceptibles d'être efficaces et collaborer à leur réalisation pratique*" *Journal officiel de la République française* (November 14, 1915), quoted in [Roussel 1989, p. 47]. An example is mentioned: that of the *obus Chilowski*, studied in [Fontanon 2005].

At the head of the Board of Inventions, Painlevé placed his good friend and younger colleague Émile Borel. Around him, Borel gathered a "technical cabinet" composed of scientists belonging to his own generation. Among them were the physicist Jean Perrin (1870–1942) who was to serve as Borel's assistant in the cabinet, the geophysicist Charles Maurain (1871–1967), the chemist André-Louis Debierne (1874–1949), and the physiologist Louis Lapicque (1866–1952). None of them was a member of the academy at the time, and only Maurain had taken part earlier in the High Commission of Inventions. Although the minister and the head of the division were mathematicians, most of the new recruits were not, a sign of a loss of influence to come for mathematics. Like the academy's commissions, the board was organized into several technical divisions to which various inventors and scientists were assigned (Figure 6). Here again, the average age was lower than in previous commissions, and mathematicians formed a clear minority. In fact, among the dozens of researchers implicated, only four mathematicians are mentioned: Hadamard, Lebesgue, Paul Montel, and Jules Drach. As described above in discussions about sound-ranging, difficulties occurring in the collaboration between Lebesgue and Borel then broke a friendship that never fully recovered.[70]

Be that as it may, the Board of Inventions examined 9,120 different projects between November 1915 and July 1916 [Anizan 2006, p. 340]. When the Armistice was signed, the board had transmitted to the various technical services of the army 781 inventions that were entirely functional and susceptible of being applied immediately [Marin 1919, p. 6]. Although tensions persisted between it, the army's technical services, and the Minister of Armament Albert Thomas, the Board of Inventions established excellent working relations with Paris laboratories in universities and *grandes écoles* (ENS, École poytechnique, the Meudon Observatory, etc.), with private donors (one of whom set up a workshop in Sèvres), with the navy (which let it use its research laboratory in Toulon), and with industrialists.[71]

As may be suspected, mathematical research was far from being prominent at the Board of Inventions. A secret report from December 1916 insisted on a small number of inventions: the development of tanks, "geophones" (used to detect sounds underground) and compressed-air horns (used for communication) both developed by Perrin, antityphic vaccines, a protection system against toxic gases developed by Lapicque at the Museum of Natural History, etc. [Roussel 1989, pp. 46–47]. However, a computing bureau was put together under Montel and Lebesgue which was used to produce new sets of firing tables and held as an instructive example of the work that was accomplished by the board: "Here is a piece of obscure and modest work, but one which was extremely useful and to which we must pay homage. Our tables today are significantly better than German tables."[72]

In January 1917, Painlevé had to resign from the government. The Board of Inventions was immediately transformed into a cabinet-level administration, called the *Sous-secrétariat d'État chargé des Inventions intéressant la Défense nationale*, which was occupied by Jules-Louis Breton (1872–1940; Figure 8). A scientist by training, Breton had worked as a laboratory assistant at the Collège de France. But, at age 26, before he could complete his scientific education, he was elected

[70]See Hélène Gispert's contribution to [Goldstein & Aubin forthcoming].

[71]On tensions, see [Roussel 1989, pp. 50ff] and [Galvez-Béhar 2008, p. 251]. On the Toulon naval laboratory, see [Soubiran 2003].

[72]"Rapport secret sur la direction des inventions," addressed to Painlevé (December 1916) [Arch. nat., 313/AP/62, p. 16].

FIGURE 8. Jules-Louis Breton. Library of Congress, Print & Photographs Division, LC-DIG-ggbain-25665.

to the Parliament as a Socialist.[73] Although, Painlevé came back to the cabinet as War Minister in March 1917, Breton stayed as Undersecretary of Inventions while his jurisdiction was greatly extended [Marin 1919, p. 2]. The Board of Inventions was reinstated to its former status on November 20, 1917, following Georges Clemenceau's replacement of Painlevé as prime minister. Breton retained his position at the head of an administration that kept all of its earlier attributions.[74]

Throughout Breton's tenure, whatever the institutional changes, the body for inventions he headed was greatly reinforced as military research and development services were placed under its direct authority. The budget for inventions increased significantly, going from 250,000 francs in 1916 to 1,431,000 francs in 1917 and 1,296,000 francs for the first half of 1918 [Roussel 1989, pp. 55–56]. But the "spirit" of the administration changed. It was now geared toward technical expertise and the development of military applications rather than pure scientific research. Great importance was put on showing consideration for inventors. The organization of research and development processes was to be as efficient as possible in terms of war purposes.[75] Meanwhile, at the end of 1917, Borel—who had consistently occupied high-level positions in Painlevé's cabinets[76]—left Paris and volunteered for active

[73]There is a recent biography of Breton [Moissinac & Roussel 2010]. On science policy in the French Parliament, see [Pinault 2006].

[74]For an indication of what was Painlevé's image in the press, see *L'Œuvre* (December 6, 1917), p. 1. The journalist Gustave Téry called him a "*brave homme de mathématicien*" (a decent man, but a mathematician) when he complained about Painlevé's inability of resisting the pressures of extremists. See also a caricature of Painlevé after the fall of his government, ibid., November 13, 1917, p. 2.

[75]See Breton's instructions of July 1917, quoted in [Marin 1919, pp. 3–5].

[76]Borel was *chef du cabinet technique* of Minister Painlevé from October 15 to December 12, 1916; then, *directeur des services techniques* at the War Ministry (Painlevé being the minister), from March 20 to September 12, 1917; and finally *secrétaire général* of Prime Minister Painlevé from September 12 to November 16, 1917.

duty in the artillery. He escaped a near death twice and was awarded a second military citation.[77]

As far as mathematicians were concerned, the institutional evolution of administrations in charge of coordinating war-related research produced dramatic changes. In a three-stage process, they were side-stepped from governmental policies regarding science and innovation. At first, old-fashioned mathematicians, who were experts in analysis and rational and experimental mechanics, had played leading roles in military and academic commissions. Then, when the Board of Inventions was put together by Painlevé and Borel, mathematical expertise became secondary to rigorous organization principles, which they still claimed to derive from their mathematical training.[78] But few mathematicians were truly active in the Board of Inventions, where natural scientists took on the leading roles. At a later stage, the coordination of war-related inventions was, by and large, taken away from active scientists including mathematicians. Many of them therefore went on active duty, like Borel, or fell back on administrative tasks. But to keep their voice in public debates—a crucial concern in a time when governments had become directly involved in scientific research—mathematicians often reinvested the persona of the intellectual, using the Academy of Sciences as a pulpit.

2.5. Back to the Academy of Sciences. When it was not occupied with unsuccessfully trying to organize the scientific and industrial mobilization of the country, the Academy of Sciences went about its usual business. Throughout the war, the mathematical activities reflected in the *Comptes rendus hebdomadaires des séances de l'Académie des sciences* (CRAS) remained substantial. Roughly 400 mathematical notices were published between 1914 and 1918, almost half of them labeled as belonging to "analysis." Coming from different generations, Paul Appell, Georges Humbert, and Gaston Julia were the most prolific authors of the period, the first one on a variety of mechanical and analytic problems, the last two on their then common theme, the arithmetical theory of forms. Like eight other students, Julia defended a doctoral thesis during the war [Leloup 2009]. But the Academy also became the natural stage on which prominent scientists developed their own analysis of the effect of war on the sciences, which were systematically reported in the press. Each year, a new president was elected who would voice concerns about the new role taken up by scientists in the total war. Over the five years of the war, three of the academy's presidents (Appell, Jordan, and Painlevé) were mathematicians.

On October 26, 1914, the annual solemn session of the five academies took place at the Palais de l'Institut. For the Academy of Sciences, Appell gave a speech on the "conflict of two civilizations." He drew a stark contrast between a Germany

[77]The first of Borel's citations, awarded on October 28, 1915, underscored his technical help in setting up a sound-ranging section on the front. His second one emphasized his qualities as leader, as well as his courage under fire: "*Officier joignant aux plus belles qualités militaires les connaissances techniques les plus étendues. Par son dévouement et son inlassable activité, a communiqué à tout son personnel le goût et le désir de l'action. Le 28 août, chargé d'occuper très rapidement une position avancée, a réussi, malgré les pertes causées par un bombardement violent, à exécuter, dans les délais fixés, une mission de tir particulièrement difficile*" (ordre no. 393 du Régiment du 9–9–1918). [Guiraldenq 1999, p. 99].

[78]In the early 1920s, Borel published many newspaper articles about the administrative, political, and economic organization of the state and international organizations collected in a book entitled *Organiser* [Borel 1925].

that was making practical use of the most recent scientific discoveries, without specifying what they were, and Allied nations fighting for freedom and justice, "in the respect of the treasures accumulated by Art and Science" [Appell 1914b]. At the end of the year, on December 21, 1914, Appell further emphasized his vision of the scientific enterprise in a time of war.[79] For Appell, war confirmed that there was a deep continuity between science and its applications:

> Almost all the developments of modern civilization take their roots in scientific research.... Thus understood, science's domain is unlimited: it reaches from the highest abstractions to the most practical applications, from the world of stars and nebulae to that of atoms and molecules; from celestial mechanics to factories, ironclads, and airplanes; [etc.].

Mathematics as well as rational mechanics occupied a special place in Appell's understanding of science and cleared new paths toward the foundation of knowledge. For the sciences, war above all provoked a moral crisis: while the search for scientific truth by "a mind enraptured by moral beauty" was the "noblest effort that was offered to a human existence," the highjacking of this ideal by impulses of specialization, domination, and utilitarianism led to "scientific barbarity." To resist this modernist anxiety, Appell counted on scientists' idealism and their voluntarism:

> What are we to think? Are science, philosophy, and religion empty words?... Have they wasted their pains, [those] who have worked so long and so hard to substitute the reign of Justice to that for brutal force? We must answer: No, a thousand times no![80]

A year later, on December 27, 1915, the geologist Edmond Perrier, as president of the academy, had other worries. Praising the savants' mobilization both on the front and in the labs and arguing for the necessity of organizing the "scientific war" at a time when the academy was being dispossessed of its leading role by the Board of Inventions, Perrier emphasized the need for scientists to take an important part in government, for only they would be able to uphold the ideals of science. "Let not science be jeopardized in the terrible adventure in which humankind is at this time entangled" [Perrier 1915]. So, when responsibility for organizing war-related scientific research was taken over by the government, a process was started in the speeches delivered at the academy by which science was disengaged at an ideological level from war.

In the following years, intellectual struggle took second stage to more practical concerns, including some concerns about the need for a new kind of internationalism. In 1915, the Board of Inventions put together an Inter-Allied Committee for Inventions. While two French officers were sent to London, delegates from Britain, Belgium, Italy, and Russia met every week in Paris. We have little specific information about their activities, but, as is well known, it fell on academicians to restructure international scientific exchanges. Already in 1916 Darboux wrote to the academies of Allied countries to raise the question. In 1917, contacts were established between Émile Picard, the physicist Arthur Schuster of the Royal Society, and the astrophysicist George E. Hale of the U.S. National Academy of Sciences;

[79] [Appell 1914c]; Appell's speech is repr. in [Appell 1922, 249–259].

[80] All quotations above are taken from [Appell 1914c, pp. 821–822].

the principle of excluding the Central Powers from new international organizations was agreed upon. As the end of the war neared, Inter-Allied scientific conferences met in London and Paris in October and November 1918 to establish the International Research Council (IRC) and various international societies, including the International Mathematical Union (IMU). On these occasions Picard took a leading role in the IRC (he was elected president).[81]

3. Effects of War on Postwar Mathematics in Paris

After the war, it was widely recognized that science had been called to play a new role in modern societies. As Appell stated in 1921: "science has ceased to be a luxury."[82] But how to apply the lessons learn from the war effort in peacetime? How should wartime Parisian mathematicians' personal and collective experiences reconfigure the professional landscape? In light of the above, it is to be expected that mathematicians would play prominent roles in turning the unprecedented scientific mobilization to peaceful goals, the economic reconstruction of France, and a new kind of scientific internationalism.

A study of the Academy of Sciences has shown that war service was most often not considered as directly relevant for the election of the seven mathematicians admitted as members in the 1920s [Gispert & Leloup 2009].[83] Borel's case is striking. To his and Painlevé's eyes, his administrative work was not to be confused with mathematical research. Reporting on Borel's candidacy, Painlevé—who as we have seen worked closely with him on the scientific mobilization—wrote: "The mathematical work of Monsieur Borel having been interrupted by the war, I have nothing to add to my report of 1912 other than a few words concerning the 18-month period from January 1913 to July 1914." Nor did Borel mention his war work in the "Supplement" to the 1912 *Notice sur les travaux* he wrote in April 1918: "From this date [August 1914] the applications that have concerned me are too different from my peacetime work to be mentioned here."[84] However, it is not the case for d'Ocagne, who in his own *Notice* emphasized his successful application of graphical computing methods to military problems (an argument Appell also used in his report on the candidate). Nor is it the case for Esclangon, who, at the end of the 1920s, included all his publications on ballistics written while he was working either at the Gâvre Commission or in Painlevé's Ministry of Inventions.

3.1. Reconstructing Mathematical Institutions. Nothing gives a stronger impression of the sense of urgency that endured right after the Armistice than to focus on Painlevé's busy schedule in the space of less than a fortnight between November 26 and December 7, 1919. His first four days were occupied with the Inter-Allied Conference of Scientific Academies in Paris, just mentioned. On the morning of November 26, Painlevé (Figure 9) opened the conference by underscoring that the principal challenge of peacetime was to maintain the links forged under the

[81]On Inter-Allied scientific relations after the war, see [Schroeder-Gudehus 1978, Kevles 1995, Rasmussen 2007, Lehto 2002], among others.

[82]Reception of Polish Marshall Pilsudski at the Sorbonne, February 5, 1921 [Appell 1922, p. 297]. For a study of scientific demobilization in France, see [Rasmussen 2007].

[83]In the order of their entrance at the Academy, they are Édouard Goursat (1858–1936) and Henri Andoyer (1862–1930) in 1919, Émile Borel in 1921, Henri Lebesgue and Maurice d'Ocagne (1862–1938) in 1922, Jules Drach (1871–1949) and Ernest Esclangon (1876–1954) in 1929.

[84]The above quotations are from [Acad. Sci. Arch., Borel's personal file] and [Borel 1918]; both are reproduced in [Gispert & Leloup 2009].

FIGURE 9. Paul Painlevé, 1920. From the National Photo Company Collection, Library of Congress, Prints and Photographs Division, LC-DIG-npcc-01680.

pressure of the war between science and industry, between science and the state, and among Allied nations. A week later, on December 2, Painlevé was ending his tenure as president of the Academy of Sciences at the Palais de l'Institut by emphasizing the moral and political lessons scientists needed to draw from their war experiences. Mathematics, Painlevé made sure to emphasize, like all other sciences, could be said to have played its part in war:

> The direct contribution of scientific research and invention to national defense was admirable. [To be solved,] problems... required the most diverse minds and the contribution of all the sciences: chemistry, mechanics, thermodynamics, optics, acoustics, electricity, meteorology.... The most abstract and most subtle mathematics was involved in the solution of sound-ranging problems and to computation of new firing tables that increased the efficiency of artillery by 25 percent.[85]

Five days afterwards, on December 7, 1918, the French government put Painlevé in charge of a subcommittee to work on the peaceful reconversion of the Board

[85]This speech is published in the *Comptes rendus de l'Académie*, [Painlevé 1918a, p. 800]; see also Painlevé's speeches on January 7 and November 11, [Painlevé 1918b, Painlevé 1918c].

of Inventions. The plan was to put together an "office for the centralization of research," the goal of which was to carry out studies for the state and to foster the development of inventions. Besides many inventors, industrialists, and military officers, the subcommittee counted Appell and Borel among its members.

Chemists and experimental scientists pushed forward with the lobbying effort undertaken during the war to establish large national laboratories (especially Le Chatelier and Charles Moureu, helped by the nationalist writer turned politician Maurice Barrès [Moureu 1924]). Industrial scientists were admitted to the Academy of Sciences in a new section devoted to "Applications of science to industry." As far as mathematics was concerned, it would seem natural to think that the experiences of World War I would have produced renewed interest in applied mathematics. In the opening speech he delivered at the Strasbourg ICM, Picard warned his colleagues against this perceived threat:

> Some say ... that in years to come applications of mathematics will be the most studied and pure theory somewhat neglected.... The times we are now living in have indeed become harder for mind workers, and the more optimistic of us sometimes ask whether our civilization will not be eclipsed. We therefore must not tire ourselves of repeating that in the final analysis the true source of all progress in the applied sciences lies in theoretical speculations [Villat 1921, p. xxviii].

The myth of pure, disintestered science—so potent, as we have seen, in the prewar period—needed to be reinvested with new meanings. In his closing speech, Picard however also argued that the world had completely changed between 1914 and 1920 and—odd in light of the above—that the mathematician now had to get out of his "ivory tower" [Villat 1921, p. xxxii]. Picard's injunctions may have seemed self-contradictory: how was one to resist utilitarianism, nurture theoretical speculations, all the while simultaneously striving to be more involved in society and industrial development? The most striking mathematical developments of the 1920s in Paris can be seen as so many attempts at resolving the conundrum.

For at least one decade after the Armistice, mathematicians retained roles of scientific leadership which they had had before the war at the university, at the Academy, in national politics, and in national and international science policy. As we have seen, Picard was a key player in the negotiations for instituting the IRC and the IMU and for the organization of the controversial Strasbourg International Congress of Mathematicians (ICM) in 1920 [Lehto 1998, Lehto 2002]. But he also participated very frequently in thesis committees of the time [Gispert & Leloup 2009]. In 1921, Appell became rector of the University of Paris. Important political figures among the Left in the 1920s, Borel and Painlevé were elected and served, respectively, as Minister of the Navy and prime minister in 1925. They also retained a crucial influence in the development of mathematics in Paris, undertaking, in their own way, lobbying efforts directed both to the state and to the academic milieu. Painlevé had a chair in fluid mechanics established for him at the Sorbonne which promoted a strong development of the mathematical approach to that field. With other Sorbonne University mathematicians, he struck an alliance with the nascent

Air Ministry to establish, in the late 1920s, well-endowed chairs and a research institute in theoretical and experimental fluid mechanics in Paris (*Institut de mécanique des fluides* (hereafter IMF)), headed by the mathematical physicist Henri Villat.[86]

As for Borel, he powerfully backed the development of statistics and mathematical physics. At the Sorbonne, leaving the chair in the theory of functions (created for him in 1909) to take Boussinesq's chair in probability and mathematical physics in 1920, Borel suggested that this old chair be transformed into a chair for theoretical physics and celestial mechanics. The proposal stirred hot debates on the Faculty Council [Gispert & Leloup 2009, pp. 76–78]. Strongly opposing the change, Lebesgue protested against the paradox that "one denie[d] to mathematics the right to be cultivated for its own sake, because it is useful." The professor of physical and experimental mechanics Kœnigs replied "there [were] many chairs for the abstract sciences and if he rallied M. Borel's request, it [was] because he [saw] that the orientation of teaching [had to be] towards the concrete sciences." At the same time, together with the central organization for statistics in France (the *Statistique générale de France*), Borel established the *Institut de statistiques de l'université de Paris* (ISUP, Institute of Statistics of the Paris University) to strengthen the teaching of statistics. In 1928, finally, Borel took advantage of Rockefeller monies to establish the *Institut Henri Poincaré* (IHP), with the explicit purpose of developing teaching and research in mathematical physics, probability theory, and mathematical statistics in Paris.[87] New links established during the war with industrial milieus, military officers, and civilian engineers also materialized by their increased membership to the French Mathematical Society (SMF): as soon as the war was over, in 1919 and 1920, up to 180 new members were admitted to the SMF, among which almost a fourth were working either for the army, as civil engineers, or even as bankers.

Before the war, as we have seen, some mathematicians had become active in politics. Besides Painlevé, who went on with his parliamentary career after the war and served once again as prime minister in 1925, Borel's case again is quite revealing [Gispert 2012]. A symbol of the mathematical renewal of the first decade of the 1900s and an influential intellectual figure, Borel belonged to a younger generation who gained access to truly powerful political positions during the war. After the Armistice, the nature of Borel's public involvement changed and, as we just saw, became more institutional. In intellectual circles, Borel took part in the foundation of the Confederation of Intellectual Workers (*Confédération des travailleurs intellectuels*) and of the French Committee for European Cooperation at the Society of Nations. The *Revue du mois* was discontinued, but Borel directed two book collections addressing general interests.[88] In Parliament, in which he was elected in 1924, Borel was heavily committed to state affairs. Vice president of the Finance Commission of the National Assembly, he used his status as a mathematician to present himself as an expert in the rational understanding of the economy. But he

[86]On fluid mechanics in interwar Paris, see [Fontanon & Franck 2006], [Aubin forthcoming], and [Mounier-Kuhn 1996]. Between 1914 and 1945, 17 doctoral theses were defended at the Sorbonne in fluid mechanics, that is 10% of the total of theses defended in mathematics in Paris during this period [Gispert & Leloup 2009, pp. 90–94].

[87]For more on ISUP and IHP, see [Siegmund-Schultze 2001, Pavé 2002, Meusnier 2006, Siegmund-Schultze 2009].

[88]"*Les Questions actuelles: études de culture générale*" (6 volumes published in 1921), as well as "*Les Questions du temps présent*" (3 volumes in 1922–23; relaunched in 1930).

also served there as a spokesman for a more general movement of scientists intent on developing public research.

In 1922, the "Office for the Centralization of Research" dreamed up by Painlevé and Borel was established under a very different guise. Named the National Office for Scientific and Industrial Research and Invention (*Office national des recherches scientifiques et industrielles et des inventions*, ONRSII), it was headed by the wartime Undersecretary for Inventions, Jean-Louis Breton. Subsuming the older CRS, the Office however focused on invention rather than on scientific research (except for physicist Aimé Cotton's great electromagnet lavishly sponsored in Bellevue). A new tax, called "the penny of the lab (*sou du laboratoire*)," was however voted on in 1925 and quickly became the main source of funding for public scientific research.[89]

At the time many expressed fear that not enough scientific personnel were being trained to replace the fallen and face the nation's new challenges. In 1923, Picard explained that financial worries were not the most important threat for scientific research:

> We have to accept that we are living through hard times, and, for that matter, for the business of science, financial worries, whatever their realities, are not the gravest; the shrinking of the number of workers in the field of disinterested science is the most worrisome point.[90]

After the war, Borel thus carried his social and cultural activities to another scale and his engagement became institutional. Many of the younger mathematicians active in war-related research in WWI also took on the task assigned to them by their elders of (re)building higher education and research institutions all over France. The interwar period witnessed a change in the relation between Paris and the provinces, characterized by a strong desire to develop in a more durable manner provincial mathematical outposts. Alone among his peers, Gaston Julia, severely disabled during the war, was able to get a professorship in Paris in 1923. Other brilliant young mathematicians were sent to the provinces with the mission of reconstructing, nay reconquering, them. There, they greatly impelled innovative teaching and new research trends and energetically renovated institutions. Albert Châtelet's example is instructive.[91] One of the few number theorists in France before WWI, he was named in 1921 dean of the Science Faculty of the University of Lille, in a region that had greatly suffered from the war. In 1924, Châtelet became rector of the university. In these roles, he established several applied science institutes for the coal industry, fluid mechanics, and agriculture. Besides Châtelet, one may mention Joseph Kampé de Fériet, also at the University of Lille, Henri Eyraud at the University of Lyons, and Maurice Fréchet, Ernest Esclangon, and Georges Valiron at the University of Strasbourg. Some of them (but not all) would come

[89] On Borel's role in Parliament, see [Pinault 2006]. The Office is often seen as an ancestor of the Centre national de la recherche scientifique (CNRS) established by Borel's friend Jean Perrin in 1939 [Picard 1991, Moissinac & Roussel 2010, Guthleben 2011]. On Cotton, see [Shinn 1993].

[90] *Il faut se résigner aux temps difficiles où nous vivons, et d'ailleurs, dans l'ordre scientifique, ce ne sont pas les préoccupations financières, si réelles qu'elles soient, qui sont les plus graves ; la diminution du nombre de travailleurs dans le domaine de la science désintéressée est le point le plus inquiétant.*" Picard to Villat, July 28, 1923 [Villat Papers].

[91] See Sébastien Gauthier's chapter in [Goldstein & Aubin forthcoming], as well as [Condette 2009].

back to Paris around WWII to exert influential positions. The war thus created a generation of skillful organizers of French higher education and the research system, who would also endeavor to rebuild international relations.

3.2. Internationalizing Mathematics. After the war, Paris naturally sought to reclaim its former status as one of the world's mathematical capitals. What set the postwar apart from earlier times, in French eyes, was the thorough disqualification of German centers and the necessity to build a strong, viable countermodel to Göttingen. Opening the Paris Inter-Allied Conference, Painlevé acknowledged that there was an "antinomy" in the exclusion of scientists of the Central Powers from international organizations:

> for there is just one geometry, just one physics; there is not a German geometry and a French geometry, and it is certainly true that, if tomorrow a German hated by us, who would have committed horrible actions, would make a discovery, if that discovery truly existed, it would be no less true.... But in the depth of mind, it is to us profoundly repugnant to sit side by side with men that have, we know it, done certain things, or accepted responsibility for them, although they now seem to feel no remorse for them.[92]

The repulsion some felt toward working with former enemies, and, for some, with the persons they felt were responsible for the deaths of their children, was quite strong. Picard, for instance, only accepted the idea that their "successors will see whether a sufficiently long time and a sincere repentance could permit mending the relations" but refused collaboration in the present (quoted in [Lehto 1998, p. 29]). On January 14, 1920, the board of the SMF accepted the reinstatement of those former members coming from enemy countries who had been excluded five years earlier, but only those formally making the demand: there was no German or Austrian member until WWII [Aubin et al. 2011, p. 191].

As is well known, the first postwar International Congress of Mathematicians was organized in the highly symbolic town of Strasbourg, just reintegrated into the French Republic. The attempt by French mathematicians to take the lead in the rebuilding of an international mathematical community verged on caricature. Mathematicians from defeated nations were of course not allowed to participate. Of the 200 delegates from 27 countries who attended the congress, 80 were French. On the board of the IMU elected on September 20, the presence of Paris mathematicians was overbearing: Jordan and Picard were elected as honorary presidents (out of 4), Appell as one of the 5 vice presidents, and Kœnigs (a mechanics professor from the Sorbonne) as general secretary. It was only the next day that nationals from former neutral countries were invited to join the IMU. Although Picard clung to the presidency of the IRC until 1931, however, the French supremacy and the exclusion

[92] "*il y a une sorte d'antinomie; car il n'y a qu'une géométrie, il n'y a qu'une physique; il n'y a pas une géométrie allemande et une géométrie française, et il est bien certain que si demain, un Allemand que nous détesterions, qui aurait commis un acte horrible, faisait une découverte, si cette découverte existait réellement, elle n'en serait pas moins vraie. . . . Mais il y a au fond de notre âme une répugnance profonde à venir demain nous trouver côte à côte avec des hommes qui ont, nous le savons, accompli certains actes, ou qui ont accepté la responsabilité de certains actes, et qui semblent aujourd'hui n'en éprouver aucun remords.*" *Conférence interalliée des académies scientifiques* (Novembre 26 to 29, 1918) [Acad. Sci. Arch., manuscrits isolés 1-J-25].

of Central Powers' scientists were already controversial, especially among American and British mathematicians.[93]

With the exclusion of most exchanges with German and Austrian colleagues, Paris tried to establish new scientific partnerships. It has to be underlined that the relations of Parisian mathematicians with their colleagues from future Allied countries had not always been especially tight before the war: reviewing in 1907 a book on set theory by William and Grace Chisholm Young, Lebesgue ironically mentioned a remark they made, according to which "as is well-known, the French are rarely well acquainted with English work" [Lebesgue 1907]. The list of SMF members shows that the number of Central and Eastern European mathematicians increased during the interwar period, and especially those from newly independent countries. The creation of the Polish journal, *Fundamenta Mathematicae*, in French and focusing on themes close to Lebesgue's analysis in particular, was perceived as a welcome testimony of this extended community of views in pure mathematics. More than ever, the citizens of these countries defended their doctoral theses in Paris, where a quarter of the students came from abroad.[94] The immigration of some mathematicians helped shape the Parisian research landscape in some areas: for example, from 1929, the director's assistant of the Fluid Mechanics Institute of the University of Paris was the Russian émigré Dimitri Riabouchinsky, who was instrumental in promoting the methods developed by Nikolay Yegorovich Zhukovsky and his students as opposed to those of the Göttingen engineer and mathematician Ludwig Prandtl.[95]

During the interwar period, Hadamard's seminar, however, was a meeting place for mathematicians coming from all over the world, personally, but above all thematically—and would welcome talks about Carl Siegel's or Emmy Noether's results in algebra and number theory as early as 1922.[96] There were a few other exceptions to the general rule of noninteraction with German scientists: we may note Albert Einstein's 1922 visit to Paris (despite the controversies it raised and the need of presenting Einstein as a Swiss citizen) and his interactions with Borel at the Society of Nations [Biezunski 1991, Guieu 1998]. The French also sent a delegation to the Second International Congress of Applied Mechanics in Zurich in 1926.

But a true reconstruction of links with German mathematics had to wait until the mid-1920s. With the help and support of their elders from the Sorbonne, the most brillant elements of the postwar generations at the ENS were granted traveling grants by the Rockefeller Foundation. Young French scientists traveled to Sweden, Switzerland, Italy, the United States, Denmark, Great Britain, Hungary, while, reciprocally, the Foundation sent young mathematicians from many countries to France [Siegmund-Schultze 2001, pp. 288–301]. Future members of the Bourbaki group overlap almost exactly with those who then chose to travel to Germany [Beaulieu 1989, Goldstein 2009]. Coming back from their study trip, these young

[93]On the International Congresses of Mathematicians after WWI and the various attempts to forge an International Mathematical Union, see [Lehto 1998, Lehto 2002].

[94]On doctoral students in Paris, see [Leloup 2009, ch. 2].

[95]On this, see Claudine Fontanon's chapter in [Goldstein & Aubin forthcoming] and [Aubin forthcoming].

[96]See announcements in the various *Bulletin de la société mathématique de France*, in the section "Vie de la société."

men would firmly oppose the idea that science could be "universal," yet specifically French. Paradoxically, the clean break from German research that was decided after the war thus favored its direct assimilation by a group of mathematicians of the postwar generations who proved to be much less critical of it than was usually the case for the previous generations.

3.3. Modernizing Mathematics. At first sight, the flurry of activities we have depicted in postwar Paris squares oddly with accounts that are usually given and emphasize its mathematical decline. Whereas Claude Chevalley emphasized wants—"the teaching of the modern developments of number theory and algebra was then totally lacking in Paris," he later wrote Arnaud Denjoy [Goldstein 2009]—others have produced accounts detailing what one could draw, as much as what was missing, from the Parisian mathematical scene. To a young mathematician from Poland like Szolem Mandelbrojt, who in no way was blind to its deficiencies, Paris still was attractive enough to settle there.[97]

The meaning of modern, modernity, and modernism in mathematics has now been largely debated.[98] The historiography usually associates it with the development of an axiomatic point of view, with the prominence of structures (like groups or fields) as the fundamental objects of investigation, with the promotion of internationalism against the claim of national particularities, and above all perhaps with the autonomy of mathematical research, that should be totally independent of the natural sciences and experience. In this understanding, the image of mathematics we have argued was dominant in prewar Paris would certainly appear as "countermodern." The approach usually described as "modernist" was pursued with great success in the postwar years in some mathematical centers in Germany, and Göttingen especially, before the wave of emigration due to the Nazi regime. From this point of view, the assimilation of these particular algebraic, structural methods by the young French mathematicians who were sent to Germany in the late 1920s has been generally understood as a key factor in the establishment of modern, Bourbakist ideals.[99]

But, before the outbreak of WWI, modernism had taken another path among Paris mathematicians. We have mentioned how the modern theory of functions was developed by Borel, Lebesgue, and Fréchet, among others. This modern theory of functions, as it was claimed to be, certainly shared many elements with the algebraic trend. These function theorists for example emphasized the need for a more rigorous foundation of basic concepts such as measure or (dis)continuity and for cutting away from banal intuition [Dugac 2003]. They never, however, argued for the self-sufficiency and complete autonomy of mathematics from the sciences. Although Borel and Fréchet devoted most of their research activity to probability and statistics thereafter, war only served to reinforce research trends that they had well started to follow in the early years of the twentieth century [Leloup 2009, ch. 5]. The new "*patrons*" of the Paris community, Montel, Borel, Hadamard, Julia,

[97]See [Mandelbrojt 1985], an interesting account of postwar mathematical Paris emphasizing both abundance and, of course, holes: "I must admit that after a few weeks [in Paris] I cried."

[98]On modernism, see the pioneering [Mehrtens 1990], as well as [Epple 1999, Sinaceur 2002, Gray 2008, Epple & Müller forthcoming]. This issue is also discussed in the first chapter to this volume.

[99]As discussed in [Leloup 2009], more general changes occur in various fields during the 1930s, with a larger assimilation of techniques from foreign mathematicians.

Villat, or Élie Cartan, as identified in [Gispert & Leloup 2009], all shared a strong analytic basis, but neither purity of methods nor extreme specialization were their priority, as witnessed by their varied posterity.

The model of Poincaré, who had emphasized the connections among the various domains of mathematics, became even more inspirational for the younger generation than it had been before the war. Rediscovering him on his hospital bed, Julia defended a doctoral thesis in number theory before working in pure analysis while underscoring the geometrical aspects of his studies. Throughout, Julia claimed to follow "precise, fruitful, clear" paths "*à la française*"—and stayed clear from any axiomatic temptation.[100] When Lebesgue reviewed Louis Antoine's thesis on topology, fittingly defended in Strasbourg in 1921, he lauded on a full page (controlled) geometric intuition as a guide in the construction of a proof. Arguing explicitly against the trend for arithmetization, he explained that by pushing the war-wounded Antoine toward his original research, war had given rise to new mathematicians whose task was to "maintain and develop French scientific production" [Lebesgue 1922]. When Denjoy, in his own *Notice sur travaux* from 1921, defended the study of singularities as the proper object of modern analysis, he linked it to a French tradition: "there will be, it seems to me, a real renunciation, from the part of French Analysis, freely to disdain a movement of ideas which it started" [Denjoy 1921, p. xiv].[101]

How could French scientists' wartime contributions be integrated into these modern currents in mathematics? Both the wish to do so and the difficulty of the task can be fittingly perceived with the example of ballistics, in a report read by Hadamard before the Paris Academy of Sciences on February 23, 1920. For him, contributions had come from "all kinds of researchers," not only mathematicians, physicists, astronomers, military engineers, but also drafted engineers previously working in industry. As a result, solutions were very diverse but often redundant: "nearly all [problems] were simultaneously solved on many sides" [Hadamard 1920, p. 437]. From a strictly mathematical point of view, however, those solutions seemed wanting: "the articles in question for the most part contribute, not to the progress of scientific knowledge concerning its principles, but rather technical modifications aiming at more or less handy applications of those principles in special practical circumstances" [Hadamard 1920, p. 437].

Contributions that seemed most interesting to Hadamard at a mathematical level also were the least useful for ballistics. The mathematicians Jules Drach and Denjoy, the latter while he was at Gâvre (Figure 10), applied Galois theory to the study of differential equation. Joseph Kampé de Fériet, author of a doctoral thesis on hypergeometric functions defended on April 16, 1915, was able, after he was sent by Borel to Gâvre, to use these functions to express the solution of the hodograph in the case when air drag was of the form of kv^n, for any positive integer

[100]For more on Julia, see [Goldstein 2009] and Catherine Goldstein's contribution in [Goldstein & Aubin forthcoming]. Quotes come from [Julia 1970].

[101]This position was supported by an idea largely shared just after the war that "French" could mean "universal." At the conference for the fiftieth anniversary of the SMF, for instance, the Belgian mathematician and president of the IMU, Charles de la Vallée Poussin, who had found refuge in various countries after the invasion of Belgium and had given courses at the Sorbonne in 1916–1917, claimed that: "The [French] Mathematical Society is not international; it is universal like the idea that it nurtures and like French thought" [SMF 1924, p. 35]; for this event and its political undertones, see [Aubin et al. 2011, p. 192].

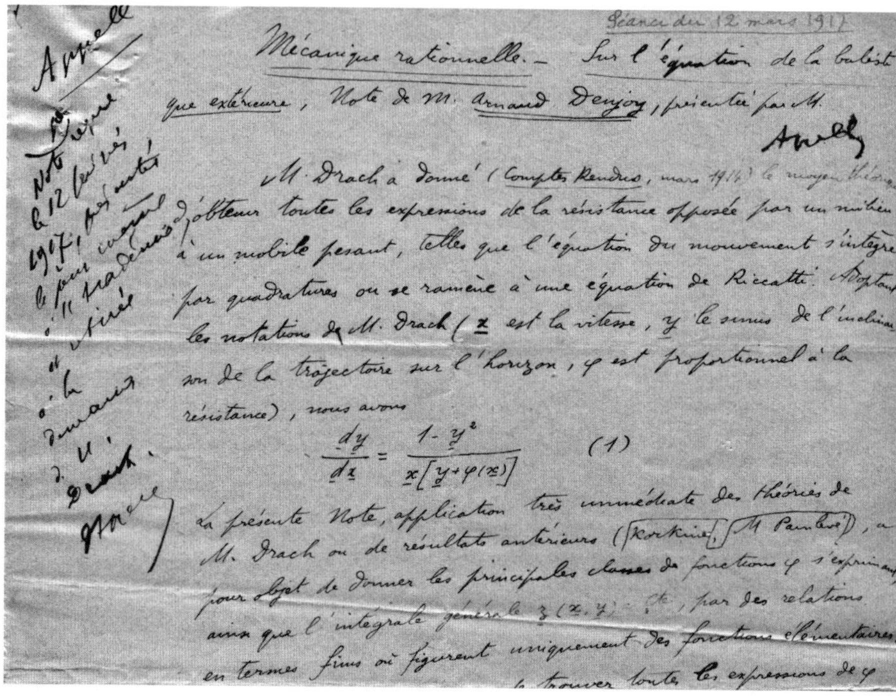
Séance du 12 mars 1917

Mécanique rationnelle. — Sur l'équation de la balistique extérieure, Note de M. Arnaud Denjoy, présentée par M. Appell

M. Drach a donné (Comptes Rendus, mars 1914) le moyen théorique d'obtenir toutes les expressions de la résistance opposée par un milieu à un mobile pesant, telles que l'équation du mouvement s'intègre par quadratures ou se ramène à une équation de Riccati. Adoptant les notations de M. Drach (x est la vitesse, y le sinus de l'inclinaison de la trajectoire sur l'horizon, φ est proportionnel à la résistance), nous avons

$$\frac{dy}{dx} = \frac{1-y^2}{x[y+\varphi(x)]} \qquad (1)$$

La présente Note, application très immédiate des théories de M. Drach ou de résultats antérieurs ([Korkine], [M. Painlevé]), a pour objet de donner les principales classes de fonctions φ s'exprimant, par des relations ainsi que l'intégrale générale $z(x, y)$ … en termes finis où figurent uniquement des fonctions élémentaires. … à trouver toutes les expressions de φ

FIGURE 10. Arnaud Denjoy's manuscript of a note titled "On the Equations of Exterior Ballistics," received by the CRAS on February 12, 1917, retracted at Jules Drach's request. With Permission of the Academie des sciences–Institut de France.

n. As detailed in David Aubin's chapter in this volume, mathematicians had played a special part in evaluating the expediency of computing procedures and allowed speeding them up considerably.

A final remark about Hadamard's report on ballistics concerns the status of mathematical research as opposed to other scientific disciplines. Following the critical assessment of the railway engineer Hippolyte Parodi and of Lebesgue, the report concurred in judging that the main scientific interest of ballistics now lay not in mathematical analysis of the equation or in computing procedures but rather in the physical analysis of the air drag problem. It was in this field that the mathematician saw "a theoretical fact … gained by Science in the course of these years of war" [Hadamard 1920, p. 442]. Studies about the effect of temperature on the velocity of sound and its impact on air drag at high velocities by the engineer Georges Darrieus and the physicist Paul Langevin were "the first step toward a truly rational aerodynamics and exterior ballistics" [Hadamard 1920, p. 443].[102]

[102]Other notable contributions to ballistics that raised interesting problems at a mathematical level include work on functions of lines [Bliss 1920], on the qualitative analysis of the equations of ballistics [Grönwall 1920] (on this, see [Gluchoff 2005]), and on approximations inspired by Poincaré's work in celestial mechanics by Kyrill Popoff (on the work of this Bulgarian mathematician and astronomer, see [Mazliak 2012]).

This "truly rational" applied (or at least applicable) mathematics embodied many French mathematicians' hopes just after the war: in order to avoid the double criticism they addressed to German mathematics before and during the war, that is, to be both too abstract and too utilitarian, they claimed that the theoretization of concrete problems alone was truly helpful for solving them efficiently. The fruitful alliance between mathematics and both rational and experimental mechanics epitomized by Painlevé's work on aviation before the war could serve as a model. As mentioned, applied mathematics was promoted by institutions and private sponsors, while many engineers working in industry had joined the SMF. After the war, this looked like an especially promising line of development for mathematics. In the ceremonies for the fiftieth anniversary of the SMF, President Poincaré thus marvelled at the way in which mathematics was "call[ed] in the factory, on construction sites, in arsenals" [SMF 1924, p. 64]. Uttered by a politician, this point of view was reinforced by all the mathematicians who spoke on the same occasion about the various applications of mechanics, the history of computing machines, and mathematics applied to engineering. Mathematicians (like Kampé de Fériet and Jules Haag) who had worked in pure analysis exclusively before the war were now producing original contributions to probability theory and fluid mechanics. Tooling up and modernizing industry appeared to them as a fitting task following their wartime engagement. But this, of course, drastically set them apart from those of their younger successors for whom pure mathematics was the mathematician's only thinkable horizon.

Conclusion

For Parisian mathematicians, just as it was for the population at large, World War I was of course a profound break at several levels: emotional, intellectual, and social. But most of these mathematicians were also convinced that the values they held before the war had prepared them well to serve as leaders of the scientific mobilization of the warring nation: in particular, their deep-rooted belief that mathematics was based on experience more than logic and axioms and that it was, and should be, closely related to other branches of science and to applications.

In the course of the conflict, however, a variety of problems occurred both at organizational and mathematical levels, and the mathematicians found that they were completely successful in their attempts at addressing them. In elite institutions like the ENS, young mathematicians suffered from an especially high number of casualties. While computing was shown to be crucial to the conduct of war, the type of mathematics that was effectively mobilized remained rather elementary (as shown in the case of sound-ranging), and, in the end, contributions from other fields of science, most notoriously chemistry, but also physics, looked perhaps more effective to the wider public. Mathematicians, nevertheless, still held influential positions at the Sorbonne, at the Academy of Sciences, or in international organizations, although Paris mathematicians would be partially side-stepped from some emerging, richly endowed decision centers in government.

In many respects, the postwar situation of mathematical Paris can nonetheless be seen as being in continuity with the war experience. Informal links activated during and because of the war often served as a basis for institutional reorganization, or at least for filling in positions created by it. It is also striking that a large part of this reorganization operated inside traditional institutions, like the

Sorbonne in particular, and not outside or independently of them. The way professoral chairs were redefined together with the teaching related to them and the way new institutes were attached to the Sorbonne (ISUP, IHP, or IMF) underlined change in continuity.[103]

Continuity with the war experience could translate, for some mathematicians, into just a slight reorientation of their prewar mathematical research and, for others, into a more obvious break. Analysis—extended from topology to functional analysis or applied differential equations—flourished after the war and Paris mathematicians were able to attract research funds and students in fields of application, such as probability theory, statistics, fluid mechanics, and mathematical physics. In this respect, Paris mathematical life in the interwar period was not a desert, nor a homogeneous place [Leloup 2009].

As is well known, a part of the generation trained after the war would on the contrary stress the vacuum it provoked and reduce the effect of WWI to the destruction of valuable lives and broken promises.[104] This emphasis on rupture rather than continuity is one of the paradoxes created by the reconstruction process which sent several important mathematicians of the war generation outside the capital for periods longer than in the past (and with more administrative duties). In the 1930s, the younger generation was encouraged to participate in the renovation process, especially by studying abroad. Some of them, then, would insist on the purification of methods and disciplinary boundaries. They would dissociate themselves from the mathematical and social values of their elders, anchor mathematics in structures, detach it from what they would consider as outside considerations, either applications or institutional activities, and show contempt for the very idea of French science [Weil 1992].

During the interwar period, however, some mathematicians of the war generation tried to further extend the bridges their experience in the conflict had convinced them they should build. The opinion is expressed, for instance, by Georges Bouligand, an ENS student just before the war, who was sent to the University of Poitiers in 1921.[105] In 1932, the very same year the future Bourbaki Claude Chevalley worked toward the elimination of analytic means from class field theory, Bouligand argued that fluid mechanics could also be seen as posing a modern set-theoretical problem:

> It is remarkable that, thus posed, the question [cavitations responsible for air bumps] connects a crucial practical interest to a high philosophical one. It indeed provides the familiar concepts of set theory with the opportunity of entering hydrodynamics and playing a very active role in it. The difficulties met in treating the problem in complete generality are especially instructive;

[103]In several towns, there were debates about the possibility of adjusting old structures or creating new ones from scratch; local relations with potential founding managers were often instrumental in the decisions; see [Mounier-Kuhn 1996, Fontanon 1998, Condette 2009].

[104]This view, which was put forward by many founders of Bourbaki, is discussed in the first chapter to this volume.

[105]Poitiers, in the center of France, was at one time perceived as a key place for the reconstruction of French mathematics and a possible French equivalent of Göttingen; see [Siegmund-Schultze 2001].

they are witness to the advances that are still required in order to harmonize the most classical parts of Fluid Mechanics with the most recent tendencies [of pure mathematics] [Bouligand 1932, p. 103].

Abbreviations

- CNRS: *Centre national de la recherche scientifique* (National Center for Scientific Research).
- AFAS: *Association française pour l'avancement des sciences* (French Association for the Advancement of Science).
- CRAS: *Comptes rendus hebdomadaires des séances de l'Académie des sciences.*
- CRS: *Caisse des recherches scientifiques* (Scientific Research Fund).
- ENS: École normale supérieure, Paris.
- ICM: International Congress of Mathematicians.
- IHP: Institut Henri Poincaré, Paris.
- IMF: Institut de mécanique des fluides (Institute of Fluid Mechanics).
- IMU: International Mathematical Union.
- IRC: International Research Council.
- ISUP: *Institut de statistiques de l'université de Paris* (Paris University Statistical Institute).
- ONRSII: *Office national des recherches scientifiques et industrielles et des inventions* (National Office for Scientific and Industrial Research and Inventions).
- SMF: *Société mathématique de France* (French Mathematical Society).
- SRS: sound-ranging sections.

Bibliography

Archival Sources

[Acad. Sci. Arch.] Archives of the Academy of Sciences, Paris.

[Arch. nat.] Archives nationales (National Archives), Paris.

[Le Verrier Papers] Le Verrier Papers, Bibliothèque de l'Institut de France.

[CDN Archives] Archives des commissions de la défense nationale, Archives of the Academy of Sciences, Paris.

[Ocagne, n.d.] Maurice d'Ocagne, *Souvenirs de Maurice d'Ocagne, membre de l'Académie des Sciences, 1862–1938*, Manuscript at the Bibliothèque de l'Institut de France, Quai Conti, Paris, Ms. 4671–4674, n.d.

[Villat Papers] Fonds Henri Villat, Archives of the Academy of Sciences, Paris, 61 J.

Published Works

[Anizan 2006] Anne-Laure Anizan, *Paul Painlevé (1863–1933) : un scientifique en politique*, doctoral thesis, Institut d'études politiques, Paris, 2006.

[Anizan 2012] Anne-Laure Anizan, *Paul Painlevé : science et politique de la Belle Époque aux années trente*, Presses universitaires de Rennes, Rennes, 2012.

[Anonymous 1900] *Le Procès Dreyfus devant la Conseil de guerre de Rennes*, 3 vols., P.-V. Stock, Paris, 1900.

[Appell 1914a] Paul Appell, *Discours du président de l'Académie des sciences*, Comptes rendus hebdomadaires des séances de l'Académie des sciences **158** (1914), pp. 17–18.

[Appell 1914b] Paul Appell, *Le conflit de deux civilisations*, Revue scientifique **52**, 2e sem. (1914), p. 173.

[Appell 1914c] Paul Appell, *Séance publique annuelle du lundi 21 décembre 1914. Allocution*, Comptes rendus hebdomadaires des séances de l'Académie des sciences **159** (1914), pp. 821–831.

[Appell 1922] Paul Appell, *Éducation et enseignement : notices et discours*, Félix Alcan, Paris, 1922.

[Appell 1923] Paul Appell, *Souvenirs d'un Alsacien, 1858–1922*, Payot, Paris, 1923.

[Aubin 1997] David Aubin, *The Withering Immortality of Nicolas Bourbaki: A Cultural Connector at the Confluence of Mathematics, Structuralism and the Oulipo in France*, Science in Context **10** (1997), pp. 297–342.

[Aubin 2009] David Aubin, Observatory Mathematics in the Nineteenth Century, in *Oxford Handbook for the History of Mathematics*, ed. Eleanor Robson & Jacqueline Stedall, Oxford University Press, Oxford, pp. 273–298.

[Aubin 2012] David Aubin, L'élite sous la mitraille : les mathématiciens normaliens "morts pour la France", 1914–1918, in *Aventures de l'analyse de Fermat à Borel. Mélanges en l'honneur de Christian Gilain*, ed. Suzanne Féry, Presses universitaires de Nancy, Nancy, 2012, pp. 681–706.

[Aubin forthcoming] *"Audacity or Precision": The Paradoxes of Henri Villat's Fluid Mechanics in Interwar France*, paper presented at a "Workshop on the History of Fluid Mechanics", Rauischholzhausen, October 15–18, 2006; preprint available at `http://hal.upmc.fr/docs/00/63/98/77/PDF/Villat_paper2010-05-14.pdf`.

[Aubin & Bret 2003] David Aubin & Patrice Bret, eds., *Le Sabre et l'éprouvette: l'invention d'une science de guerre, 1914–1939*, "14–18 Aujourd'hui," Agnès Viénot, Paris, 2003.

[Aubin et al. 2011] David Aubin, Hélène Gispert & Catherine Goldstein, Les mathématiciens français dans la Grande Guerre, in *1914–1918 Identités troublées : les appartenances sociales et nationales à l'épreuve de la guerre*, ed. François Bouloc, Rémy Cazals, & André Loez, Privat, Toulouse, 2011, pp. 183–197.

[Audouin-Rouzeau & Becker 2000] Stéphane Audouin-Rouzeau & Annette Becker, *14–18 : Retrouver la guerre*, Gallimard, Paris, 2000.

[Auvinet 2013] Jérôme Auvinet, *Charles-Ange Laisant. Itinéraires et engagements d'un mathématicien de la Troisième République*, Hermann, Paris, 2013.

[Baire 1905] Henri Baire, *Leçons sur les fonctions discontinues*, Gauthier-Villars, Paris, 1905.

[Beaulieu 1989] Liliane Beaulieu, *Bourbaki : une histoire du groupe de mathématiciens français et de ses travaux (1934–1944)*, doctoral thesis, Université de Montréal, Montréal, 1989.

[Beaulieu 1993] Liliane Beaulieu, *A Parisian Café and Ten Proto-Bourbaki Meetings (1934–1935)*, The Mathematical Intelligencer, **15**(1) (1993), pp. 27–35.

[Beaulieu 1994] Liliane Beaulieu, Questions and Answers about Bourbaki's Early Work (1934–1944), in *The Intersection of History and Mathematics*, ed. Sasaki Chikara, Sugiura Mitsuo, & Joseph W. Dauben, Birkhäuser, Basel, 1994, pp. 241–252.

[Beaulieu 2009] Liliane Beaulieu, ed., *Regards sur les mathématiques françaises entre les deux guerres*, special issue of Revue d'histoire des sciences **62-1** (2009).

[Beaunier 1909] André Beaunier, *Le Krach des mathématiques*, Le Figaro (supplément littéraire), (January 30,1909), p. 1.

[Belhoste et al. 1996] Bruno Belhoste, Hélène Gispert, & Nicole Hulin, eds., *Les Sciences au lycée. Un siècle de réformes des mathématiques et de la physique en France et à l'étranger*, Vuibert & INRP, Paris, 1996.

[Biezunski 1991] Michel Biezunski, *Einstein à Paris : le temps n'est plus...*, Presses universitaires de Vincennes, Saint-Denis, 1991.

[Bkouche 1991] Rudolph Bkouche, Variations autour de la réforme de 1902/1905, in [Gispert 1991, pp. 181–213].

[Bliss 1920] Gilbert A. Bliss, *Functions of Lines in Ballistics*, Transactions of the American Mathematical Society **21** (1920), pp. 93–106.

[Borel 1904] Émile Borel, *Les exercices pratiques de mathématiques dans l'enseignement secondaire*, Revue générale des sciences **10** (1904), pp. 431–440; repr. in *Gazette des mathématiciens* **93** (2002), pp. 47–64.

[Borel 1906] Émile Borel, *La graphologie et la méthode scientifique*, Revue du mois **2** (1906), pp. 244–250 & 366–369.

[Borel 1907] Émile Borel, *Logique et intuition en mathématiques*, Revue de métaphysique et de morale **15** (1907), pp. 273–283.

[Borel 1911] Émile Borel, ed., *De la méthode dans les sciences, deuxième série*, Félix Alcan, Paris, 1911; repr. 1919.

[Borel 1914] Émile Borel, *L'Infini mathématique et la réalité*, Revue du mois **18** (1914), pp. 71–84.

[Borel 1918] Émile Borel, *Supplément sommaire (avril 1918) à la Notice (1912) sur les travaux scientifiques*, Gauthier-Villars, Paris, 1918; repr. in *Œuvres complètes d'Émile Borel*, vol. 1, Éditions du CNRS, Paris, 1972, pp. 192–194.

[Borel 1925] Émile Borel, *Organiser*, Félix Alcan, Paris, 1925.

[Born 1978] Max Born, *My Life: Recollections of a Nobel Laureate*, Taylor & Francis, London, 1978.

[Bouligand 1932] Georges Bouligand, *Les courants de pensée cantorienne et l'hydrodynamique, ou le problème de la naissance des cavitations dans un liquide*, Revue générale des sciences pures et appliquées **42** (1931), pp. 103–110.

[Boutroux 2012] Aline Boutroux, *Vingt ans de ma vie, simple vérité. La jeunesse d'Henri Poincaré racontée par sa sœur (1854–1878)*, ed. Laurent Rollet, Hermann, Paris, 2012.

[Boutroux 1914] Émile Boutroux, *L'Allemagne et la guerre*, Revue des deux mondes **23** (1914), pp. 385–401.

[Bredin 1986] Jean-Denis Bredin, *The Affair: The Case of Alfred Dreyfus*, George Braziller, New York, 1986.

[Bret 2008] Patrice Bret, *Les laboratoires français et l'étude des munitions et matériels allemands pendant la Grande Guerre*, Cahiers du CEHD **33** (2008), pp. 7–32.

[de Broglie 1938] Maurice de Broglie, *Discours aux funérailles de Maurice d'Ocagne*, Académie des sciences, online: `http://www.academie-sciences.fr/activite/archive/dossiers/eloges/ocagne_notice.pdf` (accessed March 29, 2013).

[Charbonnier 1906] Prosper-Jules Charbonnier, *Historique de la balistique extérieure à la commission de Gâvre*, Revue maritime et coloniale, vol. 168, 1906, pp. 411–476.

[Charle 1994] Christophe Charle, *La République des universitaires, 1870–1914*, Le Seuil, Paris, 1994.

[Chickering & Förster 2006] Roger Chickering & Stig Förster, eds., *Great War, Total War: Combat and Mobilization on the Western Front, 1914–1918*, Cambridge University Press & The German Historical Institute, Cambridge, 2006.

[Christophe 1900] Christophe [Georges Colomb], *L'Idée fixe du savant Cosinus*, Armand Colin, Paris, 1900.

[Claude 1919] Georges Claude, *Politiciens et polytechniciens : Au creuset de la guerre*, self-publ., Boulogne-sur-Seine, 1919.

[Cohen 1984] Paul Michael Cohen, *Piety and Politics: Catholic Revival and the Generation of 1905–1914 in France*, University of Chicago, Chicago, 1984.

[Collective 1918] *La Vie universitaire à Paris*, Librairie Armand Colin, Paris, 1918.

[Condette 2009] Jean-François Condette, *Albert Châtelet, La République par l'école (1883–1960)*, Artois Presse Université, Arras, 2009.

[Corry 2004] Leo Corry, *Modern Algebra and the Rise of Mathematical Structures*, 2nd ed., Birkhäuser, Basel, 2004.

[Cotton 1949] Aimé Cotton, *Discours de réception de Jean Cabannes à l'Académie des Sciences* (June 9, 1949), Académie des Sciences, online `http://www.lapasserelle.com/jean_cabannes.htm` (last accessed March 29, 2013).

[Dahan Dalmedico & Bottazzini 2001] Amy Dahan Dalmedico & Umberto Bottazzini, eds., *Changing Images in Mathematics: From the French Revolution to the New Millennium*, Routledge, London, 2001.

[Darboux 1914] Gaston Darboux, *Henri Poincaré*, Revue scientifique **52**, 2nd sem. (1914), pp. 97–110.

[Daudet 1918] Léon Daudet, *La Guerre totale*, Nouvelle Librairie nationale, Paris, 1918.

[Denjoy 1921] Arnaud Denjoy, *Notice sur les travaux scientifiques de Monsieur Arnaud Denjoy*, Privat, Toulouse, 1921.

[Dierig et al. 2003] Sven Dierig, Jens Lachmund, & Andrew Mendelsohn, eds., *Science and the City*, special issue, Osiris **18** (2003).

[Dieudonné 1969] Jean Dieudonné, *Regards sur Bourbaki*, Analele universitatii Bucaresti, Matematica-mecanica **18** (1969), pp. 13–25.

[Duclert 1999] Vincent Duclert, *L'engagement scientifique et l'intellectuel démocratique : le sens de l'affaire Dreyfus*, Politix **12**, no. 48 (1999), pp. 71–94.

[Duclert 2006] Vincent Duclert, ed., *Savoir et engagement : écrits normaliens sur l'affaire Dreyfus*, ENS éditions, Paris, 2006.

[Dugac 2003] Pierre Dugac, *Histoire de l'analyse : autour de la notion de limite et de ses voisinages*, Vuibert, Paris, 2003.

[Duhem 1915] Pierre Duhem, *La Science allemande*, Hermann, Paris, 1915.

[Duhem 1991] Pierre Duhem, *The Aim and Structure of Physical Theory*, Princeton University Press, Princeton, 1991.

[Durand et al. 2013] Antonin Durand, Laurent Mazliak, & Rossana Tazzioli, eds., *Des mathématiciens et des guerres*, CNRS-Alpha, Paris, 2013.

[Durkheim 1915] Émile Durkheim, *L'Allemagne au dessus de tout : la mentalité allemande et la guerre*, Armand Colin, Paris, 1915; repr. 1991.

[Epple 1999] Moritz Epple, *Die Entstehung der Knotentheorie. Kontexte und Konstruktionen einer modernen mathematischen Theorie*, Vieweg, Braunschweig and Wiesbaden, 1999.

[Epple & Müller forthcoming] Moritz Epple & Falk Müller, eds., *Science as Cultural Practice*, vol. 2: *Modernism in the Sciences*, Akademie Verlag, Berlin, forthcoming.

[Esclangon 1925] Ernest Esclangon, *L'Acoustique des canons et des projectiles : acoustique géométrique, ondes de bouche des canons et ondes balistiques, propriétés géométriques, cas des bolides et des étoiles filantes ; acoustique physique, l'audition physiologique et l'analyse auditive, sons, bruits, détonations, infra-sons, notions auditives de direction et de distance, infra-sons engendrés par les canons et les appareils de détection, réfraction et réflexion atmosphériques des ondes sonores, zones de silence, l'acoustique et la balistique*, Mémorial de l'artillerie française, suppl., Paris, 1925.

[Fontanon 1998] Claudine Fontanon, La naissance de l'aérodynamique expérimentale et ses applications à l'aviation : une nouvelle configuration socio-technique, in *Histoire de la mécanique appliquée: Enseignement, recherche et pratiques mécaniciennes en France après 1880*, ed. Claudine Fontanon, SFHST & ENS éditions, Paris, 1998, pp. 57–88.

[Fontanon 2005] Claudine Fontanon, L'obus Chilowski et la soufflerie balistique de Paul Langevin : une recherche militaire oubliée de la mobilisation scientifique (1915–1919), in *Deux siècles d'histoire de l'armement en France : de Gribeauval à la force de frappe*, ed. Dominique Pestre, CNRS, Paris, 2005, pp. 81–109.

[Fontanon 2006] Claudine Fontanon, Painlevé et l'aviation: à l'origine de la recherche publique, in [Fontanon & Franck 2006, pp. 41–56].

[Fontanon & Franck 2006] Claudine Fontanon & Robert Franck, eds., *Paul Painlevé : Un savant en politique*, Presses universitaires de Rennes, Rennes, 2006.

[Fox 1995] Robert Fox, L'attitude des professeurs des facultés des sciences face à l'industrialisation en France entre 1850 et 1914, in *Science, Industry, and Social Order in Post-Revolutionary France*, Variorum, Adelshot, 1995, pp. 133–147.

[France 1903] Anatole France, *Histoire comique*, Calman-Levy, Paris, 1903; repr. 1921.

[France 1909] Anatole France, *L'Île des pingouins*, Calman-Lévy, Paris, 1909.

[Galison 2003] Peter Galison, *Einstein's Clocks and Poincaré's Maps: Empires of Time*, W. W. Norton, New York, 2003.

[Galvez-Béhar 2008] Gabriel Galvez-Béhar, *La République des inventeurs : propriété et organisation de l'innovation en France (1791–1922)*, Presses universitaires de Rennes, Rennes, 2008.

[Ginoux & Gérini 2012] Jean-Marc Ginoux & Christian Gérini, *Henri Poincaré : une biographie au(x) quotidien(s)*, Ellipses, Paris, 2012.

[Gispert 1991] Hélène Gispert, ed., *La France mathématique: La Société mathématique de France (1870–1914)*, SFHST & SMF, Paris, "Cahiers d'histoire et de philosophie des sciences," **34**, 1991; 2nd ed., 2014.

[Gispert 1994] Hélène Gispert, De Bertrand à Hadamard : quel enseignement d'analyse pour les polytechniciens?, in *La Formation polytechnicienne 1794–1994*, ed. Bruno Belhoste, Amy Dahan, & Antoine Picon, Dunod, Paris, 1994, pp. 181–198.

[Gispert 1995] Hélène Gispert, *La théorie des ensembles en France avant la crise de 1905 : Baire, Borel, Lebesgue... et tous les autres*, Revue d'histoire des mathématiques **1** (1995), pp. 39–81.

[Gispert 2002] Hélène Gispert, ed., *Par la science, pour la patrie : l'Association française pour l'avancement des sciences, 1872–1914*, Presses universitaires de Rennes, Rennes, 2002.

[Gispert 2012] Hélène Gispert, L'entreprise biographique à l'épreuve : écueils, défis, atouts du cas d'Émile Borel, in [Rollet & Nabonnand 2012b, pp. 139–175].

[Gispert et al. 2007] Hélène Gispert, Nicole Hulin, & Marie-Claire Robic, eds., *Science et enseignement. L'exemple de la grande réforme des programmes du lycée au début du XX*[e] *siècle*. Vuibert & INRP, Paris, 2007.

[Gispert & Leloup 2009] Hélène Gispert & Juliette Leloup, *Des patrons des mathématiques en France dans l'entre-deux-guerres*, Revue d'histoire des sciences **62-1** (2009), pp. 39–118.

[Gluchoff 2005] Alan Gluchoff, *Pure Mathematics Applied in Early Twentieth-Century America: The Case of T. H. Gronwall, Consulting Mathematician*, Historia mathematica **32** (2005), pp. 312–337.

[Goldstein 2009] Catherine Goldstein, *La théorie des nombres en France dans l'entre-deux-guerres : de quelques effets de la première guerre mondiale*, Revue d'histoire des sciences **62-1** (2009), pp. 143–175.

[Goldstein 2011] Catherine Goldstein, Un arithméticien contre l'arithmétisation : les principes de Charles Hermite, in *Justifier en mathématiques*, ed. Dominique Flament & Philippe Nabonnand, MSH, Paris, 2011, pp. 129–165.

[Goldstein & Aubin forthcoming] Catherine Goldstein & David Aubin, eds., *La Grande Guerre des mathématiciens français*, forthcoming.

[Gray 2008] Jeremy Gray, *Plato's Ghost: The Modernist Transformation of Mathematics*, Princeton University Press, Princeton, 2008.

[Grattan-Guinness 2000] Ivor Grattan-Guinness, *The Search for Mathematical Roots 1870–1940: Logic, Set Theory, and the Foundations of Mathematics from Cantor through Russell to Gödel*, Princeton University Press, Princeton, 2000.

[Grönwall 1920] Thomas Haken Grönwall, *Qualitative Properties of the Ballistic Trajectory*, Annals of Mathematics **22** (1920), pp. 44–65.

[Guiomar 2004] Jean-Yves Guiomar, *L'Invention de la guerre totale (XVIII*[e]*-XX*[e] *siècles)*, Éditions du Félin, Paris, 2004.

[Guieu 1998] Jean-Michel Guieu, *L'engagement européen d'un grand mathématicien français : Émile Borel et la "coopération européenne", des années vingt aux années quarante*, Bulletin de l'Institut Pierre Renouvin **5** (1998), online: `http://www.univ-paris1.fr/autres-structures-de-recherche/ipr/les-revues/bulletin/tous-les-bulletins/bulletin-n-05/emile-borel-et-la-cooperation-europeenne/` (accessed December 30, 2013).

[Guiraldenq 1999] Pierre Guiraldenq, *Émile Borel (1871–1956)*, Saint-Affrique, 1999.

[Guthleben 2011] Denis Guthleben, *Rêves de savants : étonnantes inventions de l'entre-deux-guerres*, Armand Colin, Paris, 2011.

[Hadamard 1920] Jacques Hadamard, *Rapport sur les travaux examinés et retenus par la Commission de Balistique pendant la durée de la guerre*, Comptes rendus hebdomadaires des séances de l'Académie des sciences **170** (1920), pp. 436–445.

[Hadamard 1937] Jacques Hadamard, *Jubilé scientifique de M. Jacques Hadamard*, Gauthier-Villars, Paris, 1937.

[Hanna 1996] Martha Hanna, *The Mobilization of Intellect: French Scholars and Writers during the Great War*, Harvard University Press, Cambridge, Mass., 1996.

[Hinman 1919] Jesse R. Hinman, *Ranging in France with Flash and Sound*, Press of Dunham Printing Company, Portland, Oregon, 1919.

[Hottin 2001] Christian Hottin, *Quand la Sorbonne était peinte*, Maisonneuve & Larose, Paris, 2001.

[Huyon 2008] Alain Huyon, *La Grosse Bertha des Parisiens*, Revue historique des armées **253** (2008), online: `http://rha.revues.org/4682`.

[IRMAR 1988] Collective, *Journée Louis Antoine : 18 novembre 1988*, Publications de l'IRMAR, Rennes, 1988.

[Julia 1970] Gaston Julia, Cérémonie de remise de la médaille de l'Institut Mittag-Leffler à M. Émile Picard, in *Œuvres de Julia*, vol. 6, Gauthier-Villars, Paris, 1970, pp. 39–50.

[Kevles 1995] Daniel Kevles, *The Physicists: The History of a Scientific Community in Modern America*, 2nd ed., Harvard University Press, Cambridge, Mass., 1995.

[Lebesgue 1907] Henri Lebesgue, Review of Young & Young, *The Theory of Set Points*, Bulletin des sciences mathématiques **31** (1907), pp. 132–134.

[Lebesgue 1922] Henri Lebesgue, *Analyse de la thèse de Louis Antoine*, Bulletin des sciences mathématiques **46** (1922), pp. 5–12; repr. in Henri Lebesgue, *Œuvres scientifiques*, vol. 5, L'enseignement mathématique, Genève, 1972, pp. 352–359.

[Lebesgue 1991] Henri Lebesgue, *Lettres d'Henri Lebesgue à Émile Borel*, Cahiers du séminaire d'histoire des mathématiques **12** (1991), 1–506.

[Leblois 1929] Louis Leblois, *L'Affaire Dreyfus : l'iniquité, la réparation, les principaux faits et les principaux documents*, Aristide Quillet, Paris, 1929.

[Lebon 1912] Ernest Lebon, *Henri Poincaré : biographie, bibliographie analytique des écrits*, Gauthier-Villars, Paris, 1912.

[Lefort 2007] Jean Lefort, *Louis Antoine, géomètre aveugle*, Pour la Science **352** (2007), pp. 16–21.

[Lehto 1998] Olli Lehto, *Mathematics without Borders: A History of the International Mathematical Union*, Springer, New York, 1998.

[Lehto 2002] Olli Lehto, The Formation of the International Mathematical Union, in *Mathematics Unbound: The Evolution of an International Mathematical Research Community, 1800–1945*, ed. Karen Hunger Parshall & Adrian Rice, American Mathematical Society, Providence, 2002, pp. 381–395.

[Leloup 2004] Juliette Leloup, *Les Dynamiques de recherche mathématique dans l'entre-deux-guerres à partir de l'étude des thèses mathématiques soutenues en France*, master thesis, EHESS, Paris, 2004.

[Leloup 2009] Juliette Leloup, *L'Entre-deux-guerres mathématique à travers les thèses soutenues en France*, doctoral thesis, Université Pierre et Marie Curie, Paris, 2009.

[Lepick 1998] Olivier Lepick, *La Grande Guerre chimique 1914–1918*, Presses universitaires de France, Paris, 1998.

[Letté 2004] Michel Letté, *Henry Le Chatelier (1850–1936) ou la science appliquée à l'industrie*, Presses universitaires de Rennes, Rennes, 2004.

[MacLeod 2000] Roy MacLeod, *Sight and Sound on the Western Front: Surveyors, Scientists and the "Battlefield Laboratory," 1915–1918*, War and Society **18** (2000), pp. 23-46.

[MacLeod & Johnson 2006] Roy MacLeod & Jeffrey Johnson, eds., *Frontline and Factory: Comparative Perspectives on the Chemical Industry at War, 1914–1924*, Springer, Dordrecht, 2006.

[Mandelbrojt 1985] Szolem Mandelbrojt, *Souvenirs à bâtons rompus, recueillis en 1970 et préparés par Benoît Mandelbrot*, Cahiers du séminaire d'histoire des mathématiques **6** (1985), pp. 1–46.

[Mansuy & Mazliak 2005] Roger Mansuy & Laurent Mazliak, *Introduction au rapport de Poincaré pour le procès en cassation de Dreyfus en 1904*, Journal électronique d'histoire des probabilités et des statistiques **1**, n°1 (2005), online: `www.jehps.net`.

[Mansuy & Mazliak 2011] Roger Mansuy & Laurent Mazliak, L'analyse graphologique controversée d'Alphonse Bertillon dans l'affaire Dreyfus. Polémiques et réflexions autour de la figure de l'expert, in *Aux origines de la police scientifique. Alphonse Bertillon, précurseur de la science du crime*, ed. Pierre Piazza, Karthala, Paris, 2011, pp. 354–372.

[Marbo 1967] Camille Marbo [Marguerite Borel], *À travers deux siècles, souvenirs et rencontres (1883–1967)*, Grasset, Paris, 1967.

[Marin 1919] Louis Marin, *Historique de la Direction des recherches scientifiques et industrielles et des inventions*, Bulletin officiel de la Direction des recherches scientifiques et industrielles et des inventions **1** (1919), pp. 1–10.

[Mariot 2012] Nicolas Mariot, *Pourquoi les normaliens sont-ils morts en masse en 1914–1918? Une explication structurale*, ARPOS Pôle Sud **36** (2012), pp. 9–30.

[Mariot 2013] Nicolas Mariot, *Tous unis dans la tranchée? 1914-1918, les intellectuels rencontrent le peuple*, Seuil, Paris, 2013.

[Mashaal 2002] Maurice Mashaal, *Bourbaki : une société secrète de mathématiciens*, Belin, Paris, 2002.

[Maurain & Pacaud 1940] Charles Maurain & Anatole Pacaud, *La Faculté des sciences de l'université de Paris de 1906 à 1940*, Presses universitaires de France, Paris, 1940.

[Mawhin 2004] Jean Mawhin, *Henri Poincaré hors de sa tour d'ivoire : Dreyfus, Galilée et Sully-Prudhomme*, Bulletin de la Classe des sciences de l'Académie royale de Belgique (6) **15**, pp. 81–102.

[Mazliak & Tazzioli 2009] Laurent Mazliak & Rossana Tazzioli, eds., *Mathematicians at War. Correspondence between Volterra and his French Colleagues during World War I*, Springer, New York, 2009.

[Mazliak 2012] Laurent Mazliak, *Une étude de trajectoire. Kyrille Popoff, les guerres et la balistique*, Almagest **3.1** (2012), pp. 78–105.

[Maz'ya & Shaposhnikova 1998] Vladimir Maz'ya & Tatyana Shaposhnikova, *Jacques Hadamard, a Universal Mathematician*, American Mathematical Society & London Mathematical Society, Providence & London, 1998.

[Mehrtens 1990] Herbert Mehrtens, *Moderne, Sprache, Mathematik: Eine Geschichte des Streits um die Grundlagen der Disziplin und des Subjekts formaler Systeme*, Suhrkamp, Frankfurt, 1990.

[Meusnier 2006] Norbert Meusnier, *Sur l'histoire de l'enseignement des probabilités et des statistiques*, Journal électronique d'histoire des probabilités et des statistiques **2**, no. 2 (2006), online: `www.jehps.net`.

[Moissinac & Roussel 2010] Christine Moissinac & Yves Roussel, *Jules-Louis Breton, 1878–1940 : un savant parlementaire*, Presses universitaires de Rennes, Rennes, 2010.

[Mounier-Kuhn 1996] Pierre Mounier-Kuhn, Un programme technologique national : La mécanique des fluides, in *Villes et institutions scientifiques : Rapport pour le PIR-VILLES*, ed. Michel Grossetti, CNRS, Paris, 1996, pp. 187–190.

[Moureu 1924] Charles Moureu, *Maurice Barrès et la recherche scientifique*, Revue des deux mondes **22** (1924), pp. 376–392.

[Nordmann 1915] Charles Nordmann, *Science et guerre*, Revue des deux mondes **30** (1915), pp. 698–708.

[Nordmann 1928] Charles Nordmann, *Notice sur les travaux scientifiques*, typographie Philippe Renouart, Paris, 1928.

[Painlevé 1906] Paul Painlevé, *L'esprit scientifique et l'esprit religieux*, Revue du mois **2** (1906), pp. 658–668.

[Painlevé 1909] Paul Painlevé, Mécanique, in [Thomas 1909–1911, vol. 1, pp. 363–409].

[Painlevé 1916] Paul Painlevé, *L'École normale supérieure et la guerre*, Revue scientifique **54** (1916), pp. 193–195.

[Painlevé 1918a] Paul Painlevé, *Séance publique annuelle du lundi 2 décembre 1918. Allocution*, Comptes rendus des séances hebdomadaires de l'Académie des sciences **167** (1918), pp. 797–810.

[Painlevé 1918b] Paul Painlevé, *Séance du 7 janvier 1918. Discours*, Comptes rendus hebdomadaires des séances de l'Académie des sciences **166** (1918), pp. 17–19.

[Painlevé 1918c] Paul Painlevé, *Élection du Maréchal Foch. Allocution*, Comptes rendus hebdomadaires des séances de l'Académie des sciences **167** (1918), p. 718.

[Painlevé & Borel 1910] Paul Painlevé & Émile Borel, *L'Aviation*, Félix Alcan, Paris, 1910.

[Paul 1985] Harry W. Paul, L'idée de recherche dans les Facultés des sciences au XIXe siècle, in *Le Personnel de l'enseignement supérieur en France aux XIXe et XXe siècles*, ed. Christophe Charle & Régine Ferré, CNRS, Paris, 1985, pp. 219–227.

[Paul 1985] Harry W. Paul, *From Knowledge to Power: The Rise of the Scientific Empire in France, 1860–1939*, Cambridge University Press, Cambridge, 1985.

[Pavé 2002] Francis Pavé, *Du concept pur aux applications et retour : les tribulations de l'Institut Henri Poincaré et de l'Institut de la Statistique de l'Université de Paris*, Annales des mines **67** (2006), online: `www.annales.org`.

[Perrier 1915] Edmond Perrier, *Séance publique annuelle du lundi 27 décembre 1915. Allocution*, Comptes rendus hebdomadaires des séances de l'Académie des sciences **161** (1915), pp. 801–819.

[Pestre 1984] Dominique Pestre, *Physique et physiciens en France, 1918–1940*, Éditions des Archives Contemporaines, Paris, 1984.

[Pestre 1997] Dominique Pestre, *The Moral and Political Economy of French Scientists in the First Half of the 20th Century*, History & Technology **13** (1997), pp. 241–248.

[Picard 1909] Émile Picard, De la science, in [Thomas 1909–1911, vol.1, pp. 1–30].

[Picard 1912] Émile Picard, *La Science et la recherche scientifique*, Revue scientifique **50–2** (1912), pp. 577–581.

[Picard 1914] Émile Picard, *La Science moderne et son état actuel*, Flammarion, Paris, 1914; repr. 1921.

[Picard 1916] Émile Picard, *L'Histoire des sciences et les prétentions de la science allemande*, Perrin, Paris, 1916.

[Picard 1991] Jean-François Picard, *L'organisation de la science en France depuis 1870 : un tour des recherches actuelles*, French Historical Studies **17** (1991), pp. 149–168.

[Picard & Pradoura 1988] Jean-François Picard & Elisabeth Pradoura, *La longue marche vers le CNRS (1901–1945)*, Cahiers d'histoire du CNRS **1** (1988); rev. version (2009) online: `http://www.vjf.cnrs.fr/histcnrs/pdf/cahiers-cnrs/picard-pradoura-88.pdf`.

[Pinault 2006] Michel Pinault, *La Science au Parlement : les débuts d'une politique des recherches scientifiques en France*, Paris, CNRS éditions, 2006.

[Poincaré et al. 1908–1909] Henri Poincaré, Gaston Darboux, & Paul Appell, Examen critique des divers systèmes ou études graphologiques auxquels a donné lieu le bordereau, in *Affaire Dreyfus, la révision du procès de Rennes : enquête de la chambre criminelle de la Cour de cassation (5 mars 1904 – 10 novembre 1904)*, vol. 3, Ligue des droits de l'homme, Paris, 1908–1909, pp. 500–600.

[Prochasson & Rasmussen 1996] Christophe Prochasson & Anne Rasmussen, *Au nom de la patrie : les intellectuels et la première guerre mondiale (1910–1920)*, La Découverte, Paris, 1996.

[Rasmussen 2003] Anne Rasmussen, Mobiliser, remobiliser, démobiliser : les formes d'investissement scientifique en France dans la Grande Guerre, in [Aubin & Bret 2003, pp. 48–59].

[Rasmussen 2004] Anne Rasmussen, *La "science française" dans la guerre des manifestes*, Mots : les langages du politique **76** (November 2004), pp. 9–23.

[Rasmussen 2007] Anne Rasmussen, *Réparer, réconcilier, oublier : enjeux et mythes de la démobilisation scientifique, 1918–1925*, Histoire@Politique. Politique, culture, société **3** (November–December 2007), online: `www.histoire-politique.fr`.

[Roche & Charle 2002] Daniel Roche & Christophe Charle, eds., *Capitales culturelles, capitales symboliques : Paris et les expériences européennes, XVIIIe–XXe siècles*, Publications de la Sorbonne, Paris, 2002.

[Rollet 1999] Laurent Rollet, *Autour de l'Affaire Dreyfus : Henri Poincaré et l'action politique*, Revue historique **298** (1999), pp. 49–101.

[Rollet 1999–2000] Laurent Rollet, *L'engagement public d'un homme de science : Henri Poincaré*, Revue des questions scientifiques **170** (1999), pp. 335-354; **172** (2000), pp. 213–239.

[Rollet 2000] Laurent Rollet, *Henri Poincaré, des mathématiques à la philosophie : étude du parcours intellectuel, social et politique d'un mathématicien au début du siècle*, Éditions du Septentrion, Lille, 2000.

[Rollet & Nabonnand 2012a] Laurent Rollet & Philippe Nabonnand, *Pour une biographie d'Henri Poincaré : le problème des sources*, Gazette des mathématiciens **133** (2012), pp. 78–93.

[Rollet & Nabonnand 2012b] Laurent Rollet & Philippe Nabonnand, eds., *Les Uns et les autres...Biographies et prosopographies en histoire des sciences*, Presses universitaires de Lorraine, Nancy, 2012.

[Roussel 1989] Yves Roussel, *L'histoire d'une politique des inventions: 1887–1918*, Cahiers pour l'histoire du CNRS **3** (1989), pp. 19–57.

[Saur & Tribout de Morembert 1965–1966] Hubert Saur & Henri Tribout de Morembert, *Ferdinand Daussy*, Mémoires de l'Académie nationale de Metz (1965–1966), pp. 13–21.

[Schiavon 2003] Martina Schiavon, *Itinéraires de la précision. Géodésiens, savants et fabricants d'instruments en France, 1870–1930 (environ)*, doctoral thesis, EHESS, Paris, 2003.

[Schiavon 2006] Martina Schiavon, *Les officiers géodésiens du Service géographique de l'armée et la mesure de l'arc de méridien de Quito (1901–1906)*, Histoire et mesure **21-2** (2006), pp. 55–94.

[Schiavon 2014] Martina Schiavon, *Itinéraires de la précision. Géodésiens, savants et fabricants d'instruments de précision en France, 1870–1930*, Presses universitaires de Nancy, Nancy, 2014.

[Schroeder-Gudehus 1978] Brigitte Schroeder-Gudehus, *Les Scientifiques et la paix : la communauté scientifique internationale au cours des années 20*, Les Presses de l'Université de Montréal, Montréal, 1978.

[Seylor 1912] Olivier Seylor, *Sur Henri Poincaré*, La Semaine politique et littéraire de Paris (July 27, 1921), pp. 7–8.

[Shinn 1993] Terry Shinn, *The Bellevue Grand Électroaimant, 1900–1940: Birth of a Research-Technology Community*, Historical Studies in the Physical and Biological Sciences, **24** (1993), pp. 157–187.

[Siegmund-Schultze 2001] Reinhard Siegmund-Schultze, *Rockefeller and the Internationalization of Mathematics Between the Two World Wars*, Birkhäuser, Basel & Boston, 2001.

[Siegmund-Schultze 2003] Reinhard Siegmund-Schultze, Military Work in Mathematics 1914–1945: An Attempt at an International Persective, in *Mathematics and War*, ed. Bernhelm Booß-Bavnbek & Jens Høyrup, Birkhäuser, Basel, 2003, pp. 23–82.

[Siegmund-Schultze 2009] Reinhard Siegmund-Schultze, *The Institute Henri Poincaré and mathematics in France between the wars*, Revue d'histoire des sciences **62-1** (2009), pp. 247–283.

[Sinaceur 2002] Hourya Benis-Sinaceur, *Modernité mathématique : quelques invariants épistémologiques*, Revue d'histoire des sciences **55** (2002), pp. 83–100.

[Sirinelli 1988] Jean-François Sirinelli, *Génération intellectuelle : khâgneux et normaliens dans l'entre-deux-guerres*, Fayard, Paris, 1988.

[SMF 1915] Société mathématique de France, *Vie de la société*, Bulletin de la société mathématique de France **43** (1915), p. 2.

[SMF 1924] Société mathématique de France, *Vie de la société*, Bulletin de la société mathématique de France **52** (1924), pp. 1–67.

[Soubiran 2003] Sébastien Soubiran, La recherche en milieu militaire : une nouvelle pratique scientifique pour les universitaires entre les deux guerres? L'exemple des marines française et britannique, in [Aubin & Bret 2003, pp. 153–167].

[Tannery 1909] Jules Tannery, Mathématiques pures, in [Thomas 1909–1911, vol.1, pp. 31–72].

[Thomas 1909–1911] P. Félix Thomas, ed., *De la méthode dans les sciences*, 2 vols., Paris, 1909–1911.

[Tollmien 1993] Cordula Tollmien, *Der "Krieg der Geister" in der Provinz: das Beispiel der Universität Göttingen 1914–1918*, Göttinger Jahrbuch **41** (1993), pp. 137–210.

[Urin 1904] Isaac Urin, *L'Université de Paris et les établissements parisiens d'enseignement supérieur (1903–1904)*, Publications de l'Université de Paris, Paris, 1904.

[Villermet 1993] Bernard Villermet, *L'École polytechnique de 1914 à 1920*, Sabix **10**, online: `www.sabix.org/bulletin/b10/sabix10.html#135`.

[Villat 1921] Henri Villat, ed., *Comptes rendus du Congrès international des mathématiciens (Strasbourg, 22–30 septembre 1920)*, Édouard Privat, Toulouse, 1921.

[Volterra 1905] Vito Volterra, *Les mathématiques dans les sciences biologiques et sociales*, Revue du mois **1** (1905), pp. 1–20.

[Volterra 1913] Vito Volterra, *Henri Poincaré : l'œuvre mathématique*, Revue du mois **15** (1913), pp. 129–154.

[Walter et al. 2010] Scott Walter, Ralph Krömer, Philippe Nabonnand, & Martina Schiavon, eds., *La Correspondance entre Henri Poincaré et les astronomes et géodésiens*, Birkhäuser, Basel, 2010.

[Weil 1992] André Weil, *The Apprenticeship of a Mathematician*, transl. Jennifer Gage, Birkhäuser, Basel & Boston, 1992.

[Wells 1901] H[erbert] G[eorge] Wells, *Anticipations of the Reaction of Mechanical and Scientific Progress upon Human Life and Thought*, Chapman & Hall, London, 1901.

[Zerner 1991] Martin Zerner, Le règne de Joseph Bertrand, 1874–1900, in [Gispert 1991, pp. 299–321].

Part II

Joining In

Italian Mathematicians and the First World War: Intellectual Debates and Institutional Innovations

Pietro Nastasi and Rossana Tazzioli

While Italian historical bibliography dealing with the Great War is extensive, secondary literature on Italian scientists' involvement in the conflict is very scarce. Compared to other European nations, Italy lacks a set of historical studies, critically and systematically organized to make better sense of the role played by its scientists—let alone, its mathematicians—during the First World War.[1] This chapter aims at providing a first approach to the question of Italian scientists' involvement in World War I. It focuses especially on mathematicians' actions, although these may not always concern mathematics exclusively nor even directly.

Our preliminary survey leads us to identify three moments where Italian mathematicians' involvement in WWI was consequential. First, during what we call the Interlude period—from the declaration of war between the Great Powers in August 1914 to Italy's decision to enter the battle on the Allied side on May 25, 1915—debates regarding intervention were intense among Italian intellectuals. In these debates, mathematicians expressed a broad range of opinions, which, as we will hint at, were significantly shaped by their prewar international networks. We will also see that while some mathematicians maintained their beliefs for the duration of the war and indeed afterwards, others drastically revised them due to the pressure of the events.

We next survey Italian mathematicians' mobilization during the active phase of WWI, that is, from May 25, 1915, until the Armistice. The first thing to point

---

Pietro Nastasi's address is Via F. Sacchetti, 7 – 00137 Rome, Italy. His email address is pgnastasi@libero.it.

Rossana Tazzioli's address is UFR de Mathématiques, Laboratoire Paul Painlevé UMR CNRS 8524, Université de Sciences et de Technologie de Lille, 59655 Villeneuve d'Ascq Cedex, France. Her email address is Rossana.Tazzioli@math.univ-lille1.fr.

[1]Omitting the large production of books by Italians soldiers on their war experiences, we mention the following: [Isnenghi 1989, Isnenghi 1993, Fabi 1994, Tomassini 1995, Isnenghi 1996–1997, Gibelli 1998, Isnenghi & Rochat 2000]. On Italian mathematicians and WWI, see [Guerraggio & Nastasi 2005, chapters 2 & 6]. Other countries seem to have been better documented: see for example [Dauben 1980, De Maria & Seidel 1980, Hartcup 1988, Roussel 1989, Pestre 1990, Aubin & Bret 2003, Goldstein 2009]. An overview of the connection between mathematics and war can be found in [Fieschi 1987, Fieschi & Paris De Renzi 1995, Booß-Bavnbek & Høyrup 2003]; see in particular [Siegmund-Schultze 2003].

out is that, as in other countries, the majority of Italian scientists became involved in the war simply as citizens, the youngest serving in active units of the army. Gradually, however, many scientists started to engage in collaborations with military institutions and developed bonds with the military and industry, which would take on a whole new dimension twenty years later. Latecomers to this scientific war, Italians were able to take cues from their allies' experience. But although they participated in international initiatives, the organization of Italian war-related research was rather slow to develop, at least at the institutional level. In the end, one remains under the impression that, although some achieved impressive results, successful instances of mathematical mobilization owed more to special circumstances than to overall planning. Even if institutions that intended to favor "inventions and research" were created during the war, scientists there often played a secondary role, in the sense that they were usually asked to draft reports on projects proposed by others (inventors, militarymen, etc.), rather than to carry out original research on their own.

Be that as it may, some mathematicians found their war experience very valuable and endeavored to shape some of their later activities accordingly. Finally, the third part of this chapter focuses on two important institutional consequences of WWI on Italian mathematics: namely, the establishment of the National Research Council (*Consiglio Nazionale delle Ricerche*, CNR) in 1923 and of the Institute for the Application of Computing (*Istituto per le Applicazioni del Calcolo*, IAC) in 1927. It is significant that the mathematicians behind the foundation of both institutions, Vito Volterra (1860–1940) and Mauro Picone (1885–1977), respectively, were also key figures who played different but exemplary roles in the mathematical and scientific mobilization of WWI.[2] In 1917–1918, Volterra had been the head of the research board set up by the Undersecretary for Arms and Munitions of the Office for Inventions and Research (*Ufficio Invenzioni e Ricerche*, UIR), while Picone—who was 25 years younger—put together a noteworthy computing bureau for ballistics attached to Italy's Sixth Army. While Picone still occupied a marginal place in the Italian mathematical community before the war, Volterra was already a central figure in Italian politics and culture. In fact, as we will see, he was one of the most ardent partisans for Italian intervention before May 1915.

In what follows, we pay attention to the local mathematical culture in two principal ways: we focus on a sample of universities—in Rome, Bologna, Turin, and Padua—during the Interlude period, and we study changing institutional settings by considering extra-academic institutions such as the Mathematical Circle of Palermo (*Circolo Matematico di Palermo*) and the Central Institute of Aeronautics (*Istituto Centrale Aeronautico*, ICA), and, in the postwar period, the creation of new institutions such as the UIR and the Italian Mathematical Union (*Unione Matematica Italiana*, UMI), where Volterra's and Picone's activities, of course, will be studied in detail. Due to the paucity of secondary literature on Italian mathematicians' involvement and role in WWI, we rely here on primary sources. Letters and reports here quoted at length are mainly kept in Italian archives. All unpublished material is provided both in its original Italian (or French) with an English translation.[3]

[2]On Vito Volterra, see: [Goodstein 2007, Guerraggio & Paoloni 2008, Paoloni & Simili 2008, Mazliak & Tazzioli 2009]. On Picone see, for example, [Fichera 1978] and [Tricomi 1977].

[3]We thank David Aubin for translating original material from the French.

1. Contributions by Italian Mathematicians to the Debate about Intervention

Although Italy was a member of the Triple Alliance with the Central Powers, Germany and Austria, it officially remained neutral at the outbreak of WWI. From the beginning of August 1914 to May 1915 when Italy finally declared war on Austria, debates between "neutralists" and "interventionists" of various hues were very heated. In the following we survey this polemic and, especially, point out various mathematicians' attitudes with respect to intervention by considering some of the most representative Italian universities.

The particular position of Italy during WWI is of course rooted in the then recent history of the country. Before 1861, Italy was divided into smaller states that were then united after a process called the *Risorgimento*, a popular movement that was both patriotic and democratic. Popular uprisings and two wars of independence with Austria led to unification in 1870 when Rome was proclaimed the capital of the new kingdom of Italy. In 1882 Italy signed the Triple Alliance together with Austria and Germany. This treaty, which guaranteed mutual support in case of aggression, was signed in spite of tense relations between Italy and Austria. Claimed by Italy, the *terre irredente* ("unredeemed lands") (Trentino, Trieste, and Fiume) indeed remained in the hands of the Austro-Hungarian Empire. Despite this bone of contention, Italy joined the Triple Alliance because it feared French aggression and because it wished to rely on Austrian and German protection to develop its own colonial empire in Africa. In 1882 Italy conquered part of Eritrea, and in 1885 Italy started a war of aggression against Ethiopia. In 1911 Prime Minister Giovanni Giolitti declared war on Libya and then on Turkey. In 1912, Italy conquered some islands in the Aegean Sea and part of Libya. Meanwhile, however, a secret agreement was signed in 1902 between Italy and France guaranteeing neutrality of the former in case of German aggression against the latter.

According to historians, Italian intellectuals generally supported Italy's intervention in WWI, as they did the colonial wars [Isnenghi 1989]. Some feared that by remaining neutral Italy was at risk of losing its status as a Great Power, as well as any hope of recovering the *terre irredente* from Austria. Intellectuals also saw interventionism as a further break with the past, fostering a progressive and anticlerical attitude. "We want to glorify war—the world's only hygiene" was famously proclaimed in the Futurist manifesto of 1909, printed on the front page of the French newspaper *Le Figaro* [Apollonio 1973, p. 22].

1.1. Roman Scientists: Volterra and Castelnuovo. Among scientists, Volterra came out strongly in favor of intervention. Succeeding Eugenio Beltrami as professor of mathematical physics at the University of Rome in 1900, Volterra had become one of Italy's most influential scholars.[4] That same year, he pronounced his celebrated *Prolusion* aimed at promoting applications of mathematics to other sciences, especially to biology and economics. He also became an ordinary member of the Academy of Lincei in 1899 and was elected Senator of the Kingdom in 1905.

Already, then, Volterra had embarked on various scientific enterprises devoted to creating stronger connections between science and industry, which would have

[4]Graduating from the *Scuola Normale* of Pisa in 1882, Volterra had been appointed professor of rational mechanics at the University of Pisa in 1883 and to the University of Turin in 1892.

important repercussions after 1915. Volterra played a prominent part in the foundation of scientific societies and institutions, such as the Italian Society of Physics (*Società Italiana di Fisica*) in 1897, the Italian Society for the Advancement of Science (*Società Italiana per il Progresso delle Scienze*, SIPS) in 1907, and the Italian Talassographic Committee (*Comitato Talassografico Italiano*) in 1910. Of special importance to this story was, as we will see, the support he lent to Arturo Crocco (1877–1968) in 1908 for the establishment of the Central Institute of Aeronautics, already mentioned.

But a crucial aspect of Volterra's prewar activities was his intense efforts to develop close relationships with French mathematics and culture. Unlike an earlier generation that had mostly looked towards Germany, he strongly encouraged cultural exchanges between Italy and France: exchanges of professors and of students, foundations of Italian institutes in France and French institutes in Italy, journals based on cooperation between the two countries, propaganda in favor of Italian culture in France and vice versa.[5] He found in Julien Luchaire (1876–1962), a professor of Italian literature at the University of Grenoble, a French colleague who enthusiastically assisted him in this project.[6] With Volterra's heavy involvement in creating closer links with France, it is no accident that none in the mathematical community argued as strongly as he did in favor of Italian participation in the war on the side of the Allies. He used his political, institutional, scientific, and social relationships, developed during the prewar period especially with France, as an instrument for strengthening support for intervention.

His exchanges of letters with Italian and foreign colleagues do not leave any doubt about his strong commitment to the cause. In a letter to Émile Borel (1871–1956), dated September 14, 1914, Volterra called the Germans "the invaders" and remarked that "France is fighting for justice and for the cause of civilization against the violence of the most brutal and odious imperialism" [Mazliak & Tazzioli 2009, pp. 36–37]. Many other letters by and to Volterra testify to his intense activity in favor of Italian intervention. This letter from Émile Picard (1856–1941) dated September 5, 1914, is particularly evocative of the extreme opinions with which Volterra constantly answered in unwavering support:

> I thank you for your warm wishes for the triumph of France over the Barbarians.... The German—I have always thought—is civilized only in appearance; in the smallest things he is coarse and without tact, and most often a compliment from a German becomes an enormous blunder. Amplify this native coarseness, and you end up with the horrors we are seeing. Furthermore, he lacks honesty and uses a philosophical coating to excuse his crimes; it is time that this immense pride be taken down and that Europe be able to breathe for the next century. All Europe should rise against the new Vandals that are only dreaming about putting the nation at its feet [Mazliak & Tazzioli 2009, p. 38].

---

[5]On this topic, see Appendices 3 and 4 of [Mazliak & Tazzioli 2009].

[6]His book [Luchaire 1965] paints an interesting picture of Italian and French cultures in the period. On Luchaire and his relationship with Italy, see [Renard 2002].

In principle, most intellectuals shared Volterra's opinion that "democratic intervention" was called for on the side of the Entente, against "German barbarity."[7] But not all Italian intellectuals were of the same opinion. The philosopher Benedetto Croce (1866–1952) expressed his discordant view in an interview published on October 13, 1914, in the *Corriere d'Italia.* The polemics concerning the privileged relations of Italian culture with French or German thought were "manifestations of the state of war." He explained:

> They are not rational questions but conflicts of passions; they do not deal with logical solutions, but with assertions of interests which, though lofty, are national, that is, particular; they are not proper reasoning but sham arguments based on imagination [Croce 1950, pp. 11–12].

Things taken for granted—like well-known statements by Volterra and other leading intellectuals insisting heavily on "German barbarity"—came under Croce's scrutiny:

> I think that when the war stops, it will be found that the soil of Europe has trembled for months or for years not only under the weight of arms, but also under that of mistakes. And Frenchmen, Englishmen, Germans and Italians will be ashamed and will beg indulgence for the verdicts they have pronounced, and will say that these were not verdicts but expressions of passion. And even more shall we have to blush who, although they were from a neutral country, spoke of "German barbarity" as an obvious thing. Among all mistakes, ... this one will hold the supremacy, since it is surely the most imposing one [Croce 1950, pp. 11–12].

According to a French witness, Croce considered himself a "*Germanofilo*" [Luchaire 1965, vol. 2, p. 22]. The Frenchmen also reported that Croce thought that to be on Germany's side was to uphold "the principle of authority" against "the abuses of freedom." But Croce was far from being alone as an Italian Germanophile. Many Italian scholars had studied and acquired specialized training in Germany; others simply liked the way its universities were organized or had high esteem for German science. While many favored Italian neutrality, few went so far as to argue for participation in the war on the Central Powers' side. "*Irredentismo*" against Austria was too widespread a feeling. But, to name just one, consider the chemist Giacomo Ciamician (1857–1922) who had studied at the University of Giessen.[8] A member of the Academy of Lincei as well as a senator, Ciamician wrote his colleague Volterra on August 6, 1914:[9]

> What a mistake this total war is! It is an event that allows thinking about nothing else. I wonder where this will lead: it is an enormous disaster, a crime against civilization.

[7]We may point out that while intervention was favored by most intellectuals, general opinion was more ambivalent. During the Interlude period, most people seemed to have backed neutrality while only a small minority was in favor of the Allies. A third position between the two called for an attack against Austria despite the treaty with the Triple Alliance. On this topic see, [MacMillan 2001] and [Rusconi 2005].

[8]Other scholars are mentioned and discussed in [Simili 1993, pp. 20–25].

[9]We note here that all letters from and to Volterra quoted in this article are in [Volterra Papers].

> I cannot believe that everybody in Europe approves—and has wanted and wished for—this destruction. I hope that there are some enlightened minds who will condemn this attempt on the lives and prosperity of the people.
>
> I would like to believe that among scholars there would be some who will not be blinded by this excessive patriotic feeling that stirs up the most ferocious feelings of race and racial hatred....
>
> But, as far as I know, no voice has spoken in this sense.
>
> Don't you think it would be opportune for us to voice such a protest in the name of offended science which knows no difference of nationality? Don't you—who, beside the great authority of your name, are well known abroad—think that it would be useful to know the opinion of the most important scholars of all countries? We hope that Italy can at least stay neutral.[10]

Ciamiacian's intuition on racial hatred is remarkable. This was a problem that became critical only after the war but was evidently rooted in prewar nationalistic feelings. Although Volterra, as we have seen, showed sympathy neither toward Ciamician's appeal to scientific internationalism nor toward his pacifist attitude and although most Italian mathematicians shared his view, there were some neutralists and even Germanophiles.

Among the Roman mathematicians who expressed anti-German views, we can also mention Guido Castelnuovo (1865–1952), then a professor at the University of Rome. With Federigo Enriques from Bologna, Castelnuovo was one of the foremost representatives of the so-called Italian school of algebraic geometry. After graduating from the University of Padua in 1886, he spent a year in Rome and then became a lecturer at the University of Turin, where he was influenced by Corrado Segre (who was in favor of Italian neutrality as we shall see in Section 1.3). In 1891 he moved to Rome where he taught until his retirement.

Though less vocal than his friend and colleague Enriques (see Section 1.2), Castelnuovo seems to have shared his attitude in favor of Italian involvement in the war. He felt—as Enriques did—that pro-German propaganda was widespread in Italy. On July 31, 1915, at a time when Italy had already declared war on Austria, although not yet on Germany, Castelnuovo wrote to Volterra:

> Concerning our relation with Germany, ambiguity unfortunately persists. Censorship, which is clearly coming from above, purges

[10]"*Che errore questa guerra generale: è un evento che non permette di pensare ad altro. Chissà a cosa potrà condurre; è un disastro immane; un delitto contro la civiltà.*

Io non posso credere che tutti in Europa approvino questa rovina e l'abbiano voluta e desiderata. Io spero che ci saranno delle menti illuminate che condanneranno questo attentato alle vite ed alle prosperità delle genti.

Io crederei che fra i cultori delle nostre scienze ve ne saranno alcuni almeno non accecati da questo esagerato spirito patriottico che fa scatenare i più feroci sentimenti di razza e l'odio di razze....

Peraltro a quanto io sappia nessuna voce s'è fatta sentire in questo senso.

Non credi tu che sarebbe opportuno farsi interpreti di una simile protesta in nome della scienza offesa, che non conosce differenze di nazionalità. Tu che oltre la grande autorità del tuo nome illustre, sei anche tanto conosciuto all'estero, non credi che sarebbe utile sentire in proposito l'opinione dei maggiori scienziati di tutti i paesi? Speriamo che almeno l'Italia possa restare neutrale" [Volterra Papers, Ciamician to Volterra, August 6, 1914].

> from newspapers the sentences coming out too harshly against the Germans.... There are many symptoms, therefore, that we want to avoid becoming Germany's enemy.
>
> Would it not be fitting that those Italian intellectuals who think more freely emphasize our war's high ideals according to which civilization should prevail in the world—ideals surely not inferior to those to which the government propaganda is nearly exclusively restricted?[11]

1.2. Interventionist Bologna: Pincherle and Enriques. Like their Roman colleagues, prominent mathematicians in Bologna seem to have come out rather strongly in favor of Italy entering the war. A professor of calculus at the University of Bologna since 1880, Salvatore Pincherle (1853–1936) was a staunch nationalist as well as an admirer of German mathematics. After his studies at the University of Pisa, he had spent the academic year 1877–1878 in Berlin, where he met Karl Weierstrass, who had a profound influence on his subsequent work in the field of functional analysis. After the war, he supported the Fascist regime and had an influential position as first president of the newborn Italian Mathematical Union (UMI) in 1922 (see Section 3.2 below), a position he kept for several years. Nationalistic as he was, he nonetheless embraced the cause of Italian intervention alongside the Entente.

At a conference organized by Volterra in 1918 in Rome in memory of the Italian mathematicians who had fallen in war, Pincherle provided what we believe to be a faithful description of the tensions among various types of nationalism in the Interlude period. Referring to the "Appeal of the 93 Intellectuals" (*An die Kulturwelt! Ein Aufruf*) signed on October 4, 1914, Pincherle wrote:[12]

> Just after the fatal day on which the Central Powers suddenly revealed a project prepared for a long time and submitted the terrified world to a storm whose horrors surpassed the imagination, the most prominent German scholars, leaders of the army of science which was believed not to know any national border, threw off their masks just as did their rulers; and a famous Manifesto—which would have announced the true bankruptcy of science, had the ideas inspiring it prevailed—claimed that doctrines count only to the extent that they help the hegemonic ideas that German militarism was going to translate into reality.

---

[11] "*Purtroppo l'equivoco nei nostri rapporti colla Germania persiste. La censura, evidentemente ispirata dall'alto, va sopprimendo nei giornali le frasi troppo aspre verso i tedeschi; ... vi sono insomma molti sintomi che non vogliamo inimicarci la Germania.*

Non sarebbe il caso che quella parte degli intellettuali italiani che pensa più liberamente, mettesse in rilievo le alte idealità della nostra guerra, riguardo alla civiltà che dovrebbe prevalere nel mondo, idealità certo non inferiori a quelle, a cui la propaganda governativa si è quasi esclusivamente ristretta?" [Volterra Papers, Castelnuovo to Volterra, July 31, 1915].

[12] Also known as the Fulda Manifesto (after Ludwig Fulda, a well-known Jewish writer), the "Appeal" was signed by 93 high representatives of German art and scholarship and argued for their patriotic commitment in a very assertive tone. Illustrious physicists, mathematicians, chemists signed the "Appeal" (such as Max Planck, Wilhelm Wien, Felix Klein, Walther Nernst, Fritz Haber, and others). The Manifesto was reinforced by two others, the first one signed by three thousand faculty members of German universities and the second one by the chancellors of twenty-two of them, all upholding the unassailable virtues of German culture. See [Gratzer 2000], in particular, Chapter 8: "Science, Chauvinism and Bigotry."

> And if in those days the large majority of our country's scientists were absolutely convinced that Italy could not stay away from the Great War, this was not only due to the intuition that neither now nor ever could Italy reclaim those territories that belonged to it by incontestable right, nor because the indefensible borders imposed by unjust treaties had to be corrected, nor because of the natural sympathy [we feel] toward the oppressed and menaced nations. It was also because the German Manifesto's ethical stance was against what was for us the supreme ideal of science; we had cultivated and venerated science on the condition that it should not become the monopoly of a race or the tool of a brutal power. We therefore felt that, should the enemy's evil plans prevail, everything we hoped and expected from science would have gone to waste.
>
> In the heart of the noblest among our young brothers, this conviction was translated into action; they abandoned their favorite studies, the School where they could have continued without disgrace to render their services, in order to take up arms: and if they fell in battle, they left the most beautiful example and the most shining memory to their School. Among these noble minds we commemorate today several of our most famous mathematicians; and it is my task to evoke—alas with inadequate words—the figure of one of them, Eugenio Elia Levi [1883–1917] [Roma 1918, p. 27].

Pincherle's colleague at the University of Bologna, Federigo Enriques (1871–1946) also supported Italian intervention on the Allies' side. After studying in Pisa, Enriques had joined the University of Bologna in 1896, where he taught projective and descriptive geometry. Later, in 1923, he would move to the University of Rome, where he joined his friend and brother-in-law Castelnuovo.

Enriques shared his feelings about Italian intervention in his correspondence with Xavier Léon, the director of *Revue de métaphysique et de morale*.[13] On August 25, 1914, Enriques wrote:

> I was in Switzerland, in Zurich, when the storm broke.... I spent hours in anguish before Italian neutrality was proclaimed; coming home to Italy, I understood that any other decision from the government would have been impossible, since the common feeling of Italians from all classes and all parties was unanimously against the aggressors.
>
> If you could see the anxiety with which we wait here for news about the war and how the entire population hopes for France's salvation, you would be moved. For that matter, we feel here that this assertion of independence vis-à-vis Germany will be paid for dearly in the case of a German victory.

[13]Letters from Enriques to Léon and some drafts from Léon to Enriques are in [Manuscrits Léon]. They are reproduced in [Quilici & Ragghianti 1989].

> We are now ready to be attacked, too. But if peace cannot be kept, let us be on the side of civility and right! This is the intimate thought of the whole Italian people whose calm and pacifism in fact merely hide our wish also to contribute ourselves—when the time will have come—to the enterprise of liberation.[14]

Polemics about Italy's position in the war caused personal trouble for Enriques, when he was forced to leave the editorship of *Scientia*, the scientific and philosophical journal he had founded in 1907 and edited with Eugenio Rignano (1870–1930). Rignano had launched an inquiry, finding it inadmissible that a scientific journal—such as *Scientia*—should shut itself up in an ivory tower and remain impassive toward the tragic events of the war. The inquiry was meant to achieve clarity about "the causes and factors of the war" but soon slid into the expression of personal opinions. At a moment when Italy had already entered the war, Rignano had no hesitation writing about the imperialistic aims of its British ally and about the former's co-responsibility in bringing about the war. When Enriques asked Rignano to withdraw the opinion piece and Rignano refused, Enriques left the editorship of the journal with a clamor—only to resume it after Rignano's death in 1930.

Concerning Italian intellectuals' opinion towards the war, Enriques wrote to Léon on February 4, 1915:

> No one knows quite what to think about this war. You know where our sympathies and aspirations lie; there are but a small number of people who do not share them (unfortunately Hegelian philosophy[15] has not adopted an honorable position, it is among the small number of German sympathizers). But I believe that the Italian press fails to give a proper idea of the situation when it seems to be busy only with the alliance treaty. No one is quite sure about our obligations, but it is to be feared that the government itself is not entirely free, except in the case where the Germans themselves would be foolish enough to attack our rights and become a menace [Quilici & Ragghianti 1989, p. 318].

1.3. Neutralist Turin and Padua: Segre, Severi, and Levi-Civita. We have already observed that some Italian intellectuals—though the minority—were neutralists. Even among mathematicians, not all agreed with the majority opinion championed by Volterra. As an example, we mention Corrado Segre (1863–1924) at the University of Turin. One of the most influential geometers of his time, Segre is considered to be the founder of the Italian school of algebraic geometry. In a letter to Volterra, Carlo Somigliana (1860–1955) suspected Segre of pro-German neutralism. According to Somigliana, Segre's point of view was that war and its consequences should be ignored, except for "the due respect towards Germans" [Volterra Papers, Somigliana to Volterra, May 15, 1916].

When Segre learned that the French mathematician Jacques Hadamard wished to come to Italy to lecture on the theory of differential equations and waves, he believed that his true motive was to meet with mathematicians—such as Volterra—who were pro-war activists. That indeed happened some months later in May

[14]Enrique to Léon (August 25, 1914); repr.[Quilici & Ragghianti 1989, p. 317].

[15]The reference to the "*philosophie hégélisante*," dishonored because of its pro-German ideas, was surely directed at Croce, about whom we have already spoken.

1916. We are informed of Hadamard's stay in Italy through his correspondence with Volterra, who had told Painlevé about his wish to get Hadamard to Rome for a lecture series. After some hesitation due to reluctance to interrupt his "war work," Hadamard came to Rome and gave six lectures on "Partial Differential Equations and Related Problems" (*Équations au dérivées partielles et problème aux limites qui s'y rattachent*) in the first two weeks of May 1916. Of course Hadamard's lectures were also organized in order to tighten the bonds between Italian and French mathematicians. He moreover intended to discuss the problem "of the very strong opposition with which the pro-Allied movement is met by the Jewish population of the United States."[16] Informing Volterra of actions taken by the French Jewish community to change this view, Hadamard expressed the wish that Volterra put pressure on the Italian government too.

Together with Castelnuovo and Enriques, Francesco Severi (1879–1961) was the third leading figure of Italian algebraic geometry. After initially promoting Italian neutrality, Severi adopted a position similar to that of Volterra and Enriques. Severi had also studied with Segre at the University of Turin. In 1904, he was professor of geometry at the University of Parma and a year later he moved to the University of Padua. Later, in 1922, Severi would be appointed professor of algebraic geometry at the University of Rome. Thanks to his support of the Fascist regime, his academic and political career took a new turn: in 1923 he became chancellor of the University of Rome, in 1929 a member of the Fascist Academy of Italy, and in 1939 founded the National Institute of Higher Mathematics (*Istituto Nazionale di Alta Matematica*, INDAM).

In Padua, Severi was already occupying administrative positions. He was director of the School for Engineers, the town councillor in charge of education, a member of the commission for urban planning, and president of the municipal society for water and gas. A member of the Official Socialist Party,[17] Severi was a pacifist. He quickly changed his mind, however, to become a partisan of what Italian historiography has called "revolutionary intervention," a term connoting the interventionist stance of various branches of the revolutionary movement (socialists, anarchists, syndicalists, ...) [Caraccioli 1968].

In the newspaper *L'Adriatico* from March 9, 1915, Severi described his reasons for abandoning a neutralist position in favor of support for Italian war participation, although the Socialist Party's position was still firmly neutralist:

> I hope and believe that in this crucial moment the attitude of the governing bodies of my party is an expression of that state of anguished tension, which any Socialist of reason or heart feels, between the ideal imperatives of his own faith and his perception of the unavoidable needs of the present; and [that the party's attitude is] not the outcome of a thoroughly considered and methodically pursued plan for action.
>
> But if it is so, and if it is also true, as I think, that the Socialist Party, as a political organization, could never support an

[16]Letter from Hadamard to Volterra dated April 3, 1916, in [Volterra Papers] reproduced in [Mazliak & Tazzioli 2009, p. 80]; see also Appendix 2 in ibid.

[17]In the 14th Congress of the Italian Socialist Party taking place in Ancona in 1914, the revolutionary branch of the Party defeated the revisionist branch and became the "Official" Socialist Party.

> intervention in war, I would find it rather preferable if the Socialist protest against the war remained, in any case, restricted to the purely ideal field, and if it was at the same time acknowledged that the actual situation is inescapable and cannot be modified today, precisely because it follows from social conditions that our Party cannot suddenly change....
>
> A less absolute attitude from our party's leadership would also be very important from a political point of view, since it would leave every party member free to evaluate the actual elements of the situation according to his own conscience as an Italian citizen and, at the same time, *allow each of us* to keep spreading Socialist propaganda to the masses by pointing out the monstrous disasters that follow from a capitalistic organization of society.
>
> I am convinced of the necessity of an Italian intervention on the side of the Triple Entente.... It is true that the causes of this war are properly capitalistic, but we cannot ignore that because of the brutal violations of the natural rights of nations of which Germany is guilty, that because of the existence of many unresolved questions, and finally that because it is in *the interest* of some belligerent states, and especially of England, that minor nationalities be respected, ... that because of all this, the war has on the whole developed into a battle between two conflicting visions of rights and forces that shall prevail in the modern world [Severi 1915].

Severi's colleague at the University of Padua, Tullio Levi-Civita (1873–1941) also was a Socialist. Appointed professor of mechanics at Padua in 1897, Levi-Civita had studied there under Gregorio Ricci and Giuseppe Veronese. In 1918 he would move to Rome where he taught until 1938, when the Racial Laws were promulgated and he was forced to retire. One of the best Italian mathematicians of the period, Levi-Civita also enjoyed a high reputation abroad due to his research work on the three-body problem, tensor calculus, hydrodynamics, and general relativity theory.

Contrary to Severi, Levi-Civita remained a pacifist throughout the war. This considerably affected his friendship with Volterra: their correspondence, usually very amicable, became extremely formal during the war. As far as we know, Levi-Civita never showed hostile feelings towards his German colleagues before, during, or after the war. From March to May 1915, he exchanged very interesting letters with Albert Einstein. Their main discussion concerned general relativity and strongly contributed to the correct expression for the gravitational field equations obtained by Einstein in November 1915. But Levi-Civita expressed his neutralism and pacifism whenever there was an occasion, and especially during the postwar period (see Section 3.1 below). He wrote to Arnold Sommerfeld (1868–1951) on December 9, 1920:

> I have always been, and not only in science, a convinced internationalist; and, on the basis of this ideal, I consider *au-dessus de la mêlée* all nationalisms without exception....

> However we agree on an essential point—and I am very pleased about it. We both believe that scientific and personal relationships between scholars from every country in general, and between us in particular, must not be disturbed in any case by circumstances or memories of national or state divergences.[18]

Levi-Civita's general hatred of war and of its awful consequences is well expressed in a letter to George David Birkhoff (1884–1944) dated August 23, 1916:

> As you may easily imagine, we are feeling in Europe much more than in America the depressing influence of war on scientific activities and collaboration: efforts, aspirations, young energies, and, generally, all forms of energy are mobilized by present circumstances; and unfortunately no symptom yet leads us to expect any time soon any kind of appeasement capable of insuring justice and reestablishing fraternity among the nations.[19]

In the end, however, Italian pro-war intellectuals were successful and, on May 23, 1915, Italy officially declared war to Austria. Fifteen months later, on August 27, 1916, it also declared war on Germany.

2. The Mobilization of Italian Mathematicians

2.1. French Mathematicians' Reaction to Italian Intervention. Understandably, French mathematicians reacted to the Italian intervention with great enthusiasm. In an undated letter to Enriques, Léon lauded the "admirable" Italian decision to enter the war in spite of its dangers and horror. "There will not be," he added, "a *German peace* founded on an Empire extending from the Pyrenees to the Urals: this is morally and historically impossible" [Quilici & Ragghianti 1989, p. 311, original emphasis]. On May 26, 1915, Picard wrote:

> Your recent letter gave me a glimpse of the great event that just took place [the Italian declaration of war]. In France, we have seen with infinite joy Italy enter the great struggle of Civilization against Barbarity. The enemy is still strong, but he is wounded, and one is allowed to think that the staunch resistance it is putting up will suddenly weaken. Here, we consider that this gigantic struggle much surpasses the greatest wars of history. Enough ruins would have been accumulated and enough blood shed for Europe to enjoy peace for a long time to come [Mazliak & Tazzioli 2009, p. 55].

In a letter to Volterra dated July 1, 1915, Hadamard clearly saw that Italian intervention in the war on the side of France should be the occasion for strengthening cooperation between mathematicians from both countries:

---

[18] "*Io sono sempre stato, non soltanto in scienza, un internazionalista convinto, e, in base a tale idealità, considero au dessus de la melée tutti indistintamente i nazionalismi....*

Però in un punto essenziale — e me ne compiaccio vivissimamente — ci troviamo in pieno accordo: nel convincimento che i rapporti scientifici in genere e quelli personali tra gli studiosi d'ogni paese, e tra noi due in particolare, non debbano essere comunque turbati da contingenze o ricordi di divergenze nazionali o statali" [Sommerfeld Papers, Levi-Civita to Sommerfeld, December 9, 1920].

[19] Letters from Levi-Civita to Birkhoff [Birkhoff Papers]. Courtesy of Andrea B. Goldstein. The letters from Birkhoff to Levi-Civita are published in [Nastasi & Tazzioli 2000, pp. 199–220].

> I have wanted to write to you for a long time, since I have learned through [Joseph] Pérès, and M. Richet, and many others how much pain and enthusiasm you gave to the great movement that is currently taking hold of Italy. The admirable élan of sympathy shown by our youth through the sacrifice of their life would have to be doubted if we refrained from honoring the occasion of remembering and tightening the links we have woven at a time when we were merely suspecting the *boche* Barbarity [Mazliak & Tazzioli 2009, p. 57].

"As for me," Hadamard added, "I make myself useful as I can—this is not much—by lending my help to a certain number of technical researches" (ibid.). Did, like the French, Italian intellectuals, and mathematicians in particular, also contribute to the war effort? We may say that, in spite of their great enthusiasm, Italian mathematicians—with the notable exception of Volterra and Picone—generally contributed at a personal, rather than formal, level.

2.2. Initial Stages of the Scientific Mobilization in Italy. All mathematicians who joined the army—and several indeed volunteered—were immediately put into the reserves. Most made contributions to problems of exterior ballistics and of the related field of telemetry.[20] All of them were—at least at first—absolutely incompetent in ballistics; many of them certainly knew so-called Galilean ballistics (external ballistics in empty space), whose first elements were taught in high schools and developed in the university course of mechanics. However, they were skilled mathematicians, able to face mathematical questions and to work out useful solutions. As Pincherle noted in his obituary of Eugenio E. Levi [Roma 1918], solid theoretical training easily allowed one to serve as a technician. This was true at least in the case of Leonida Tonelli (1885–1946), Luigi Tenca (1877–1960), Francesco Giacomo Tricomi (1897–1978), and E. E. Levi himself.[21] Significantly, however, even if the latter was one of the best scholars in complex analysis (according to [Dieudonné 1977, p. 114]), he was modestly employed before he was killed on active duty as a military engineer.

We also point out that none of them—at least if we are to judge from the silence of those working on this topic—asked for Levi-Civita's opinion. This is somewhat surprising when it is remembered that the latter might have been one of the very few mathematicians actually engaged in "terminal ballistics," before the war, with a study of the penetrating power of bullets in solid bodies [Levi-Civita 1906].[22] The only hint to that effect comes from Picone's commemoration of Levi-Civita:

> I knew Tullio Levi-Civita in his hospitable house in Padua, during the war 1915–18.... I visited him from Schio, during the night, in a motor-lorry used for supplies.... He always received me cordially and studied with me some ballistic problems that I submitted to him. Out of these [visits], I always drew much

[20]On the Italian contributions to ballistics (mainly due to Fubini, Picone, Severi, and Volterra), see [Tanzi Cattabianchi 1977, Tanzi Cattabianchi 1981, Tanzi Cattabianchi 1988].

[21]War memories of some Italian mathematicians may be found in [Tenca 1959, Terracini 1968, Tricomi 1967].

[22]Other interesting results in ballistics produced by Levi-Civita can be found in the following books he wrote in collaboration with the physicist Ugo Amaldi: [Amaldi & Levi-Civita 1923–1927] and [Amaldi & Levi-Civita 1935].

> profit in view of my various ballistic assignments. One day, he lent me a Brunswiga calculating machine which I had found on his desk.... This allowed us to make, in a short time, the long, difficult computations necessary for compiling the new firing tables.... I do not know where the calculating machine is now. Levi-Civita had the delicacy, natural to him, never to ask for it back [Picone 1974, p. 21].

Generally speaking, Italy's military and institutional organization was very poor when it entered the war, compared with the other great powers. This was especially true with regards to the integration of science and technological innovation. In 1894, France for example had set up a military commission for examining inventions (*Commission des inventions intéressant l'armée de terre et de mer*, [Roussel 1989]). Before the war the physicist Éleuthère Mascart (1837–1908) and then Jules Violle presided over the commission; the hydrodynamicist Joseph Boussinesq (1842–1929) and the mathematician Paul Appell (1855–1930) were among its members. In 1914, at the very beginning of the war, the commission changed its name to High Commission for Inventions Related to National Defense (*Commission supérieure des inventions intéressant la défense nationale*) and was reinforced by the presence of technicians, engineers, politicians, and members of the Academy of Sciences. It was headed by the mathematician Paul Painlevé (1863–1933)—replaced by Borel when the former became Minister of War—and leading mathematicians (such as Borel, Hadamard, Henri Lebesgue, and Paul Montel) and physicists (such as Aimé Cotton, Paul Langevin, Jean Perrin, and Pierre Weiss) worked for it. The commission had access to university and industrial laboratories and was actively engaged in several problems involving the application of science to national defense (gas protection systems, improvement of firing tables, sound location of enemy batteries, detection of submarines, and so on).

Contrary to France, Italy seemed to have made no effort to improve the scientific organization of military innovation between August 1914 and its entry into the war. This much is acknowledged in a book about Italian "military culture":

> Just before [WWI], anxiety for renewal could be noticed in military organizations; it also concerned [military] culture: the formalization of practice had been replaced by lively interest in the military disciplines. For other reasons—financial difficulties, systematic oppositions on the part of certain currents of public opinion, governmental weakness—this much-needed process of renewal in the Italian Army was not as radical and deep as would have been necessary to live up to our increasing status as great power [Landogna 1937, vol. 3, p. 125].

Indeed, in spite of rapid progress in artillery techniques and doctrine made manifest during the first bloody months of WWI, old firing tables elaborated by General Francesco Siacci (1839–1907) thirty years earlier were still in use. As Picone wrote in his papers on this subject [Picone 1934], in numerous cases Italian artillery fired at soldiers from their own side because they did not have sufficiently advanced firing tables. Later official documents also confirmed the "scarcity and inadequacy of our artillery" [Landogna 1937, vol. 3, p. 127].[23]

[23] About French ballistics and how Siacci's methods were replaced by new ones, see David Aubin's contribution to this volume. The chapters by June Barrow-Green and by Thomas

Although the Italian mobilization of scientific resources thus took place in a generally disorganized manner,[24] local structures were set up to foster technical and military innovation, such as the institutions created by Picone and Volterra. The National Committee for Inventions relating to War Materials (*Comitato Nazionale per le invenzioni attinenti al materiale di guerra*) was founded in July 1915 and officially recognized in March 1916 when it was delegated to represent Italy on the Inter-Allied Committee for Inventions (*Comité interalliés des Inventions*). As we shall see, it was only between the end of 1916 and the beginning of 1917 that, thanks to Volterra, the Inventions and Research Office (UIR) was created on the French model.

Volterra traveled to France several times in order to visit the analogous French Commissions and also promoted Marie Curie's visit to Italy. Some years before the war, Volterra had asked Curie—without success—to take part in meetings of the Italian Society for the Advancement of Science. Named director of UIR in 1917, Volterra renewed his invitation to Curie on behalf of the Italian government and this time she accepted. For nearly three weeks, from July 30 to August 18, 1918, she toured Italian deposits of radium for medical and military uses. In the end, a report was drawn up, and it was decided to constitute a new Commission for Radioactive Materials attached to Volterra's UIR.[25]

But Volterra was not dealing only with international scientific relations. He got his own hands dirty working on technical devices, as can be gathered from the following event reported by Giovanni Sansone (1888–1979), professor of analysis at the Universities of Florence and Pisa:

> During the First World War, I had the good fortune of being personally introduced to [Volterra] in April 1918 at Pero di Piave, where [as a reserve officer] I directed a "Cotton-Weiss phonotelemetric section," in charge of locating Austrian batteries positioned on the other side of the Piave along an eight-kilometer front. He served as Reserve Major in the Corps of Engineers; he had obtained instruments like mine from France and came to check whether the results obtained by my section were comparable to French ones, and indeed they were [Sansone 1977, p. 30].[26]

Sansone's words clearly show that Volterra was in close scientific contact with his French colleagues, for example those involved in the computing bureau set up by Lebesgue and Montel. In Italy, the physicists Augusto Righi (1850–1920) and Antonio Garbasso (1871–1933), together with Severi, were the main scientists involved in phonotelemetric measurements. On this subject a heated controversy

---

Archibald, Della Dumbaugh, and Deborah Kent, respectively, discuss ballistic research in Great Britain and the U.S.

[24]There were however some exceptions to the unsystematic nature of Italian scientific mobilization: the Institute of Pharmaceutical and Toxicological Chemistry of Naples directed by Arnaldo Piutti (1857–1928) and the Polytechnic of Milan whose director was Giuseppe Colombo (1836–1921). For their war work, see the 1919 report of the General Director of the Ministry of Public Instruction, Giovanni Filippi, [Filippi 1919, pp. 345–346, pp. 355–356]; on Colombo, see [Lacaita 2007].

[25]See [Picone & Bilinski 1969, p. 19]. There is an unpublished report by Volterra that documents Curie's trip to Italy [Volterra Papers].

[26]For a detailed study of sound-ranging techniques in France and Italy, see [Schiavon 2003a] and [Schiavon 2003b].

between French and Italian scientists broke out, which documents how the latter failed to be integrated into a complex military-scientific coordinated ensemble.[27]

2.3. The Central Institute of Aeronautics. Another domain where Italian scientists were able to establish close relations with their foreign colleagues from Allied countries was in the rising field of aeronautics, where Italy had an impressive early start.[28] The first courses of aeronautics to be given in Europe actually took place at the *Istituto Centrale Aeronautico* (ICA), founded in 1908 by Arturo Crocco (1877–1968) with Volterra's decisive help.[29] A wind tunnel had already been built under Crocco's supervision in 1903—together with a British counterpart, it was the first of its kind in Europe.

Despite a promising start, Italy however failed to adopt ambitious plans for pursuing aviation, focusing instead on the development of more efficient dirigibles. A first experimental airship was built in 1906 by Crocco, together with Ottavio Ricaldoni and Antonio Munari. Two years later the first military dirigible in Italy (the so-called *P.1*) was produced in plants for aeronautical constructions set up in Rome and in nearby Vigna di Valle employing 128 civilians, 32 tailors, and 326 military workers. Once again, Crocco was in charge of the theoretical supervision of the construction while Ricaldoni oversaw its technical aspects. A hydroplane built by Crocco himself was used as a theoretical basis for the *P.1*, which was also known as the "little dirigible" ("*P*" stood for *piccolo*). The Roman plant then put other airships into production, such as the *P.2* and the *P.3* (used in the military invasion of Tripolitania in 1911), the *M.1* (the "medium" one), and then the *P.4*, *P.5*, and *M.2*. The institutional picture was completed by the Aeronautic Observatory set up by Lieutenant Cristoforo Ferrari in Vigna di Valle in 1909, which started to launch sounding balloons three years later. A second—and far superior—wind tunnel was built in 1914 by Crocco. Able to reach velocities as high as 200 km/h, this wind tunnel functioned until 1935 when Crocco oversaw the establisment of Guidonia, which came to be known as the "air town."

Volterra played a prominent role in the development of the aeronautical sector, not only as an influential political figure, but also as an experimenter.[30] In 1911 he served on a commission established to improve aerial navigation headed by Lieutenant-General Giuseppe Valleris [Volterra Papers, Italy's Minister of War to Volterra, November, 1911]. This commission was active until 1922, with some changes due to the war. Volterra was moreover called to take part in various other committees and commissions set up during the war, such as the committee for coordination and improvement of the aerological service.

[27]On the polemic and, generally, on Italian and French phonotelemetric measures during WWI, see [Schiavon 2003a, vol. 2, chap. 7].

[28]An overview on the history of aerodynamics is in [Anderson 1997]: on wind tunnels in particular, see pp. 296–299. For another case study of aeronautics research in prewar Europe, see Laurent Rollet and Philippe Nabonnand's contribution to this volume.

[29]The ICA was closely followed by the Paris *École nationale supérieure d'aéronautique* established in 1909. See [Siegmund-Schultze 2003, p. 71]. More about French aerodynamic research at this time may be found in [Fontanon 2003], [Fontanon 2011], and Claudine Fontanon's contribution to [Goldstein & Aubin forthcoming], as well as [Aubin forthcoming].

[30]See [Volterra 1916]. Concerning Volterra's experimental work for the compilation of firing tables, see [Crocco 1961, pp. 26–27].

When war broke out, this entire organization was put to the service of defense. The Volterra Archives holds various technical documents showing that the mathematician was heavily involved in the programs. New airships had to be designed. New firing tables needed to be computed for dirigibles. Volterra traveled through the country to carry out experiments and test new materials. The war also intensified international cooperation. It seems that Volterra and Crocco wished to enroll British help in the hope of scaling up their dirigible program.[31] Writing from London in April 1916, Crocco reported on the progress of this collaborative program:

> Here, they welcome our "*G*" [the "big" one] in the most flattering way. They have a need for dirigibles ... and lack the technical expertise for building them.
>
> Therefore, they have the best possible opinion of this project, which, although it has yet to be completed, presents itself as having a firm basis and extraordinary aeronautical efficiency.
>
> However, we shall need to change the objectives we suggested in Italy .... Here, they do not want *offensive*, but rather *reconnaissance* [airships]. The English mind, moreover, does not lack inertia. Nothing is therefore decided as to whether it will be built, even if everything leads us to believe that it will, and—since they already have all the proper equipment—that it will probably be built in England.
>
> Waiting for these decisions—which are compounded by the English government's difficulties in satisfying Italian requests for machine-guns, I do not know for how long I will have to stay here. I therefore think with nostalgia about the whole program in progress there [in Italy]: I miss the intellectual stimulation I derived from our daily exchanges of ideas.... [32]

Meanwhile, Volterra was actively working on setting up a more suitable institutional environment for the pursuit of war-related scientific research back in Italy.

2.4. The Office for Invention and Research. For Volterra, France was the obvious model for organizing the scientific mobilization of Italy. But in France, as in Italy, even if some invention evaluation boards existed before the war, the mobilization of the scientific resources of the country was at first mostly a matter

[31] On the history of Italian aeronautics during WWI, see [Ferrari 2004, Di Martino 2005, Miana 2007].

[32] "*L'accoglienza qui fatta al nostro "G" non poteva essere più lusinghiera. Hanno bisogno di dirigibili (come rileverà dall'accluso articolo) e non hanno alcuna preparazione tecnica per costruirli.*

Hanno fatto quindi il miglior viso possibile a un progetto che, sebbene non ancora completo, si presenta tuttavia impostato su basi solide e prevede un rendimento aeronautico eccezionale.

Però occorrerà cambiare lo scopo che noi ci eravamo proposto in Italia, come rileverà dal citato articolo. Qui non vogliono offensiva, ma servizio di esplorazione. Ed inoltre la mentalità inglese non è sprovvista d'inerzia. Nulla è perciò ancora deciso circa la sua costruzione, sebbene tutto lasci supporre che si farà, e, poiché sono già attrezzati, si farà probabilmente in Inghilterra.

Nell'attesa di queste decisioni, complicate dalla difficoltà da parte del governo inglese di soddisfare alle richieste italiane di mitragliatrici, io non so prevedere quanto tempo mi toccherà di rimanere quassù. Penso quindi con nostalgia a tutto il programma che si sta svolgendo costaggiù: sento la mancanza prolungata di quell'appoggio intellettuale che nel nostro scambio giornaliero di idee, guidava la mia mente" [Volterra Papers, Crocco to Volterra, April 2, 1916].

of personal initiatives. In November 1915, however, the mathematician Painlevé, who beginning in August 1914 headed an invention evaluation committee, as already mentioned, became Minister of Public Instruction, of Fine Arts, and of Inventions Concerning National Defense. Painlevé immediately established a Directorate of Inventions (*Direction des inventions intéressant la défense nationale*) and placed another mathematician, Émile Borel, at its head.[33] Taking inspiration from Britain, which had already set up its own Board of Invention and Research (BIR) in July 1915, Painlevé's ambition was to institutionalize, to rationalize, and to strengthen the mobilization of French scientific resources for military purposes.

On December 27, 1915, Volterra wrote to Borel (who had just lost his adoptive son, Fernand Lebeau):

> I learned that you were militarized [sic] and that you are now head of the study board of inventions that is part of the Ministry of Public Instruction. I think that you will be very busy with the work of this board which is very important. For several months, I have been part of the Military Aeronautical Institute. I am carrying out several experimental and computational tasks there. We work with great ardor to prepare all that is necessary for this crucial branch of the army [Mazliak & Tazzioli 2009, p. 66].

On January 4, 1916, Borel replied expressing the wish that official collaboration be established between various Allied boards of invention: "I hope that ... we will achieve that coordination of efforts that is the main strength of our enemy and against which the liberal principles themselves of the three great Western Powers are putting up obstacles" [Mazliak & Tazzioli 2009, p. 68]. The following October, while emphasizing that scientific relations with Britain were very tight, Borel lamented the fact that relations with Italy were less developed and announced his intention of visiting Italy [Mazliak & Tazzioli 2009, pp. 95–97]. At the end of November, it was Volterra's turn to visit him in Paris.

At the beginning of 1917, Volterra was finally in position to create a research council in charge of overseeing war-related scientific research at a national level. A letter (dated February 27, 1917) from the Undersecretary for Arms and Munitions to the Minister of War provides details about the tasks of the new body which was clearly organized along the lines adopted by the Entente and especially by France:

> In a personal letter from November 2 of this year [1916], His Excellency the Minister of Public Instruction drew the attention of Your Excellency to the advisability, pointed out to him by Mr. Emilio Borel, Head of Cabinet of the French Ministry of Inventions, to send an Italian professor there to study how scientific institutes and university laboratories are associated with the military administration with regard to inventions; he suggested entrusting Senator Professor Volterra with this task.
>
> Acquiescing to this suggestion, Your Excellency assigned this specific task to the said Professor Volterra, who, when his mission was concluded, addressed to Your Excellency a detailed report on the French organization of everything related to national defense, the conclusion of which was that it would be advisable to

[33]On this, see the chapter by David Aubin, Hélène Gispert, and Catherine Goldstein in this volume.

> Italy to mobilize the scientific strength of the country and make use of various scientific institutes' important resources in order to lighten the burden on military technical plants and offices.
>
> Agreeing to this advisability, Your Excellency invited Professor Volterra to design a project expressing the aims and competences of the new Office that would have to be created for this purpose, and indicating the public and private institutes on which the Office might draw for scientific or experimental assistance.
>
> In consequence, Professor Volterra presented to Your Excellency a meticulous project for the new Central Office....
>
> Your Excellency having in principle accepted these proposals and agreeing with my suggestion that purely scientific work be clearly distinguished from practical proposals that have the potential of being immediately implanted, I therefore suggest that a new office for the study of inventions be created on the basis of the present Department of Inventions of the Undersecretary [of Arms and Munitions]. This Office would receive the task of communicating to the specialized offices placed under the authority of the undersecretary all inventions that are strictly military in character or that directly concern artillery, engineering, or aeronautical materials, merely keeping a statistical trace of them, while it would communicate directly with the various nonmilitary scientific institutes all matters that concern the analysis and technical evaluation of all other inventions that are susceptible of contributing to national defense. [...]
>
> And since Professor Volterra has already had the occasion of observing the French organization of such matters at first hand, I should suggest, if Your Excellency agrees with what was said above, that Professor Volterra—who better than anybody else could live up to the important task which would thus be entrusted to him—be called upon to head the Office to be created.[34]

The papers left by Volterra show that his involvement in the organization of the UIR thus created was intense [Volterra Papers]. These documents allow us to reconstruct the way in which the new institution was created, its purpose, the

---

[34]We quote here from the copy of the letter sent to Volterra: "*Sua Eccellenza il Ministro della Pubblica Istruzione con lettera personale del 2 Novembre u.s.* [1916] *faceva presente all'E.V. l'opportunità, fattagli rilevare dal Sig. Emilio Borel Capo di Gabinetto del Ministro delle Invenzioni in Francia, di inviare colà un Professore Italiano per prendere conoscenza dei rapporti esistenti fra gli Istituti Scientifici e i Laboratori Universitari con le Amministrazioni Militari in materia di Invenzioni; e proponeva di affidare tale incarico al Senatore Prof. Volterra.*

V.E. in adesione a tale proposta, affidava l'incarico specifico al detto Prof. Volterra, il quale, compiuta la sua missione, indirizzava a V.E. stessa una dettagliata relazione sulla organizzazione francese per tutto ciò che riguarda le invenzioni interessanti la difesa Nazionale, concludendo con l'opportunità di mobilitare anche in Italia le forze scientifiche del Paese e di utilizzare le risorse non trascurabili dei numerosi Istituti Scientifici, e ciò con sollievo degli Stabilimenti ed Uffici Tecnici Militari.

Convenendo su tale opportunità V.E. invitava il Prof. Volterra a formulare un progetto in cui fossero meglio precisati gli scopi e le attribuzioni del nuovo Ufficio che si sarebbe a tal fine

difficulties it faced, and the way it was abolished. Consider, for example, an undated memorandum Volterra addressed, as head of the UIR, to the Undersecretary for Arms and Munitions:

> The task of this Office, which was established with the purpose of collecting and examining inventions concerning national defense, has very adequately been expanded so as to encompass interesting scientific and technical issues relevant to the same class of problems. This second task ... is rather wide, since modern warfare is more or less directly using all of the country's industrial resources and requires that the performance of each resource be driven to the maximum. The problems that must be resolved if this aim is to be attained can be grouped in the following three categories:
>
> (1) Problems of purely industrial and technical character....
> (2) Problems suggested by practical application, but that bear on questions of scientific interest, and which it is therefore adequate to study from a general or comprehensive point of view. From their solution, mostly derives an extension of scientific conquests beneficial to technical practice, from which other applications may spring, more or less associated with the one from which the problem arose originally.
> (3) Problems of scientific character, whose elements belong to the field of pure research and that ought to be studied independently from practical applications that may derive from them.
>
> ... Problems belonging to the third category can conveniently be studied in university laboratories; those belonging to the second [category] in laboratories equipped with workshops and administered by the state; and those belonging to the first [category] in industrial settings. In general, the study of these problems asks for experiments. What is therefore needed is the following:

---

dovuto costituire, e dal quale fosse risultato di quali Istituti Pubblici e privati ci si sarebbe potuti avvalere pel concorso scientifico o di indole sperimentale.

Il Prof. Volterra di conseguenza presentava a V.E. un progetto organico del nuovo Ufficio Centrale da istituire....

V.E. avendo in massima accettato siffatte proposte e convenendo a mio avviso tener presente ciò che è lavoro scientifico da ciò che è proposta pratica e quindi di immediata attuazione, se adatta allo scopo, proporrei costituire un nuovo ufficio per lo esame delle proposte di invenzioni, il quale potrebbe trarre il suo nucleo costitutivo dall'attuale Reparto Invenzioni di questo Sottosegretariato, e che dovrebbe avere il compito di trasmettere senz'altro ai diversi uffici speciali del Sottosegretariato stesso quelle invenzioni di stretto carattere militare o concernenti materiali di Artiglieria, Genio, e Aeronautica sì da conservarne presso di sé la sola traccia statistica, mentre dovrebbe mettersi in diretta relazione coi vari Istituti Scientifici Civili per quanto riguardi lo studio e l'esame tecnico di tutte le altre invenzioni che interessano la difesa del Paese. [...]

E poiché il Prof. Volterra ha già avuto occasione di vedere da vicino l'organizzazione data in Francia a tale materia, proporrei, nel caso V.E. accedesse a quanto sopra è detto, che fosse chiamato a dirigere il costituendo Ufficio lo stesso Prof. Volterra, il quale meglio di ogni altro potrebbe rispondere all'importante compito che gli verrebbe in tal modo affidato" [Volterra Papers, cartella VI].

(a) to have experienced staff for conducting experimental research relative to various subject matters;
(b) to have conveniently equipped laboratories at one's disposal;
(c) to possess adequate financial means.[35]

As soon as it was estabished, the UIR changed rapidly and was increasingly turned into a body for scientific consulting. Already divided into various sectors (the Mathematical, Physical, Chemical, Electrical Branches...), the UIR also set up several special technical commissions focusing on nitrogen, zinc, potassium, helium, etc. As Volterra explained to the undersecretary in an undated letter, these commissions were modeled on those existing in other countries and were useful for quickly getting a specialist's opinion when needed [Volterra Papers, cartella VI].

To develop his project and increase the scientific competitiveness of the UIR, Volterra required adequate funding. With this goal in mind, he wrote the undersecretary on June 18, 1918, to explain the structure of the UIR:

> The Office for Invention and Research was created for the modest task of evaluating inventors' proposals, but (following the example set by similar institutes created by other warring countries) it became a body for scientific and industrial consulting that all ministries, and the ones concerned with military affairs especially, trust with the assessment of technical and practical questions, the pursuit of scientific research, and the conception of new industries relevent to the country's defense and economy.

---

[35] "*Il compito di questo Ufficio, sorto con lo scopo di raccogliere ed esaminare le invenzioni relative a problemi interessanti la difesa nazionale, si è molto opportunamente allargato associandovi quello delle ricerche di carattere scientifico e tecnico interessanti la stessa categoria di problemi. Questo secondo compito, ora indicato anche nel nome dell'Ufficio, è assai vasto, perché la guerra moderna, utilizzando più o meno direttamente tutte le risorse industriali del Paese, esige che il rendimento di ciascuna sia spinto al massimo grado*

I problemi da risolvere per il raggiungimento di questo fine possono essere catalogati secondo i tre tipi seguenti:

(1) *Problemi di carattere puramente tecnico-industriale. Per studiarli si deve scegliere il metodo che più direttamente possa condurre al risultato pratico, ed ogni ricerca relativa, principale o sussidiaria, è limitata al campo dell'applicazione pratica, che si ha in vista.*

(2) *Problemi, il cui enunciato è suggerito dalla previsione di una pratica applicazione, ma che hanno attinenza con questioni di interesse scientifico, ed è quindi opportuno di studiare da un punto di vista generale e comprensivo. Dalla loro risoluzione col vantaggio della tecnica si associa per lo più un'estensione dalle conquiste scientifiche, da cui altre applicazioni possono germogliare più o meno connesse con quella da cui il problema ha tratto origine.*

(3) *Problemi di carattere scientifico, i cui argomenti appartengono al campo della ricerca pura, e che debbono essere studiati indipendentemente dalle applicazioni pratiche che possono derivarne....*

I problemi della terza categoria si studiano convenientemente presso i Gabinetti Universitari; quelli della seconda presso i Laboratori forniti di officine dipendenti dalle Amministrazioni dello Stato; quelli della prima presso gli Stabilimenti Industriali. Lo studio di questi problemi esige in generale l'esperimento. È pertanto necessario:

(a) *avere il personale esperto nella condotta di ricerche sperimentali attinenti ai vari argomenti;*

(b) *disporre di Laboratori convenientemente attrezzati;*

(c) *di possedere mezzi finanziari adeguati*" [Volterra Papers, cartella VI].

The Office thus essentially consists of two divisions: Inventions, assigned to a special group of technicians; and Research, assigned on a case-by-case basis to special commissions appointed with well-defined aims.

The Office belongs to the Ministry of Arms and Munitions and is placed directly under the authority of the undersecretary, according to what was decided when it was created and confirmed to me by Your Excellency when I paid him a visit yesterday.

However, the Office has established relations with various other ministries, in particular with the Ministry of Naval Affairs (through a section directed by Admiral Avallone and by General Valsecchi from the Naval Engineers Corps) as well as with the analogous offices of the Allied nations. It therefore has representatives in Paris, London, and Washington, D.C., while the British government has a representative at the Office, and France and the United States will be represented soon.

As said, the [business of the] Division of Research is assigned to commissions specially appointed for the purpose at hand whenever the Ministry of Arms and Munitions, whether from an initiative of its own or after agreement with other ministries, request some particular theoretical or practical investigation from the Office.

Several of these Commissions have already completed their task and submitted recommendations that have been translated into ministerial decisions. Let me recall the work done by the Commission for Raw Materials regarding the production of sulphuric acid whose recommendations were implemented through agreement between the Ministry of Arms and Munitions and the Ministry of Agriculture; let me also mention the work of the Commission for Bomb Welding which, ... through its practical studies and instructions, favored the development of electric welding and put together an exhibition in Milan to convince manufacturers of the advantages of the new system.[36]

---

[36] *L'Ufficio Invenzioni e Ricerche nacque col modesto compito di esaminare le proposte presentate dagli inventori, ma questo (seguendo l'esempio dei consimili Istituti creati dalle altre nazioni belligeranti) divenne un organo di consulenza scientifica ed industriale a cui tutti i Ministeri, ed a modo speciale quelli militari affidano l'esame di questioni tecniche e pratiche, l'esecuzione di ricerche scientifiche, l'approntamento di progetti di nuove industrie interessanti la difesa e la economia del paese.*

Esso è quindi essenzialmente diviso in due servizi, quello delle Invenzioni affidato ad uno speciale corpo di Tecnici e quello delle Ricerche che viene affidato caso per caso a speciali Commissioni nominate con scopi ben determinati.

L'Ufficio fa parte del Ministero Armi e Munizioni e dipende direttamente dal Sottosegretario, secondo quanto fu stabilito dalla sua costituzione e mi è stato confermato dalla E.V. in occasione della mia visita di ieri.

L'Ufficio ha però relazioni coi vari altri Ministeri e specialmente con quello della Marina (a mezzo di apposita Sezione diretta dall'Ammiraglio Avallone e dal Generale Valsecchi del Genio Navale) nonché cogli Uffici analoghi delle Nazioni alleate. Ha perciò suoi rappresentanti a Parigi,

A few months after this letter was written, the Allied nations had won the war and the UIR was dissolved. But this experience was not lost on Volterra, who endeavored to establish a new institution with the aim of perpetuating the collaboration of pure and applied scientists among themselves and with the military and industry. This was an institution that postwar Italy needed in order to reemerge on more solid ground and with a stronger national industry. In this sense, the wartime UIR sowed the seed of the CNR, with which we shall deal in Section 3.2.

2.5. The Sixth Army. "[N]ot only is mathematics beautiful, it may also be useful" [Picone 1972, p. 7]. This is what a mathematician remembered thinking when, as a young man, he was mobilized on the Trentino front and was quickly able to solve a pressing ballistic problem assigned to him by his superior. Then a recent graduate from the *Scuola normale superiore* of Pisa, Mauro Picone was 25 years younger than Volterra. His experience of WWI was therefore very different from that of the overseer of the Italian scientific mobilization, but it arguably had a stronger influence on his future life and career.

Before the outbreak of WWI, Picone, who had studied with Luigi Bianchi (1856–1928) and Ulisse Dini (1845–1918) among others, was a young promising mathematician. He could already count numerous significant publications to his credit that contributed to the theory of ordinary and partial differential equations. Remaining in Pisa as Dini's assistant untill 1913, he then accepted an assistant position in rational mechanics and analysis at the Turin Polytechnical School.[37]

The portion of the front Picone was sent to, in the Trentino, was a mountainous area. He served under Colonel Federico Baistrocchi who would later play a prominent political role in the Fascist regime. At the time, however, Baistrocchi was facing serious trouble with his artillery. Medium- and large-caliber artillery were made useless in this mountainous zone, due to failings in Siacci's old firing tables. Even if summarily corrected, the tables—derived from his classical treatise on ballistics [Siacci 1888]—produced catastrophic results (Figure 1). In Picone's account, Baistrocchi asked for the young mathematician's help as soon as he heard he had joined his unit (Figure 2). He specifically requested new firing data in order to help destroy enemy strongholds. And, after about one month of intense computing, Picone successfully concluded his task. "After this success, one can imagine how I came to see mathematics differently" [Picone 1972, p. 7].

Picone drafted a note "On the Firing of Medium- and High-Caliber [Cannons] in the Mountain" [Picone 1917] which explained his achievement. His contribution to

---

a Londra, a Washington, mentre presso l'Ufficio è rappresentato il Governo Inglese e saran presto rappresentati quelli di Francia e degli Stati Uniti.

Il Servizio Ricerche è affidato, come già dissi, a speciali Commissioni, che vengono appositamente costituite ogni qualvolta il Ministero delle Armi e Munizioni o per sua iniziativa o per accordo con altri Ministeri decide di affidare all'Ufficio una speciale indagine teorica o pratica.

Varie di queste Commissioni hanno già espletato l'opera loro, presentate da proposte che furono tradotte in provvedimenti ministeriali. Ricorderò l'opera della Commissione per le materie prime occorrenti alla produzione dell'acido solforico, che propose i provvedimenti tradotti poi in pratica d'accordo tra i Ministeri delle Armi e Munizioni e dell'Agricoltura; così pure ricorderò l'opera della Commissione per la saldatura delle bombe che ha studiato profondamente tale problema in vista specialmente della saldatura delle bombe ed ha favorito lo sviluppo della saldatura elettrica con studi e con indicazioni pratiche, indicando in Milano apposita esposizione per convincere gli industriali della convenienza del nuovo sistema" [Volterra Papers, cartella VI].

[37] After the war, Picone taught successively at the Universities of Cagliari, Catania, Pisa, and Naples, where he remained from 1925 to 1932, when he moved to his definitive chair in Rome.

FIGURE 1. The original caption of this picture reads: "A battery of Italian '75's' in action on a plateau overlooking the Isonzo River, which for a long time constituted a formidable barrier against the Italian attempts to reach the Carso Plateau." Source: *The War of the Nations: Portfolio in Rotogravure Etchings Compiled from the Mid-week Pictorial*, `http://hdl.loc.gov/loc.gdc/sgpwar.19191231`. Library of Congress, Serial and Government Publications Division.

the war effort is best summarized in a report probably written by the mathematician Guido Fubini (1879–1943), then a professor at the University of Turin. Recounting the story of Picone's encounter with Baistrocchi, the report describes the technical problem artillerymen faced with Siacci's firing tables. It states the assignment given to Picone:

> Find the inclination of the artillery piece such that the smallest possible charge will be needed to reach the target, overcome resistance, and activate the fuse. Determine the inclination and the load; compute the adequate changes to be given either to the load or the initial angle to adjust for the varying y-coordinate of the point of impact, for instance by 100 m[eters].
>
> By using the firing tables computed by Picone, this problem could always and easily be solved in the field, even by people ignorant of ballistics.[38]

[38] "*Determinare quell'inclinazione da dare al pezzo, per la quale la carica occorrente a portare il proietto sul segno, a superare gli ostacoli frapposti e a far funzionare la spoletta, sia la minima possibile. Determinare l'inclinazione e la carica, calcolare, per l'aggiustamento del tiro, la*

FIGURE 2. Artillery Command of the Italian 6th Army Corps, in the war zone (August 1917). Mauro Picone—bearded—stands second from left in the middle row. Courtesy of the Archivio Storico dell'Istituto per le Applicazioni del Calcolo.

To achieve his end, Picone created a small computing bureau in a ramshackle wooden barracks in the Dolomites. In this makeshift installation, one can see the root of the Institute for the Application of Computing (IAC) which Picone would establish.[39] He once remembered how, towards the end of 1916, a chance encounter with Volterra had made a strong impression on him and reinforced his conviction that such an institution could render great service to science and industry. Waiting to be received by his superior officer, Captain Volterra shared some thoughts about the war and about tactical and technical problems with his young colleague, who remembered them forty years later in the obituary he wrote for Volterra [Picone 1956, p. 3285]. Volterra, Picone recalled, spoke of the necessity of a closer collaboration between science and technology, which would not only immediately increase the Italian Army's military capacities but also be instrumental in the progress of peacetime Italy in industry and health. Picone went on:

> Thus did I find a first—authoritative and extremely encouraging—support for the idea which had come to me when I faced the problems that plagued our heavy artillery between the lofty peaks and

---

variazione da far subire o al peso della carica o all'inclinazione del pezzo, per far variare, alla costante quota del bersaglio, l'ascissa del punto d'arrivo del proietto, ad esempio di m. 100.

Colle tavole di tiro che dovrebbe costruire il Picone questo problema potrebbe prontamente essere sempre risoluto in batteria, anche da chi ignorasse la Balistica" [Volterra Papers, Report annexed to a letter from Segre to Volterra, January 13, 1917].

[39]We analyse the consequences and the development of these ideas below in Section 3.3.

> abyssal valleys of our Alps—the idea of an Institute for the Application of Computing (which I was able to establish some ten years later at the University of Naples) designed to mobilize the powerful instruments of numerical computing for the quantitative mathematical analysis of the problems of experimental science and of technology [Picone 1956, p. 3285].

In the following passages from a commemoration in honor of Antonio Signorini, Picone shared some vivid recollections of the war experience that had led to this idea:

> [Signorini] also devoted himself to ballistics. He did so, intensely in the years 1917–1922, driven at first by the needs of our artillery as they arose during the war of 1915–1918, and later, in 1946, by his position as a consultant for [IAC] which I then directed. He dealt with both exterior and interior ballistics, which he studied after the defeat of Caporetto, from 1917 to 1918, at the artillery headquarters of the Sixth Army [which was] operating in the Asiago Plateau. He was called there, on my proposal, to join a group of officers who were on the ground so as to be able to step in immediately and usefully whenever the solution of frequent and unexpected problems of a mathematical nature was needed for the tactical use of our medium- and large-caliber artillery. This group consisted of *il Nostro* [Signorini], the mathematicians Alessandro Terracini, Domenico Mattioli, Gino Cecchini, Arturo Cecconi, the engineer Brusini (a brave refugee from Trieste whose first name I unfortunately do not remember) and the undersigned, and I cannot refrain from adding that it was this group's useful work in the application of mathematics to the solution of the problems of field artillery that gave rise to the idea for the foundation of [IAC] in 1927 as part of my chair of infinitesimal calculus at the University of Naples.
>
> Artillerymen in charge of medium- and large-caliber pieces were complaining about the loss of initial velocity ... due to the wear and tear of the pieces, which caused the projectiles, when inserted into the bore, not to sink as much as in unused ones; as a consequence the combustion chamber increased in volume considerably. Well, [Signorini] was able to compute the loss incurred by the initial velocity imparted to the projectile at the blast of the explosive in the enlarged chamber, as a function of the (easily measurable) advance of the projectile in the bore; this loss being known, standard formulae of exterior ballistics could be applied to determine the angle correction affecting the inclination of the piece in order for the staff to be able to use the firing tables that were at their disposal to reach their target. After the war, [Signorini] published a thermodynamical, analytical, and numerical study of this problem in a paper entitled "*Calcolo della perdita di velocità iniziale dovuta al logorio dell'anima*", in vol. IV from 1919 of the *Rivista d'Artiglieria e Genio.* It should be noted that this beautiful article presents a new method for the approximate determination

> of one-dimensional integrals which has a remarkable approximating efficiency and which, in modern numerical analysis, is known as "Signorini's method."[40]

In an unpublished typewritten report from "Catania, December 1919," Picone further described the way in which his methods were implemented in the Sixth Army:[41]

> In February 1918 artillery command of the Sixth Army was taken over by General Roberto Segre, one of the most learned and most brilliant artillery officers of our army.
>
> General Segre was one of the most vigorous supporters of my firing methods.... He wanted without delay that all calibers [that is, all models of cannons] in our artillery be given rational firing tables calculated on the scientific basis I had laid down.
>
> He provided me with all the means I requested. At the army's headquarters, an office for ballistic studies was set up, which was later known to have an analogue in Paris, at the Artillery Depot of the French Army. Eminent mathematicians like Borel, Hadamard, Lebesgue, and Montel worked in this Paris office.
>
> In our small office, located in the attic of a countryside farm, I worked together with the collaborators I was able to assemble: Professor Terracini (Lieutenant of the Engineers), Dr. Cecconi (Artillery Lieutenant), Engineer Brusini (Artillery Lieutenant), Dr. Mattioli (Second Lieutenant of the Infantry), and later, toward the end of October 1918, Professor Signorini (Artillery Lieutenant), whom I managed to take out of a position where he could not achieve anything. We had five calculating machines, [human] computers, draughtsmen.
>
> The office worked day and night. Due to this, I list among my work the publications of the artillery headquarters of the Sixth Army.... The artillery of the Sixth Army and of some of the neighboring armies were thus provided with all the technical firing advances that had been achieved by the Allies, sometimes achieving progress that, in many respects, could stand comparison with the methods used by the Allies.[42]

---

[40] [Picone 1964, p. 397]. See also [Signorini 1919] and [Signorini 1922].

[41] Presented to the Competition of Analysis for the University of Cagliari, this report is in the *Archivio Storico dell'Istituto per le Applicazioni del Calcolo "Mauro Picone."* We thank its director, Michiel Bertsch, for providing us with a copy of the document.

[42] "*Col febbraio 1918 assunse il Comando dell'Artiglieria della VIa Armata il Generale Roberto Segre, uno dei più colti e più brillanti ufficiali d'artiglieria del nostro esercito.*

Il generale Segre fu uno dei più fervidi fautori dei miei metodi di tiro.... Egli volle che si procedesse senza indugio a dotare ogni calibro della nostra artiglieria di razionali tavole di tiro compilate sulle nuove basi scientifiche che io avevo posto.

Egli mi fornì di tutti i mezzi che io richiedevo. Al Comando dell'Armata si istituì un vero e proprio ufficio di studii di Balistica, che si seppe poi avere un analogo a Parigi, presso il Deposito d'Artiglieria dell'esercito francese. In questo ufficio di Parigi lavoravano insigni matematici, come il Borel, l'Hadamard, il Lebesgue, il Montel.

Nel nostro ufficietto, sito in una soffitta di una fattoria di campagna, lavoravamo io e i collaboratori che avevo potuto ottenere: il Prof. Terracini (Tenente del Genio), il Dott. Cecconi (Tenente d'Artiglieria), l'Ing. Brusini (Tenente d'Artiglieria), il Dott. Mattioli (S. Tenente di

In this official report, Picone described an efficient team working with the support of the Italian Army and developing methods able to compete favorably with those of other Allied nations.[43] Things were however somewhat messier. In the letter Segre wrote to Volterra on January 13, 1917, to which was appended a report we have already quoted and assigned to Fubini, the difficulties Picone encountered at a military and a practical level come to the fore:

> In consequence of the excellent results deriving from Picone's computations in campaigns at the *Alpe di Cosmagnon* and the *Dente del Pasubio* last summer, Dr. *Picone* ... was asked to compute here, during the winter, new tables for the firing of medium- and large-calibre [cannons] in the mountains and was given access to the resources of the nearby shooting range of Ciriè. This assignment was made as official as possible: Headquarters, Ministry of War, and so on. And the work is supposed to be completed by the end of March.
>
> Well, at the shooting range of Ciriè, after [receiving] many compliments, Picone (1) was informed that he had to wait for the appropriate ministerial authorization *before firing any shot* that might be needed for his experiments and to check his computations; (2) was also forbidden to examine results of any prior experiments [performed on the site] without a new agreement with the Ministry; (3) was not even allowed to consult books, without General Arlorio's indirect intervention from Turin. And I'm not even speaking about the fact that no draughtsman and no computer were assigned to him!
>
> *Fubini* himself took care of Picone's problems and also suggested some changes to his method. Among other things he found a formula for adjusting fire which is in general valid and may be used to replace usual correction coefficients which are invalid when the difference in height [between site of fire and target] is too large.... Colonel Bianchi from [Turin] (the professor of ballistics on whom everything concerning artillery seems to depend...) objected that [Fubini's formula] was too complicated....[44]
>
> Obstructionism in Ciriè, obstructionism in Turin with Colonel Bianchi. They do not want any help from those who can deliver it! And this is our Italy [whose fate is at stake]!

Fanteria), e, più tardi, verso la fine di ottobre del 1918, il Prof. Signorini (Tenente d'Artiglieria), che riuscii a strappare da una situazione, nella quale egli non poteva rendere nulla. Ebbi cinque macchine calcolatrici, calcolatori, disegnatori.

Il lavoro era diurno e notturno in questo ufficio. A questo lavoro si devono le pubblicazioni del Comando d'Artiglieria della VIa Armata elencate fra i miei titoli. ... All'Artiglieria della VIa Armata e a parte delle Armate limitrofe, si fornirono così tutti i perfezionamenti tecnici del tiro che erano stati conseguiti presso gli alleati, realizzando anche, in molti punti, progressi notevolissimi, di fronte ai metodi in uso presso gli alleati."

[43]For an extensive bibliography on external ballistics, see [Adhémar 1934].

[44]Artillery Colonel Giovanni Bianchi (1861–1917) was indeed an expert in exterior ballistics. See [Bianchi 1910].

> Neither Picone nor Fubini has asked me to write you. I alone resolved to do it. May you act to improve the state of things![45]

What can we conclude from this fascinating sample of published and unpublished material? How can contradictory documents be assessed? More research would be required to provide a more complete understanding of the war work of Italian mathematicians, but the documents here exhibited are enough to hint at the reasons why his wartime experience had a much stronger impact on Picone than on any other mathematician involved, as he was, in war research. Fubini and Severi, Oscar Chisini (1889–1967), and Pietro Teofilato (1879–1952) had war experiences that do not differ much from Picone's. They solved a variety of ballistic and telemetric problems created by the rapid development of artillery and by the need to adapt firing tables to the special geographical conditions encountered in certain phases of WWI.[46] But after the war they all went back to pure mathematics.

Picone, in contrast, recalled his work for the Sixth Army as a life-changing experience. As we have seen, he concluded from this that mathematics was "not only beautiful" but "also useful." This was a hard-won conviction that was forged during decisive years for the determination of his career path and scientific character. To be truly useful, moreover, mathematics had to be backed up by the proper organization of research resources. For Picone, therefore, the realization that mathematics may be useful would not lead to the bracketing of his wartime experience. On the contrary, Picone would pursue, even as a civilian, the military experience of appropriating the right resources to make mathematics useful. As we shall see, the establishment of the IAC was the concrete expression of an intuition born in the bunkers of WWI.

3. Some Effects of WWI on the Italian Mathematics Community

3.1. Italian Mathematicians and Ostracism. After the Armistice, mathematics was engulfed by the seaside transformation in international relations among

---

[45] "*Il Dr Picone ... in seguito agli ottimi risultati che i suoi calcoli avevano dato l'estate scorsa nelle azioni contro l'Alpe di Cosmagnon e contro il Dente del Pasubio, ebbe l'incarico di compilare, qui, nell'inverno, nuove tavole di tiro per il tiro in montagna di medii e dei grossi calibri, valendosi dei mezzi che si trovano nel vicino poligono di esperienze d'artiglieria di Ciriè. Questo incarico è dato con tutte le formalità possibili: Comando supremo, Ministero della Guerra, ecc. E il lavoro deve essere ultimato entro la fine di marzo.*

Orbene al poligono di Ciriè, dopo molti complimenti, il Picone: (1) fu avvertito che per le esperienze di rettifica e di controllo dei calcoli avrebbe dovuto, per ogni colpo, attendere di avere un'apposita autorizzazione ministeriale. (2) ebbe persino il divieto di consultare le vecchie esperienze, senza una nuova corrispondenza col Ministero. (3) nemmeno libri poté avere, se non quando ricorse all'intervento indiretto, a Torino, del generale Arlorio. Non parlo poi del fatto che non gli si diede nemmeno un disegnatore o un calcolatore!

Fubini s'interessò ai problemi del Picone, e gl'indicò anche qualche modificazione nel metodo. Trovò, fra l'altro, una formola per l'aggiustamento dei tiri, valida in generale, da sostituire ai soliti coefficienti di correzione non validi nei tiri con forti dislivelli ... Il colonnello Bianchi di qui (prof. di Balistica, da cui pare dipenda tutto ciò che riguarda l'Artiglieria...) gli obbiettò che era troppo complicata....

Ostruzionismo a Ciriè, ostruzionismo a Torino col colonnello Bianchi. Non vogliono aiuti da chi potrebbe darli! E si tratta della nostra Italia!

Non sono né Picone, né Fubini che mi han detto di scriverti. Sono io solo che ho pensato di farlo. Chi sa che tu non possa agire perché le cose procedano meglio!" [Volterra Papers, Segre to Volterra, January 13, 1917; translation contains original emphasis].

[46] On telemetric problems, see [Teofilato 1920].

European states. Generational issues left a deep mark on the mathematical communities that had suffered the most from casualties in the trenches. In France, in particular, the new Bourbaki generation argued that they were called to fill in the void left by those fallen for the motherland.[47] In Italy, new national schools were established, sometimes with a strong and aggressive agenda, especially under the Fascist regime. But in fact, very few Italian mathematicians had died during WWI and most influential mathematicians before 1914 still played a prominent role in the postwar reconstruction.

Institutions and people were physically and morally wounded by the terrible experience they had just lived through. The happy dream of turn-of-the-century modernism was shattered. People now feared the lapse back to barbarism through the subversion of consolidated powers and customs. Before the war they had dreamed of universal progress made possible by scientific discoveries; now they were awakening to a new reality where there were heated discussions about whether scientific meetings and cooperation with defeated countries would be allowed to take place as if nothing had happened. This was a hard blow to internationalism, which had characterized generations of Italian mathematicians after unification.

It is well known that scientists from Allied nations decided to institute new kinds of international cooperation embodied by the International Research Council (IRC) and the International Mathematical Union (IMU), from which German, Austrian, and Turkish colleagues were excluded. The first International Congress of Mathematicians to take place after the war was held in 1920 in the symbolic location of Strasbourg, a town that France had just recovered from Germany.[48] In a letter to Volterra, Picard explained the criteria that should be applied for inviting mathematicians to the Strasbourg Congress: "Invitations—addressed to scholars from Allied and neutral nations—will be personal. We do not intend to invite anyone among the neutrals who was too openly against our cause during the war" [Mazliak & Tazzioli 2009, p. 139]. Volterra was in complete agreement with the hard line. "It is with very great pleasure that I have seen the choice of this town [Strasbourg] as the seat of the next Congress of Mathematicians" [Mazliak & Tazzioli 2009, p. 141]. In a letter to Borel, Volterra added: "German militarism has not been knocked down at all; given the occasion it will always try to reassert itself.... I have no confidence in what they still want to do" [Mazliak & Tazzioli 2009, p. 142].

While most French mathematicians seemed adamant about it, Italian mathematicians were more divided about the measure of exclusion than Volterra's militant attitude would seem to indicate. Contrary to what Volterra claimed in many of his letters to Picard, Borel, and Hadamard, Italy was not unanimously against Germany before, during, and after the war. The Mathematical Circle of Palermo, for one, had a different point of view. After Picard, as a member of the Mathematical Circle, asked for German members to be expelled, Max Noether, recognizing that

[47]The Bourbaki generation had often argued this; see, especially, [Dieudonné 1978]. As is discussed in the introduction of this volume and in the chapter by David Aubin, Hélène Gispert, and Catherine Goldstein, recent historiography is more circumspect about this claim; see also [Leloup 2009, Beaulieu 2009, Goldstein 2009, Aubin 2012].

[48]The next congress took place in Toronto in 1924. Only in 1928 in Bologna would German mathematicians be welcomed to the congress. On the International Congresses of Mathematicians, see [Lehto 1998, Albers et al. 1986]; on the international scientific community in the 1920s, see [Schroeder-Gudehus 1978].

the position of its director Michele De Franchis (1875–1946) was a difficult one, tendered his resignation. In a letter to Corrado Segre from October 7, 1919, De Franchis however asserted that "the *Circolo* does not and will not distinguish mathematicians according to nationality or race" [Brigaglia & Masotto 1982, p. 377]. De Franchis refused to accept Max Noether's resignation and never expelled German members from the *Circolo*. In the end, he prevailed since Picard, Borel, Klein, and Hilbert belonged to the directing board of the circle in the postwar period.[49]

An ardent believer in the internationalism of science, as we have seen, Levi-Civita also opposed any kind of ostracism. In the interwar period, he devoted much effort to the organization of the International Congresses of Applied Mechanics in collaboration with Theodore von Kármán (1881–1963), the director of the Aerodynamic Institute in Aachen. In a letter he sent to Levi-Civita on April 2, 1922, von Kármán explained that his intention was to organize a European scientific meeting on fluid mechanics.[50] In spite of its great interest from a theoretical point of view and of its even greater importance from of a technological one, this domain which attracted mathematicians, physicists, and engineers was not constituted as a single research area. In order to overcome political problems arising from conflicting nationalistic beliefs, von Kármán suggested that the meeting be organized in a rather informal way—he would directly contact his colleagues from Germany, Austria, and former neutral countries (including his mentor Ludwig Prandtl and Carl Oseen from Sweden), while Levi-Civita would invite "Romanic and English" scientists. Levi-Civita's choice as a partner was based on scientific and personal grounds: indeed interested both in pure mathematics and in its applications to physics and engineering, Levi-Civita was moreover highly esteemed as one of the best Italian mathematicians and as an unwavering internationalist. Levi-Civita enthusiastically accepted von Kármán's invitation, suggesting that the congress be considered as the "personal business of some scholars," very few in number but excellent in quality [Levi-Civita Papers, Levi-Civita to von Kármán, April 21, 1922]. He invited his colleagues who were specialists in fluid mechanics, including Volterra and Gaudenzio Fantoli (1867–1940), a professor of hydraulics at Milan's Polytechnic. As they were both members of the Italian Committee of the IRC whose statutes excluded the former Central Powers, they expressed strong opinions against any kind of international congress of that sort.

But such negative answers were not enough to cancel the congress. In September 1922, as planned, some thirty scientists met in Innsbruck. Most came from Germany and Austria, but among the former Allies, Italy was the only country to be represented. All Italian scientists present, Modesto Panetti, Giulio De Marchi, Bruto Caldonazzo, and Umberto Cisotti, were Levi-Civita's friends and pupils. In Innsbruck, other international congresses were planned on general subjects of applied mechanics—and not only fluid dynamics. The series of International Congresses of Applied Mechanics was launched. The next one took place in Delft in April 1924 and was organized by Jan Burgers (1895–1981) and Cornelius Benjamin Biezeno (1888–1975).[51]

---

[49] For more details, see [Brigaglia & Masotto 1982, pp. 373–378].

[50] The letters to Levi-Civita are found in [Levi-Civita Papers].

[51] On the International Congresses of Applied Mechanics, see [Battimelli 1996].

Between Volterra's and Levi-Civita's extreme attitudes, a continuous spectrum of opinions existed. Some wished to renew old ties established with German colleagues before the war. One of them was Luigi Bianchi, who had been Volterra's teacher at the University of Pisa. While he was at the University of Munich from 1879 to 1881, Bianchi had befriended Adolf Hurwitz (1859–1919). In a letter to Bianchi written on January 23, 1883, Hurwitz had even called those Munich years "the best time in my life."[52] When WWI broke out Bianchi however expressed very different feelings:

> About the Germans, my thought is always the same—indeed, if possible, I hate them more than before, due to the accumulated misdeeds of this horrible war. It is my firm hope, and also my conviction, that they will eventually have to pay dearly for them.[53]

But in spite of the hatred of the Germans he expressed in 1915, Bianchi was soon ready to rehabilitate German scholars in the name of his love of science. Just before the Strasbourg Congress of 1920, Bianchi wrote to Volterra:

> You know how much I dislike the German people who have turned the world in such a nice way that old men like me will never be allowed to know better times! But I do believe that, in the very interest of science, it is preferable (if possible) to mitigate hostilities and not to ostracize the scientific production. Now and for many more years, it is clearly impossible to have personal relationships with the Germans; but I would like things to be organized in such a way as to ease rather than impede scientific, and especially mathematical, exchanges. I think you share my opinion and hope that the resolutions of the congress will ease a task that is already difficult in itself and that it is better not to embitter. This could only succeed in harming the interests for which we especially care.[54]

These examples show that, Volterra's opinion notwithstanding, Italian mathematicians' attitude toward ostracism was rather nuanced. Most indeed found it challenging to manage their relationships with their German and Austrian colleagues and everyone had to find the personal solution that fitted his or her own conscience and circumstances. But on the whole, we may say that Volterra's line

---

[52]Letters from Hurwitz to Bianchi are published in [Bianchi 1959, pp. 74–110] and stop in 1906. The above quotation is on p. 89 of [Bianchi 1959]

[53]"*Riguardo ai tedeschi, il mio pensiero è sempre lo stesso; anzi, se è possibile, li odio più di prima, in ragione dei misfatti accumulati durante questa orribile guerra. Che finiranno per pagarli cari è la mia ferma speranza, ed anche la persuasione*" [Volterra Papers, Bianchi to Volterra, March 3, 1915].

[54]"*Tu sai quanto io sia avverso ai tedeschi che hanno conciato il mondo in questa bella maniera, che ai vecchi come me non sarà dato di vedere tempi migliori! Però io credo che, nell'interesse stesso della scienza, convenga piuttosto attenuare se è possibile, le ostilità, e non dare loro l'ostracismo nei riguardi della produzione scientifica. Per ora, e per molti anni ancora, è evidentemente impossibile avere coi tedeschi relazioni personali; però io amerei che le cose si mettessero in modo da facilitare anziché ostacolare lo scambio di tutto quanto riguarda la produzione scientifica, in particolare la matematica. Credo che anche tu sarai dello stesso parere, e spero che le risoluzioni del Congresso saranno per facilitare un compito già arduo per sé e che non conviene inasprire. Ciò non potrebbe che riuscire a danno degli interessi che ci stanno massimamente a cuore*" [Volterra Papers, Bianchi to Volterra, March 2,1920].

nevertheless prevailed not only at a moral level, but also at the level of the institutionalization of Italian mathematics in the interwar period. We now turn to innovative scientific institutions established in interwar Italy by Volterra and Picone.

3.2. The Birth of the National Research Council. As noted earlier, the links between science, industry, and the military seem to have been weak in Italy before 1914, especially when compared with other western nations. But in every country the war had led scientists to increase their involvement in society. In 1937 Arnaud Denjoy (1884–1974) gave a speech at the *Réunion internationale des mathématiciens* that provides a good account of this postwar trend:

> The prestige of science was increased by the war. This is a bitter consideration. But the splendid benefit it had previously generously granted men had moved them much less than the ruins and disasters that were lavished with profusion on the people of Europe by science-rooted technology. Hitherto indifferent and disdainful with respect to science in its useful and benevolent guise, humanity was seized by consideration and respect in front of the science that generated terrible and fatal effects....
>
> Innovation lay in the conception of scientific research seen as an independent public service, whose funding from the state was justified by its object alone, without having to complement it with the practicalities of definite applications. On its part, private enterprise has also distributed much more than before [to scientific research] [Denjoy 1939, pp. 2–3].

When Denjoy pronounced these words, Italy was under Fascist rule and had entered an official regime of economic and intellectual "autarchy." But just after the war, it seemed to be the perfect embodiment of the views expressed by Denjoy. At the first meeting of the Italian Society for the Advancement of Science (SIPS) to take place after the war, in Trieste in 1921, the physicist Orso Mario Corbino (1876–1937), who was behind the exceptional school of physics founded by Enrico Fermi in Rome, delivered an inspirational speech as Minister of Public Instruction:

> Among those to whom we should express our gratitude, let us also include science students and school teachers. Even though the technical deficiencies of Italian schools are at all levels significant, no one can deny its educational efficiency in the face of the admirable example set by the Italian youth on the battlefield. All social classes covered themselves with glory during the war, but in no other class was the disregard for danger and death so great and serene as in our student body....
>
> Behind the heroic soldiers, ever more arms and resources were produced by Italian industry which was conferred more power due to its closer links with scientists. And *the whole country became engaged in the pursuit of scientific progress*, whose extreme importance they now felt.
>
> But, if science was the primordial factor in the Entente's victory thanks to its powerful means of destruction, the world should

> not forget that science alone has the power to heal it from the terrible ruins brought by the war.... And science will be able, I believe, to repair the damages that it was forced to inflict to bring about the triumph of justice in the world [Corbino 1921, pp. LXXXIX–XC, original emphasis].

An even more destructive war that broke out less than twenty years later was to show that Corbino was dreadfully mistaken. But in the immediate aftermath of the war, it was a comfort to believe in Corbino's speech. Volterra, like Corbino, wished to believe in the positive value of science. Volterra's scientific and political stature emerged much reinforced from WWI. In what we have called the Interlude period, between August 1914 to May 1915, he had strongly come out in favor of Italian intervention, and when Italy joined the struggle on the side of the Allies, he had heavily invested himself in military activities, in war-related scientific research, and above all in the promotion of cultural initiatives meant to strengthen the social implications of science beyond military applications. In 1916 he had for example founded the Italian Association for the Cooperation between Allied and Friendly Countries (*Associazione italiana per l'Intesa fra i Paesi alleati e amici*) in order to favor social, political, and cultural exchanges, as well as exchanges of teachers and students between institutions and universities of the Allied countries. Volterra was convinced that an industrialized country needed to tighten the interaction between science and industry. Science, in his view, was no longer the product of just a few scientists, as he pointed out in a speech he gave to honor Henri Poincaré at the Rice Institution in 1915: it was now a collective enterprise. And as such the development of scientific research needed to be organized and coordinated on national and international levels.

Volterra had increased his personal stature as a mathematician as well. One might even say that it was his own conception of mathematics that was reinforced by the war experience since he had always been a fervent supporter of an abstract and highly innovative brand of mathematics that nevertheless always remained directed toward applications. In the 1920s, he was the main representative of Italian mathematics in most international organizations and meetings. His position in the Italian scientific community and his international network made him a natural reference point for military and state authorities with regard to interactions with the scientific community.

It is therefore hardly surprising that, in the years 1918–1919, Volterra was the main Italian interlocutor of the British physicist Arthur Schuster (1851–1934), the Belgian astronomer Georges Lecointe (1869–1929), the American astronomer George Ellery Hale (1868–1938), and the French mathematician Picard in the promotion of the IRC. The IRC was a project concocted by Hale with a double objective: in the U.S., the IRC was to confer an international dimension to the National Research Council and would help turn it into a permanent institution; abroad, the IRC would serve to export the American model and coordinate the refoundation of international scientific organizations around the scientists from victorious nations. The IRC was moreover meant as a way of pursuing in peacetime the kind of intellectual and technoscientific cooperation that had taken place between the Allies. It would also develop the kind of international collaboration that used to characterize solar physics before the war, a field in which Hale had played a significant role. During the Inter-Allied Conference that took place in Brussels in July 1919, where

FIGURE 3. Portrait of Vito Volterra, 1929. With permission of the Académie des sciences–Institut de France.

Hale was confirmed as president of the IRC, Volterra was elected vice president, a position he kept until 1928.

In the immediate postwar period, Volterra thus assumed a major role on the international scientific scene. He was moreover appointed president of the International Bureau of Weights and Measures (*Bureau international des poids et mesures*), a position he kept until his death. Under Volterra's presidency, the general secretary was the Swiss physicist Charles Édouard Guillaume, who won the Nobel Prize in 1920. Under their leadership, the seat of the Pavillon de Breteuil in Sèvres was expanded, thanks to funding from the Rockefeller Foundation (1929–1931), and new metrological units were established in electricity and photometry. Therefore, it is not surprising that Volterra's colleagues called him "Mr. Italian Science," a nickname which captures well his role and, at the same time, says something more about the weakness of the Italian community which he represented (Figure 3).[55]

In Italy, however, Volterra faced some setbacks when he tried to set up the National Research Council (CNR) as the Italian section of the IRC. His project came to a halt due to the tensions we have already mentioned concerning Italian attitudes with respect to colleagues from defeated nations. Political and social instabilities in postwar Italy also hindered Volterra's project. But the CNR was not only the expression of the fundamental agreement between Volterra's views and international trends for reorganizing scientific research; it also represented the logical continuation of Volterra's directorship of the Office for Inventions and Research (UIR) during WWI. In a letter dated July 3, 1919, the Undersecretary of Arms and Munitions had already informed Crocco that all documents and instruments in possession of the UIR when it was dissolved right after the war's end were destined to the still nonexistent CNR:

> I am pleased to inform Your Lordship that I have ordered the closure of the service, i.e. [the UIR], on July 15 and that all scientific material, the library, the archive, the machinery, and the furniture of the Office, will be taken by this Institute [Crocco's Central Institute of Aeronautics], under the condition that everything shall be the property of the "National Research Council" once it is established.
>
> I do not think that difficulties can arise from this; in any case, I ask you to send me the remarks and suggestions you may have, especially regarding the liquid-air plant which has recently been seized in Trento and which will surely be ruined if inactive, and regarding the completion and publication of the reports concerning nitrogen, zinc, pyrite ashes, potash, and other physical topics that the [UIR] leaves in a state of near completion.[56]

[55]See [Goodstein 2007, p. 3]; the author refers to a letter from the American astronomer George Ellery Hale to John T. Carty, July 3, 1918, where he wrote that Volterra was the "leading spirit" of his country.

[56]"*Mi pregio comunicare alla S.V. di aver disposto che l'Ufficio Invenzioni e Ricerche cessi le sue funzioni dal 15 corrente, e che tutto il materiale scientifico, la biblioteca, l'archivio, il macchinario ed i mobili dell'Ufficio medesimo vengano presi in consegna da codesto Istituto, con la condizione che quando sarà costituito il 'Consiglio Nazionale di Ricerche' tutto quanto sopra passerà in proprietà del Consiglio medesimo.*

Non credo che potranno esservi difficoltà per questo; ad ogni modo prego presentarmi le opportune osservazioni e proposte: in modo speciale per quello che riguarda l'impianto per aria liquida di recente catturato a Trento, impianto che certo si deteriorerebbe se rimanesse inoperoso; e per quello che riguarda la ultimazione e la pubblicazione a stampa delle relazioni relative all'Azoto, allo Zinco, alle ceneri di Pirite, alla Potassa, nonché di alcune altre sopra argomenti di Fisica che l'Ufficio Invenzioni lascia in corso di avanzata redazione" [Volterra Papers, the Undersecretary of Arms and Munitions to Crocco, July 3, 1919; copy of the letter forwarded to Volterra by Crocco].

While the establishment of the CNR faced resistance, Volterra had nevertheless succeeded in establishing the Italian Mathematical Union (UMI) in the spirit of the Brussels conference that had instituted the IRC. In other words, the UMI was not an expression of the Italian mathematical community (as in the case of other Italian professional societies, of physicists and chemists, for instance), but the local branch of the International Mathematical Union.[57]

Finally approved on November 18, 1923, the statutes of the CNR underscore its national role: it was asked to coordinate and rationalize research activities by developing special fields; to be in touch with various national institutions regarding scientific questions; and to manage or to institute research laboratories.[58] On January 12, 1924, Volterra was unanimously elected president of the CNR. This is perhaps the moment when his reputation reached its apex. A year before, he had been elected president of the Academy of Lincei, after spending three years as its vice president. He was the grey eminence of the UMI (the president was Salvatore Pincherle), which served as the mathematical committee of the CNR, and was the most influential figure of the SIPS.

The statutes of the CNR betrayed its main promoter's wild hopes. Aimed at rationalizing and developing Italian scientific organizations, the CNR positioned itself in competition with traditional academies while insisting on tightening its connections with the industrial world and encouraging public support for scientific research. Previously, scientific research had been free from outside intervention (aside from economic pressure), though fragmented by the structure of the university network as well as by personal ambitions. Volterra's aim was to transform this system by orienting scientific research towards the solution of societal problems and by supporting the integration of science with industry.

But the establishment of the CNR turned out to be Volterra's last victory. Two years later, Volterra indeed broke with Mussolini's Fascist government, leaving him in the role of the progressive but isolated old lion.[59]

3.3. The Foundation of the Institute for the Applications of Computing. We have already seen how, during WWI, Picone directed a team in charge of computing firing tables to be used in the Alpine areas, which apparently were successfully put to use on the front. For his exceptional merits, Picone was promoted first to the rank of lieutenant and then captain. Above all, as we have seen, his success in this domain had made him aware of the great promise of computing and numerical analysis. This new awareness conditioned the rest of his professional life. Already during the war, Picone dreamed of a new institute where numerical analysis would be applied to experimental and technological research. As he explained in his autobiography:

> When the war ended, I returned to the university lecture halls, and even though I was immediately absorbed by the necessity of

[57]For the history of the UMI, see for example [Sansone 1974]. On Volterra's role more specifically, see [Guerraggio & Nastasi 2005, chapter 3].

[58]For the history of the CNR and, in particular, its first years, see [Simili 1993] and [Paoloni & Simili 2001].

[59]Volterra remained an influential figure in the international community especially through his role in the Rockefeller Foundation, where he supported projects for changing the seat of the International Bureau of Weights and Measures and for creating the Institut Henri Poincaré in Paris. On this, see [Siegmund-Schultze 2001].

> reconquering my position on the battlefield of pure science, which I had lost during the three years of the war, I never stopped thinking of mathematics as a powerful tool for the experimental sciences and for technology and of an organization of things which might allow the mathematician to intervene in a more opportune way in problems of a strictly mathematical nature that had blocked the progress of these sciences and their applications, including industrial applications.
>
> Already in these early years of reconquered peace—alas so transitory!—I was struck by the idea of creating an institute in which mathematicians, equipped with the most powerful numerical computing instruments, would be able to collaborate with experimental scientists and with technicians to reach concrete solutions of their numerical problems....
>
> Hence the use of calculators, even by the mathematician, hence the design of laboratories for the mathematician, who could no longer be portrayed as the abstract isolated thinker needing only paper and pencil for his work. The mathematician had to get out of his office and mingle with the crowd of those who seek to unravel the mysteries of nature and win possession of hidden treasures.
>
> Since those years, I have been advocating for my ideas with an inexhaustible tenacity among my friends and in my teaching, as well as in scientific and industrial circles. But these ideas progressed with extreme sluggishness! Today more than ever, the resistance of almost all mathematicians to their progress seems inexplicable [Picone 1972, p. 9].

With "inexhaustible tenacity," Picone recalled, and it is true. But in his campaign in favor of a computing institute he also showed that he was endowed with great foresight and political skills. In 1923, recounting to the members of the *Circolo Matematico di Catania* the part scientists had played in WWI, Picone concluded with a most lucid assessment of the future role of science and technology in wars: "We must think of the future! I hope that the warning I voice, even if it comes from my humble level, may reach high up: In a future war, the armies that will win will be those with the best technological preparation. Future wars will be wars among scientists [*le future guerre saranno guerre fra scienziati*]," [Picone 1934, p. 29]. To reach as high up as the Italian government, Picone sent a copy of his speech to the "Duke of victory" Armando Diaz (1861–1928), Minister of War of Mussolini's first government.[60] But his time was yet to come.

In 1925 Picone moved to Naples, where he created a small, one-chair institute with funds from the Bank of Naples (*Banco di Napoli*). In his memoir he called it an "embryonic Institute of Computing, equipped with machines of modest computing power, but sufficient to start the experiment" [Picone 1972, p. 12]. His first disciple was Renato Caccioppoli (1904–1959), who was immediately recognized as one of the best Italian analysts.

In 1932, when Picone accepted a chair in Rome, the small institute followed and was attached to the CNR as the Institute for the Application of Computing

[60] Diaz's answer, dated April 23, 1923, is reproduced by Picone in the reprint of his speech [Picone 1934].

(IAC).[61] According to his long-held beliefs, the IAC was to develop projects of civil, mechanical, and electronic engineering to which to apply numerical analysis—and mathematics in general. It was to devise new methods for the solution of problems submitted to it as well as to contribute to the development of analysis itself.[62] The IAC was a hierarchical structure at whose head was a director. The director was assisted by a vice director, some assistants and researchers, and about ten graduate computers and draughtsmen. The success of the IAC was quick and impressive—it was consulted by the Ministry of Aeronautics, the Ministry of the Army, and the Ministry of the Navy, and it launched collaborations with various industrial firms in civil engineering and electric production. Around fifty consulting contracts were signed on average every year. Activities at the IAC gave rise to about 250 publications during the period 1927–1940 covering many different fields, such as pure analysis, numerical computation, rational mechanics, civil engineering, the theory of elasticity, hydrodynamics, aerodynamics, etc. On an administrative level, the IAC, where scientific research was organized outside the academic environment, was a great innovation on the Italian mathematical landscape. It was able to hire young mathematicians by creating various professional positions. Mathematics (both pure and applied) had entered the world of consulting.

Picone's extraordinary skills as teacher and manager were essential for theIAC's success, but he also was an innovative researcher. For the first time on the Italian mathematical stage, numerical approaches became prominent. To Picone, the proofs of an existence theorem and of the uniqueness of the solution were not enough. Now new constructive procedures for computing explicit solutions were to be devised. In other words, to find the numerical algorithm, to prove its convergence, and to compute the approximate error were now fundamental questions to be treated with the highest level of rigor. Of course, Picone was not the first to face similar questions (not even in Italy). However, he was the first mathematician who explicitly took on problems emerging from other fields and to gather in the same institute different skills coming from various branches of pure, applied, and computational mathematics.

Picone's success—it must be emphasized—owed much to his unwavering support of the Fascist regime and to his close connection to the military establishment.[63] The institute was well known and appreciated by high-ranking Fascist authorities who supported it financially. In this connection, Picone wrote in 1938 to the president of the CNR, to whom he presented the IAC's scientific activity:[64]

> The highest task for [Italy's] autarchy assigned by the Duce to the [CNR] can be faced if the mathematical instrument provides industries with procedures not only for the minimizing of materials

[61]On the history of the IAC, see [Nastasi 2006], as well as the accounts of its founder [Picone 1938] and [Picone 1953]. On Picone's engagement in numerical computations, see [Fichera 1986].

[62]See [Picone 1968] and Picone's "Sulla necessità per il progresso delle Scienze sperimentali e matematiche della creazione di un Istituto Centrale di Calcolo," unsigned report in the name of the "*Comitato Matematico del CNR*," 1929 [IAC Archives].

[63]In a letter to Gentile (June 5, 1923) Picone warmly congratulated him for his adhesion to the regime, which, he wrote, would pour "new pure blood into the robust veins of the Party that has rebuilt and renewed the country" [Guerraggio & Nastasi 1993, p. 185].

[64]Picone's unpublished report here quoted is to be found in [IAC Archives].

used in construction without compromising stability and power, but also for the maximizing of time savings.

... [The t]echnician cannot be requested to have the high mathematical skills on which the computations required by the task mentioned above are based; in the present Institute, the technician may thus find the scientific assistance and the full understanding of his needs and objectives....

Industial technical offices must not fear that this collaboration might erode their own importance and necessary autonomy; on the contrary, once the adequate computing method for a special purpose has been established, ... this method will remain in their exclusive possession; absolute discretion about the method will be kept by the Institute; and industrial technicians will be freely able to apply the method or modify it in any way the company will think beneficial....

A large part of the research carried out by the Institute comes from the ministries of national defense: the stability of aeronautic apparatuses, aerial bombing, stabilization instruments for underwater torpedoes, the firing of artillery at great distances, military radio communications, etc., all have provided ground for long and laborious researches, completed successfully with the help of the mathematical resources and the numerical computational means that belong to the Institute.

And the National Institute for the Application of Computing, which is today at the orders of His Eminence, Marshal of Italy Pietro Badoglio, aims at soundly maintaining its efficiency and even increasing it, also in order to always be able to answer any call for the power of Italian arms.[65]

---

[65] "*L'altissimo compito autarchico assegnato dal DUCE al Consiglio Nazionale delle Ricerche può essere affrontato, se, da parte delle industrie, si richiede anche allo strumento matematico di pervenire a ridurre al minimo il materiale impiegato nelle costruzioni, senza comprometterne la stabilità e la potenza, conseguendo altresì tutta la possibile economia di tempo.*

D'altra parte non è da richiedere al tecnico il possesso di quelle elevate cognizioni matematiche che sono di fondamento ai calcoli che impone il compito ora accennato, onde la creazione di questo Istituto nel quale il tecnico potrà trovare tutta l'assistenza scientifica possibile e la piena comprensione delle sue necessità e delle sue mete....

Gli uffici tecnici delle industrie non possono temere la collaborazione indicata come tendente a sminuirne l'importanza e l'autonomia necessaria, anzi, una volta stabilito il metodo di calcolo adatto per le particolari ricerche nell'indirizzo sopraddetto, tale metodo, con la più assoluta riservatezza da parte dell'Istituto, sarà di loro completa proprietà e potrà essere liberamente applicato e modificato a puro ed unico vantaggio del successo della missione ad essi affidata nell'industria alla quale appartengono....

Molta parte delle ricerche compiute dall'Istituto provengono dai Ministeri della difesa nazionale: la stabilità delle costruzioni aeronautiche, il tiro di bombardamento da aereo, gli apparecchi di stabilizzazione del siluro subacqueo, il tiro delle artiglierie a grandi distanze, le radio-comunicazioni militari, ecc., hanno fornito materia di lunghe e laboriose ricerche, felicemente condotte a termine, con l'ausilio dei mezzi matematici e di calcolo numerico in possesso dell'Istituto.

E l'Istituto Nazionale per le Applicazioni del Calcolo, oggi agli ordini di S. E. il Maresciallo d'Italia Pietro Badoglio, si propone di mantenere integra ed in aumento la propria efficienza anche per poter sempre rispondere a qualsiasi appello per la potenza delle armi italiane" [IAC Archives, ibid.].

Picone adhered to all aspects of the Fascist ideology without exception—including the racial laws. He took part in the UMI meeting in December 1938 when the Italian mathematical community as a whole decided to accept Fascist racial policy. Right after the meeting, Picone wrote Wacław Šierpinsky (1882–1969) to point out that it was "pressing" that Aryan scientists work together to show that science could progress without Jewish participation. Picone also mentioned that in the Academy of Lincei "the percentage of Jewish members is very high."[66] Picone nevertheless offered unmistakable tokens of friendship and support for the Jewish mathematicians Fubini, Alessandro Terracini (1889–1968), and Guido Ascoli (1887–1957). Picone's faith in the regime and in Mussolini was so strong that he refused to cancel a joint lecture series with Fabio Conforto (1909–1954) in Germany during the month of July 1943. After the Fascist regime fell, on September 8, 1943, Picone distanced himself from it and started to convert "his" institute to peacetime work. But this is another story [Nastasi 2006].

Conclusion

We hope to have thrown some light on Italian mathematicians' role in World War I—a subject that has hitherto almost completely escaped the attention of historians. We have argued that most Italian mathematicians played a clear and unambiguous role: politically they were against the Central Powers from August 1914 to the early 1920s even when Italy was not yet or no longer at war with them, and they soon sided with the Entente, arguing for Italian intervention, developing scientific collaborations with scientists from Allied nations, and enforcing the ban decided upon by international scientific organization in 1918. This was the case of Volterra, Picone, Castelnuovo, Enriques, Severi, and many others. As we documented, no one in the mathematical community was as strongly involved as Volterra in favor of Italian intervention in the war on the side of France, Great Britain, and Russia.

But other mathematicians (such as Levi-Civita and Segre) had a different attitude and did not want Italy to enter the war. The polemics between supporters of Italian intervention and neutralists—reviled as "Germanophiles"—were harsh. However, when war was declared, all of them, either in favor of or opposed to intervention, joined together for the "defense of the fatherland"—with the sole exception, as far as we know, of Levi-Civita, who remained a committed pacifist. Segre's case is emblematic: considered a Germanophile before the war broke out, he soon addressed a moving letter to Volterra expressing his deep nationalistic feeling that he concluded by the call: "*Viva l'Italia*" [Volterra Papers, Segre to Volterra, May 27, 1915].

During the war, mathematicians fought in the Italian Army (generally as officers) to defend their country and often became close collaborators of military organizations. We have in particular seen how Volterra and Picone, respectively, created two institutions attached to the Ministry of War: the Office for Inventions and Research and the Ballistic Office attached to the Sixth Army. For both of them, the war experience was fundamental and life-changing: it showed them the way in which closer links between mathematics, science, industry, and the military were to be woven. In the interwar period, such experience was drawn upon to establish new institutions: the CNR for Volterra and the IAC for Picone. Both institutions can, as we have seen, be interpreted as the natural continuation of their war work.

[66] This letter was published in [Guerraggio et al. 2007, pp. 57–58].

But we should point out that the majority of Italian mathematicians showed little interest in the further development of such applied mathematics.

After the war another polemic arose in Italy between the promoters of ostracism against German and Austrian science and those who wanted to reintegrate them into international meetings and organizations. In Italy, supporters of intervention were for the most part in favor of ostracism. In his 1920 *Saggi scientifici*, Volterra reprinted some of his well-known lectures held at the International Congresses of Paris (1900) and Rome (1908) [Volterra 1920]. In this reprint, he carefully erased all references to German mathematicians, including those to Felix Klein.[67] Other mathematicians had a more international spirit and wished to reestablish scientific collaboration with Germany. Among them, Levi-Civita played a large role in the creation of the International Congresses of Applied Mechanics with German colleagues. Similarly, De Franchis refused to exclude German mathematicians from the Mathematical Circle of Palermo.

In the spring of 1914, an international meeting took place in Palermo for the celebration of the thirty-year anniversary of the Mathematical Circle, organized mainly by Segre and Volterra. Many illustrious mathematicians and physicists took part in the celebrations, spending a week together in Palermo, most of them coming from Germany, France, and, of course, Italy.[68] About one month later, war tore down personal and scientific relationships generally for many years. Many of them would resume these relationships with the "enemy" only at the International Congress of Mathematicians in Bologna in 1928, fourteen years later. This may serve to underscore the absurdity of World War I.

Abbreviations

- BIR: Board of Invention and Research.
- CNR: *Consiglio Nazionale delle Ricerche* (National Research Council).
- IAC: *Istituto per le Applicazioni del Calcolo* (Institute for the Application of Computing).
- ICA: *Istituto Central Aeronautico* (Central Institute of Aeronautics).
- IMU: International Mathematical Union.
- INDAM: Istituto Nazionale di Alta Matematica.
- IRC: International Research Council.
- MIT: Massachusetts Institute of Technology.
- SIPS: *Società Italiane per il Progresso delle Scienze* (Italian Society for the Advancement of Science).
- UIR: *Ufficio Invenzioni e Ricerche* (Office for Inventions and Research).
- UMI: *Unione Matematica Italiana* (Italian Mathematical Union).

Bibliography

Archival Sources

[Birkhoff Papers] George David Birkhoff Papers, Harvard University Archives.

[IAC Archives] Archives of the Istituto per le Applicazioni del Calcolo "Mauro Piccone," Rome.

---

[67]More on this is in [Guerraggio & Nastasi 2008, p. 143].

[68]Among the attendees, one notes Edmund Landau representing the Göttingen Mathematical Society, William Sedgwick from the Society of Arts of MIT, and Edward Van Vleck from the American Mathematical Society. Unable to attend, Camille Jordan and Émile Picard wired their kindest regards. See [Nastasi 1988].

[Manuscrits Léon] Manuscrits Xavier Léon. Bibliothèque Victor Cousin, Bibliothèque interuniversitaire de la Sorbonne, Paris, MSVC 358-369.

[Levi-Civita Papers] Tullio Levi-Civita Papers, Accademia Nazionale dei Lincei, Rome.

[Sommerfeld Papers] Arnold Sommerfeld Papers, Archives of Deutsches Museum, Munich.

[Volterra Papers] Vito Volterra Papers, Accademia Nazionale dei Lincei, Rome.

Published Works

[Adhémar 1934] Robert d'Adhémar, *La Balistique extérieure*, Mémorial des sciences mathématiques **65**, Gauthier-Villars, Paris, 1934.

[Albers et al. 1986] Donald Albers, Gerald L. Alexanderson, & Constance Reid, *International Mathematical Congresses: An Illustrated History, 1893–1986*, Springer, New York, 1986.

[Amaldi & Levi-Civita 1923–1927] Ugo Amaldi & Tullio Levi-Civita, *Lezioni di Meccanica Razionale*, 2 vols., Zanichelli, Bologna, 1923–1927.

[Amaldi & Levi-Civita 1935] Ugo Amaldi & Tullio Levi-Civita, *Nozioni di balistica esterna*, Zanichelli, Bologna, 1935.

[Anderson 1997] John David Anderson, *A History of Aerodynamics and Its Impact on Flying Machines*, Cambridge University Press, Cambridge, 1997.

[Apollonio 1973] Umbro Apollonio, ed., *Documents of 20th-Century Art: Futurist Manifestos*, trans. Robert Brain, Robert Willard Flint, John C. Higgit, & Caroline Tisdall, Viking Press, New York, 1973.

[Aubin 2012] David Aubin, L'élite sous la mitraille : les mathématiciens normaliens "morts pour la France", 1914–1918, in *Aventures de l'analyse de Fermat à Borel. Mélanges en l'honneur de Christian Gilain*, ed. Suzanne Féry, PUN-Éditions universitaires de Lorraine, Nancy, 2012, pp. 681–706.

[Aubin forthcoming] David Aubin, *"Audacity or Precision": The Paradoxes of Henri Villat's Fluid Mechanics in Interwar France*, paper presented at a "Workshop on the History of Fluid Mechanics", Rauischholzhausen, October 15–18, 2006. To appear.

[Aubin & Bret 2003] David Aubin & Patrice Bret, eds., *Le Sabre et l'éprouvette: l'invention d'une science de guerre, 1914–1939*, "14–18 Aujourd'hui," Agnès Viénot, Paris, 2003.

[Battimelli 1996] Giovanni Battimelli, *Tullio Levi-Civita e i congressi internazionali di meccanica applicata*, Rivista di Storia della Scienza **4** (1996), pp. 51–80.

[Beaulieu 2009] Liliane Beaulieu, *Regards sur les mathématiques françaises entre les deux guerres. Introduction*, Revue d'histoire des sciences **62-1** (2009), pp. 9–37.

[Bianchi 1910] Giovanni Bianchi, *Corso teorico pratico di balistica esterna*, Pasta, Turin, 1910.

[Bianchi 1959] Luigi Bianchi, *Opere*, vol. 11: Corrispondenza, ed. UMI & CNR, Edizioni Cremonese, Rome, 1959.

[Booß-Bavnbek & Høyrup 2003] Bernhelm Booß-Bavnbek & Jens Høyrup, eds., *Mathematics and War*, Birkhäuser, Basel, 2003.

[Brigaglia & Masotto 1982] Aldo Brigaglia & Guido Masotto, *Il Circolo Matematico di Palermo*, Dedalo, Bari, 1982.

[Caraccioli 1968] Alberto Caraccioli & al., *Il trauma dell'intervento, 1914–1919*, Vallechi, Florence, 1968.

[Corbino 1921] Orso Mario Corbino, *Discorso di S.E. Orso Mario Corbino*, Atti della Societá Italiana per il Progresso delle Scienze (September 1921), pp. LXXXIX-XCI.

[Crocco 1961] Arturo Crocco, Socio dell'Accademia Nazionale dei Lincei, in *Vito Volterra nel I centenario della nascita, 1860–1960*, Problemi attuali di scienza e di cultura **51**, Accademia Nazionale dei Lincei, Rome, pp. 26–27.

[Croce 1950] Benedetto Croce, *L'Italia dal 1914 al 1918. Pagine sulla guerra*, Laterza, Bari, 1950.

[Dauben 1980] Joseph W. Dauben, *Mathematics and World War I: The International Diplomacy of G.H. Hardy and Gösta Mittag-Leffler as Reflected in Their Personal Correspondence*, Historia Mathematica **7** (1980), pp. 261–288.

[De Maria & Seidel 1980] Michelangelo De Maria & Robert W. Seidel, *Lo scienziato e l'inventore: l'inizio dell'integrazione sistematica fra scienza e industria in USA durante la prima guerra mondiale*, Testi & Contesti **4** (1980), pp. 5–32.

[Denjoy 1939] Arnaud Denjoy, *Aspects actuels de la pensée mathématique*, Bulletin de la Société mathématique de France **67** (1939), pp. 1–12.

[Dieudonné 1977] Jean Dieudonné, *Panorama des mathématiques pures : le choix bourbachique*, Gauthier-Villars, Paris, 1977.

[Dieudonné 1978] Jean Dieudonné, ed., *Abrégé d'histoire des mathématiques: 1700–1900*, 2 vols., Hermann, Paris, 1978.

[Di Martino 2005] Basilio Di Martino, *I dirigibili italiani nella grande guerra*, Ufficio storico aeronautica militare, Rome, 2005.

[Fabi 1994] Lucio Fabi, *Gente di trincea*, Mursia, Milan, 1994.

[Ferrari 2004] Paolo Ferrari, ed., *L'aeronautica italiana: une storia del Novencento*, Franco Angeli, Milan, 2004.

[Fichera 1978] Giuseppe Fichera, *Mauro Picone*, Atti dell'Accademia delle Scienze di Bologna (13) **5** (1978), pp. 245–263.

[Fichera 1986] Giuseppe Fichera, Mauro Picone, un pioniere dell'analisi d'oggi, in *Atti del Convegno celebrativo del centenario della nascita di Mauro Picone e di Leonida Tonelli (Roma, 6–9 maggio 1985)*, Atti dei Convegni Lincei **77**, Accademia nazionale dei Lincei, Rome, 1986, pp. 75–88.

[Fieschi 1987] Roberto Fieschi, *Scienza e guerra*, Editori Riuniti, Rome, 1987.

[Fieschi & Paris De Renzi 1995] Roberto Fieschi & Claudia Paris De Renzi, *Macchine da guerra: gli scienziati e le armi*, Einaudi, Turin, 1995.

[Filippi 1919] Giovanni Filippi, *Le Università e gl'Istituti d'istruzione superiore in Italia durante la guerra*, Bollettino ufficiale del Ministero della Pubblica Istruzione **47** (1920), pp. 329–384.

[Fontanon 2003] Claudine Fontanon, Paul Painlevé et l'aviation : aux origines de l'étatisation de la recherche scientifique, in *Paul Painlevé (1863–1933): un savant en politique*, ed. Claudine Fontanon & Robert Frank, Presses Universitaires de Rennes, Rennes, pp. 41–56.

[Fontanon 2011] Claudine Fontanon, La carrière scientifique de Gustave Eiffel : un long parcours en aérodynamique (1892–1922), in *L'ingénieur entrepreneur. Les centraliens et l'industrie*, ed. Jean-Louis Bordes, Annie Champion, & Pascal Desabres, Presses Université Paris-Sorbonne, Paris, pp. 333–356.

[Gibelli 1998] Antonio Gibelli, *La Grande Guerra degli italiani, 1915–1918*, Sansoni, Florence, 1998.

[Goldstein 2009] Catherine Goldstein, *La théorie des nombres en France dans l'entre-deux-guerres : de quelques effets de la première guerre mondiale*, Revue d'histoire des sciences **62-1** (2009), pp. 143–175.

[Goldstein & Aubin forthcoming] Catherine Goldstein & David Aubin, eds., *La Grande Guerre des mathématiciens français*, forthcoming.

[Goodstein 2007] Judith R. Goodstein, *The Volterra Chronicles: The Life and Time of an Extraordinary Mathematician (1860–1940)*, American Mathematical Society, Providence & London Mathematical Society, London, 2007.

[Gratzer 2000] Walter Bruno Gratzer, *The Undergrowth of Science: Delusion, Self-deception, and Human Frailty*, Oxford University Press, Oxford & New York, 2000.

[Guerraggio et al. 2007] Angelo Guerraggio, Maurizio Mattaliano, & Pietro Nastasi, *Mauro Picone e i matematici polacchi: 1937–1961*, Accademia Polacca delle Scienze, Rome, 2007.

[Guerraggio & Nastasi 1993] Angelo Guerraggio & Pietro Nastasi, *Gentile e i matematici italiani: Lettere 1907–1943*, Bollati Boringhieri, Turin, 1993.

[Guerraggio & Nastasi 2005] Angelo Guerraggio & Pietro Nastasi, *Italian Mathematics Between the Two World Wars*, Birkhäuser, Basel, 2005.

[Guerraggio & Nastasi 2008] Angelo Guerraggio & Pietro Nastasi, *Roma 1908: il Congresso Internazionale dei Matematici*, Bollati Boringhieri, Turin, 2008.

[Guerraggio & Paoloni 2008] Angelo Guerraggio & Giovanni Paoloni, *Vito Volterra*, Franco Muzzio Editore, Rome, 2008.

[Hartcup 1988] Guy Hartcup, *The War of Invention: Scientific Developments, 1914–18*, Brassey's, London, 1988.

[Isnenghi 1989] Mario Isnenghi, *Il mito della Grande Guerra*, Il Mulino, Bologna, 1989.

[Isnenghi 1993] Mario Isnenghi, *La Grande Guerra*, Giunti-Casterman, Florence, 1993.

[Isnenghi 1996–1997] Mario Isnenghi, *I luoghi della memoria*, 3 vols., Laterza, Rome & Bari, 1996–1997.

[Isnenghi & Rochat 2000] Mario Isnenghi & Giorgio Rochat, *La Grande Guerra 1914–1918*, La Nuova Italia, Florence, 2000.

[Lacaita 2007] Carlo G. Lacaita, *Università e impresa*, Annali di Storia delle Università Italiane, **12** (2007), 17 pp., online `http://www.cisui.unibo.it/annali/12/testi/01Lacaita_frameset.htm`.

[Landogna 1937] Francesco Landogna, ed., *Corso di Cultura Militare*, 3 vols., Morano Editore, Naples, 1937.

[Lehto 1998] Olli Lehto, *Mathematics without Borders: A History of the International Mathematical Union*, Springer, New York, 1998.

[Leloup 2009] Juliette Leloup, *L'entre-deux-guerres mathématique à travers les thèses soutenues en France*, doctoral thesis, Université Pierre et Marie Curie, Paris, 2009.

[Levi-Civita 1906] Tullio Levi-Civita, *Sulla penetrazione dei proiettili nei mezzi solidi*, Atti dell'Istituto Veneto di Scienze, Lettere ed Arti **65** (1906), pp. 1148–1158. Repr. in *Opere matematiche*, 6 vols., ed. Accademia nazionale dei Lincei, Zanichelli, Bologna, 1954–1973, vol. 2, pp. 505–513.

[Luchaire 1965] Julien Luchaire, *Confession d'un français moyen*, 2 vols., Olschki, Florence, 1965; reproduced from first ed. Sagittaire, Marseille, 1943.

[MacMillan 2001] Margaret Mac Millan, *Peacemakers: The Paris Conference of 1919 and Its Attempt to End War*, J. Murray, London, 2001.

[Mazliak & Tazzioli 2009] Laurent Mazliak & Rossana Tazzioli, eds., *Mathematicians at War: Volterra and his French Colleagues in World War One*, Springer, New York, 2009.

[Miana 2007] Paolo Miana, *I bombardiero caproni nella grende guerra*, Mecchione ed., Varese, 2007.

[Nastasi 1988] Pietro Nastasi, *Documenti della vita del Circolo Matematico di Palermo*, Rendiconti del Circolo Matematico di Palermo **20** (1988), suppl.

[Nastasi 2006] Pietro Nastasi, *I primi quarant'anni di vita dell'Istituto per le Applicazioni del Calcolo "Mauro Picone"*, Bollettino dell'Unione Matematica Italiana, sez. A., (2006), pp. 1–244.

[Nastasi & Tazzioli 2000] Pietro Nastasi & Rossana Tazzioli, eds., *Aspetti scientifici e umani nella corrispondenza di Tullio Levi-Civita (1873–1941)*, Quaderni Pristem **12**, Palermo, 2000.

[Paoloni & Simili 2001] Giovanni Paoloni & Raffaella Simili, eds., *Per una storia del Consiglio Nazionale delle Ricerche*, 2 vols., Laterza, Rome & Bari, 2001.

[Paoloni & Simili 2008] Giovanni Paoloni & Raffaella Simili, Vito Volterra and the Making of Research Institutions in Italy and Abroad, in *The Migration of Ideas*, ed. Roberto Scazzieri & Raffaella Simili, Science History Publications, Sagamore Beach, pp. 123–150.

[Pestre 1990] Dominique Pestre, La première guerre mondiale et ses conséquences sur la science en France, Talk delivered at the Conference on "Scienza, Tecnologia, Istituzioni in Europa (1900–1920): Vito Volterra e l'origine del CNR," Rome, November 27–28, 1990.

[Picone 1917] Mauro Picone, *Sul tiro dei medii e dei grossi calibri in montagna*, Rivista dell'Artiglieria e Genio **35** (1917), pp. 5-15.

[Picone 1934] Mauro Picone, *L'Artiglieria italiana nella guerra mondiale*, Tipografia del Senato di G. Bardi, Rome; repr. from a lecture held in Catania in 1923 and first published in *Esercitazioni di Matematica del Circolo Matematico di Catania* **3** (1923), pp. 1–31.

[Picone 1938] Mauro Picone, *L'Istituto Nazionale per le Applicazioni del Calcolo nel quadriennio 1933/37*, Pubbl. dell'INAC **27**, CNR, Rome, 1938.

[Picone 1953] Mauro Picone, *Sull'opera matematica dell'Istituto Nazionale per le Applicazioni del Calcolo nel decorso quarto di secolo della sua esistenza*, Atti del IV Congr. dell'U.M.I. (Taormina, October 25–31, 1951), Cremonese, Rome, 1953, vol. 1, pp. 27–44.

[Picone 1956] Mauro Picone, *Commemorazione di Vito Volterra pronunciata a Palermo il 15 settembre 1956, cinquantesimo anniversario della Società Italiana per il Progresso delle Scienze*, Ricerca Scientifica **26** (1956), pp. 3277–3289.

[Picone 1964] Mauro Picone, *Commemorazione di Antonio Signorini*, Atti e Memorie dell'Accad. Petrarca di Lettere Arti e Scienze di Arezzo **37** (1958–1964), pp. 378–402.

[Picone 1967] Mauro Picone, *Presentazione di pubblicazioni dovute alla guerra 1915–1918*, Rendiconti delle sedute della Cl. Sci. Fis. Mat. Natur. Accad. Lincei **42** (1967), fasc. 1, 7 pp.

[Picone 1968] Mauro Picone, *Presentazione di pubblicazioni riguardanti l'attività dell'Istituto per le Applicazioni del Calcolo, dal 1927, anno della sua fondazione, al 1960, in cui fu sottratto alla direzione del suo ideatore*, Rendiconti dell'Accademia dei Lincei **44**(4) (1968), pp. 1–10.

[Picone 1971] Mauro Picone, *L'Artiglieria italiana nella prima guerra mondiale*, in Conferenze 1970–1971, Scuole di Applicazione (Turin), 1971, pp. 67–98.

[Picone 1972] Mauro Picone, *La mia vita*, Bardi, Rome, 1972.

[Picone 1974] Mauro Picone, *Discorso tenuto nel Convegno Internazionale per la celebrazione del centenario della nascita di Tullio Levi-Civita (17–19 dicembre 1973)*, Tip. Bardi, Rome, pp. 20–22.

[Picone & Bilinski 1969] Mauro Picone & Bronislaw Bilinski, *Marie Sklodowska-Curie in Italia. Nel centenario della nascita (1867–1934)*, Ossolineum, Wroclaw, Warsaw, & Cracow, 1969.

[Quilici & Ragghianti 1989] Leana Quilici & Renzo Ragghianti, *Il Carteggio Xavier Léon: corrispondenti italiani con un'appendice di lettere di Geroges Sorel*, Giornale critico della filosophia italiana **9** (1989), pp. 295–368.

[Renard 2002] Isabelle Renard, *"Il Grenoble": Il primo Istituto francese al mondo*, Antologia Vieusseux **7** (2002), pp. 1–30.

[Roma 1918] *Onoranze a Luciano Orlando, Ruggiero Torelli, Eugenio Elia Levi, Adolfo Viterbi, professori di matematica nelle Università italiane caduti in guerra*, Seminario Matematico della Facoltà di Scienze dell'Università di Roma, June 22, 1918, Bertero, Rome, 1918.

[Roussel 1989] Yves Roussel, *L'histoire d'une politique des inventions: 1887–1918*, Cahiers pour l'Histoire du CNRS **3** (1989), pp. 19–57.

[Rusconi 2005] Gian Erico Rusconi, *L'azzardo del 1915: come l'Italia decide la sua guerra*, Il Mulino, Bologna, 2005.

[Sansone 1974] Giovanni Sansone, *Le attività dell'Unione Matematica Italiana nel primo cinquantennio della sua fondazione*, Bollettino dell'Unione Matematica Italiana **2** (1974), pp. 7–43.

[Sansone 1977] Giovanni Sansone, *Algebristi, analisti, geometri differenzialisti, meccanici e fisicimatematici ex-normalisti del periodo 1860–1929*, Scuola Normale Superiore di Pisa, Pisa, 1977.

[Schiavon 2003a] Martina Schiavon, *Itinéraires de la précision : géodésiens, artilleurs, savants et fabricants d'instruments en France, 1870–1930 (environ)*, doctoral thesis, EHESS, Paris, 2003.

[Schiavon 2003b] Schiavon, Martina, Des savants-officiers entre science, armée, État et industrie de précision : les géodésiens du Service géographique de l'Armée, 1887–1920, in [Aubin & Bret 2003], pp. 60–73.

[Schroeder-Gudehus 1978] Brigitte Schroeder-Gudehus, *Les scientifiques et la paix : la communauté scientifique internationale au cours des années 20*, Les Presses de l'Université de Montréal, Montréal, 1978.

[Schwartz 2001] Laurent Schwartz, *A Mathematician Grappling With His Century*, trans. Leila Schneps, Birkhäuser, Basel & Boston, 2001.

[Severi 1915] Francesco Severi, *L'opinione di un socialista ufficiale*, L'Adriatico (March 9, 1915).

[Siacci 1888] Francesco Siacci, *Balistica*, Casanova, Turin, 1888.

[Siegmund-Schultze 2001] Reinhardt Siegmund-Schultze, *Rockefeller and the Internationalization of Mathematics Between the Two World Wars*, Birkhäuser, Basel & Boston, 2001.

[Siegmund-Schultze 2003] Reinhardt Siegmund-Schultze, Military Work in Mathematics 1914–1945: An Attempt at an International Persective, in [Booß-Bavnbek & Høyrup 2003], pp. 23–82.

[Signorini 1919] Antonio Signorini, *Un teorema di confronto in Balistica esterna e alcune sue applicazioni*, Rendiconti del Circolo Matematico di Palermo **43** (1919), pp. 357–393.

[Signorini 1922] Antonio Signorini, *Sulla velocità minima*, Rendiconti dell'Accademia Nazionale dei Lincei **31**(2), pp. 101–104.

[Simili 1993] Raffaella Simili, ed., *Scienza, tecnologia e istituzioni in Europa*, Laterza, Bari, 1993.

[Tanzi Cattabianchi 1977] Luigi Tanzi Cattabianchi, *I contributi di Mauro Picone alla Balistica razionale*, Rivista Matematica dell'Università di Parma **3** (1977), pp. 357–389.

[Tanzi Cattabianchi 1981] Luigi Tanzi Cattabianchi, *I contributi di Guido Fubini e di Francesco Severi ad alcuni problemi di balistica esterna*, Atti dell'Accademia delle Scienze di Torino, Cl. Sci. Fis. Mat. Natur. **115** Suppl. (1981), pp. 217–233.

[Tanzi Cattabianchi 1988] Luigi Tanzi Cattabianchi, *I contributi di Vito Volterra alla balistica da aeromobili*, Rivista Matematica dell'Università di Parma **14** (1988), pp. 87–105.

[Tenca 1959] Luigi Tenca, *Matematici combattenti*, Archimede **11**(4) (1959), pp. 186–190.

[Teofilato 1920] Pietro Teofilato, *Sulla determinazione dei centri sonori*, Atti della Pontificia Accademia dei Nuovi Lincei *73* (1920), pp. 185–189.

[Terracini 1968] Alessandro Terracini, *Ricordi di un matematico*, Ediz. Cremonese, Rome, 1968.

[Tomassini 1995] Luigi Tomassini, *L'Italia nella Grande Guerra*, Fenice, Milan, 1995.

[Tricomi 1967] Francesco G. Tricomi, *La mia vita di matematico attraverso la cronistoria dei miei lavori*, Cedam, Padua, 1967.

[Tricomi 1977] Francesco G. Tricomi, *Mauro Picone (1885–1977). Cenni commemorativi*, Atti della Accademia delle Scienze di Torino **111** (1977), pp. 573–576.

[Volterra 1916] Vito Volterra, *Metodo di calcolo degli elementi di tiro per artiglieria aeronautica*, Rendiconti dell'Istituto Centrale Aeronautico **5** (1916); repr. in *Opere matematiche*, 5 vols., ed. Accademia nazionale dei Lincei, Ediz. Cremonese, Rome, 1954–1962, vol. 4, pp. 200–248.

[Volterra 1920] Vito Volterra, *Saggi scientifici*, Zanichelli, Bologna, 1920.

[Weil 1992] André Weil, *The Apprenticeship of a mathematician*, trans. Jennifer Gage, Birkhäuser, Basel & Boston, 1992.

A Mobilized Community: Mathematicians in the United States during the First World War

Thomas Archibald, Della Dumbaugh, and Deborah Kent

There are many stark contrasts between the American and the European experiences of the Great War. Before the war the United States, although militarily involved in the Pacific Ocean and the Americas, was only occasionally interested in European affairs. The war provided economic opportunity for an initially neutral U.S., and the brief involvement in hostilities, though very important for the Allied victory, was nowhere near as costly as for the European combatants. The 116,000 deaths of U.S. soldiers, about half of them in combat, can hardly compare with the French loss of about 3.5% of its population in military deaths alone, and of course there was no damage to property on U.S. soil [Leland & Oboroceanu 2010, p. 2]. But the U.S. emerged from the war with a new role in international affairs, one which accorded it recognition as a major player politically and economically. Mathematics and science were of course touched by these transformations, though the direct impact on the main themes of mathematical research may have seemed minimal to some historians. Yet the community that reconstituted itself following the rapid mobilization and demobilization entered the 1920s with a renewed self-confidence, a new national awareness, and a feeling that mathematics and mathematicians were a national resource with a role to play in the shaping and carrying out of national aims. Individuals, research groups, and institutions saw new collaborative potential and new fields in which mathematics could find a place, among them government and industry.

Thomas Archibald's address is Department of Mathematics, Simon Fraser University, British Columbia, Canada. His email address is tarchi@sfu.ca.

Della Dumbaugh's address is Department of Mathematics, University of Richmond, Richmond, VA 23173. Her email address is ddumbaugh@richmond.edu.

Deborah Kent's address is Department of Mathematics, Drake University, Des Moines, Iowa 50311. Her email address is deborah.kent@drake.edu.

In doing the research for this chapter, T. Archibald and D. Kent were supported in part by the Social Science and Humanities Research Council (Canada) and by Simon Fraser University, support which is gratefully acknowledged. D. Kent also thanks the American Philosophical Society for its Franklin Grant support. We would like to thank David Aubin for providing materials from the G. D. Birkhoff Papers held at the Harvard University Archives.

This chapter owes a considerable debt to earlier writings on aspects of the role of U.S. mathematics in the war effort. Many of these works however have tended to focus primarily on local and specific issues. David Alan Grier has looked in some detail at the activities of Oswald Veblen at the Aberdeen Proving Ground (APG) and related matters [Grier 2001, Grier 2005a]. The late Loren Butler Feffer cogently considered aspects of Veblen's research agenda and the role of applications in U.S. mathematics in ways that focussed partly on the war [Butler Feffer 1997, Butler Feffer 1998]. The general issue of science in the war effort forms an important part of the classic 1978 study by Daniel Kevles, *The Physicists*, and Kevles also looked in particular at the role of the astronomer George Ellery Hale during the war [Kevles 1968]. One of the authors of this chapter has devoted considerable attention to Leonard E. Dickson, an important figure in U.S. mathematics during the war years [Fenster 1999a, Fenster 1999b]. The volumes edited by Peter Duren provide a good deal of raw data on the matter, much of it previously published elsewhere [Duren 1988]. And Alan Gluchoff's recent and excellent technical survey of ballistics in the period gives substance to a number of the general remarks in our study [Gluchoff 2011].

The U.S. entry into the war did not occur until early April 1917. President Woodrow Wilson, formerly president of Princeton University, preferred to remain neutral, but ongoing German submarine attacks on U.S. merchant shipping to Britain and intelligence that the Germans were encouraging Mexico to go to war against the U.S. provoked action. The late entry into the war and geographic distance from the conflict characterize the rapid mobilization of American forces in the First World War. These factors similarly shaped the activities of American mathematicians engaged in the war.

The aim of this chapter is to explore the effect of the war on mathematics and mathematicians in the U.S., as well as the role of American mathematics in the conflict. To understand the interrelations between mathematics and the First World War in the context of the U.S., this chapter considers mathematics before, during, and after World War I, with particular attention to academic institutions and governmental and private agencies and foundations. We survey the national, educational, and scientific contexts for mathematics in the U.S. before, during, and after the war, at times focusing on certain individuals and institutions to illustrate specific mathematical activity in the U.S. during WWI. What emerges is a picture of a mathematical community broadly involved in war-related educational efforts, war research, and the conduct of the war at all levels. One impact of this activity on the mathematical community was to forge a more unified community; another was the establishment of government/military/academic collaborations. We describe the main features of the interface between the mathematical community and the war effort in a way that surveys the existing material and introduces a certain amount of new archival evidence.

A few institutions will serve our story as focal points. It is not a coincidence that Harvard University, Princeton University, and the University of Chicago play leading roles in attempting to delineate the responses of the mathematical community to the war. In fact, their leadership is intertwined in complex ways, hinging on the activities of certain figures who themselves had leadership roles in the different arenas of mathematical activity. George David Birkhoff, Veblen, and Dickson are not the only important figures, but their names will come up repeatedly. As research leaders, they are emblems of the coming of age of mathematics in the U.S., while playing important roles in bringing together a complex U.S. community. They were also important in the linking of American with European mathematics both via the content of their work and teaching and through their personal activity with European mathematicians. They, and others from these institutions, performed crucial roles in the creation and management of new infrastructure for mathematics, such as the National Research Council (NRC), the American Mathematical Society (AMS), and the Mathematical Association of America (MAA). Indeed, while we focus on the universities, we also attempt to identify aspects of the wartime activity of trans-local associations such as the AMS, the NRC, the Carnegie Institution of Washington (CIW), and the Rockefeller Foundation. Faculty and administrators from these schools also figure prominently in the temporary structures that arose in connection with military conduct of the war, as we shall see. These universities are also important by virtue of their networks of alumni and, not incidentally, by the relatively large amount of information we have about them from both primary and secondary sources. This is not to say that other universities and colleges did not play important roles, quite the contrary. However the institutional leadership that was already manifest before the war was reinforced by both the conduct of the war and events in its immediate aftermath.

1. Before the War

The development of the American university system dramatically influenced the emergence of an American mathematical community during the last quarter of the nineteenth century. It is important to realize that this is a "system" in certain practical senses, sharing degree structures and to some extent admission and graduation requirements, but individual institutions are highly independent and there was and is no national system of higher education in the U.S. There is a very large number of institutions that range in size, mandate, religious affiliation, and programs, and these are highly dispersed geographically. An American mathematical community had come into existence only in the last quarter of the nineteenth century and had begun to take shape and grow in the first quarter of the new century [Parshall & Rowe 1994]. Structures for a professional mathematics community began taking shape as a small but growing group of mathematicians had actually begun to receive their training in America. The introduction of teaching and research as part and parcel of the American university experience—and, thus, the opportunity for graduate training at home—came with the opening of Johns Hopkins University. When Hopkins opened in 1876 it aimed to rival the best institutions in Europe. The trustees at Hopkins hoped to launch the first school in America to emphasize not only teaching but also research and the training of future researchers. Instead of emulating Harvard University, Yale University, or Princeton University, with traditions rooted in the English college system for undergraduate education, they modeled the school on the German seminar system. In mathematics, James Joseph Sylvester led the department, trained graduate students, and launched a specialized mathematics journal, though the momentum he created could not be sustained after Sylvester returned to England in 1883 [Parshall 2006, especially ch. 9].

By 1900, nearly 200,000 miles of railroad bound together a United States that stretched more than 3,000 miles across.[1] Settlement patterns in the western portions of the U.S. often hinged on the location of the terminal points of the railroad. Chicago played a key role first as railhead, then as rail hub, as it experienced population growth from 112,172 in 1860 to 1,099,850 in 1890. Fortunes ballooned for some as new railroad tracks linked the East and West Coasts. Among a handful of others, John D. Rockefeller, an oil magnate, and Andrew Carnegie, a powerful captain of the steel industry, acquired staggering wealth from the late nineteenth century progress and industrial development. Sizable fortunes such as theirs contributed to a new wave of philanthropic initiatives in America. In particular, both Rockefeller and Carnegie made substantial investments to advance American education, contributing to a second wave of research institutions. Rockefeller's new University of Chicago benefitted from a combination of location, population growth, philanthropic activity, and foresightful trustees and would, in fact, play a crucial role in the development of American mathematics.

Opening in 1892, the University of Chicago began its operations in mathematics with a commitment to the training of researchers. Under the chairmanship of Eliakim Hastings Moore, a small but determined group of mathematicians established a flourishing, successful Ph.D. program. Moore handpicked German colleagues Oskar Bolza and Heinrich Maschke to bolster the program. Modeled after the German research university and built, in part, by Maschke and Bolza, Chicago embraced the seminar as one of its most effective modes of teaching and training future researchers [Parshall & Rowe 1994, ch. 9]. The mathematics department at Chicago offered a viable opportunity for aspiring American mathematicians who would otherwise have had to travel to Europe for their training. L. E. Dickson earned one of the first Ph.D.s in mathematics at Chicago. As a representative of a new kind of American mathematician, one both trained and employed at research institutions in the U.S., Dickson provides us one insight into American mathematics during the First World War.

Like the University of Chicago, and Johns Hopkins University before it, Princeton and Harvard rapidly emerged as institutions that shared a simultaneous commitment to teaching, research, and the training of future researchers, a novel combination for American higher education at the time [Parshall & Rowe 1994, Birkhoff 1988, Aspray 1988]. This profile was soon to become widespread, as was the vogue for German training for American mathematicians. In contrast to the University of Chicago, Princeton and Harvard enjoyed geographical locations on the Eastern Seaboard in some of the most populous regions in America. Location, in fact, would be a critical component of successful higher education venues, particularly in the late nineteenth and early twentieth centuries. When Father Edward Soren founded Notre Dame University in 1841, for example, he viewed its "central location" in South Bend, Indiana, as one of its selling points.[2] Universities on the

[1]The transcontinental railroad usually refers to a line over the Rocky Mountains between the Midwest and the Pacific Ocean. Completed in 1869, the lines from east and west met in Utah. While there are many sources on general American history, we have relied on [Berky & Shenton 1966].

[2]Despite his advertisement, South Bend has never really been centrally located, then or now, especially when compared to New York City, Boston, or even Chicago. With its deliberate addition of a graduate program in the 1930s, Notre Dame University points to another wave in the history of American education, and the history of American mathematics in particular, that is, the

Eastern Seaboard, not just Harvard and Princeton but (notably) Yale University, Columbia University, and Brown University, benefitted from an accessible location and long tradition in attracting both faculty and students.

The American mathematicians and the developments in higher education they instigated echoed the call for progress and growth championed by President Theodore Roosevelt and embraced by the country generally.[3] For the mathematical community, the advantage of growth manifested itself in the form of an increasing recognition from the Europeans, an increase in the training of future researchers on American soil, and a changing definition of what it meant to be a mathematician.

A remarkably large number of the leaders of research mathematics in the U.S. in the early years of the twentieth century were German-trained. The Göttingen of Felix Klein and David Hilbert figured prominently in many curricula, as did Berlin. A few examples will indicate the pervasiveness of this tendency. At Harvard, William Fogg Osgood and Maxime Bôcher had both studied under Klein. Charles L. Bouton had been a student of Sophus Lie in Leipzig. Julian Lowell Coolidge had trained with Eduard Study at Bonn. And Edward Vermilye Huntington had been a student of Ernst Schröder at the Kaiser-Wilhelm University in Strasbourg in 1900 (when it was part of Germany). At Chicago, Moore had spent a study year in Berlin, and Bolza and Maschke had been students of Klein. At Cornell the interest in German mathematics was sufficient that in 1897 and 1898 lectures were given in German by John Tanner (on Heinrich Weber's *Lehrbuch der Algebra*) and Virgil Snyder, a Klein student, on line geometry. Many others with no direct initial contact with Germany worked with these and other supervisors who had: illustrious examples are Veblen and G. D. Birkhoff, who both studied under Moore's direction [Parshall & Rowe 1994, Cornell Sun 1898].

The creation and expansion of the AMS over the first 15 years of its existence illustrate the growth and maturing of the mathematical community. In December of 1888, six men met to found the New York Mathematical Society. Their number grew to 16 the following year and to 23 by 1890. Convinced of the importance of a scholarly venue for the exchange of mathematical research, they founded the *Bulletin of the New York Mathematical Society*. Challenged by the country's westward-moving boundary and prompted, in part, by the prospect of increased revenue, the New York Mathematical Society worked towards a more national purview. Thus in 1894 the New York Mathematical Society made its debut as the American Mathematical Society at the national meeting of the American Association for the Advancement of Science [Archibald 1938, pp. 3–9]. It would take some time for the geography of the society to catch up with the name. As Garrett Birkhoff, G. D. Birkhoff's son and himself a mathematician, would later put it,

> [P]rogress was accompanied by some growing pains; ... with growth came factionalism. The leadership of the AMS was divided on two issues: regional balance, and concern with mathematical education. In particular, the Chicago Section felt neglected by the AMS. Although it had produced the lion's share

---

transition from teaching to research in the "second tier" schools in America. David Zitarelli, for example, has begun to address this question with his study of Lehigh University [Zitarelli 2007].

[3]The ebullient Theodore Roosevelt had assumed the U.S. presidency in 1901. Reelected in 1904 to a four-year term, Roosevelt built his presidency and, to a certain extent, the country itself on the premise of change as he portrayed himself and his country as adolescents in search of adventure.

> of the outstanding American-educated Ph.D.s and the center of population was steadily moving westward, meetings were usually held to suit the convenience of the Eastern Ivy League faculties, rather than that of University of Chicago, Northwestern University and the "big ten" middle Western state universities [Birkhoff 1976, p. 41].[4]

The founding of the Mathematical Association of America (MAA) in 1916 addressed a demand for an enhanced role for mathematical education within the burgeoning AMS, reflecting the large number of members who taught at colleges (with little or no research component) rather than universities with a professorial research mandate. Regional balance formed a much more pressing—and challenging—issue. The introduction of the Chicago Section in 1896 and the San Francisco Section six years later ostensibly began to address the geographical issues. On the one hand, in the first twenty years of the society, all annual meetings were held in the East, sectional meetings were not regarded as meetings of the society, and only two AMS presidents came from outside the Eastern Seaboard. On the other hand, the group of University of Chicago mathematicians "was full of enthusiasm, fertile in new ideas, and impetuous to see them in operation" [Archibald 1938, p. 8].

"[E]ven with wise counselors both in the East and in the Middle West," Raymond C. Archibald, from Brown University, observed,

> there was a distinct chance for misunderstandings and cleavage to arise. That nothing serious in this regard did develop is due in part to E. H. Moore's fine influence; but older members agree that the role played by Edward B. Van Vleck during the critical years 1906–1920, after he had arrived at the University of Wisconsin, was of no little importance. As a trusted friend of prominent men in both the East and Middle West, and in constant contact with them, he was a frequent interpreter to each of the thought and aspirations of the other. Thus, with others also contributing, differences were composed and cooperation became more intimate [Archibald 1938, p. 8].

In the end, then, careful leadership and communication eased the New York Mathematical Society into a truly national society. Once again, E. H. Moore assumed a critical role in this transition.

While the society addressed various concerns on the collective level, the definition of what it meant to be a research mathematician, only an isolated phenomenon in America until the very end of the nineteenth century, began to take a firmer shape as the new century dawned. In the late nineteenth century, the role of the professional mathematician came to encompass a commitment to teaching, research, and the training of future researchers. For some this role expanded further in the early

---

[4]The "Ivy League," the term referring to a league in collegiate football, consists of Brown University, Columbia University, Cornell University, Dartmouth College, Harvard, the University of Pennsylvania, Princeton University, and Yale University. All of these with the exception of Cornell were founded before the American Revolution. The corresponding western group usually includes the University of Chicago, Northwestern University, Purdue University, and the Universities of Illinois, Indiana, Iowa, Michigan, Minnesota, Ohio and Wisconsin, though there has been some variation in the actual composition of the football league.

twentieth century to include negotiating with philanthropic organizations and, as we will see after the war, also with governmental and industrial entities.

The early twentieth century brought the American mathematical community more than new students and new sources of funding, however. It also brought war. Wilson won the presidency in 1912 by a plurality rather than a majority vote. The Republican Party divided its vote among William Taft and Theodore Roosevelt, and Eugene Debs, the Socialist candidate, received more than one million votes. When the war began, Wilson was distracted by domestic affairs. He urged the country to maintain neutrality with regard to European affairs, seen by many as a divisive and costly trap. Wilson urged a spirit of neutrality, and his reelection to the presidency in 1916 was almost universally viewed as a mandate for peace [Leuchtenberg 1966, p. 1112].

America was a country of immigrants, some of whom were called home in various European drafts, some of whom had divided interests, but many of whom had strong feelings about the war. Although America did not enter the war until April 1917, an important segment of public opinion increasingly regarded U.S. involvement as inevitable. Ernest W. Brown, AMS president from 1915–1916, issued a call to American mathematicians in his retiring presidential address on December 28, 1916. The English-born Brown had invested more than forty-five years of research in lunar theory, with twenty-five years of his career on the faculty at Yale University in New Haven, Connecticut. The vast majority of the address considered the relations of mathematics to the natural sciences, and indeed this was the theme described in the title. However, the final page of his 16-page address adopted a prophetic tone, noting both the threat the war presented to Europe and the need to assess whether mathematics in the U.S. could carry European achievements forward.

> It is true that here we are happily removed from the devastation which has overtaken Europe, but the lack of stimulus from abroad cannot fail to have its effect. We need at this time to take stock of the resources of ourselves and of our Society, both mental and physical, to find whether they are sufficient for what may come and more than all to conserve the heritage which we have received from the old world so that it may later be returned alive and productive.
>
> The possibility of great changes coming rapidly and with little warning must be faced. We cannot therefore be content to sit back and await events. The future of research in pure science is in danger as never before and the less practical the science the greater the danger. New economic and social forces are at work in every direction and may call many of us to activities which may stop temporarily and perhaps permanently the work and thought which have in the past most fully occupied us. Two generations of steady and continuous development have tended to make us settle in grooves of thought and habits of mind in which great upheavals have no place.
>
> To such changes at least a proper mental attitude is necessary. Action may not be needed now but preparation for action

> is essential and the chances of the future should be in the mind of every individual if the right course is to be pursued when the time for action comes [Brown 1917, pp. 229–230].

Four months before America entered the war, Brown understood the distinct possibility of war and what it might mean for American mathematicians who neglected what Brown called for, a taking stock of its resources for foreseen challenges.

Brown did not propose explicit actions to enhance preparedness. However in the same time period, before the entry of the U.S. into the war, the question of the role of government in science became linked to issues related to the overseas conflict. This led to a call for a scientific advisory body to the national government. The seeds for such a formal advisory scientific entity had been planted when the National Academy of Sciences (NAS) was organized in a time of war in 1863. By the early 1900s, the NAS had, at best, a tenuous relationship with the government. On the one hand, the expansion of the government's own scientific bureaus had virtually rendered the academy's help superfluous. On the other hand, many NAS members viewed their membership as a satisfying honor, a "blue-ribbon" society as one described it (Edwin Grant Conklin in [Kevles 1968, p. 430]), rather than something that implied a role in national service.

George E. Hale, the distinguished American astronomer, entertained more ambitious ideas for the NAS. Hale's increasing concern for the general advancement of science in America recognized the growing interest of American businesses in science [Kevles 1968, pp. 428–429]. His study of Europe's scientific academies also helped Hale understand the importance of government patronage of science. As an astrophysicist, Hale also possessed a keen awareness of the benefits of what would now be termed interdisciplinary research. Moreover, Hale felt group efforts—backed by a body with scientific authority—stood a better chance of securing financial support than individual applications for funding. From Hale's perspective, the NAS was the natural venue to implement these ideas. International events set the stage for him to act on this plan in April 1916.

Before the U.S. officially entered the war, many academics joined or attempted to join the forces in Europe. They volunteered to drive ambulances, enlisted in Canadian regiments, or attended summer training camps in the hope of gaining an officer's commission. Meanwhile, the U.S. began to respond officially to the escalating tension. On April 19, 1916, President Wilson publicly called for cessation of the German practice of employing submarines to sink all ships in enemy waters, without warning. The day after President Wilson issued this ultimatum, members of the NAS personally delivered a message to the president that the academy was at his service as an advisory board for scientific matters. Wilson's acceptance of their offer led to the creation of the National Research Council (NRC) in June–July 1916 [Kevles 1968, p. 431].

The NRC aimed to bring education, industry, science, and government together to promote the national welfare.[5] While Hale's ideas served as the impetus for the formation of the NRC, the physicist Robert Millikan, vice chairman of the NRC,

[5] Written shortly after he earned his Ph.D. at Princeton for his work on physics in America, [Kevles 1968] focuses on the development of the NRC within Hale's overall aims for American science. As the title of his article suggests ("The Genesis of the National Research Council and Millikan's World War I Work" [Jewett 1948]), the physicist Frank B. Jewett describes the creation of the NRC within the context of Millikan's contributions. Together, they form a comprehensive look at the NRC at the time of its inception.

actually ran the day-to-day activities in Washington, D.C. [Jewett 1948]. In addition to his sizable administrative duties in the running of the NRC, Millikan would eventually meet with French, English, and Italian war representatives, addressing issues on topics as diverse as airplanes, tanks, and poison gases, and, perhaps most importantly, organized antisubmarine research.

While the NRC's efforts touching on military issues seemed largely defensive in the prewar period, discussion of mobilization was in progress in early 1917, in the context of Wilson's recent re-election, his efforts to distinguish German and Allied war aims in favor of the latter, and the continuation of unrestricted submarine warfare by Germany. On February 26, 1917, President Wilson requested permission from congress to arm U.S. merchant ships, and on March 20 the cabinet voted for war. In late March of 1917, the War College Division, a planning arm of the Army General Staff, issued a report calling for a force of one million men. It estimated it would take at least ten months to train a force of 500,000 and ship them to Europe [McCarthy 1992, ch. 3]. President Wilson then called for a national army, to be "raised and maintained exclusively by selective draft." One week later, on April 6, 1917, congress voted to declare war on Germany, thus bringing the U.S. into the First World War.

2. The United States Enters the War

2.1. A Mathematical Community Mobilized in the War Effort. When the United States entered the war, the urgent circumstances prophesied by Brown came to pass. The first U.S. troops reached France late in June 1917. Just a week later, Brigadier General John Pershing requested an army of one million men, a figure he would soon change to over three million [Pershing 1919, p. 8]. A frenzied mobilization followed, and this brought with it changes throughout the educational and research communities as well as in society more generally.

In the process of mobilization, the primary directors of operations were of course military men, among whom we find no mathematicians. This brings up the question of the role of the U.S. Military Academy (West Point) and Naval Academy (Annapolis), where those training for positions as officers would acquire basic skills, including the mathematical knowledge needed for artillery and navigation. These courses were fixed, and it is likely that the textbooks for them may have found use in the soon-to-materialize desperate attempts to bring nonprofessional soldiers up to speed on such matters. However, we have no evidence that faculty or students at these institutions took on any exceptional role in these efforts—unsurprisingly, since they were in great demand for active service of other kinds.

At the time, the U.S. had no developed system of wartime science and only a fledgling NRC. The necessities of mobilization required resources that exceeded the capabilities of either the government, military, or industry alone, including arms, ships, and supplies. The flurry of activity involved a new kind of collaboration between industrial, military, and governmental entities. The many arrangements undertaken to put the formerly neutral nation on the road to a full fledged war operation almost all occurred between April 1917 and the Armistice of November 11, 1918. Factories were repurposed to manufacture equipment and airplanes for the navy. Some companies, such as the Carr Fastener Company, became exclusively dedicated to military supply. The U.S. manufactured at least forty three-thousand-ton wooden ships that were sent to the French. One factory produced thirty-two steel minesweepers for the British forces. Angier Mechanical Laboratories developed and produced expandable cartridges for Browning machine guns. Canadian Car and Foundry was central to design and construction of new TNT plants. The U.S. Treasury Department hired geologists to manage shipments of war minerals such as manganese, phosphate, salt, and gypsum. The DuPont Chemical Company worked in collaboration with American University (in Washington, D.C.) to develop mustard gas, while other chemists at DuPont hoped to process smokeless gunpowder. The Bureau of Mines was heavily involved in helium production for use in dirigibles. Academic researchers of various kinds collaborated with industrial personnel and government aims in the war effort.[6]

Although the organization of research universities in the U.S. was based primarily on the German system and scholars of the two countries had close cultural ties before 1914, American college students and professors were among the strongest supporters of the war. "Of all my colleagues at Harvard," G. D. Birkhoff wrote to the Italian mathematician Tullio Levi-Civita just before the U.S. entered the war in 1917, "only two not of pure German blood and German born incline toward the other side [i.e. the Central Powers]; and even these keep very quiet." His brother Ted had earlier mentioned to him that Bolza had become quite isolated at the University of Chicago by taking a stance in favor of the Central Powers: "he would rather sacrifice for the German army than teach mathematics, and the '*kunst und wissenchaft*' [art and science] seem useless and superfluous luxuries to be discarded at a time like this, without a regret."[7] Eager American students and faculty who overran college recruiting stations seemed to share, if not Bolza's sympathies, at least the intensity with which he engaged in the struggle, be that at the expense of art and science.

Both faculty and students were subject to the draft, which in the first instance touched all men 21–31 years of age. Unmarried men were normally in the first category to serve. Many students and faculty fell into this category. Princeton University, for example, has claimed that it sent a larger fraction of its student body to train for the war than any other university, though how this could be known at the time is unclear. The fact that in June 1917, of 1,407 undergraduates, 1,394 were either in active service or taking military courses suggests that Princeton's ranking in this regard would be near the top [Collins 1932, p. xxii]. In addition to these students, 138 Princeton faculty served in uniform during the war. Many in the postsecondary mathematics community from across the U.S. participated in the war effort in many different capacities. Some engaged in similar war activities as their academic and nonacademic contemporaries including volunteer ambulance and relief service abroad, Red Cross or Young Men's Christian Association (YMCA) participation, both home and abroad, enlistment in state guards, volunteer civilian government work, and service in the U.S. Army or Navy. In addition to Princeton, faculty and students who served came from the University of Chicago, Harvard, the University of Wisconsin, Rice University, the University of

[6]On the American scientific war effort, see [Yerkes 1920]. The Harvard War Records Office Forms also provide interesting information on the industrial mobilization. On helium production, see [Aubin 2003].

[7]Both letters in [Birkhoff Papers]: HUG 4213.2, Box 3, Birkhoff to Levi-Civita, March 7, 1917 (facsimile of originals in California Institute of Technology Archives Acc#7129); Box 7, Ted Birkhoff to G. D. Birkhoff, December 16, 1914. As is well known, Bolza abandoned his mathematics for fifteen years, devoting himself instead to Indian theology.

Texas, Brooklyn Polytechnic, Cornell University, Catholic University (Washington D.C.), the University of Michigan, the University of Pennsylvania, the University of Illinois, and the University of Minnesota. Examples of mathematicians involved nonmilitarily include Louis L. Silverman of Cornell University, who was attached to the Massachusetts Board of Public Safety, and Arnold Dresden of the University of Wisconsin, who went to France with the Red Cross.[8]

The case of Marston Morse, who had completed a noted Ph.D. dissertation on "Recurrent Geodesics on a Surface of Negative Curvature" under Birkhoff's supervision at Harvard in the spring of 1917 and would later be appointed at Harvard and at the Institute for Advanced Study in Princeton, is enlightening. Morse joined the Ambulance Service, on June 24, 1917, and was immediately sent to France. There, he experienced the Spartan living conditions of the fighting soldier, slept in underground hideouts with African troops and German prisoners under constant shelling, took part in battle at the Chemin-des-Dames, near Verdun, was moderately gased near Rheims in August 1918, and was awarded the Croix de Guerre for his service. But Morse felt he might play a more useful part in war than as an "automobiliste":

> Although the ambulance work is great, still I think I could be of more use in some branch of the service where some mathematical training is necessary. In particular, there will be an opportunity for me to apply in several weeks, for the rather rare chance of going to the American artillery school over here. If I can match my mathematical training against other men's experience I might obtain an opening.[9]

Raymond W. Brink—another student of Birkhoff's who went on a fellowship to study mathematics in France during the academic year 1916–1917!—agreed that Morse was "too fine a lad to be fed to the cannon." Nonetheless, Morse's heroic stay in France was not a complete loss to mathematics. In July 1917, he found time to send a revised copy of his thesis to Birkhoff from the front and was concerned with its publication. "The parts changed are for the most part in the section on Recurrent Geodesics." He read mathematical books, requesting volume 2 of Édouard Goursat's *Cours d'analyse* from his mother on March 26, 1918, and considered attending Jacques Hadamard's course on celestial mechanics in 1919 (although it is not clear whether he was able to do so).[10]

Because of this flurry of military activity outside the confines of academia, mathematicians often turned their attention away from mathematics. Serving as battery captain in the First Indiana Field Artillery, one of Birkhoff's many young

---

[8][AMS Notes, vol. 24, pp. 106, 163, 317, & 413]; [AMS Notes, vol. 25, pp. 93 & 139–140]. The *Notes* section of the *Bulletin of the American Mathematical Society* provides a good source for an overview of the extent to which the postsecondary mathematics community in the U.S. participated in the war. A list was also compiled and published by the MAA in 1919, which included 178 names of "known mathematicians of the United States in national service occasioned by the war" [Rothrock 1919]. This list also includes graduate students.

[9][Birkhoff Papers, HUG 4213.2, Box 4, Morse to Birkhoff, December 1, 1917].

[10]Materials used for the paragraphs above come from Morse's correspondence with his mother [Debru 2001] and with his supervisor [Birkhoff Papers, HUG 4213.2]; see Box 3, Raymond W. Brink to Birkhoff, January 21, 1917, and Morse to Birkhoff, July 5, 1917, both quoted above; and in Box 4, Morse to Birkhoff, September 10; December 1, 1917; and March 8, 1918. About Brink's travel to wartime Paris, see the introduction to this volume.

mathematical correspondents wrote: "At present my mind has been quite removed from mathematics on account of the war situation.... I like military work, but fear it will make it very difficult to return to mathematics, particularly since the prospects are for a long war."[11] The young mathematician William L. Hart was teaching a few hours of artillery and equitation in Fort D. A. Russell, Wyoming, but was also in charge of supplies and clothing. He lamented:

> I have had no time for mathematics so far. I am in hopes of getting a little opportunity for reading and shall pick some easy French math book if I can find one. I haven't seen a thing in our practical artillery which demands math beyond trigonometry. So I get no practise in mathematical thinking.... It certainly will be interesting to see how the war affects our research men. I'll be very sorry if mathematicians drop everything on account of the war.[12]

In the early stages of U.S. engagement, specifically mathematical work was indeed hard to find. Oswald Veblen from Princeton believed strongly in supporting the war effort but was frustrated by his initial failure to find opportunity for mathematical service, as he wrote in his letter to Birkhoff, on May 19, 1917, quoted in the introduction to this volume. At the time, however, Birkhoff apparently was already "at work upon a very pretty problem in mathematical physics of some importance for war reasons."[13] The physicist Owen W. Richardson, who had left Princeton for King's College London in 1914, discouragingly replied to Veblen's inquiry "as to what there was to do for mathematicians in war [that] there is practically nothing that requires real mathematics." Veblen added:

> From other sources I know of an important application of the problem of Apollonius (circle tangent to 3 given circles) but the real scientific interest in this is physical. I am not expecting to find any considerable math[ematica]l interest in my new [assignment].[14]

And indeed if the following letter transmitted to Birkhoff by the Harvard Registrar Office can be taken as representative, American seamen were only asking for solutions to trivial problems: "I will appreciate it greatly if you will turn the problem noted below to some authority in the Department of Mathematics for solution."

> This problem, called, I am informed, the Harvard problem, has thus far defied the efforts of a number of my fellow officers of solution and we are interested to see a demonstration and an explanation of the principles involved in the solution of this and others of a similar nature:[15]

[11][Birkhoff Papers, HUG 4213.2, Box 3, Kenneth R. Williams to Birkhoff, May 29, 1917].

[12][Birkhoff Papers, HUG 4213.2, Box 3, Hart to Birkhoff, October 19, 1917].

[13][Birkhoff Papers, HUG 4213.2, Box 3, Birkhoff to Levi-Civita, May 10, 1917 (facsimile of originals kept in California Institute of Technology Archives Acc#7129)].

[14][Birkhoff Papers, HUG 4213.2, Box 3, Veblen to Birkhoff, September 10, 1917]. This is probably related to sound-ranging; for more on this problem, see the introduction to this volume.

[15][Birkhoff Papers, HUG 4213.2, Box 4, Capt. H. R. Searles to the Registrar, July 12, 1918]. The problem as quoted has an obvious solution with $X = 2$ and $Y = 3$ and thus reduces to a one-variable cubic equation; in the general case, it amounts to the intersection of two parabolas, as already known to Descartes.

$$
\begin{aligned}
X^2 + Y &= 7 \\
X + Y^2 &= 11.
\end{aligned}
$$

Be that as it may, mathematicians did gradually engage in technical work and their mathematical training became viewed as a direct resource and as a valuable addition to training recruits. Those who were accepted into the armed forces often served as officers in various services, including Army Ordnance, the Signal Corps, Field, Coast, and Heavy Artillery, Supplies, and Accounts. Many mathematicians participated in the war effort in ways related to their education and few in the mathematical community seem to have seen the frontlines. The support provided by mathematicians to the war effort then could take the form of mathematical research, educational activities, and potentially other forms. To these we now turn.

2.2. Mathematical Research in the War. The two largest entities engaged in war-related mathematical research were intertwined. These were the ballistics research unit at Aberdeen and the Ordnance Department of the War Department in Washington, D.C. These had specific connections at the leadership level to Princeton and Chicago. The Aberdeen unit, directed by Veblen, undertook practical experimental tests. A group around Forest Ray Moulton of the University of Chicago was based at Army Ordnance in Washington and directed some of the activity at Aberdeen. Veblen and Moulton were thus in regular contact, and in this sense, due to his connections with Birkhoff and others, Veblen served as a node in a Chicago-Princeton-Harvard network at this period and later on, and this is one way in which members of these leading research institutions engaged the war in a mathematical sense. Beyond Aberdeen and Ordnance, some individual mathematicians were engaged elsewhere, like at the Naval Experimental Station, or with the Fuel Administration. Still other mathematicians continued their normal academic work, despite wartime upheaval.

2.2.1. *Oswald Veblen at Aberdeen.* Certainly one of the most important arenas of mathematicians' involvement was munitions design and testing, and the Aberdeen Proving Ground (APG) was the central location and institution for this work in the American context. Since 1880, the Army Ordnance office had used a proving ground located in Sandy Hook, New Jersey. In the early twentieth century, new, more powerful, munitions exceeded the range there and came dangerously close to New York waterways and development. Because of its scale and perceived importance, one historian has referred to this institution as the "Manhattan Project of its time." The APG consisted of 35,000 acres of Chesapeake Bay Shoreline and cost $73 million to build [Grier 2005a, p. 254]. The Aberdeen facility was an up-to-date successor of earlier installations: a nearby Naval Ordnance facility at Indian Head (Figure 1) was also in use for war work. These Maryland proving grounds, accessible by barge from the Potomac River yet remote enough for military test sites, were also sites of training.[16] The APG also served as the last stop in the production lines for artillery, shells, gun carriages, and explosives and was thus involved in coordination with industry.

---

[16]For example, some members of the U.S. Marines stationed at the Naval Proving Grounds in Indian Head, Maryland, received military instruction and training, including a course on ballistics and firing data given by French Army Captain Roulleau de la Roussière. [Harvard Mathematics Papers, UA V 561.8, Box 1911–1920, Correspondence 1919, S. W. Bogan, June 19, 1919].

FIGURE 1. Navy gun being tested at the Indian Head Proving Ground, Maryland, in 1919. Library of Congress, Prints & Photographs Division, photograph by Harris & Ewing, LC–DIG–hec–12662.

Six months after lamenting a lack of mathematical war work, on November 20, 1917, Veblen received his assignment to active service at Sandy Hook Proving Ground. The route leading to this assignment is unclear, which is unfortunate since the precise methods by which these decisions were taken are in general rather obscure. Various kinds of personal connection are not ruled out, but we have no evidence to present. In early January of 1918, Veblen took command of the office of experimental ballistics at its new APG location. Born in June 1880, Veblen completed high school at the age of 13, earned his first B.A. degree in 1898 at the State University of Iowa and his second from Harvard in 1900 [Veblen Papers, Box 33], [Montgomery 1963, p. 27]. He studied under the influential E. H. Moore at the University of Chicago and received his Ph.D. at the age of 23 in 1903. Woodrow Wilson, Princeton University's president at the time, hired Veblen in 1905 to strengthen the mathematics department there. Veblen was appointed to a full professorship in 1910 and was later awarded the Henry Burchard Fine Professorship in 1926.[17] Given that Veblen's early research focused on geometry and, at the outbreak of the war, on algebraic topology (or analysis situs), ballistics research at the APG was a departure for him in a way that it would not have been for, say,

[17]In 1932, he joined the faculty of the newly created Institute for Advanced Study. Veblen invested considerable energy not only in the advancement of mathematics at Princeton, but also in America at large. He served as AMS president in 1923–1924, roughly the same time he began working to secure external funds for mathematics [Butler Feffer 1998]. Veblen also had a particular penchant for encouraging younger mathematicians.

Birkhoff. On the other hand, it is closer to Veblen's later interests, as we discuss below [Ritter 2011].

Research at the APG fell under the subdiscipline of exterior ballistics, modeling the flight of the projectile released from a gun. The end product of exterior ballistics was the all important range table that provided the distances guns could propel a shell. These distances, or ranges, often depended on muzzle velocity, angle of the barrel, and type of shell [Grier 2001, p. 924].[18] On the ranges at Aberdeen, ordnance officers would test new weapons before deploying them to arsenals or, ostensibly, to the units of the American Expeditionary Force (AEF).[19] This was not glamorous work. Fitzhugh S. Carmichael's "History of the Range Firing Section of the Proof Department of the Aberdeen Proving Ground" described the work candidly:

> Range firing at Aberdeen began in February, 1918, with a 2795 mountain gun. This gun was fired for range to impact on land. The firing was done whenever a fresh snow storm had obliterated previous marks on the ground. Captain (later Major) Veblen, who conducted this firing, would go down the range on horseback while the firing was suspended for meals, identify the shell holes and place distinctive marks in them to enable them to be found by the surveyors, who made the actual measurements at the range [Veblen Papers, Box 33].

This primitive testing, "more picturesque than satisfactory," took place as decidedly real projectiles intended to destroy and damage were fired at various points in Europe [Veblen Papers, Box 33]. "The work was interesting," summarized the young mathematician Charles N. Harkins, a Harvard graduate student, after he had left Aberdeen, "but conditions were rather uncomfortable in many ways."[20] Even with data in hand, the computation of the range tables proved a formidable challenge. Veblen used a differential equation model that had been developed in the 1880s by the Italian ballistician Francesco Siacci.[21] The APG data went to Sandy Hook, New Jersey, where a group of humans created the tables. "In spite of the best efforts of all concerned," as Carmichael described this process in his report, "it was found very awkward to do the computation at a distance from the actual firings." In March 1918, Veblen ultimately traveled to Sandy Hook to sort out "numerous questions" related to the technical aspects of the data. The relatively small collection of records on Veblen's work at the APG capture the "leadership" skills, "personal qualities," and "intellectual" prowess the lieutenant colonel observed in his assessment of Veblen at the time of his discharge [Veblen Papers, Box 33]. It

[18]James W. Alexander, Veblen's former student at Princeton, worked on questions related to interior ballistics during WWI. He once asked Veblen if he had "any 'dope' on interior ballistics." Alexander captured a certain reality when he described his approach to Veblen. Alexander used "the Navy theory ... which has the advantage of being simple and inaccurate instead of complicated and inaccurate" for his work comparing muzzle pressure in French and American guns [Veblen Papers, Box 33, Alexander to Veblen, February 14, 1918].

[19]In fact, as Grier points out, the (human) computers that helped calculate the range tables grew restless, in part, because they realized only a small part of their computations would actually reach Europe. The American Expeditionary Force primarily used French and British guns, and troops and supplies had first priority on American transport ships [Grier 2005b, p. 151]. Only at the very end of the war did American guns actually reach the frontlines.

[20][Birkhoff Papers, HUG 4213.2, Box 3, Charles N. Harkins to Birkhoff, December 28, 1918].

[21]A further development on Siacci's method can be found in David Aubin's contribution to this volume dealing with the Gâvre Commission, the French equivalent of Aberdeen.

TABLE 1. Some Aberdeen Proving Ground mathematicians.

Name	**Institution/ Location**
Albert A. Bennett	University of Texas
Hans F. Blichfeldt	Stanford University
Gilbert Ames Bliss	University of Chicago
Fitzhugh Lee Carmichael	Princeton graduate student
Charles R. Dines	Dartmouth College
Percy Austin Fraleigh	Cornell University
Haig Galajikian	Princeton University
Harry Merrill Gehman	University of Pennsylvania
Bennington P. Gill	College of the City of New York
Thomas Hakon Gronwall	New York, NY
Charles N. Harkins	Harvard graduate student
James S. Mikesh	Hibbing Minnesota Junior College
Howard H. Mitchell	University of Pennsylvania
William H. Roever	Washington University (St. Louis)
Irwin Roman	Northwestern University
Clarence Albert Shook	Harvard graduate student
Oswald Veblen	Princeton University
Norbert Wiener	University of Maine
Frank Edwin Wood	University of New Mexico
William Hammond Wright	University of California

was precisely this same collection of skills that made Veblen a powerful force in American mathematics.

Veblen's office of experimental ballistics was an active part of the military, not a laboratory where intellectuals could evade military work. Veblen found most academic mathematicians adjusted easily to the restrictions of army life. Veblen's job at the APG also included finding and recruiting mathematical talent for the work. David Grier reports that as many as 60 mathematicians worked at least part-time with Veblen.[22] Through Rothrock's and Grier's works, we can account for more than one-third of these mathematicians by name, recorded in Table 1.

2.2.2. *Moulton and the Ordnance Office.* Not long after Veblen was called to Sandy Hook, and shortly thereafter to APG, Forest Ray Moulton, a professor of astronomy at the University of Chicago, was contacted to assist the government school of artillery practice at Fort Sill, Oklahoma. By March 27, 1918, Moulton was on his way to the War Department in Washington, D.C., to help them with what he described as "large and important problems." Moulton records his official date of enlistment as April 16, 1918. The president of Carnegie Institution of Washington (CIW) Robert S. Woodward had apparently recommended him for this army position and he plainly told Moulton that he expected him to live up to his "commendations."[23]

[22] Perhaps Grier is also counting the "human calculators" that engaged in work at the proving ground.

[23] All quotes in the above paragraph are from [CIW Papers]. See Moulton to Woodward, January 3, 1918; Moulton to Woodward, March 27, 1918; Woodward to Moulton, April 2, 1918.

Prior to entering Army Ordnance, Moulton had pursued work on periodic orbits and a new lunar theory and was thus a colleague of Yale's E. W. Brown. Moulton enjoyed a truly luxurious longstanding relationship with the CIW from 1908–1923. During this time, the CIW paid Moulton between $1,800–$2,000 per year for "salary in order to be relieved from teaching duties" [CIW Papers, Application for CIW Grant, January 9, 1908]. As a comparison, in 1919, the University of Chicago paid Moulton $2,888 per year for his salary. Thus very early in the century—and continuing through the war—Moulton enjoyed (what was then) substantial support for his research.[24]

The role of Woodward, president of the Carnegie Institution, in making specific recommendations for war work requires further research, but a few observations may be made [Kevles 1978, pp. 82–90]. Woodward had been dean of pure science at Columbia, issuing from two of the largest federal science organizations, the Coast and Geological Surveys. As president of Carnegie since 1904 he had chosen to support the view of the trustees that large projects were preferable targets for funding. One of these trustees, Charles Walcott, had formerly led the U.S. Geological Survey and, with a reputation for practicality, also served on the board of Hale's new NRC, as did the director of the National Bureau of Standards, another service that was to employ mathematicians. The long service of these men in Washington and their deep acquaintance with the scientific community via their involvement with funding made them knowledgeable potential advisors in the good use of staff with sophisticated skills.

In particular, Moulton and Woodward had corresponded since 1906. Moulton's move to Washington appears to have been at Woodward's suggestion. A letter from Moulton to Woodward of March 27, 1918, announced his intention to move: "About ten days ago I was in Washington at the request of the War Department, to confer with them regarding mathematical questions which have arisen. They are face to face with large and important problems, and have insisted strongly that I give them help. I have decided to do so, and shall leave for Washington for the period of the war, within a few days." Woodward replied to Moulton on April 2, 1918: "I am in receipt of your letter of March 21 [note: he means March 27] and am glad to know that you will be able to come to Washington to give assistance to the army in this time of need. You may be interested to know that this appears to be a sequel to recommendations made some time ago to army officials. I am glad my commendations have been followed and shall now expect that you will live up to them" [CIW Papers].

Moulton's prewar specialty in celestial mechanics was more relevant to war work than that of most mathematicians. The Washington group in the Ordnance office included James W. Alexander (Princeton), Thomas Buck (University of Chicago), William Duncan MacMillan (University of Chicago), Dunham Jackson (Harvard), Joseph F. Ritt (Columbia University), Caroline E. Seely (Columbia University), Herman Lyle Smith (Princeton), and Harry S. Vandiver (Philadelphia). We may note once again the presence of representatives of Chicago with three members, Princeton with two, and Harvard with one.[25]

---

[24]Interestingly, his subject was not designated as mathematics or astronomy by the CIW, but, rather, as geology, perhaps because of the close relation in the American context between the U.S. Geological Survey—the "biggest science" of the nineteenth century—and mathematics.

[25]Vandiver had contacts with Harvard, having published a paper with Birkhoff, and after his stint as a Naval Reservist wound up on faculty at Cornell [Lehmer 2007].

Several kinds of experimental work were involved in ballistics investigations. One type is exemplified by the proving ground experiments at Aberdeen, guided by Veblen who "carried out the orders of the Washington office with vigor and success," according to Moulton, but who also brought imagination, insight, and initiative to the task. The aim of these experiments was to determine the dependence of range on projectile shape and weight (for the case of exterior ballistics) and on variations in the rifling of the barrel. These experiments were designed in Washington and involved the production of modular projectiles with different nose shapes, bodies, and bases. The cylinders forming the bodies were threaded inside so that different weights could be screwed in. This meant not only that many different kinds of projectiles could be tested but that the centers of gravity and moments of inertia could be altered while leaving the outer form constant. Finally, artillery pieces with varying pitches of rifling were designed and tested [Moulton 1926].

2.2.3. *Research Activity Elsewhere.* Aberdeen and Washington were not the only loci of war-related research. Professors Oliver Dimon Kellogg (University of Missouri, later Harvard) and Max Mason (University of Wisconsin) were engaged in government service at the Naval Experimental Station at New London, Connecticut. Kellogg had requested and received a one-year leave of absence from the University of Missouri in order to work chiefly on submarine defense problems[26] [AMS Notes, vol. 25, p. 93]. Mason, whose thesis had been on existence theory for ordinary differential equations, was involved in the invention of the hydrophone, a device for the detection of submarines [Butler Feffer 1997, p. 73], [Kevles 1978, pp. 122–124]. Both Mason and Kellogg had studied in Germany under Hilbert.

Warren Weaver, of Throop College of Technology in Pasadena, California, served in the Science and Research Division of the Signal Corps, the beginning of an extended career linking the military with applied mathematics. Weaver had been a student of Mason's at the University of Wisconsin and had just come to California in 1917; Throop was to become California Institute of Technology in 1920. Likewise, Charles Wiggings Cobb, a professor at Amherst College in Massachusetts, was granted a year's leave to "enter the aviation work of the government" [AMS Notes, vol. 24, p. 163]. David F. Barrow of Yale University "entered the government service" [AMS Notes, vol. 25, p. 139]. Joseph E. Rowe of Pennsylvania State University is described as undertaking mathematical research for another agency, the National Advisory Committee for Aeronautics in Washington. Industrial work of a military nature also received support from the academic community: William Raymond Langley of Yale University was granted leave in order to work in the ballistic division of the DuPont Chemical Company in Delaware [AMS Notes, vol. 25, p. 139].

Some leaves were more limited in term. George Edwin Howes was absent from Williams College, in Williamstown, Massachusetts, for more than a year to serve as a member of the Fuel Administration. He spent a three month sabbatical in the work of the administration, but "when it was time for the College to open in the fall it was thought that I might be of somewhat more service here than at Williamstown." In January, Howes could not "see the date of ending my work here, but I am sure I ought to stay until the late spring at least, and possible

[26][Harvard Mathematics Papers, UA V 561.8, Box 1911–1920, Correspondence 1919, Kellogg to W. F. Osgood, April 11, 1919].

until early summer" [Harvard Mathematics Papers, UA V 561.8, Box 1911–1920, Correspondence 1919, G. E. Howes to W. F. Osgood, January 4, 1919].

Some mathematical faculty were sent on missions to Europe, the nature of which is now not clear. Professor Julian Lowell Coolidge of Harvard, having received a commission as a major in ordnance, was sent to Europe as head of a scientific mission. In point of fact he wound up at headquarters of the AEF. Similarly, Griffith C. Evans, a Harvard Ph.D. of 1910 then at Rice University in Texas, who began in the Signal Corps, moved into Army Ordnance and undertook a "special mission" [AMS Notes, vol. 25, p. 189]. The reasons for this "upward mobility" are not clear and may reflect technical requirements, academic pedigree, family connections (the Lowells being extremely prominent in Boston), all or none of these.

More prosaic support concerned the important task of keeping track of operational statistics. Harvard's Edward V. Huntington took a leave and went to do statistical work under the chief of staff in Washington, a leadership role in which he was joined by John Roger Musselman of the University of Illinois and probably others. Huntington was commissioned as a major. President of the MAA in 1918, Huntington was unable to attend the meeting of December 1918 due to work commitments, despite the end of hostilities. In a telegram to that assembly,

> [he] threw a vivid light on his experiences in certain lines of statistical activity by the remark that "there is a great future for mathematics in this country, especially *accurate arithmetic*" [his emphasis] [AMS Notes, vol. 25, p. 92].

Mathematicians from U.S. institutions were thus involved in a very broad range of war-related work, ranging from overseas military and intelligence service to training activities, and various features of civil defense. War-related mathematical research took place both directly within the military and alongside it. The work clearly had very little to do with the research specialties of the participants, at least in many cases. Trained in Strasbourg, Huntington, for example, made contributions both before and after the war in algebraic logic. William Fogg Osgood was a German-trained specialist in complex analysis, Coolidge a geometer. In war research, the emphasis was on ballistics, aeronautics, and issues related to naval warfare (notably antisubmarine activity). At least some of the special missions were for the purposes of coordinating certain aspects of this research. Given the size of the mathematics community in the U.S. in 1917–1918 this was certainly a broad participation, an experience widely shared. It was not as all-encompassing as that in France and Germany: we learn from these notes that 49 members of the Hamburg Mathematical Society out of 121 were in the military in 1916 [AMS Notes, vol. 23, p. 152].

Institutions for scientific research were likewise strongly developed after the entry of the U.S. into the war. One of the most significant of these developments concerns the NRC and antisubmarine work. While the U.S. Navy understood the offensive role of the submarine, it had not given much thought to the submarine as a defensive weapon. As chairman of the Antisubmarine Committee, Millikan worked closely with the chief engineer of the navy to develop methods to detect an enemy submarine. This antisubmarine work and, in fact, all the initial work of the NRC took place while America was at war. A peacetime NRC had never existed.

2.2.4. *Mathematical Research Outside the War Effort.* Not all mathematicians participated actively in the war effort, and we have no way to compare their rate of participation (say) with that of the general population, with other academics, and so on; nor can we parse the participation quantitatively as regards research mathematics versus education. If we find ourselves at least mildly surprised that pure mathematicians turn to applied war work with effectiveness, it is due in part to our lack of ability to grasp the various dimensions of their drives to thus engage, and sources that help us to explain individual decisions of a patriotic nature are few. Nonetheless, given the widespread apparent enthusiasm for this involvement and the success with which it was taken up, we may also find ourselves surprised by those who continue as before with pure or applied research that seems to have little or no immediate relation to war work. We will here mention two people, L. E. Dickson and Joseph Slepian, not as representative, but as indicative of a phenomenon that also needs to be understood better in this context.

Leonard Dickson is an example of a mathematician who took on no direct military activity. Dickson kept a steady beat as a research mathematician at his academic home, the University of Chicago, throughout the war. Born in January 1874, Dickson was 40 when WWI broke out. Dickson drew his first breath on land claimed by a Civil War veteran in Iowa and moved with his family to a small town in Texas, when he was four.[27] Dickson earned one of the first Ph.D.s in mathematics at the University of Chicago in 1896, and after a time abroad and a couple of short appointments at other universities, he spent the vast majority of his professional career, 1900–1940, on the faculty at the University of Chicago. In this position, he would write more than 300 manuscripts and 18 books, advise 67 doctoral students, and serve as AMS president. He also edited some of the major U.S. mathematical journals, including the *Transactions of the American Mathematical Society* and the *American Mathematical Monthly*. From 1911 to 1923, Dickson also found the time in his pure mathematical career—which included finite field theory, invariant theory, the theory of algebras, and number theory—to write a three-volume, 1,500-page historical account of the theory of numbers.

It took Dickson more than a decade to bring his monumental historical study of number theory to an end.[28] Aside from mentioning in 1917 that "conditions in Europe" would influence the demand for this work and grumbling in 1919 about his "fixed, prewar salary," the war made little impact on this (or any other) work of Dickson's [CIW Papers, Dickson to Woodward, January 22, 1917; March 25, 1919]. Although Woodward continued to support Dickson's historical study throughout the war, he also mentioned the wartime activity of the CIW. In June 1917, he commented that "[t]he Institution is now much preoccupied with affairs of the war; but most of our work is going on without serious interruption or delay" [CIW Papers, Woodward to Dickson, June 15, 1917]. By March 1919, Woodward cited the "pressure of complications arising out of the war" as preventing him from traveling to Chicago to meet with Dickson [CIW Papers, Woodward to Dickson, March 31, 1919]. In 1927, Dickson acknowledged other ramifications of the war on American mathematics. When asked for his advice as to whether Duke University should launch a new research journal, Dickson urged the publication of books in advanced mathematics instead. "Practically no such books have appeared in America

[27]The community activism of his parents took root in young Dickson and later expressed itself in the form of his commitment to the American mathematical community. Dickson's older brother, Tracy Dickson, invested this same commitment in a career devoted to American military service.

[28]For more on his *History of the Theory of Numbers* [Dickson 1919–1923], see [Fenster 2005].

since the war without being financed by some University or the Research Council" [CIW Papers, Dickson to Carmichael, May 3, 1927]. The *Duke Mathematical Journal* was subsequently founded in 1935.

Joseph Slepian provides another example of a researcher whose mathematical career appears largely unaffected by the war. Slepian completed his Ph.D. in mathematics at Harvard in 1913, with Birkhoff as supervisor, and devoted a year to post-doctoral studies at the University of Göttingen and the Sorbonne before becoming a mathematics instructor at Cornell University. After a year, he joined the Westinghouse Electric Corporation and quickly advanced to its research department, where he would invent a device called the ignitron and earn 204 patents, mainly related to the electric power industry, developed during a lifetime career there [Fowler 2003, Furfari 2000]. Slepian also participated in the Manhattan Project during World War II and worked with both the Office of Scientific Research and Development and the War Production Board.

At the outbreak of World War I, though, Slepian hoped to join the military with his brother but was unfit for service due to diabetes. In October 1918, Slepian wrote to his former professor at Harvard describing his experiences with mathematics in the industrial context and his role as a consultant among the engineers. "Naturally," he wrote to Osgood, "a large part of this consultation is on mathematical problems, usually of a rather elementary nature not beyond the elementary calculus. Most always, too, the data of the problem are not sufficiently constant or determinable to justify other than approximate mathematical solutions. Occasionally, however, problems come up which require more intense study and higher mathematics." Slepian further specified that he did find "higher mathematics exceedingly useful in considering problems of a general nature."[29] Slepian worked to educate his engineer colleagues at Westinghouse about the power of rigorous mathematical methods. In the winter of 1917, he taught them a course on differential equations and, throughout his years there, he organized many other courses on topics including vector analysis, theory of electricity and magnetism, the kinetic theory of gases, and conduction of electricity through gases [Fowler 2003, p. 248]. But it does not appear that the war shaped any of this work directly.

2.3. Mathematical Education and Training. Many mathematicians, mobilized or otherwise, engaged in educational efforts related to the war. Roland G. D. Richardson of Brown University wrote in 1918 that all physically fit college men "must look forward to being called by their country to some arm of service" and, therefore, every college must recognize the "imperative necessity" of technical preparation [Richardson 1918, p. 321]. Colleges and universities responded in various ways to the call of the country's war program. Some institutions, such as Harvard and the University of Pennsylvania, for example, lent their facilities to the government for instruction and housing of enlisted men. Naval issues were in the forefront of their collective imagination. This led to a variety of short courses in naval topics as part of intense efforts to equip new soldiers with the critical basics of modern warfare. These educational initiatives, whether on campus, in military camps, or in civilian institutions like the YMCA, were among the most common for people involved in mathematical education at any level. Let us review some of these initiatives.

[29][Harvard Mathematics Papers, UA V 561.8, 1911–1920, Slepian to W. F. Osgood, October 23, 1918].

2.3.1. *University Course Programs for the Navy, Air Force, and Artillery.* Regular naval units were formed on many campuses, such as Brown University, Princeton, Yale, and the Universities of Michigan, California, and Washington. The Massachusetts Institute of Technology hosted its own Naval Aviation Attachment. Some students were trained in engineering, while others pursued commissions. The bulk of instruction for these students came from the regular faculty members, not military personnel. At the University of California, for instance, the course of study designed to prepare students for naval service included plane and spherical trigonometry, map construction, analytic geometry, nautical astronomy, navigation, oceanography, marine meteorology, as well as seamanship, naval regulations, physical education, and naval history. The navigation and nautical astronomy courses conformed to naval procedure and relied on both Nathaniel Bowditch's *Practical Navigator* (a venerable object first published in 1802) and the *American Nautical Almanac.* In addition to this plan of study designed for college students, the University of California also offered extension courses in navigation and nautical astronomy taught by college faculty to about 800 men. Armin Otto Leuschner of the University of California was delegated to administer navigational instruction for the U.S. Shipping Board on the Pacific Coast, notably by finding suitable instructors; others in many locations participated directly in navigational training, as we discuss in detail below.

A government school of aeronautics was founded at Princeton, for example, commanded by Major Dana H. Crissy of the U.S. Military Academy at West Point. The War Department had enabled the creation of such 3-year military colleges for schools with a Reserve Officers' Training Corps. The Aviation College consisted in a 3-year program approved by the Army General Staff in which the students were to undergo military training prior to being sent to "aviation centers" such as San Diego and Pensacola [New York Times, June 4, 1918].

Another nationwide training effort concerned artillery. By the time of the Armistice, the Central Officers Training for Field Artillery, located at Fort Zachary Taylor in Kentucky, had on its mathematical staff Lester R. Ford of Harvard, Theophil Henry Hildebrandt and Raymond Walter Barnard of Michigan, Joel David Eshleman of Rochester, Temple Rice Hollcroft of Columbia, Charles E. Wilder and Chester Henry Yeaton of Northwestern, and Harold Eicholz Wolfe of Indiana. Similarly, Paul R. Rider of Washington University in St. Louis went to the coast artillery training camp at Fort Monroe, Virginia, presumably as an instructor. While considerable training activity took place, only some of these efforts entailed release from the teacher's usual institution.

2.3.2. *Military Field Schools at Chicago.* Mathematics instructors being in short supply, some mathematical opportunities opened up for women during the war. The case of Alice Bache Gould, who had prepared a Ph.D. thesis on "Brodian geometry" with E. H. Moore at Chicago, may serve to illustrate this. The commanding officer of the Naval Reserve School of the Chicago Municipal Pier requested that the Departments of Mathematics of the University of Chicago and of Northwestern University conduct courses in mathematics and navigation for enlisted men waiting for their call to the Municipal Pier School, a branch of the Ensign School in Pelham Bay Park, New York [Barnett 1918, p. 328]. From the University of Chicago, E. H. Moore and former mathematics Ph.D. student Gould offered four three-week courses which met for three hours daily in the summer of 1918.

Gould had spent the early war years volunteering as an assistant in espionage to the naval attaché office in Spain. For various personal reasons, she returned to the U.S. in the summer of 1918. Armed with significant mathematical training and superior Spanish language skills, Gould set her mind on serving the war effort with these. She drew on extensive personal contacts in hopes of finding a technical position, but this quest proved difficult because no one quite knew what to do with her. Her Chicago supervisor invited her to help the volunteer efforts of the mathematics department at the University of Chicago in navigation courses for enlisted men in the Reserve Officer Training Corps (ROTC). Moore also supposed that there would be an opening for Gould to work with Moulton since most men in mathematics capable of helping there would be draft age. Moore referred Gould to Moulton, who wrote her with an invitation to enter the engineering division of the Ordnance Department as a computer but warned that it would "be routine and monotonous and the pay very moderate."[30] In the end, Gould opted to teach navigation for the navy men in Chicago.

About 550 students enrolled in the four courses while they also participated in infantry drill and swimming. The instruction was based on Bowditch's *Practical Navigator* and a text in preparation by Gould about great circle sailing. Topics included dead reckoning, determination of latitude, longitude, and time. An additional week was added for signaling, taught by navy instructors rather than Chicago mathematicians. Members of the department of mathematics from Northwestern University similarly offered four-week courses with theoretical instruction in mathematics—most likely trigonometry—and astronomy from civilians supplemented by drills from petty officers.

2.3.3. *The Students Army Training Corps (SATC).* The SATC was part of a nationwide program designed, partly as a student-retention plan, by the Committee on Education and Special Training of the War Department. Alumni newsletters document the activities of the SATC, which served as a central entity to mediate military training. One hundred and fifty-seven educational institutions participated by contract with the War Department. The colleges and universities provided instruction and space, while the SATC offered free college education, room and board, plus \$30 pay per month. Participating institutions transformed campus buildings into military barracks, mess halls, commissary stores, and medical offices. In some cases, they added new buildings or remodeled laboratories for military use (at New York University and Vanderbilt University in Nashville, Tennessee, for example). The specific courses offered varied from school to school. About 500 colleges and universities offered SATC programs with a total enrollment from all military and collegiate sectors at around 179,000 participants [Dickson 1919, p. 295].

A Committee on Education and Special Training was formed in the War Department in early 1918 with the end in mind of coordinating various types of education and training efforts [Committee 1919]. The philosopher Ralph Berton Perry of Harvard was the executive secretary of this committee for much of its life. In September of 1918, this committee produced a descriptive circular for "the guidance of educational institutions which are giving instruction in mathematics to members of

[30][Gould Papers, Moulton to Gould, June 24, 1918]. Information in the above paragraph comes mostly from [Gould Papers]; see, especially, Robert W. Willson to Gould, June 23, 1918; and E. H. Moore to Gould, June 18, 1918.

the S.A.T.C." The primary general program called for three terms of twelve weeks each, with nine hours of instruction per week, in algebra and trigonometry, analytic geometry, and elementary calculus. An additional twelve weeks of calculus, four weeks of probability, thirty-six weeks of navigation based on Bowditch's *Practical Navigator*, and twelve weeks of analytical mechanics were necessary for students preparing to be specialists in certain branches of the service. The detailed course descriptions repeatedly stressed that exclusively practical mathematics should be taught. In analytic geometry, for example, "the chief aim of this course is to enable the student to use graphic and analytic methods for the solution of simple concrete problems of military or engineering character." With regard to analytical mechanics, "it should be borne in mind that the purpose of this course is to teach mechanics rather than mathematics." The committee advised further that "there should be a large number of easy problems, and all complicated algebraic and trigonometric reductions should be avoided."[31]

SATC instructors remarked on the high rate of absenteeism due to the influenza epidemic, but also on the high motivation of the students since better-qualified students would have a chance at a higher rank. At the University of Illinois, roughly 2,100 students in the SATC were taught trigonometry. The students were divided into two groups (engineers and nonengineers). The engineers received four hours weekly of instruction, while nonengineers received three hours. The extra hour concentrated on slide rule use and the application of trigonometry to military and nautical problems. This invites the question of what the other students learned as applications. At Oberlin in Ohio, a much smaller school, about 100 students did a course on map-reading and map-making. The instructor there remarked on the desirability of such concrete applications in regular courses, and this was a recurring theme. Uses of mathematics in the solution of practical concrete problems were consistently emphasized, rather than the mathematical development of the theory.

On October 18, 1918, George Porter Paine wrote to Osgood, then at Harvard, from Madison, Wisconsin, bemoaning the "temporary havoc" that both the influenza epidemic and the SATC worked on the University of Wisconsin, where "chaos reigned supreme."[32] Not only was the university unprepared for the new crop of freshmen in 1918, but the already heavy instructional hours increased further "from 75 to 100 per cent above normal, while salaries remain as before."[33] The heavy demands of SATC work, including several evening classes, combined with illness to force all faculty efforts exclusively to actual classroom instruction. The mathematics department at Wisconsin struggled to maintain even that. Paine commiserated with Osgood, describing war instruction in Madison as "extremely unsatisfactory to us all because of the wholly inadequate preparation of a majority of the students" and because of the management of the classes. Paine concluded by saying how much "the men here, in the mathematics department at least, will

[31]The quotations in the preceding paragraph come from [Harvard Mathematics Papers, UA V 561.8, Box 1911–1920, Correspondence 1919, War Department Special Descriptive Circular, September 19, 1918].

[32]Paine was a Harvard alumnus (AB 1905, AM 1907) who completed a Ph.D. in Wisconsin in 1918. It seems likely that Osgood had been a professor of his, with whom he maintained a personal correspondence [Blue Hen 1925, n.p.].

[33][Harvard Mathematics Papers, UA V 561.8, Department of Mathematics, Correspondence 1919, G. P. Paine to W. F. Osgood, October 18, 1918].

be glad when the S.A.T.C. is disbanded." Whether Paine's experience was typical of the situation elsewhere is not clear, but evidently the additional instruction imposed a necessity for new arrangements wherever it occurred.[34]

2.3.4. *The YMCA and other civilian organizations.* The Young Men's Christian Association, the Red Cross, the Knights of Columbus, and Hostess Houses similarly sought to provide support and encouragement to military men through various services, including lessons in navigation and radio through its general education program. These are "grassroots" institutions, loosely coordinated at best, though here too we see participation from leading universities. In particular, the YMCA was active in the effort of providing education for the war, especially the navy. The YMCA's mathematical efforts were overseen by a committee chaired by George Abraham Miller of the University of Illinois. John W. Young of Dartmouth College also served as a director of the mathematical instruction organized by the YMCA's National War Work Council.

As a result, mathematical work at the Great Lakes Naval Station, for example, operated under YMCA auspices. Approximately 800 of the 40,000 enlisted men at Great Lakes Naval Station attended voluntary mathematics courses, some in preparation for the Ensign School entrance examination and others to aid their aviation studies. The YMCA courses covered only topics in arithmetic, algebra, and trigonometry, but still they faced difficulties in finding competent instructors [Barnett 1918, p. 326]. Isaac Albert Barnett, who had completed his Ph.D. at the University of Chicago in the spring of 1918, spent that summer supervising mathematics with Henry Blumberg of the University of Illinois, before taking a job at Washington University in St. Louis. Based on this experience, Barnett and Blumberg published "A Short Course in Trigonometry for Navy Men." The YMCA similarly offered lessons in navigation and radio at naval camps throughout Massachusetts. Mathematics students at Harvard received credit for these courses towards their concentration.[35]

2.3.5. *The War and Education at Harvard.* This intense effort in military teaching and research of course impacted the life of mathematics departments across the country. We focus on the Harvard Division of Mathematics to illustrate how this also had an effect on postwar reconstructions. On March 24, 1917, the chairman of the committee on military affairs issued recommendations for all Harvard graduates and students anxious to serve their country in the war. The suggestions included joining the ROTC, or serving in a branch where expert knowledge would be of service, such as aviation, chemistry, engineering, navigation, topography, or medicine. For these, enrolling in a military course was recommended to master military vocabulary. Specialists were encouraged to offer their services through their own schools or departments. Harvard's committee on military affairs, in cooperation with a central committee in Washington, planned to refer these specialists accordingly.[36]

[34][Harvard Mathematics Papers, UA V 561.8, Department of Mathematics, Correspondence 1919, G. P. Paine to W. F. Osgood, November 12, 1918].

[35][Harvard Mathematics Papers, UA V 561.8, Department of Mathematics, Correspondence and Papers, 1911–1920, Application for Peirce Instructorship, March 21, 1919; M. Crosby to W. F. Osgood, February 4, 1919], [Harvard Ayer Papers, UA V 596.10, U.S. Naval Radio School, Records of Commander Nathaniel F. Ayer, 1918-1921, 1942, Correspondence II, Mrs. H. C. Read to G. Daniels, October 27, 1918].

[36][Harvard Mathematics Papers, UA III.10.194.10, Committee for Military Affairs—Forms].

By the last months of the war the Harvard mathematics department had been brought to a complete standstill. Coolidge, stationed abroad at the AEF headquarters, wrote to Osgood in October of 1918 to "appeal to [his] courage and tenacity for the great work that lies ahead." He added:

> For the moment nothing is possible. There is no college and no mathematical department. But that does not mean that there is no problem for us to be facing right now. I read in the paper today of the capture of Ostende [in Belgium]. I hear from the intelligence section that Zeebrugge [also in Belgium] is ours; I gather from various sources that the German morale has collapsed. Therefore I hazard the guess that hostilities will be over by Christmas. The conclusion from this premise is that this is to be the only year when there is no Harvard college, that the gov[ernmen]t will cease to draft many men after next spring, and that the first return to quasi normal conditions will come in Cambridge next autumn."[37]

Although Coolidge could not speculate on his own return to Harvard, he focused keenly on the future of the mathematics department there. He also speculated at length about ways to build up the department after the war.

As Osgood explained in his report to Harvard president Ira N. Hollis, it was "somewhat indecent to dwell at any great length on what the Division [of Mathematics at Harvard] has done" with respect to making serious contributions to mathematical science in the past year. Osgood submitted statistics about research production and publication over the previous twenty-five years and explained that the falling off between 1915 and 1919 "is largely attributable to the war and to death. Two members of the Division died during the last year named [Gabriel Marcus Green and Maxime Bôcher], two were in military service, and a fifth lost nearly two academic years in the same way. Those that remained were obliged to bear a very serious burden of teaching, and were scarcely well placed to do fruitful scientific work."[38]

Despite great effort, they encountered tremendous difficulty finding qualified instructors to help with the burden of teaching. Hollis of Harvard wrote to ask Dunham Jackson to fill a post at Harvard. Jackson had graduated from Harvard in 1908 and completed a masters there in 1909. He then studied in Göttingen, where he completed his Ph.D. in 1911 under the direction of Edmund Landau. Jackson became an instructor at Harvard, and in 1916 an assistant professor. Starting in 1918, he worked as captain in the Ordnance Department, doing methods of numerical integration with Moulton until 1919. In February of that year, D. Jackson wrote to Osgood with sympathies for the heavy strain of present emergencies, but not solutions. "I can only tell Hollis," he wrote, "that similar inquiries are coming in from all sides, and that the demand for instructors continues greatly to exceed the supply; I do not know of a single available one at present."[39] Obtaining leave

[37] [Harvard Mathematics Papers, UA V 561.8, 1915–1916–1917–1918–1919 folder, Coolidge to Osgood, October 19, 1918].

[38] [Harvard Mathematics Papers, UA V 561.8, Report to the Visiting Committee, Osgood to Hollis, April 27, 1920]. G. M. Green was a young graduate from Columbia University (with a thesis on differential geometry in 1913) who died of influenza in 1919.

[39] [Harvard Mathematics Papers, UA V 561.8, Dunham Jackson to Osgood, War Department letterhead, February 27, 1919].

was impossible for Jackson, but sworn affidavits were sent to Rexford Tucker and Bancroft H. Brown (a Birkhoff student) so they could urgently request immediate discharge and teach mathematics at Harvard. In the fall of 1919, William Casper Graustein—a Harvard graduate of German descent who had trained with Study at Bonn before serving in the Ordnance Department at Aberdeen—was called back "to help in the congenial task of rehabilitating the Division of Mathematics" at Harvard [Coolidge 1941, p. 343].[40]

3. After the War

In 1919, the young Bancroft H. Brown, later at Dartmouth College, wrote to G. D. Birkhoff: "I am very anxious to continue my studies. I don't consider these two years as by any means wasted, but I feel that I have done my share, and will be glad to get back into harness again."[41] But the experience of the war was to shape what getting into harness again would mean for many researchers. While the war seemed to have little impact on Leonard Dickson's mathematical researches, it did influence the topic of his retiring address as AMS president. Dickson's address was scheduled for December 27, 1918, and, for all but the final weeks of its preparation, the world was at war. He intended for his "Mathematics in War Perspective" to introduce his audience to military and naval instruction in France, England, and the U.S. The Armistice ironed a wrinkle into Dickson's plans for his address, however, since he would now speak to an audience in peacetime. Consequently, he emphasized, as E. W. Brown had done two years earlier, that national preparedness in wartime depends on national preparedness in peacetime, particularly relative to mathematical training.

"Let it not become possible," Dickson urged his audience, "that thousands of young men shall be so seriously handicapped in their army and navy work by lack of adequate preparation in these subjects" [Dickson 1919, p. 289]. It was in his opening remarks, however, where Dickson revealed his true distance from the frontlines. "An army officer of high rank," Dickson began,

> now facing the problems involved in stopping the huge war machine which he had helped to build, recently remarked to me that this getting out of war is far more trouble than getting into it. The Armistice has put me in the same boat and, for the purposes of this address, came a few weeks too soon. I had already put myself under obligations to numerous friends, including two in England and France, for furnishing me authoritative information on the role of mathematics and its applications in the war. While this information is fortunately no longer needed for its initial purpose, it bears on the timely question of the kind of preparedness which the nation should adopt [Dickson 1919, p. 289].

[40][Harvard Mathematics Papers, UA V 561.8, Bancroft Brown to Osgood, September 18, 1919; M. Morse, April 17, 1919; January 2–16, 1919. Telegrams to/from Rexford S. Tucker, Corporal, 20th Trench Mortar Battery, Camp Jackson, SC; from Osgood to Graustein, February 15, 1919 , and Graustein's reply, February 18, 1919, on APG paper: "—Lt. William C. Graustein invited to teach at Harvard for the spring term."]

[41][Birkhoff Papers, HUG 4213.2, Box 7, Bancroft H. Brown to Birkhoff, February 10, 1919].

While the question of how, say, America mobilized more than four million men and then disbanded the operation less than two years later certainly merits attention, no German, French, Italian, or British citizen—mathematician or otherwise—would have ever mentioned that the Armistice came too soon, even in jest.

Many years later Veblen would echo Dickson's (and others') point about the critical role of preparedness in peacetime when he wrote in support of Dickson's brother, Colonel Tracy Dickson, for an appointment as Chief of Ordnance of the U.S. Army.[42] "You will recall discussions which we had years ago," Veblen wrote to F. R. Moulton in December, 1929,

> about the proper organization of the Ordnance Department in time of peace, how we felt that in peace time the important thing was the proper development of all the technical devices, both theoretical and practical, so we should not again see the unfortunate confusions and lack of preparation which we observed at the beginning of the last war. My experience in the mud of Aberdeen impressed these ideas upon me perhaps even more than your broader experiences at Washington did upon you [Veblen Papers, Box 33, Veblen to Moulton, December 16, 1929].

Veblen's comparison of mud vs. no mud experiences emphasizes his point about peacetime preparation. It also suggests a distinction among some actors between those who engaged in intellectual endeavors related to the war and those who engaged in field operations. Nevertheless, Veblen was in the mud just outside of Baltimore, not Verdun.

After the war, Veblen was not part of the immediate demobilization. With the possibility of an armistice on the horizon, the U.S. Army sent Veblen to Europe to visit the British, French, and Italian ballistics laboratories before they were dismantled [Grier 2001, p. 929]. During his four-month tour, Veblen visited the Gâvre Proving Grounds and also the British munitions facility at Woolwich. He paid special attention to the discussions he had with John E. Littlewood and Vito Volterra, both of whom had undertaken ballistics research.

Veblen became active in American mathematical institutions after the war, serving as AMS president, working to build a new mathematics faculty at Princeton, participating in the first faculty at the Institute for Advanced Study, and, eventually, serving on the Applied Mathematics Panel of the Office of Scientific Research and Development (OSRD) in WWII. Veblen's experience as a team leader during the war also may have been a factor in transforming his way of doing mathematics, so that on his return to Princeton he no longer worked as a solitary mathematician but as the leader of an energetic and innovative research group in what was for him a new interest, differential geometry. Ritter in fact argues persuasively that Veblen left Aberdeen convinced of the effectiveness of organizing research in such a way that a relatively large group of researchers works collectively on a common program. More than this, the ballistics work highlighted an idea already coming to the fore in other contexts, namely that the interactions between physics and mathematics were of central importance for contemporary research agendas [Ritter 2011, pp. 150–155].

---

[42] Dickson asked Veblen to write on his brother's behalf [Veblen Papers, Box 4, Dickson to Veblen, November 26, 1929].

3.1. Mathematics and Military Research in the Postwar Years. After the end of the war, some in the American mathematical research community wished to take stock of the war-related research that was conducted. For one thing, the mathematical sophistication of exterior ballistics had increased significantly, and mathematicians were intent to stay associated with it [Gluchoff 2011]. An example will underline this point. In December of 1918, at the joint meeting in Chicago of the MAA and the AMS, a joint special session on mathematics and the war presented to a general audience a diverse collection of papers that provides some picture of the effects of wartime involvement on mathematics [AMS Notes, vol. 25, pp. 241–242]; [MAA Meeting 1919]. Bennett, one of the "Aberdeen men," gave an overview of some general aspects of ballistics. He noted the importance of the motion of the projectile both inside and outside the barrel of the weapon. Inside the barrel, the question of the effects of uniform versus variable pitch of the rifling posed problems in classical mechanics. As for outside the gun, he discussed the question of nutation and precession of the projectile, the effects of air resistance, in particular the perturbing effect resulting from proximity of the velocity of the projectile to that of sound. However, Bennett appears to have said little beyond describing the general character of the problems, mentioning the importance of probability, statistics, mechanical quadrature, and graphical methods in their solution.

A more vivid example was given by William D. MacMillan in a speech on "Ordnance Problems," describing the magnitude of the problems involved. MacMillan, formerly at Chicago, was part of the Moulton group at Army Ordnance in Washington.

> A plane or dirigible is at an unknown height, is moving in unknown direction with an unknown velocity, the force and direction of the wind and some other such data are incompletely known. The artillery officer has six seconds within which to make the necessary observations and compute the required direction and elevation of fire [MAA Meeting 1919, p. 101].

MacMillan pointed out that while progress had been made on simplifying the calculations, the users of these methods were still only attaining one hit in 15,000 rounds. On the bright side, MacMillan "good-humoredly" added that a practiced gunner who did not bother with these methods hit such targets one in about 3,000 or 4,000 shots!

The account of Philip Langdon Alger, titled "Practical exterior ballistics," also published in [MAA Meeting 1919], is more illuminating about the relationship of such research to broader research practice. Langdon had graduated from the Massachusetts Institute of Technology (MIT) in electrical engineering in 1915 and later served as lieutenant in the Ordnance Department. His account shows an awareness of contemporary British ballistic work, as well as a good understanding of relevant recent literature internationally. Alger pointed out that in the old methods it was usual to determine experimentally a constant for each projectile and employ the Newtonian assumption that air resistance is proportional to the square of velocity. Tables by Otto and Euler worked on this Newtonian assumption, though more recently artillery tables employing third-power laws (Bashforth) and fourth-power laws (Zaboudski, Lardillon) had been devised. He then mentions that following experimental work by Krupp, Mayevski, and Hojel, a discontinuous law of air resistance had come into use, where the term used would depend on the velocity. He likewise mentions the fact that high-angle fire, frequent in WWI, meant that taking the density of air as constant was likewise not suitable. In any case, in this context, the numerical solution of differential equations became important. Methods of Otto, Didion, and Siacci are described as "very satisfactory for use in conjunction with Mayevski's laws."[43]

More recent research, according to Alger, took this further:

> ...the modern use of high-angle fire has rendered necessary new methods for exact computation. This need has been met by the assumption of a smooth empirical law for air resistance and by the introduction of methods of solution of these equations by numerical integration. Such methods have long been in use but others have recently been developed in England by Littlewood and in America by F. R. Moulton. Improvements and extensions of the American method by Gilbert Ames Bliss, Capt. A[lbert] A. Bennett, and T[homas] H. Gronwall have much reduced the labor of computation required [MAA Meeting 1919, pp. 101–102].

In another contribution, Hans F. Blichfeldt, a Danish-born mathematician from Stanford who had studied group theory with Sophus Lie in Leipzig, used a variation of Siacci's method for expressing by quadratures the trajectory associated with low velocity high-angle fire connected with mortars in the trenches. Speaking on ballistics at both the joint AMS-MAA meeting of 1918, where he reported on the effect of the rotation of the earth and its curvature on the path of a projectile, and at the Toronto International Congress of Mathematicians (ICM) in 1924, William Henry Roever, now of Washington University in St. Louis, kept his interest in the field alive for a few years and would return to it during WII [Roever 1924, Roever 1943].

The end of the war however led to a reshaping (and in some cases abandonment) of the research programs associated with it, and the return of mathematicians to their usual careers induced various changes. "I think that the ballistics work will be in large part abandoned soon," wrote Roland Richardson at the end of 1918.[44] APG experiments appear to have ended, temporarily anyway, by then. Mason and Kellogg stayed at Aberdeen for a time. Veblen was in Europe and all the other mathematicians had already left. Moulton's group, however, did remain involved in related research. As Moulton pointed out, proving ground experiments could not isolate individual external effects, and wind-tunnel experiments thus came into play. These continued for several years after the war, in conjunction with industrial cooperation. Moulton's description is informative about the conduct of this military-industrial interaction:

> Unfortunately the power required to give velocities above that of sound in cylinders of considerable cross-section is very great. It was learned, however, that the General Electric Company was

[43]For a further discussion of air resistance laws used in ballistics and shooting procedures, see David Aubin's contribution dealing with the Gâvre Commission in this volume. High-angle shooting in anti-aircraft gunnery and high-angle shooting in mountainous areas are also discussed, respectively, in June Barrow-Green's and Pietro Nastasi and Rossana Tazzioli's contributions to this volume.

[44][Birkhoff Papers, HUG 4213.2, Box 3, Richardson to Birkhoff, December 31, 1918].

FIGURE 2. Bureau of Standards wind tunnel, around 1918–1921. Library of Congress, Prints & Photographs Division, LC–DIG–npcc–20523.

planning to test at Lynn, Massachusetts, in the early autumn of 1918, some huge blowers for use in blast furnaces. Accordingly, [Moulton] inquired of Dr. Sanford A. Moss [of the General Electric Company], respecting the possibility of making ballistic experiments during the process of making the blowers. At a conference in Washington in July, 1918, Dr. Moss promised the full co-operation of the General Electric Company. Immediately general plans for the experiments were worked out in conjunction with Professor (then Captain) Albert A. Bennett[45]; an appropriation of as much as $100,000 as might be required was secured to prosecute the experiments; upon application to Director [Samuel W.] Stratton, of the Bureau of Standards, Dr. L[yman] J. Briggs was assigned to work on the experiment; and on the first of October the transfer of Professor (then Major) G[eorge] F[errie] Hull from the Signal Corps to the Ordnance Department was secured and he was assigned to co-operate with Dr. Briggs, and to be officially in charge of the practical work of carrying out the experiments. These experiments have proved most difficult, as was expected, and have been continued over several years [Moulton 1926, pp. 3–4].

[45]Bennett was a Princeton Ph.D. from 1915 who had spent a year in Göttingen. He was to return to the APG in WWII [Mitchell, s. v. Bennett].

This is an interesting constellation of circumstances. General Electric was developing blast furnaces for the steel industry, but a spin-off benefit was the possibility of acquiring wind-tunnel data. Assembling even a small team involved the cooperation, not only of various units in the armed forces, but also of the National Bureau of Standards and private industry. It was a type of interaction that was to become very common in the future.

At some point after the war, Moulton returned to Chicago. In 1925 he completed his book *New Methods in Exterior Ballistics*, which appeared the following year. This contained the results of roughly seven years of work which had begun when he became head of the Ballistics Branch of Army Ordnance in April 1918. Nor had he been alone in this effort:

> United States army and naval officers have been studying ballistics, mathematics, and physics under [my] direction at the University of Chicago during the entire period since the close of the war [Moulton 1926, p. 1].

Moulton appears to have taken it on himself to wholly recast ballistics, in particular noting the importance of combining theory and experiment. All but the last chapter, treating the motion of a rotating projectile, were complete in essence by the end of the war.

Moulton's book is almost wholly mathematical, although attesting to the engineering side of the research the book contains one condensation wave photo. Two of the chapters are on numerical integration, where a very basic scheme is used to demonstrate how to compute trajectories, essentially a forward Euler method applied to the system: the system is first converted into a first-order system by the usual Lagrangian method. Moulton sets the numerical quadrature problem up as one in which the earth is flat and immobile, taking rotation and curvature into account separately using what he terms "differential variations" to keep the differential equations simple. A chapter is devoted to this method, which is probably the most important part of the book practically. The problem of accuracy of the numerical integration method is at first waved off by noting that where no observational checks are available (always the case with shell trajectories, as opposed, say, to celestial mechanics) "reliance must be placed on the smooth progression of the higher differences of the functions involved" [Moulton 1926, p. 78]. However, later an entire chapter is devoted to the validity and error in these approximations.

The influence of this book, and Moulton's ballistic research program, on mathematical research seems to have been very limited, at least in the immediate wake of these events. Moulton mentions one University of Chicago Ph.D. thesis by Lawrence Wainwright, directed by him. The preface rather suggests that things had already ground to a halt by 1926 as far as the program of comparing the theory to practical results.

The account by Moulton identified the National Bureau of Standards as a further government site where war-related mathematical research found a home in the years following WWI. A key figure here is Lyman Briggs, a scientifically versatile figure trained as a physicist. Briggs's war work was not limited to wind tunnels. He had also been instrumental in designing the "stable zenith" gyroscope, a device useful for aiming guns at sea. This was tested aboard the battleships USS Arizona and USS Mississippi in October 1918 and was later adopted as standard by the navy on all its battleships [Landa & Nimmo 2003]. Briggs reported on his

FIGURE 3. Charles Buckles Falls, "Learn to make and test the big guns—better yourself, enlist and learn a trade in the Ordnance Department" Recruiting poster showing firing at Aberdeen Proving Ground, 1919. Library of Congress, Prints & Photographs Division, POS-US .F35, no. 5 (C size).

research in the *Proceedings of the International Congress of Mathematicians* in Toronto [Briggs 1924]. In 1924 Briggs was the chief of the Division of Mechanics and Sound of the bureau, and his report describes a spectrum of activities at the bureau. The first of these concerns the experimental testing, and preparation of buckling tables for, compression members. One side of this program concerned light columns and girders for aircraft use and was done in response to requests by the Bureau of Aeronautics of the Navy Department. Ballistics was another area of endeavor, and Briggs describes various measurement devices associated with testing at Aberdeen, including projectile velocity measurements, pressure measurements in the breech, and air resistance measurements. Briggs himself was involved in the last of these, and General Electric is credited for supplying blowers. Fabric tension meters for airships and the hydrophone for use in locating ships are described in the same report.

The 1924 Toronto Congress remains unusual among the ICMs for the extent to which applied mathematics and in particular directly military issues were discussed. The dominance of pure mathematics reasserted itself in Bologna in 1928. Subsequent meetings have occasionally seen controversy concerning the applicability of mathematics to war aims. However, the direct communication of mathematics with military applications was to take a markedly different direction during and after WWII, notably in the Cold War context. Much of the research activity undertaken during the First World War was done with very specific and urgent objectives that included the engineering of devices, among them guns, shells, and submarine detection instruments. Even if often qualitative rather than being given in the kind of detail that permits a thorough understanding of the state of the art, postwar publications such as those reviewed here were symptoms of an awareness of the importance of war uses of mathematics, an awareness that did not die instantly with the war's end.

3.2. Postwar Changes in Education and Training. Even after the Armistice, wartime activities continued to have a ripple effect on students and institutions. The war necessarily interrupted the studies of enlisted students. Following the war, they petitioned for credit in mathematics for studies pursued abroad or for aviation experience. Students were granted mathematics credit in many cases, for example, for eight months service in the Lafayette Flying Corps of the French Army, for courses in radio communication, and studies undertaken at the French Artillery School in Fontainebleau. The Administrative Board of the University Extension Office also added an extra session of summer school "primarily for Harvard undergraduates who are behind in their work on account of war service."[46]

3.2.1. *Curricular Changes in University Mathematics.* Considerations of war service and the mathematics involved prompted discussions of curricular changes. The advent of the war had shut down research communication on aeronautics, which was widely seen as an extremely promising military tool. In 1918, Edwin Bidwell Wilson of MIT published a short note on the mathematics of aerodynamics [Wilson 1918]. After a very brief survey of some of the major issues in basic aerodynamics, Wilson noted:

> The right kind of mathematics [for aerodynamics] is the good old traditional Cambridge, England, type, ... a type of math which in this country, owing perhaps to our preponderance of German-trained mathematicians, has all too little prestige.

Wilson concluded with a call to reconsider curriculum without prejudice and, apparently, with a break from history. This address was given in December 1916, but the call to examine curriculum, as well as methods, came forward once again in connection with the educational problems posed by the war effort. Wilson was not alone in believing that U.S. universities did not devote enough attention to applications. This had been a major theme in the address of E. W. Brown, the outgoing AMS president, discussed earlier [Butler Feffer 1997, Parshall 2000]. Brown, however, had not been the first AMS president to take up this issue of applied mathematics. Moore and Woodward each devoted at least a portion—if not all—of their retiring presidential addresses to the topic of pure vs. applied mathematics in the very early years of the twentieth century.

The effect of Wilson's call is not clear, though at Harvard it appears to have gone unheeded. In the decade following the war, only one thesis among Harvard's graduate students involved an obviously war-related topic. Lewis E. Ward wrote

[46] [Harvard Mathematics Papers, UA V 561.8, Department of Mathematics, Correspondence 1919, A. F. Whittem to W. F. Osgood, December 20, 1918]. The above paragraph is also based on informations gathered in [Harvard Mathematics Papers, UA V 561.8, Department of Mathematics]. See, especially, Correspondence and Papers, 1911–1920, J. F. Rolfe to W. F. Osgood, July 19, 1919; M. Crosby to W. F. Osgood, February 4, 1919; and W. F. Osgood to L. S. Mayo, May 27, 1919; Box 1911–1920, 1915–1916–1917–1918–1919, Memo for Prof. Osgood, May 21, 1919.

"Computation in Trajectories in the Theory of Ballistics," in December 1920. He relied on the French translation of Siacci's *Balistique extérieure*, Prosper-Jules Charbonnier's *Traité de balistique extérieure*, Carl Cranz's *Lehrbuch der Ballistik*, and James M. Ingalls's *Ballistic Tables (Artillery Circular "M")*, "as well as numerous blue print pamphlets very kindly loaned me by Professor Graustein [who had worked at Aberdeen]; and Lieutenant R. S. Hoar's book on exterior ballistics in blue print form, which was very kindly loaned me by the author."[47] The following spring, Harvard considered the possibility of having Roger S. Hoar give a course in ballistics. The Division of Mathematics discussed the proposition at length and, afterwards, "[i]t was the feeling of the Division that such a course would be given more appropriately under the Department of Military Science."[48]

In the spring of 1919, the Harvard faculty discussed introducing a program in military science under the ægis of the newly formed Department of Military Science and Tactics. Colonel Robert C. Goetz solicited particular input from Osgood, chair of mathematics, on the program:

> I realize fully there is only an absolute minimum of Mathematics required. I have the feeling that if more is put in at this time it might be discouraging to a large number of students. However, I believe that if I can get the students interested in gunnery and in the mathematical computations in Ordnance and Artillery Firing, many will realize the necessity of a thorough working knowledge of Mathematics. From the interest aroused in the practical application of the theoretical work and by proper advising, I believe many students will take higher courses to equip themselves in Mathematics.[49]

No response from Osgood to this prognosis has been located. There does not appear to have been substantial curricular change at Harvard due to the war, at least not in the direction of increasing applied mathematics.

In other institutions, there is evidence that new mathematics courses were introduced as a result of the war. Henry Lewis Rietz of the University of Iowa for example suggested the value of a course in statistics. In February and March of 1918, Rietz had somehow been involved in a reorganization of the Quartermaster Corps, the army entity in charge of the details of distribution of supplies. Here he had an opportunity to see at first hand what kind of statistical methods were being used in the War Department: as he diplomatically puts it, the experience convinced him that a course might well be given in a department of mathematics [MAA Meeting 1919, p. 99] which would be suitable for such work. He devised such a course and offered the first half of it by the end of 1918. This statistics course was given to 73 people, 6 hours weekly, and covered tabulation of data, frequency distributions, and the basics of probability; graphical methods; and averages. He

---

[47][Harvard Mathematics Papers, UA V 561.255 Mathematics Department, Minor Thesis Cards, c. 1902–1941, December 1920, L. E. Ward, "Computation of Trajectories in the Theory of Ballistics"]. *A Course on Exterior Ballistics* by Roger Sherman Hoar was published as an Ordnance textbook by the Government Printing Office in 1921.

[48][Harvard Mathematics Papers, UA V 561.3, Minutes of Meetings , Vol. II, 1920–1924, March 1, 1921]. Note that R. S. Hoar later became a successful science-fiction writer under the pseudonym of Ralph Milne Farley.

[49][Harvard Mathematics Papers, UA V 561.8, Box 1911–1920, Correspondence 1919, R. C. F. Goetz to W. F. Osgood, April 4, 1919].

FIGURE 4. U.S. President Calvin Coolidge at the National Academy of Sciences and Research Council, April 28, 1924. Library of Congress Prints and Photographs Division LC-F8-30602.

did not have time, as it turned out, to cover averages of different kinds (as he put it) to his satisfaction. He gave an example of the kind of question he hoped students could solve: estimate the relative frequency of there being more than 3 inches of monthly rainfall in Iowa City (based on some unspecified data).

3.2.2. *The National Research Council.* Once the Armistice was signed, Hale set out to make the NRC a permanent coalition of science and government. Striking a careful balance between federal support without government control, Hale sought an Executive Order to ensure an "enduring legal foundation" for the NRC [Kevles 1968, p. 432]. After various negotiations, President Wilson signed the (revised) Executive Order on May 10, 1918, thereby guaranteeing the existence of the NRC in peacetime. One significant outcome of WWI, this coalition (Figure 4) between science and government is perhaps more striking in the American than in the European context because of its novelty as an effective association in the U.S. context; the earlier envisaged role of the NAS had been less effective and did not endure.

The NRC made two substantial contributions to American mathematics in the immediate postwar years. First, the NRC provided desperately needed funds for publications in mathematics. The NRC—and the General Education Board of the Rockefeller Foundation, for that matter—gave generously to support the publication of the *Transactions of the American Mathematical Society* and to establish a revolving book fund for the publication of mathematical books. Second, at Veblen's urging, the NRC established fellowships in mathematics to fund "the elite of the mathematical intellectuals" in the U.S. by providing opportunities for further research in optimal conditions [Archibald 1938, p. 15].

In October 1919, representatives of mathematics in the Division of Physical Sciences investigated the place of mathematics within the NRC and the status of applied mathematics in the U.S. After careful discussion, the division addressed a statement to mathematics departments throughout the U.S.[50] This statement observed that undergraduate training currently did not equip students "to consider the application of mathematics to physical problems except of the most elementary kind." Yet needs for research in dynamics, optics, sound, and related mathematical subjects increased dramatically due to wartime developments in air, surface, and underwater transportation. The NRC statement asserted that "applied mathematics has been underemphasized in America" but also acknowledged researchers' reluctance to abandon their present pure mathematical investigations in pursuit of new and unfamiliar directions. Instead of advising such drastic departure, the NRC called mathematics departments to serve the country by teaching, encouraging, supporting, and directing their students to undertake problems in applied mathematics. The NRC advocated a return to methods that based research questions on contact with branches of physical science and additionally urged mathematics departments to stimulate "more students of good ability to prepare themselves for careers in applied mathematics, so that the present and probably increased future demand in this subject may be met."[51] A detailed assessment of the effect of this call is beyond the scope of this chapter, but there is some evidence that the effect was slight and that the view was not shared by the NAS, for example. In the mid-1920s E. B. Wilson complained to Harlow Shapley, the Harvard astronomer, that pure mathematicians at the NAS and at Harvard had blocked applied mathematics [Reingold 1988, p. 196].

While Hale may have envisioned an interdisciplinary organization that linked science and government, certain aspects of the peacetime NRC do not seem to have coalesced with Hale's vision. In 1920, for example, Leonard Dickson chaired the Committee on Algebraic Numbers of the NRC. That a committee on algebraic numbers existed already suggests a reshaping of Hale's original aims. Together with Howard H. Mitchell and George Eric Wahlin, Dickson authored the "Report on the theory of algebraic numbers" which, combined with "the report of Hilbert [on algebraic number fields, in the *Jahresbericht*] give a full direction to the literature of that part of the theory of numbers which is not being covered in Dickson's *History of the Theory of Numbers*" [Carmichael 1923, p. 327]. A report on the theory of algebraic numbers seems a long way from Hale's idea of interdisciplinary studies or national security and welfare.

Conclusion

The rate at which events unfolded as the United States engaged the conflict is remarkable. The formerly neutral nation launched a full-fledged war operation

[50][Harvard Mathematics Papers, UA V 561.8, 1911–1920, Correspondence, etc., 1919–1920, The Status of Applied Mathematics in the United States, October 20, 1919]. Osgood reviewed this document for Harvard and responded in April of 1920 by analyzing the needs of mathematics. This letter has not been located but was referred to the chairman of the Division of Physical Sciences. [Harvard Mathematics Papers, UA V 561.8, 1911–1920, Correspondence, etc., 1919–1920, C. E. Mendenhall to W. F. Osgood, October 22, 1919 and R. M. Yerkes to W. F. Osgood, May 4, 1920].

[51][Harvard Mathematics Papers, UA V 561.8, 1911–1920, Correspondence, etc., 1919–1920, The Status of Applied Mathematics in the United States, October 20, 1919].

in April 1917 and started the process of massive demobilization shortly after the Armistice of November 11, 1918. This chain of events impacted institutions and individuals alike, as well as training and research in the U.S. The current chapter has specifically considered, in the American context, the effects of the war on mathematics and mathematicians as well as the lesser influence mathematics had on the war.

The rapid mobilization of mathematicians, whether in educational roles, in research, or in intellectual leadership, parallels the very rapid mobilization of American forces across the board. It took a little over a year for U.S. soldiers to actually enter battle, in May 1918, and American materiel was not to reach the front until the very end. Although the conflict generated a spike of interest in ballistics, aeronautics, and other war-related mathematics, the timeline of events minimized the direct influence of American mathematics on the war. Most, in fact, came in the context of local efforts at low-level mathematical training for military purposes. The eventual hope afforded by the U.S. commitment was a key contribution. Like the rest of the country, academic mathematicians had stood behind the national effort to bring troops and weaponry to the aid of the Allies. This large and apparently nearly unanimous participation from academic quarters would have been particularly reassuring to their British and French counterparts. It contributed one piece of the overarching American support that was one decisive factor in winning the war.

Although the immediate impact of American mathematics on the war was minimal, the fact of U.S. involvement in the war did mark mathematics in America. The national effort to aid the Allies led both institutions and individuals to specific challenges and questions that resonated in the postwar years. Specific institutions found themselves bound to the service of the state and its goals: the newly formed NRC and the Carnegie Institution of Washington learned that negotiation with government and the military was part of their mandate and that the government's aims could be decisive in their decision-making as well as in what they could successfully accomplish. There were obvious lessons here for the future for all parties about the potential of the role of the academy in public affairs and in defense. Accompanying these lessons in a growing community were increased links between different elements of the community, particularly the leading research universities, and a renewed institutional basis that was to shape future developments through the AMS, MAA, government, and private philanthropy.

These lessons were clearly underlined by individual experiences such as those of Veblen and the "Aberdeen men."[52] Many American mathematicians and scientists left their academic positions to aid the U.S. in World War I. Veblen measured the distance traveled by projectiles while soldiers fired them on the other side of the Atlantic. Millikan worked to develop methods to detect submarines while the Germans flooded the Atlantic with submarines. The demands of frontline warfare were met with initiatives that took place, in part, more than 3,000 miles away. The new "economic and social forces" E. W. Brown had mentioned in his retiring presidential address did, indeed "stop temporarily ... the work and thought" [Brown 1917, p. 229] which had occupied Veblen, Millikan, and others in the preceding decade.

[52]The expression is used by Benjamin Fife in a letter to Birkhoff, February 12, 1919 [Birkhoff Papers, HUG 4213.2, Box 3].

Not all American mathematicians abandoned their habits, though, and in some cases the support for the war effort in Europe was more theoretical than concrete. As an idea (and project) conceived before World War I, sustained throughout the war, and published in 1919, 1920, and 1923, Dickson's *History of the Theory of Numbers* serves as one example of how America continued in its "rapid mathematical progress ... with only a brief pause" during the years from 1914–1920 [Birkhoff 1976, p. 42]. G. D. Birkhoff's work on the three-body problem and on dynamical systems also has its origin in these years [Aubin 2005].

However, for some of the many researchers and educators in the mathematical sciences who *did* participate in the war effort, their applied mathematical experiences continued after the war ended. The basis for further development was thus kept alive during this period. Veblen, for example, remained a consultant for the Aberdeen Proving Ground after the war and was instrumental in persuading John von Neumann to take on a consulting role there in 1937 [Price 1988, p. 267]. Von Neumann is described as having "learned from Robert Harrington Kent the related theories of shock and detonation" and became an expert on these matters. After his 1915 M.A. at Harvard, Kent had joined the Ordnance Department and was responsible in France for the preparation of firing tables used by the AEF. Staying at Aberdeen after the war, Kent went on to become "almost surely, the country's only 'complete' ballistician." The "outstanding carrier of the torch of science in the Army," Kent was a direct connection between Veblen's and von Neumann's experience in ballistics [Simon et al. 1971, pp. 96 & 104].

Mature mathematicians overwhelmingly returned after WWI to the pure-mathematical practice in which they had apprenticed and begun their careers before war-related activity, but others retained interests in the kind of associations they had learned through their military efforts. Max Mason became director of natural sciences for the Rockefeller Foundation in the late 1920s, later moving to its presidency. Warren Weaver succeeded him beginning in 1932 in the natural sciences director's post. Weaver went on to head the Applied Mathematics Panel of the OSRD from 1943 to 1946, also providing a direct link between American scientific activities in the two wars. The work of Lyman Briggs at the National Bureau of Standards likewise draws our attention to the continued interests of government in the uses of mathematics for military technology. Briggs was to become director of this bureau in the 1930s. Industry, too, was introduced to mathematical contacts in this period, as we have seen through the work of Moulton and Slepian. Future investigation of such associations and the kind of mathematical work they supported is highly desirable. In any case, it is clear that at the onset of WWII the U.S. no longer found it problematic to determine what to do with mathematicians.

The U.S. mathematical research community had developed from European roots, drawing particularly on German training and models. American mathematicians continued to seek contact with European colleagues through their writings, as well as through involvement in projects like Klein's and Meyer's *Encyklopädie der mathematischen Wissenschaften*. The extent to which these contacts were interrupted during the longer war, and during the U.S. involvement, is unclear. As for the resumption of broader international contacts in the postwar period, many American mathematicians appear to have drawn on their personal allegiances to German universities. This connection was to reassert itself after the war through the active role taken by Hale and others in advocating the reopening of international scientific meetings to German and Austrian scholars. American discomfort at the failure of these moves led to their revocation of an offer for the International Congress of Mathematicians to meet in New York in 1924. Continued American agitation on this matter was to prove key to the reintegration of German and Austrian mathematicians into the international community, notably the 1928 Bologna ICM [Lehto 1998].

Abbreviations

- AEF: American Expeditionary Force.
- AMS: American Mathematical Society.
- APG: Aberdeen Proving Ground.
- CIW: Carnegie Institution of Washington.
- ICM: International Congress of Mathematicians.
- MIT: Massachusetts Institute of Technology.
- MMA: Mathematical Association of America.
- NAS: National Academy of Sciences.
- NRC: National Research Council.
- OSRD: Office of Scientific Research and Development.
- ROTC: Reserve Officer Training Corps.
- SATC: Student Army Training Corps.
- YMCA: Young Men's Christian Association.

Bibliography

Archival Sources

[Birkhoff Papers] George David Birkhoff Papers, Harvard University Archives.

[CIW Papers] Carnegie Institution of Washington Archives, Carnegie Institution of Washington, Washington, D.C.

[Gould Papers] Alice Bache Gould Papers, Massachusetts Historical Society.

[Harvard Ayer Papers] U.S. Naval Radio School, Records of Commander Nathaniel F. Ayer, 1918–1921, Harvard University Archives.

[Harvard Mathematics Papers] Division of Mathematics, 1911–1920, Harvard University Archives.

[Veblen Papers] Oswald Veblen Papers, Archives of the Library of Congress, Library of Congress, Washington, D.C.

Published Works

[AMS Notes] Anonymous, *Notes*, Bulletin of the American Mathematical Society **21** (1914–1915) to **25** (1918–1919), passim.

[Archibald 1938] Raymond C. Archibald, *A Semicentennial History of the American Mathematical Society, 1888–1938*, American Mathematical Society, New York, 1938.

[Aspray 1988] William Aspray, The Emergence of Princeton as a World Center for Mathematical Research, 1896–1939, in [Duren 1988, vol. 2, pp. 346–366].

[Aubin 2003] David Aubin, La guerre du froid, la défaite de l'inventeur: bombes à oxygène liquide et production d'hélium, in *Le Sabre et l'Éprouvette: l'invention d'une science de guerre 1914–1939*, ed. David Aubin & Patrice Bret, Noesis/Agnès Viénot, Paris, 2003, pp. 105–116.

[Aubin 2005] David Aubin, George David Birkhoff, Dynamical Systems (1927), in [Grattan-Guinness 2005, pp. 871–881].

[Barnett 1918] Isaac Albert Barnett, *Mathematical Instruction at the Great Lakes Naval Station*, American Mathematical Monthly **25** (1918), pp. 326–328.

[Barry 2004] John M. Barry, *The Great Influenza: The Epic Story of the Deadliest Plague in History*, Penguin Group, New York, 2004.

[Berky & Shenton 1966] Andrew S. Berky & James P. Shenton, eds., *The Historians' History of the United States*, vol. 2, G. P. Putnam's Sons, New York, 1966.

[Birkhoff 1976] Garrett Birkhoff, Some Leaders in American Mathematics: 1891–1941, in [Tarwater 1977, pp. 25–78].

[Birkhoff 1988] Garrett Birkhoff, Mathematics at Harvard, 1836–1944, in [Duren 1988, vol. 2, pp. 2–58].

[Blue Hen 1925] *The Blue Hen* [Student Yearbook], Delaware College, University of Delaware, 1925.

[Briggs 1924] Lyman E. Briggs, Research in Mechanics and Sound at the Bureau of Standards, in *Proceedings of the International Congress of Mathematicians*, ed. John Fields, University of Toronto Press, Toronto, 1924, vol. 2, pp. 393–403.

[Brown 1917] Ernest W. Brown, *The Relations of Mathematics to the Natural Sciences*, Bulletin of the American Mathematical Society **23** (1917), pp. 213–230.

[Butler Feffer 1997] Loren Butler Feffer, *Mathematical Physics and the Planning of American Mathematics: Ideology and Institutions*, Historia Mathematica **24** (1997), pp. 66–85.

[Butler Feffer 1998] Loren Butler Feffer, *Oswald Veblen and the Capitalization of American Mathematics: Raising Money for Research, 1923–1928*, Isis **89** (1998), pp. 474–497.

[Carmichael 1923] Robert D. Carmichael, Review of L. E. Dickson et al.'s *Algebraic Numbers*, American Mathematical Monthly **30** (1923), pp. 327–328.

[Collins 1932] Varnum Lansing Collins, *Princeton in the Great War*, The Office of the Secretary, Princeton University, Princeton, N.J., 1932.

[Committee 1919] Committee on Education and Special Training Advisory Board, *Committee on Education and Special Training: A Review of its Work during 1918*, War Department, Washington D.C., 1919.

[Coolidge 1941] Julian Lawrence Coolidge. *William Caspar Graustein—In memoriam*, Bulletin of the American Mathematical Society **47** (1941), pp. 343–349.

[Cornell Sun 1898] *Mathematics Department*, The Cornell Daily Sun, **19**, no. 48, November 15, 1898.

[Debru 2001] Claude Debru, *Harold Marston Morse: un mathématicien dans la première guerre mondiale*, Gazette des mathématiciens **87** (2001), pp. 9–15.

[Dickson 1919–1923] Leonard Dickson, *History of the Theory of Numbers*, 3 vols., Carnegie Institution of Washington, Washington, 1919, 1920, & 1923.

[Dickson 1919] Leonard Dickson, *Mathematics in War Perspective*, Bulletin of the American Mathematical Society **25** (1919), pp. 289–311.

[Duren 1988] Peter Duren, ed., with the assistance of Richard Askey & Uta Merzbach, *A Century of Mathematics in America*, 3 vols., American Mathematical Society, Providence, 1988.

[Fenster 1999a] Della Dumbaugh Fenster, *Why Dickson Left Quadratic Reciprocity out of the History of the Theory of Numbers*, American Mathematical Monthly **106** (1999), pp. 618–626.

[Fenster 1999b] Della Dumbaugh Fenster, *Leonard Dickson's History of the Theory of Numbers: An Historical Study with Mathematical Implications*, Revue d'histoire des mathématiques **5** (1999), pp. 159–179.

[Fenster 2003] Della Dumbaugh Fenster, *Funds for Mathematics: Carnegie Institution of Washington Support for Mathematics from 1902–1921*, Historia Mathematica **30** (2003), pp. 195–216.

[Fenster 2005] Della Dumbaugh Fenster, Leonard Dickson, History of the Theory of Numbers, in [Grattan-Guinness 2005, pp. 833–843].

[Fowler 2003] T. Kenneth Fowler, *Joseph Slepian, 1891–1969*, Biographical Memoirs of the National Academy of Sciences **83** (2003), pp. 240–251.

[Furfari 2000] Frank Anthony Furfari, *Joseph Slepian: Scientist, Engineer, Inventor*, IEEE Industry Applications Magazine **6**, no. 6 (Nov./Dec. 2000), pp. 14–19.

[Gallian 2006] Joseph Gallian, *Contemporary Abstract Algebra*, 6th ed., Houghton Mifflin Company, Boston, 2006.

[Gibson 1998] Campbell Gibson, *Population of the 100 Largest Cities and Other Urban Places in the United States: 1790 to 1990*, Population Division Working Paper No. 27, Population Division, U.S. Bureau of the Census, Washington, D.C., June 1998.

[Gluchoff 2011] Alan Gluchoff, *Artillerymen and Mathematicians: Forest Ray Moulton and Changes in American Exterior Ballistics, 1885–1934*, Historia Mathematica **38** (2011), pp. 506–547.

[Grattan-Guinness 2005] Ivor Grattan-Guinness, ed., *Landmark Writings in Western Mathematics, 1640–1940*, Elsevier, Amsterdam, 2005.

[Grier 2001] David Alan Grier, *Dr. Veblen Takes a Uniform in the First World War*, American Mathematical Monthly **108** (2001), pp. 922–931.

[Grier 2005a] David Alan Grier, Dr. Veblen at Aberdeen: Mathematics, Military Applications and Mass Production, in [Walton 2005, pp. 253–270].

[Grier 2005b] David Alan Grier, *When Computers Were Human*, Princeton University Press, Princeton, 2005.

[Jewett 1948] Frank B. Jewett, *The Genesis of the National Research Council and Millikan's World War I Work*, Reviews of Modern Physics **20** (1948), pp. 1–6.

[Kevles 1968] Daniel J. Kevles, *George Ellery Hale, the First World War, and the Advancement of Science in America*, Isis **59** (1968), pp. 427–437.

[Kevles 1978] Daniel J. Kevles, *The Physicists: The History of a Scientific Community in Modern America*, Alfred A. Knopf, New York, 1978.

[Landa & Nimmo 2003] Edward R. Landa & John R. Nimmo, *Soil History: The Life and Scientific Contributions of Lyman J. Briggs*, Soil Science Society of America Journal **67** (2003), pp. 681–693.

[Lehmer 2007] Derrick Henry Lehmer, Vandiver, Harry Schultz, in *Complete Dictionary of Scientific Biography*, Scribner's, New York, 2007, vol. 18, pp. 957–958.

[Lehto 1998] Olli Lehto, *Mathematics Without Borders: A History of the International Mathematical Union*, Springer, Heidelberg, 1998.

[Leland & Oboroceanu 2010] Anne Leland & Mari-Jana Oboroceanu, *American War and Military Operations Casualties: Lists and Statistics*, Congressional Research Service, Document 7-5700, Feb. 26, 2010.

[Leuchtenberg 1966] William Edward Leuchtenburg, The Perils of Prosperity, in [Berky & Shenton 1966, pp. 1106–1118]. Extr. from *The Perils of Prosperity 1914–1932*, Chicago, University of Chicago Press, 1958.

[MAA Meeting 1919] William DeWeese Cairns, *Third Annual Meeting of the Mathematical Association of America*, The American Mathematical Monthly **26, 3** (1919), pp. 91–107.

[McCarthy 1992] Michael J. McCarthy, *"Lafayette, We Are Here": The War College Division and American Military Planning for the AEF in World War I*, Ph.D. Thesis, Marshall University, May 1992.

[Mitchell] Martha Mitchell, *Encyclopedia Brunoniana*, Brown University, online: `http://www.brown.edu/Administration/News_Bureau/Databases/Encyclopedia`, accessed on November 18, 2013.

[Montgomery 1963] Deane Montgomery, *Oswald Veblen*, Bulletin of the American Mathematical Society **69** (1963), pp. 26–36.

[Moulton 1926] Forest Ray Moulton, *New Methods in Exterior Ballistics*, University of Chicago Press, Chicago, 1926. Reprinted under the title *Exterior Ballistics*, Dover, New York, 1962.

[New York Times] *Princeton Makes War Training Compulsory*, New York Times, June 4, 1918.

[Parshall 2000] Karen H. Parshall, *Perspectives on American Mathematics*, Bulletin of the American Mathematical Society **37** (2000), pp. 381–405.

[Parshall 2006] Karen H. Parshall, *James Joseph Sylvester: Jewish Mathematician in a Victorian World*, Johns Hopkins University Press, Baltimore, 2006.

[Parshall & Rowe 1994] Karen H. Parshall & David Rowe, *The Emergence of the American Mathematical Research Community (1876–1900): J. J. Sylvester, Felix Klein, and E. H. Moore*, American Mathematical Society, Providence & London Mathematical Society, London, 1994.

[Pershing 1919] John J. Pershing, *Final Report of Gen. John J. Pershing, Commander-in-Chief, U.S. Expeditionary Forces*, U.S. Government Printing Office, Washington, 1919.

[Price 1988] G. Bailey Price, American Mathematicians in World War I, in [Duren 1988, vol. 1, pp. 267–268].

[Reingold 1988] Nathan Reingold, Refugee Mathematicians 1933–1941, in [Duren 1988, vol. 1, pp. 175–200].

[Richardson 1918] Roland G. D. Richardson, *Courses in College in Preparation for the Navy*, American Mathematical Monthly **25** (1918), pp. 317–321.

[Ritter 2011] Jim Ritter, Geometry as Physics: Oswald Veblen and the Princeton School, in *Mathematics meets Physics*, ed. Karl-Heinz Schlote & Martina Schneider, Harri Deutsch, Frankfurt, 2011, pp. 145–179.

[Roever 1924] William H. Roever, Derivation of the Differential Equations of Motion of a Projectile Regarded as a Particle, in *Proceedings of the International Congress of Mathematicians*, ed. John Charles Fields, University of Toronto Press, Toronto, vol. 2, p. 766.

[Roever 1943] William H. Roever, *Effects of the Earth's Rotation on the Range and Drift of a Projectile*, Science **97** (1943), pp. 115–116.

[Rothrock 1919] David Andrew Rothrock, *American Mathematicians in War Service*, American Mathematical Monthly **26** (1919), pp. 40–44. Repr. in [Duren 1988, vol. 1, pp. 269–273].

[Simon et al. 1971] Leslie E. Simon, Frank E. Gribbs, & Serge J. Zaroodny, *Robert Harrington Kent, 1886–1961*, Biographical Memoirs of the National Academy of Sciences **51** (1971), pp. 94–117.

[Tarwater 1977] Dalton Tarwater, ed., *The Bicentennial Tribute to American Mathematics, 1776–1976*, The Mathematical Association of America, n.p., 1977.

[Walton 2005] Steven A. Walton, ed., *Instrumental in War: Science, Research, and Instruments between Knowledge and The World*, Koninklijke Brill, Leiden, 2005.

[Wilson 1918] Edwin Bidwell Wilson, *The Mathematics of Aërodynamics*, American Mathematical Monthly **25** (1918), pp. 292–297.

[Yerkes 1920] Robert M. Yerkes, ed., *A New World of Science: Its Development during the War*, The Century Co., New York, 1920.

[Zitarelli 2007] David Zitarelli, *Straddling Centuries: The Struggles of a Mathematician and His University to Enter the Ranks of Research Mathematics, 1870–1950*, Historia Mathematica **34** (2007), pp. 271–288.

Part III

Moving On

Debating the Place of Mathematics at the École polytechnique around World War I

Jean-Luc Chabert and Christian Gilain

When considering the evolution of mathematics and the various occupations of mathematicians around the time of the First World War, a study focusing on the École polytechnique would seem especially relevant.[1] Here was indeed both a military school and a scientific institution of higher education where the teaching of mathematics has always been highly valued. In all its administrative structures, mathematicians found themselves side by side with career officers. Often used to denote the École, its students, or its graduates (both being called "polytechnicians"), the letter "X" can be seen as an allusion both to the traditional symbol of the unknown in an algebraic equation and to the crossing cannons displayed on its coat of arms, a reminder of the privileged links the École polytechnique always had with artillery (see Figure 1).[2]

In their various capacities as officers, engineers, or savants, polytechnicians played a considerable role in World War I. Two of the highest commanders of the French and Allied forces during the war were graduates of the École. Joseph Joffre (X 1869) and Ferdinand Foch (X 1871) had heard Charles Hermite's lectures on

---

Jean-Luc Chabert's address is Université de Picardie, LAMFA, 33 rue Saint-Leu, 80039 Amiens, France. His email address is jean-luc.chabert@u-picardie.fr.

Christian Gilain's address is Université Pierre et Marie Curie, Institut de mathématiques de Jussieu-Paris Rive Gauche, 4 place Jussieu, 75252 Paris Cedex 05, France. His email address is christian.gilain@imj-prg.fr.

We would like to thank Marie-Christine Thooris, director of the *Centre de ressources historiques* of the École polytechnique, Palaiseau, France, as well as Olivier Azzola, archivist, who have greatly eased our work, providing us in particular with most of the illustrations included. This chapter was translated from the French by David Aubin, whom we also thank for his helpful suggestions for improving this text.

[1] As far as the history of science is concerned, no work appears to have adequately dealt with the École polytechnique during WWI. In the most complete reference on the history of the École [Belhoste et al. 1994], the Great War figures in only marginally. Some contributors to this book indeed selected the year 1914 as a cut-off, but the overall periodization adopted simply jumped over WWI, embedding it in a continuum titled "From one defeat to the next" [1870–1940]. There is, however, an interesting masters thesis specifically devoted to the institutional history of the École polytechnique during WWI [Villermet 1992], until now only partly published [Villermet 1993], but it does not deal directly with questions concerning the history of science.

[2] The origin of this traditional use of the letter "X" is not known precisely but has been part of the École's oral tradition since the nineteenth century. See [Lévy & Pinet 1894] and [Mattatia 1994].

analysis.[3] Among polytechnicians, the death toll was high: 883 of them died during the conflict, including 260 who were still students at the time (that is, 18% of the classes concerned). However, our aim here is not to review the range of polytechnicians' wartime occupations, but rather to examine the way in which the École polytechnique functioned as an institution during the war. We believe that this exercise can better illuminate mathematics' special place, as well as mathematicians' roles, at the École and, more broadly, in French society.[4]

While the mathematical disciplines—and above all analysis—always figured prominently in the polytechnician curriculum, this special status was always disputed [Gilain 1989, Belhoste 2003]. The terms of the debate, however, varied over the years. Initially, the École polytechnique explicitly stated that it had a dual goal. The École would, on the one hand, diffuse current scientific knowledge in the mathematical, physical, and chemical sciences as well as in the graphical arts and, on the other, provide the particular technical training needed by engineers entering the public service, as civilian or military officers. Already under attack, the general scientific goal was officially abolished in 1832.[5]

At the start of World War I, the situation had not changed and the École's mission was officially restricted to the basic training of engineers before they went to Special Application Schools (the *Écoles d'application*, such as mining, civil engineering, or officer's training schools). In a 1908 memo Joffre himself for example strongly underscored that the teaching of the École polytechnique should be geared towards the applications that one will make of them later: "mathematical analysis is a means to an end and not an end in itself. The course should be restricted to elements needed for other courses at the École and for applications this branch of mathematics receives in engineering."[6] Joffre's memo notwithstanding, debates went on about the place that should be reserved for theoretical science in the basic training of high-level engineers, as well as the content of mathematical curricula and the best-suited mathematical teachers.

More broadly, at the start of the twentieth century, debates were many about the École's role in French society as well as about the way in which mathematics had to be taught—at the École or not. Terry Shinn, [Shinn 1980], has argued that those debates were characteristic of a struggle between, on the one hand, "modernist" reformers who supported an inductive pedagogy based on the solution of concrete problems and who wished to develop the study of the physical sciences at the expense of the mathematical sciences and, on the other, conservatives attached to the privileges of abstract science and the deductive method. While admitting that some tentative changes were occurring before the outbreak of WWI, Shinn

[3]Contrary to Ivy League schools in the United States, it is customary to identify polytechnicians by the year they were admitted to the École, preceded by the letter "X" for obvious reasons.

[4]Wartime activities of polytechnicians are examined in the catalog of a recent exhibition on "Une grande École dans la Grande Guerre" [Thooris 2004]. On the place of polytechnicians in the French mathematical community before WWI, see [Gispert 1991, pp. 25–28 and 65–66].

[5]Only at the end of the 1950s was scientific research admitted as a career path for polytechnicians on a par with the others [Grison 1989, pp. 16–17], [Dahan Dalmedico 1994, pp. 315–317], [Dahan Dalmedico 1995, pp. 287–289].

[6]"*L'Analyse mathématique est un moyen et non un but. Le cours doit être réduit aux développements nécessaires pour l'intelligence des autres cours de l'École et pour les applications qu'on en tire dans le domaine de l'art de l'ingénieur*" [Joffre 1908].

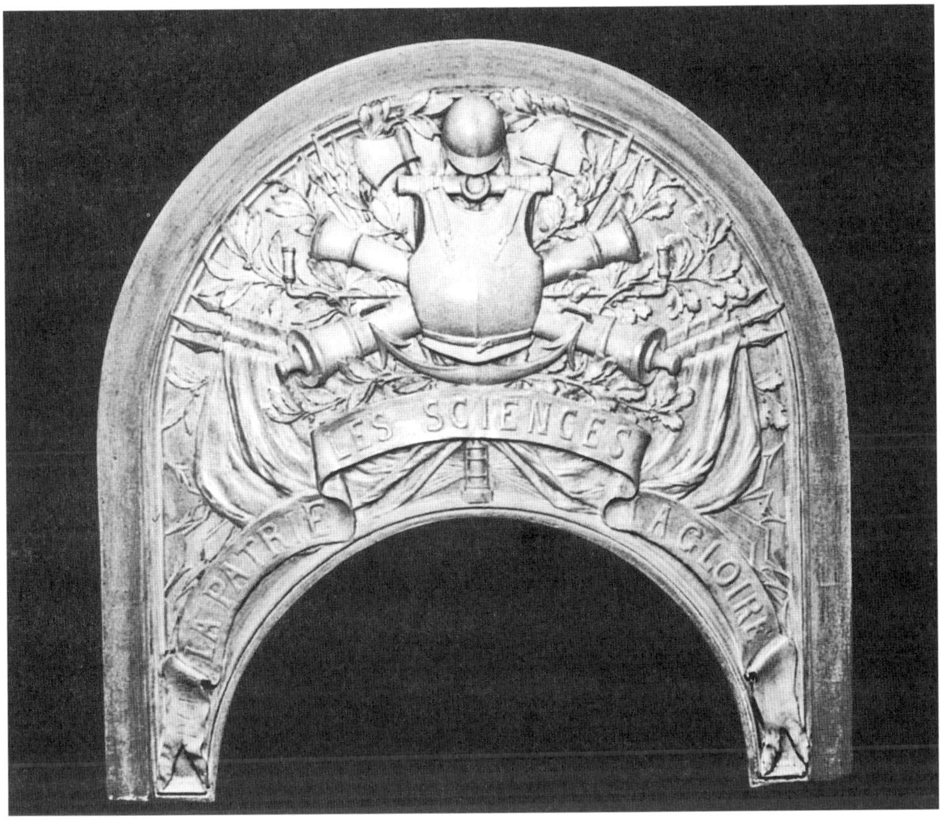

FIGURE 1. "For the Fatherland, Science, and Glory": a version of the coat of arms of the École polytechnique, 1912. © Collection École polytechnique.

concludes that conservatism won the day. Insisting on the "mixture of rigidities and evolutions" that prevailed in the prewar period, Antoine Picon has put forward a more balanced view [Picon 1994, p. 144]. But he also concluded that, on the whole, the École polytechnique was "sinking" at the time and that it remained backward-looking in the face of contemporary scientific progress and new social and industrial needs.

Our detailed study has led us to suggest a different picture with regard, in particular, to the mathematicians' "modernism" at the École polytechnique. Looking for some "WWI effect" on the way such questions were approached at the École, we show that at first glance continuity seemed to have characterized to a considerable degree the mathematical curricula taught there as well as the place of mathematics in the general polytechnician course of study. But we argue that overall continuity was a self-conscious choice in the face of severe criticism. This choice was made possible by the de facto alliance struck between mathematicians and the military establishment of the École polytechnique, which came out reinforced in the postwar period. In order to see this, we will broach for each time period (before, during, and after the war) three main themes: (1) the mathematical courses needed to prepare for the École as well as that taught once there; (2) initiatives by various

FIGURE 2. A study room of the École polytechnique, from a postcard produced just before WWI. Personal Collection.

organizations and personalities, including in wartime, to reform the content of the polytechnician training deemed too abstract; and (3) the profile of the École's professors of mathematics, focusing mainly on recruitment, the debates they brought about, and what they reveal about the École's institutional position with respect to mathematics and mathematical training.

1. Before the War

1.1. Mathematical Teaching Programs. Before the war, the École polytechnique was prominent in the elitist system of higher education (the *grandes Écoles*) put in place at the time of the French Revolution. In the context of the École polytechnique, the mathematical program did not stand alone; rather, it can be taken to refer to one of several interrelated entities:

- the program of the entrance examinations, or *admission program*[7];
- prior to that, the curriculum of the preparatory *lycée* classes (the *classes de mathématiques spéciales* that the students took after they had taken the final examination in secondary education, the famous *baccalauréat*) where pupils were taught what was needed to succeed on the entrance examinations of the *grandes Écoles* (mostly, but not only, the École polytechnique and the École normale supérieure, hereafter ENS);

[7]Formally, the French system of higher education distinguished between an entrance examination, where candidates with grades higher than a certain level are admitted, and an admission *concours* where candidates are ranked and only a fixed number of them are admitted. In the following, we nevertheless refer to the *concours d'admission* of the École polytechnique as the admission examination.

- the teaching curriculum during the two years of training at the École polytechnique itself, called the *internal teaching program*;
- and after that, the teaching program in the Special Schools where students completed their training in a specific branch of engineering.

In the following, we will pay attention mostly to the admission and the internal teaching programs. In principle these programs were to be revised every ten years, but revisions have rarely taken place so regularly in practice. The revision process followed several steps: first discussed and approved by the École polytechnique's *conseil d'instruction* (CI), programs were then discussed and voted on by the *conseil de perfectionnement* (CP) before they were sent for approval to the War Ministry.[8] These two boards were the main governing bodies of the École polytechnique. At the CI, the professors and examiners (*examinateurs*) sat together with representatives of the central military administration of the École, while the CP included only some professors together with representatives of various ministries and representatives of the Special Schools. Ad hoc committees were often moreover formed in order to formulate specific propositions. Let us now look at the discussions surrounding both programs in the prewar years.

1.1.1. *The Admission Program.* In 1902 there was a comprehensive reform of teaching programs in secondary education (the *lycées*) [Belhoste et al. 1996]. As a consequence, the Ministry of Public Instruction decided that the preparatory class programs needed to be adapted. A committee including, among others, the chemist Marcellin Berthelot and the mathematicians Gaston Darboux and Paul Appell adopted the new program for the *mathématiques spéciales* class in 1904. This program was deemed "maximal," in the sense that it was designed to be a preparation for all the *grandes Écoles*—not just the École polytechnique. Each *grande École* was supposed to determine the content of its admission examination among the materials covered by the preparatory classes. As far as the École polytechnique was concerned, the consequence of this reform was to take away from it the de facto control it used to have on the program of the preparatory classes. One thus notes that the new program reduced the place of analytic geometry, which according to Appell's report had been allowed to expand unreasonably due to the influence of the entrance examination of the École polytechnique.

Forced to adapt to the reform, the École simply adopted the maximal program as its own admission program. In May 1913, however, the CP decided to reduce the admission program for 1914, in particular by eliminating some aspects that seemed too advanced theoretically (the notions of cut and of incommensurable number, the proofs of the d'Alembert theorem, some theorems of Apollonius, etc.). To see how the dialectic between applications and theory was played out, it might be useful to look at the case of rational mechanics as discussed by the CP in February 1914. Reiterating a proposal he had already made a year earlier, Léon Lecornu, one of the two professors of mechanics at the École, and the chemist Henry Le Chatelier argued for the suppression of mechanics from the admission program. In Le Chatelier's view, since they were taught by mathematicians, preparatory courses of

[8] In the following, we will use the abbreviations PV of the CP (resp. CI) to denote the minutes of the proceedings of the CP (resp. CI) and AEP to denote the Archives of the École polytechnique.

mechanics were too mathematical and therefore not very useful.[9] Lecornu thought that this inadequate introduction to the field before his own course was counterproductive. Lucien Poincaré however expressed his fear that this suppression might be "misinterpreted by the public and seen as a sign that the École polytechnique was returning to abstract mathematics."[10] In the end, the CP decided to keep mechanics in its admission program.

1.1.2. *The Internal Teaching Program.* "Mechanics," one of the École's physics professors, Alfred Cornu, had once written, "must be the soul of our teaching."[11] Indeed, in the constant tension between the needs for advanced theoretical teachings and the insistence on applications immediately useful for engineering practice, mechanics could be made to fulfill both goals. It was one of the five disciplines that benefited from more than 60 lessons over the two years of polytechnician training (together with analysis, chemistry, physics, and drawing).[12]

The 1904 reform discussed above led to a modification of the internal teaching program at the École adopted in September 1909. This was the context of Joffre's intervention quoted earlier. The general who was then in charge of overseeing the military schools in France had specified: "The École polytechnique is not just an institution of high scientific culture. Its well-defined role is *to train students for the various Special Schools* of the civil service."[13] Some professors at the École reacted strongly to this view, notably the professor of analysis Camille Jordan:

> A second-rate engineer (and such will always be the most numerous) barely needs rather restricted knowledge to do his work correctly. A compendium can be all the science he needs. But in order not to lose its standing, our country must draw on more than those useful but modest servants; it must be able to draw on a few elite engineers capable of giving new impulses to their profession.[14]

---

[9]Le Chatelier was reported as saying: "*les professeurs de mathématiques ne peuvent enseigner la mécanique parce qu'ils la connaissent mal ; c'est aux physiciens qu'il appartient de l'enseigner*" [AEP, PV du CP, vol. 11, p. 24]. Himself a polytechnician, Le Chatelier was professor of chemistry at the *École des Mines* in Paris. A staunch defender of "industrial science," he rejected atomism. Twice a candidate for a teaching position at the École polytechnique in 1894 and 1897 and twice rejected, he became a professor at the *Collège de France* and at the Sorbonne. In 1912, he was named by the Académie des sciences as one of its two representatives to the CP of the École polytechnique [Letté 2004].

[10]"*mal interprétée dans le public et regardée comme un retour de l'École polytechnique vers les mathématiques abstraites*" [AEP, PV du CP, vol. 11, p. 25]. Lucien Poincaré was Henri's first cousin; he held a doctorate in the physical sciences and then occupied a high ranking position for higher education at the Ministry of Public Instruction [*directeur de l'enseignement supérieur*].

[11]He went on: "Thus the course of analysis ... must be adjusted ... to the needs of the course of mechanics" [Cornu 1896, p. 899].

[12]To complete the picture, let us note that three other disciplines were taught in 40 to 60 lessons (geometry, German language, and history and literature), while some others were taught in less than 40 lessons (social and political economy, architecture, astronomy, and geodesy).

[13]"*L'École polytechnique n'est pas un simple établissement de haute culture scientifique. Son rôle, bien défini, est de* former des Elèves pour les Ecoles d'application *des différents services publics*" [AEP, III3a/6, original emphasis].

[14]"*Un ingénieur médiocre (et ce sera toujours le plus grand nombre) n'a besoin pour son service que de connaissances assez restreintes. Un aide-mémoire peut lui tenir lieu de science. Mais sous peine de déchoir, notre pays doit posséder en dehors de ces serviteurs utiles mais modestes, quelques ingénieurs d'élite, capables de faire faire un progrès à leur profession*" [AEP, III3b/7].

Such protest notwithstanding, the preamble of the École's internal teaching program adopted in September 1909 reminded that:

> The teaching of the École polytechnique has for its immediate objective the training of students for successfully following the teaching of the Special Schools; it embraces all the scientific knowledge that can be necessary for an engineer or an officer.... Any theoretical development must necessarily come with numerical examples and applications.[15]

We review, in the domains of mathematics, the principal modifications to the programs that came out of this reform. First of all, to the course of analysis (34 lessons in the first year; 31 in the second year), important changes were made. While all developments concerning elliptic functions were eliminated from the program, emphasis was put on all applications of the calculus relevant to mechanics and physics. Debates about whether or not elliptic functions should be taught at the École are interesting because they provide a good illustration of the École's hesitations with respect to the balance between theory and applications. Already at the end of the 1860s, Joseph Bertrand was teaching it in the course of analysis, and in 1875 the theory of elliptic functions was officially included in the internal program. Later, in the 1890s, Jordan argued in favor of keeping it in his course by emphasizing that these functions "arise in a mass of problems in mechanics, physics, astronomy, and even in various questions directly linked to the engineer's practice" [Gispert 1994, p. 189]. Jordan went on to mention the theory of waves, the effects of rolling, the trajectory of projectiles, the arches of bridges, the study of springs, etc.

While the theory of elliptic functions disappeared from the official program in 1909, Jordan, Georges Humbert, and Jacques Hadamard kept it in their course syllabi for second-year students.[16] The ambiguity of the place of elliptic functions at the École polytechnique was strikingly put to the fore in November 1916 in an official speech by Moïse-Emmanuel Carvallo (1856–1945). A specialist of rational mechanics, who was "director of studies" (*directeur des études*) from 1909 to 1921,[17] Carvallo (see Figure 3) brought up elliptic functions in his speech as an example of advanced mathematics with practical applications: "the integration of the problem of rotations in a solid body leads to elliptic functions, which expand the domain of periodic functions studied in trigonometry" [Carvallo 1916, p. 2].[18]

But if elliptic functions were half-heartedly eliminated from the official program, the theory of second-order partial differential equations was given a key role in the course of analysis. Generally speaking, applications to physics were systematically added to the program. Fourier series, the Stokes formulas, potential theory, ordinary differential equations, and, above all, partial differential equations

---

[15]"*L'enseignement de l'École polytechnique a pour objet immédiat de préparer les élèves à suivre avec fruit celui des Écoles d'application; il embrasse toutes les connaissances scientifiques qui peuvent être nécessaires à un ingénieur ou un officier.... Toute théorie de quelque importance est obligatoirement accompagnée d'exemples numériques et d'applications*" [AEP, III3a/7].

[16]Humbert: one lesson in 1909–1910, then 3 lessons in 1911–1912; Jordan: 5 lessons in 1910–1911. During the war, Hadamard would also devote a few lessons to elliptic functions.

[17]At the École polytechnique, the director of studies was responsible for the daily organization of teaching. He was elected by the CP among candidates who often were civilian members of the École's teaching staff.

[18]For more details about the circumstances of Carvallo's speech, see below.

FIGURE 3. The director of studies of the École polytechnique Moïse-Emmanuel Carvallo. © Collection École polytechnique.

were illustrated by problems relevant to magnetism, telegraphy, the diffusion of heat, acoustics, and hydrodynamics. As we can see, although highly theoretical, the evolution of the program of analysis never lost touch with applications and, in particular, reflected the great contemporary development of research in the field of partial differential equations [Gispert 1994, pp. 194–196].

Second, considering the course of astronomy and geodesy (17 lessons in the second year), changes were even more drastic. Before 1904, the teaching of this course had occupied 28 lessons, of which 18 were devoted to astronomy proper and 2 to error theory. In 1904, Octave Callandreau, who had been teaching the course since 1893, died. War Minister Louis André[19] suggested that the course of astronomy be abolished and that geodesy be taught in a Special School only to officers who would have a need for it. This proposal would make room for social economics, which the government wished to introduce into the curriculum.

Protest from professors was immediate, and Henri Poincaré, whose prestige was immense at the time, accepted a professorship in order to save the course of

[19]General André (X 1857) had earlier been commander of the École; see Section 1.2 below.

astronomy [Crépel 1994, pp. 210–212]. Appointed by special decree, Poincaré wrote a program in 15 lessons, of which 2 were devoted to error theory. The course of geodesy was, for its part, replaced by social economics. In fact, it was not Poincaré, but a *répétiteur*, Constant Rozé, who taught that course. In 1907, there was a total of 21 lessons, of which 3 dealt with geodesy.

But when Poincaré left his position in 1908, he was replaced by Robert Bourgeois (ranked first by the CP ahead of Maurice d'Ocagne who had been placed on top by the CI). A career officer and a world-renowned geodesist, Bourgeois was a representative of a generation of officer-savants [Schiavon 2003]. As a result, the program of this course shifted toward geodesy. Of Bourgeois's 17 lessons, 7 were devoted to astronomy, 2 to error theory, and 8 to geodesy. Bourgeois's course was adopted without change as the official program of 1909; the end result was a definite shift toward the practical.

Finally, as far as the course of geometry (26 lessons in the first year; 12 in the second year; all 38 lessons taught during the first year starting in the fall of 1913) was concerned, program modifications seemed smaller but with considerable symbolic significance. Stereotomy, that is, the cutting of wood and stone, was a historic application of descriptive geometry taught at the École polytechnique since its foundation by Gaspard Monge. It even figured in the official title of the geometry course. In 1909, despite the isolated resistance of the *examinateur* of geometry Louis Picquet, graphical calculation was introduced at the expense of stereotomy. This process was accelerated by the death of the professor of geometry, Paul Haag, and the nomination of d'Ocagne in 1912.

Looking at the evolution of programs and courses in the prewar years, one gets the impression that mathematicians were far from being ossified conservatives. On the contrary, they seemed anxious to design courses combining high-level theory, engagement with current research, and practical applicability. We now turn our attention to the professors themselves.

1.2. Mathematics Teaching Faculty. The curriculum vitae of hired teaching personnel, we argue, is an important element in the evaluation of the policy of the École polytechnique relative to the roles assigned to it by the government. In mathematics especially, the choice between a polytechnician and a *normalien*, a graduate from the École normale supérieure, often gave rise to heated discussions at the *conseils* (CP and CI). Among polytechnicians, one should note, there was a great disparity with regards to their research activities.

Let us consider the whole faculty in the mathematical sciences (analysis, geometry, mechanics, astronomy, and geodesy) with permanent positions.[20] In 1914, while the places where the analysis faculty had made their study was mixed, the situation was very different for the other branches of the mathematical sciences where polytechnicians dominated (see Table 1).

As far as hiring was concerned, the process generally required three formal votes by (1) the scientific subcommittee corresponding to a specific position, (2) the CI, and (3) the CP, whose decision was final.[21] It is instructive to follow the

[20]Among the permament faculty, there were professors and various levels of teaching assistants (in reverse hierarchical order): *répétiteurs titulaires*, *répétiteurs adjoints*, and *répétiteurs auxiliaires*.

[21]In fact, the minister could reverse a decision of the CP, but this seldom happened.

TABLE 1. Faculty with permanent positions in mathematics in 1914. The following abbreviations are used in the tables: PR = professor; *Rép. tit.* = *répétiteur titulaire*; *Rép. adj.* = *répétiteur adjoint*; *Rép. aux.* = *répétiteur auxiliaire*; Dr = doctorate in the mathematical sciences; *agr.* = *agrégé* (certified high school teacher); *Ing.* = *ingénieur* (engineer); mil. = military officer; Acad = member of the Academy of Sciences; X = polytechnician; N = *normalien.*

Analysis:

PR:	Georges Humbert (X - *Ing.* - Dr - Acad 1901), Jacques Hadamard (N - Dr - Acad 1912).
Rép. tit.:	Édouard Goursat (N - Dr - Acad 1919), Gabriel Kœnigs (N - Dr - Acad 1918).
Rép. adj.:	Auguste Boulanger (X - *agr.* -Dr), Paul Lévy (X - *Ing.* -Dr - Acad 1964).
Rép. aux.:	Ernest Vessiot (N - Dr - Acad 1943), Jacques Chapelon (X - *Ing.* -Dr).
Total:	4 N and 4 X (all Dr.).

Mechanics:

PR:	Léon Lecornu (X - *Ing.* -Dr - Acad 1910), Paul Painlevé (N - Dr - Acad 1900).
Rép. tit.:	Georges Simart (X - mil. - Dr), Jacques Jouguet (X - *Ing.* -Acad 1930).
Rép. adj.:	Alfred Liénard (X - *Ing.*), Alphonse Marbec (X - *Ing.*).
Rép. aux.:	Jean Réveille (X - mil. - Dr), Charles Plâtrier (X - mil. - Dr).
Total:	1 N and 7 X.

Geometry:

PR:	Maurice d'Ocagne (X - *Ing.* - Acad 1922).
Rép. tit.:	Raoul Bricard (X - *Ing.*).
Rép. adj.:	Joseph Fouché (X - *agr.*).
Total:	3 X.

Astronomy and geodesy:

PR:	Robert Bourgeois (X - mil. - Acad 1917).
Rép. tit.:	Maurice Hamy (astronomer - Dr. - Acad 1908).
Rép. adj.:	Henri Noirel (X - mil.).
Total:	2 X and 1 astronomer.

whole process when the extant archives allow it. In order to see whether the hiring process was affected by the war, let us examine first a few earlier instances of hiring.

The fact that Émile Picard was not chosen for the professorship in analysis in 1895 has often been seen as a confirmation that the École preferred one of its alumni to a brilliant mathematician from the ENS. In fact, the whole process indicates that the situation was more complex. On May 7, 1895, the majority of the CI indeed chose Picard by 12 votes to 11, only to see, three days later, the CP overturn this choice by a 13-to-7 margin in favor of Humbert. The minutes of both meetings show that General André (1838–1913), the commander of the École, as well as General Charles-Nicolas Peaucellier (1832–1913), chairman of the CP, both strongly supported Picard.[22] It was the directors of the Special Schools sitting on the CP who voted against the *normalien*.

In 1905, on the contrary, it was the *normalien* Paul Painlevé, already a professor at Sorbonne, who was selected as the new mechanics professor, over Prosper-Jules Charbonnier who was not only a polytechnician but also an engineer in the Naval Artillery.[23] The margin by which Painlevé was elected was this time larger at the CP (17 to 5) than at the CI (14 to 8). Once more, we note that the commander of the École, General Jules-Albert Corbin (1844–1906), favored the *normalien* Painlevé. The director of studies, Ernest Mercadier (1836–1911), an electrical engineer and a polytechnician who had opposed Picard in 1895, supported Painlevé, saying that the latter had long been associated with the École and could be considered a "polytechnician ... by adoption" [AEP, PV of the CP, vol. 10, p. 170]. Later, as we shall see, Painlevé had as minister the opportunity to pay the École back. Such a case of the "adoption" of a nonpolytechnician is however quite rare.

In 1911, Jordan retired from his chair of analysis. Two main candidates vied to replace him: Hadamard (a *normalien* and professor at the Collège de France) and Edmond Maillet (a polytechnician with a doctorate in the mathematical sciences and professor at one of the most prestigious Special Schools, the *École nationale des ponts et chaussées*, hereafter ENPC). On November 23, the CI leaned heavily in favor of Hadamard (15 to 5). The minutes show that Maillet was backed by Lecornu and by the physics professor Auguste Lafay. For his part, Hadamard got the support of Painlevé, Carvallo, and Humbert. Harshly criticizing Maillet's course at the ENPC, Humbert added:

> Over the last 10 years, M. Hadamard has published many more results about physical reality than M. Maillet who is an engineer. [Hadamard] is therefore wholly capable of orienting his lectures in this direction. Lastly, M. Hadamard is an indisputable master in the field of "partial differential equations" whose development is desired by the École [AEP, PV of the CI, vol. 11, p. 53].

On December 9, the CP also elected Hadamard (13 to 7). Not surprisingly Maillet's candidacy was supported by Auguste Kleine, who directed the ENPC. Emphasizing the fact that Maillet was a polytechnician, he stated that "as an engineer Maillet perhaps has a practical side that M. Hadamard is lacking" [AEP, PV of the CP, vol. 10, p. 332]. Henri Poincaré and, once again, Humbert argued in favor of Hadamard, the latter reiterating that even if he was not an engineer, Hadamard's work had a more practical inclination than Maillet's, especially as far

---

[22]See also the letters from Charles Hermite to Gösta Mittag-Leffler, April 9 and June 17, 1895, in *Cahiers du Séminaire d'histoire des mathématiques*, vol. 10, 1989, pp. 29–33.

[23]On Charbonnier's role in ballistic research in France, see David Aubin's contribution to this volume.

as elastic wave theory was concerned.[24] Hadamard's hiring, one notices, was in perfect coherence with the new program for analysis adopted two years earlier in 1909.

1.3. The Place of Mathematics in the Polytechnician's Curriculum. After 1870, the expansion of the university system in France posed a threat to the key role hitherto assumed by the École polytechnique in higher education. In 1904, the dean of the Science Faculty in Paris, Appell, wrote that contrary to the practical schools that dispensed training "directly useful to an art or a profession," the École polytechnique was "a purely scientific institution providing general education, like a university [*Faculté*]. It appears as an exclusive *Faculté des sciences*." Appell concluded that the École was completely expendable: "If the École did not exist, the universities would organize themselves without difficulty in order to fulfill its function" [Appel 1904, pp. 292–295].

Not surprisingly, many expressed different opinions regarding the respective roles of the university and of the École. In front of a Senatorial Commission, General Charles-Alexandre-Louis Kreitmann (1851–1915) belabored the case of elliptic functions (see p. 281 above), which he acknowledged as an advanced scientific theory, but one that "wholly lacked of applications: none have been found yet. So, we will devote only 5 lessons to them and some projects even completely dispense with them. On the contrary, a university can devote 20 to 25 lessons to elliptic functions, because it is a very beautiful theory."[25] Should a mathematical theory that seemed to have no application—however beautiful it might be—deserve to be included in the training of high-level engineers?

In 1908, a new actor entered the debates—the Society of Friends of the École polytechnique (*Société des amis de l'École polytechnique*, hereafter SAEP). Incorporated "to contribute to the prosperity of the École polytechnique and its preservation at the forefront of advanced scientific teaching," the SAEP immediately set up a permanent Education Committee in charge of examining the programs of the École both at a general level and at the level of its main courses.[26]

Under the impulse of Le Chatelier, the Education Committee of the SAEP soon launched an offensive in order to replace the second-year course of geometry by a course of "physico-mechanics" intended as an example of industrial science. SAEP president, the hydraulic engineer Louis Cuvinot, wrote to the Minister of War to suggest that the course proposed by Le Chatelier would tighten the link between the École polytechnique and the Special Schools. A mediator between the "perfect" sciences and engineering practice, industrial science would teach students

[24]The author of several publications in algebra and number theory (including his doctoral thesis in 1892), Maillet had also worked on hydraulic problems, but he can hardly be considered an analyst in the sense that the École usually meant it. On Hadamard, see Jean-Luc Chabert and Christian Gilain's chapter in [Goldstein & Aubin forthcoming].

[25]"*Les fonctions elliptiques constituent une haute théorie scientifique, mais cette théorie manque totalement d'applications: on n'en a pas encore trouvé. Dès lors, nous y consacrons cinq leçons en tout, et encore un projet de révision qui est en préparation la supprime-t-il complètement. Au contraire, une faculté des sciences peut consacrer aux fonctions elliptiques vingt ou vingt-cinq leçons parce que c'est une très belle théorie*" [AEP, III3a/7].

[26]For the statutes of the SAEP, see the *Bulletin de la société des amis de l'École polytechnique*, no. 1, April 1909, p. 11. The Education Committee included 13 members, Humbert and Le Chatelier among them; see *Bulletin de la société des amis de l'École polytechnique*, no. 2, February 1910, pp. 16–17.

to "deal with real objects and not just with mathematical concepts." At the time, Haag's death had put the question of his replacement as professor of geometry on the table. Finally, the Minister of War, following the decision of the CP, replied negatively to Cuvinot, and d'Ocagne was called to fill the position.

Denunciations of the abstract character of the École polytechnique's teaching, focusing especially on mathematics, were common. Months before the outbreak of war, General Frédéric-Antoine Chapel (1849–1932) wrote a report following his inspection of the École in May 1913 (the report was dated February 1, 1914) lamenting the fact that polytechnicians were not active enough in the field of inventions. It is significant that looking for an explanation, Chapel focused on the teaching dispensed by the École:

> I see no other explanation than that the general orientation of our studies is a little too prone to letting abstraction prevail over realization; formula over fact; the blind grinding of the X over geometric vision; work of memory over effort of conception and imagination; more concerned with putting the Past in equations than to uncover the unknown of the Future; lastly, favoring the deductive method at the expense of intuition.... [27]

The mathematical training of future engineers was also debated at an international level. Founded at the Rome International Congress of Mathematicians in 1908, the International Commission for Mathematical Instruction (ICMI) held its congress in Paris in April 1914. Professor Paul Staeckel from Heidelberg University in Germany presented a synthesis of an inquiry into the mathematicial training of engineers in some twenty countries. Emphasizing the model and the worldwide influence of the École polytechnique on all mathematical teaching in general and that of engineers in particular, Staeckel claimed: "The truly novel idea was that a uniform theoretical education was needed for all the young men intent on joining a civilian or military body" [Staeckel 1914, p. 307].

That same evening at a meeting of the French Society for Civil Engineers (*Société des ingénieurs civils de France*, hereafter SICF),[28] d'Ocagne summarized the conclusions of the congress of the ICMI. Pointing out that one clear consensus was not to limit the mathematical training of engineers to basic and classical notions, d'Ocagne argued that this was the condition for the adaptability of engineers to the rapidly changing technological landscape. "We do not know," he said, "what tomorrow may have in store for us. It may be that such theory which

---

[27] "*Je n'en vois d'autre explication que dans l'orientation de nos études en général un peu trop propres à faire prévaloir l'abstraction sur la réalisation; la formule sur le fait; la trituration aveugle de l'X sur la vision géométrique; le travail de mémoire sur l'effort de conception et d'imagination; plus préoccupé de mettre le Passé en équation que de dégager l'inconnue de l'Avenir; enfin favorisant la méthode déductive au dépend de l'intuitive, au nom d'une prétendue infaillibilité de la première que l'histoire des sciences pourtant est loin de ratifier*" [Chapel 1914, p. 7].

[28] The SICF had been founded in March 1848 by engineers from the *École centrale des arts et manufactures*, hereafter ECAM. As opposed to polytechnicians, most of whom became public servants or military officers after their studies, the engineers trained at the ECAM (called *centraliens*) worked in industry. The SICF's members, whose relation with polytechnicians was often tense, wanted the state to recognize their status [Jacomy 1984]. As we shall see below, the SICF was sometimes involved in debates about the mathematical training of engineers.

today seems purely abstract could be useful in some area of technology."[29] The geometry professor singled out imaginary numbers that had in less than fifty years become the indispensible tool of "electro-technicians." D'Ocagne then addressed the question of who should teach mathematics to engineers. If Staeckel had unsurprisingly concluded that mathematicians were best suited to this task, d'Ocagne—while expressing his agreement with Staeckel about the necessary mathematical skill needed—underscored, in front of his audience of civil engineers, that knowledge of engineering needs acquired through professional experience was also highly desirable.

Even so, the discussion that followed d'Ocagne's lecture shows that his position was controversial. A professor at the ENPC and an engineer who had pioneered the use of reinforced concrete, Charles Rabut (1852–1925) cited "the abuse of computation [*calcul*] in the applied science." His opinion was opposed to d'Ocagne's on both points. Rabut expressed the wish:

> (1) that, except for basic elements, future engineers' mathematical training be provided, not at the university [*en Sorbonne*], but in technical schools by engineers with practical experience;
> (2) that this training be limited to what is strictly necessary, the first principle of this teaching being to subsume the deductive method under the experimental method and *never compute what can be measured.*[30]

When the war broke out in 1914, would the "Sacred Union" put an end to those debates for the duration of the conflict? Contrary to what has been argued elsewhere [Shinn 1980, p. 173], we will now see that this was far from the case.

2. The École polytechnique in Wartime

When the army was mobilized on August 1, 1914, the scientific training of the class of 1911 had just been completed at the École polytechnique. At the time, in addition to their two-year study course, polytechnicians indeed had to spend one full year in the army. But the law of August 7, 1913, had abolished the requirement, so that students in the classes of 1912 and 1913 had together completed their first year of studies at the École. These cohorts involved 220, 231, and 270 students, respectively.

As far as the class of 1914 was concerned, the recruiting process was stopped halfway through. Indeed the written part of the examination was completed but not the oral part. All candidates, assimilated to students of the École polytechnique, were mobilized, mainly in the artillery. The suspension of the examination process would hang over the next years as a recurrent problem for the École and the War Ministry. On August 9, 1914, 228 candidates with a grade of 13/20 or higher at the oral examination were officially admitted to the École. Only in 1916 would 190 other candidates be officially admitted among those who had been successful in the written examination [Bourgoignon 1917b].

[29] Minutes of discussions at SICF are published after Staeckel's report in *L'Enseignement mathématique*, vol. 16, 1914, pp. 344–356. D'Ocagne's quote is on p. 346.

[30] *L'Enseignement mathématique*, vol. 16, 1914, p. 350 (emphasis by Rabut).

The buildings occupied by the École were emptied and turned into a military hospital. Some laboratories were used for war purposes. Carvallo, for example, turned his workshop into a periscope factory. On December 26, 1914, the question of whether classes should be resumed was addressed by a new CP put in place under the directorship of General Chapel.

> The board, mindful that, except for rare exceptions due to reasons of health, all the students of the three classes of the École are in the army, [and mindful] that there is no purpose in taking them away from [the army] for studies which, while necessary to the intellectual development of an elite, are not immediately and directly useful to national defense, decides unanimously that the courses of the École polytechnique cannot be resumed under the present circumstances.[31]

The nearly total halt to which the École was brought lasted for almost two years. In 1915, the entrance examination was canceled. The next year, only 70 candidates were received, half of whom were sent directly to the front. Courses started again in November 1916, but the audience was much smaller than usual. In his inaugural speech, on November 6, 1916, the vice commander of the École Jean-Antoine-Henri Bourgoignon (see Figure 4) testified to the evolution of the mood from immediate urgency to the sense that the future nonetheless was to be prepared. Having taken part in the "sublime effort of the French people" and in "the victorious resistance on the Marne, on the Yser, at Verdun," polytechnicians were now assuming a different role (see Figure 5).

> The struggle goes on, in a manner that is still bloody, but the enemy is entrenched. He is under attack. The worrying anguish has vanished. Social life can resume. We are planning for the future.... Your presence here is not only the presage but the constatation that the heavy menace that weighed on the destinies of France is now definitely discarded [Bourgoignon 1916, p. 2].

Under these circumstances, the way in which the course of study at the École polytechnique was described by Director of Studies Carvallo is revealing. In his "Introductory conference on the teaching of the École polytechnique," Carvallo recalled the law of 25 *frimaire* Year VIII of the Revolutionary calendar (December 16, 1799)—which had reorganized the École polytechnique—to reaffirm its initial dual goal and to insist on the value of theoretical teaching. The science taught at the École which is part of the common knowledge necessary for engineers in all the Special Schools, Carvallo said, "is general and philosophical; engineering science is positive and useful. This double character is responsible for both the attraction and the difficulty of the École's teaching" [Carvallo 1916, p. 1]. Let us now see whether the reaffirmation of the initial dual goal of the École was translated into practice.

---

[31] "*Le Conseil considérant que, sauf de rares exceptions motivées par des raisons de santé, tous les élèves des trois promotions de l'École sont aux armées; qu'à aucun égard il ne convient de les en éloigner pour des études qui, si elles sont nécessaires au développement intellectuel d'une élite, ne sont pas immédiatement et directement utilisables pour la Défense nationale, est à l'unanimité d'avis que la reprise des cours de l'École polytechnique ne doit pas être envisagée dans les circonstances actuelles*" [AEP, PV of the CP, vol. 11, p. 44].

FIGURE 4. Jean-Antoine-Henri Bourgoignon, vice commander of the École polytechnique. © Collection École polytechnique.

2.1. Mathematical Teaching Programs. In preparation for the reopening of the École, d'Ocagne proposed in the summer of 1916 to reorganize the teaching of geometry.[32] In the new program, stereotomy vanished almost completely, leaving only two lessons on stone cutting while the standing of graphical calculation was increased. A world leader in this domain, d'Ocagne seized the occasion to introduce his own specialty, nomography—that is, the use of curve systems that allow for the reading of the numerical results of computations directly from the graphs [Tournès 2000]. "Nomography," he emphasized on July 4, 1916, "is the part of

[32]D'Ocagne's proposal was successfully examined and unanimously accepted (except for Louis Picquet, *examinateur* of geometry) by the geometry subcommittee on June 27, by the CI on July 4, and finally by the CP on July 22.

FIGURE 5. A polytechnician dressed as a "*poilu*" pays a last tribute to fallen soldiers while he is shown the way toward an industrial society by the French Republic. Engraving by Paul-Albert Laurens on the cover of the Album for the classes 1916 and 1917 of the École polytechnique. © Collection École polytechnique.

the course that has found the largest applications in [illegible] war." And he cited the building and use of airplanes, the binding of heavy artillery pieces at the Schneider factories, etc. Graphical integration needed to be further emphasized since "it is nowadays of capital importance notably for the rapid determination of heavy projectile trajectories" [AEP, PV of the CI, vol. 11, pp. 214–215].

Otherwise, the École's mathematical teaching seems to have remained in accordance with the official programs [AEP, X2C7]. Between November 1916 and January 1917, Hadamard thus devoted 34 lessons to his course of analysis for a group of 35 first-year students. We note that 3 out of the 33 second-year lessons from October 1917 to January 1918 were devoted to elliptic functions, even though they had officially been abolished from the curriculum in 1909 on the grounds of their being too theoretical.

2.2. Civil Engineers and Polytechnicians. Contrary to what might have been expected, prewar debates about the nature and role of mathematical education from an engineering point of view did not stop—far from it. At a meeting of the SICF, on November 3, 1916, postwar needs in terms of engineers' higher education were once again addressed by distinguished guests. The main event was a speech by Léon Guillet. A *centralien* engineer, Guillet was a professor of metallurgy at his *alma mater*, the ECAM. A close friend of Le Chatelier [Letté 2004, p. 146], he was mobilized during the war, serving in war-related industry first in Puteaux and later in Saint-Nazaire. In February 1916, he was attached to the Ministry of Trade and Industry and put in charge of technological matters [Fontanon & Grelon 1994, vol. 1, p. 614]. Before Guillet's talk, the Minister of Trade and Industry himself, Étienne Clémentel, drew lessons from the war to establish the parameters of the debate: "In the future, a tight collaboration between science and industry will be required: science must get down from its ivory tower; after the distresses of the present tragedy it would be criminal for science to remain speculative [and] not to understand that it needs to penetrate industry."[33]

Guillet then presented his assessment of the new needs in terms of engineering education. Taking his cue from foreign models, and notably the German one, he made the case for the following reform:

- the suppression of the class of *mathématiques spéciales* whose programs were not, according to him, adapted to engineering;
- the introduction of an age limit between 17 and 18 years old for admission in the schools, so that engineers would be available at a young age for the job market;
- the restriction of the place of mathematics in engineers' training in favor of classical culture;[34]
- and, finally, changes in teaching methods: among the latter, he insisted on reducing the number of examinations in order to develop the students' initiative, on tightening professors' links with industry, on limiting the number of lectures, etc.

The École polytechnique clearly was the main target of Guillet's attacks: "the exaggerated role of the mathematical teaching—which only takes place in France—is due to the influence of the École polytechnique in the design of educational programs" [Guillet 1916, pp. 676–677]. Since all preparatory classes were modeled on those giving access to the École polytechnique, this bad influence was seen to affect on the whole system of higher education in engineering.[35] In a long series of remarks, Dean Appell expressed a view that was in total agreement with Guillet's. It would be a great service to France, he said, "to put engineering training back on its true basis which is both scientific and practical [and] to supress ... all privileges that exist in no other country."[36]

At the beginning of 1917, the SICF addressed the issue of its *Bulletin* containing those remarks to General Fernand-Alexandre Curmer (1854–1937), who had been named commander of the École polytechnique on March 6, 1916 (see Figure 6).

[33] *Bulletin de la Société des ingénieurs civils de France*, no. 10, October–November 1916, p. 619.

[34] Many at the École polytechnique also emphasized the importance of classical culture. This position does not allow for the discrimination between two camps, one "conservative" and one "modernist", as has been suggested [Shinn 1980, pp. 101–121]. Shinn's position in terms of social segregation has been criticized [Picon 1994, p. 153].

[35] We note that even if these recommendations were inspired by foreign examples, they were contrary to those voiced at the ICMI in 1914 (see above).

[36] *Bulletin de la Société des ingénieurs civils de France*, no. 10, October–November 1916, p. 633. This spectacular alliance between Guillet and Appell hides divergent positions: the latter in fact believed that the training of engineers should be provided by universities and not special schools.

FIGURE 6. General Fernand-Alexandre Curmer, commander of the École polytechnique. © Collection École polytechnique.

Refusing an invitation to debate at the next meeting of the SICF, Curmer wrote a memo in which he refuted Guillet's claims about the École. Regarding the claim that an exaggerated place was allotted to mathematics in preparatory schools due to the École's alleged influence, Curmer recalled that Appell himself had been a member of the expert committee responsible for the 1904 program of classes in *mathématiques spéciales*.[37] On June 16, 1917, some members of the CP adopted a more conciliatory position. According to its president, General Chapel, the defense of the École's position was no excuse for dispensing with a lucid examination of

[37]This 9-page typed memo bears the handwritten notation: "*Note remise à quelques personnalités—anciens élèves de l'École*" [AEP, III3a/8, file SICF, no. 4 bis].

criticisms and propositions addressed to it. One may recall that as inspector of the École, Chapel had insisted on reducing the mathematical content of polytechnician training [AEP, PV of the CP, vol. 11, p. 120].

The chronology of the events that followed is complex, but important in order to grasp the relationship between the stakes of the debate. In July 1917, the Minister of Trade and Industry, Clémentel, asked the SICF to formulate specific recommendations to reform the training of engineers. Clémentel then forwarded these recommendations to the War Minister, Painlevé, requesting that an interministerial commission be formed to address the issue [AEP, III3a/8, file SICF, no. 9 bis]. By that time, Painlevé, who (as we have seen) had been hired as a professor of rational mechanics by the École polytechnique in 1905, was as Minister of War directly responsible for the École. Painlevé therefore consulted the CP of the École on the matter. After long discussions of the issue in the CP on September 1 and 15, a majority emerged in favor of seriously discussing this type of reform only after the war. By then principles, "perhaps wholly new and different from the present" ones, might have emerged on which to base the training of engineers and officers [AEP, PV of the CP, vol. 11, p. 138]. On this basis, Painlevé, acting both as Minister of War and now as the head of the government (*Président du Conseil*), responded negatively to Clémentel's wishes on October 22.

A further attempt was made by the SAEP in November 1917 after power had changed hands and after Georges Clemenceau had succeeded Painlevé as the head of the government. Once again, negative governmental response was apparently suggested by Curmer himself.[38]

In a book published before the end of the war, Guillet publicized his ideas while avoiding any discussion of the case of the École polytechnique—"since it does not train engineers" [Guillet 1918, p. 63]. But in the preface written by Le Chatelier, the attack was explicit. Speaking of the course programs of the École, Le Chatelier wrote:

> M. Guillet is right to criticize the École for leaving too large a place for mathematics. The unexpected developments of physics and chemistry, the growing influence of both sciences on the progress of industry should have been reflected by an evolution in teaching. Nothing of the sort was done: mechanics and physics increasingly are taught as if they were annexes of mathematics. The experimental method [and] the art of making precise measurements are neglected [Guillet 1918, Le Chatelier's preface, p. 16].

There seems to be a political motivation behind the denunciation of engineering training as too theoretical and too mathematical in spirit. To reform the teaching of the École polytechnique in this way was synonymous with tightening its connection with industry and with increasing at the expense of the state the weight of the private sector in the training of engineers. In the preface to Guillet's book, Le Chatelier was explicit about this when he pointed to the "more liberal organization" being experienced by the Ministry of Trade, which had just asked various

[38]This is, in any case, what can be inferred from a letter dated January 11, 1918, written by Clemenceau. On this typed letter, one indeed finds the handwritten comment: "*rédaction proposée par le général Curmer*" [AEP, III3a/8, file SICF, no. 21].

scientific and industrial societies to name representatives to the boards of the *Conservatoire national des arts et métiers* (CNAM). "This example could be imitated and expanded" [Guillet 1918, p. 17].

The alliance between Painlevé and Curmer, between the mathematician at the head of the government and the military man in command of the École polytechnique, was crucial in defeating the "reformers." On November 8, 1918, Curmer inaugurated the new academic year by referring, just like Carvallo two years earlier, to the law of 25 *frimaire* Year VIII. He asked: "What is the École polytechnique?" Public opinion, he said, has it that this is a place "where a lot of mathematics is done ... and from where one comes out ready for any career [*d'où l'on sort n'importe quoi*].... However simplistic it may seem, this definition in reality is not far from the truth" [Curmer 1917, p. 1]. He went on:

> Some, even among ex-students, claim that the teaching that the École provides is not always absolutely useful in practice. This may have the appearance of truth. But, as Monsieur [Alexandre] Millerand rightly said in 1912 when as Minister of War he visited the École, "from this, does not follow that this teaching is useless baggage, since as far as intellectual matters are concerned, it is not right to omit the superfluous which is a *highly necessary thing.*" On the road that you are taking, you will have the surprise of retrieving the superfluous at the very moment when its usefulness would perhaps seem to be the furthest away: most often you will feel its effect and its help will come to you at an unconscious level [Curmer 1917, p. 2, emphasis in original].

From this speech, one gets the feeling that the École polytechnique had successfully withstood the assault of the SICF. If the war had taught the social need for an effective industrial sector able to organize quickly to produce highly technological goods in large quantities, it had also shown that an elite with advanced scientific and mathematical training was indispensable to the modern warring state.

2.3. Year 1918. Such lessons were for the postwar period. In the meantime, the École polytechnique still needed to train soldiers, and to train them fast. The selection process in 1918 was moved forward to March to allow young men scheduled to be mobilized on April 15, 1918, to have the chance to take the examination before being sent out to the front (which is what happened to most of the 160 accepted candidates). Preparatory classes were shortened to 5 months. On June 16, 1917, the CP therefore decided to reduce the admission requirements for the entry examination in 1918.[39] The CP made sure to emphasize that those changes were no more than a timely response to exceptional circumstances and had nothing to do with the SICF's requests.

For a time, in July 1918, it seemed that the German offensive might force an evacuation of the École to Montpellier.[40] But the situation was soon reversed and courses started in October 1918. At this time, victory seemed certain. In his inauguration speech, Vice Commander Bourgoignon could thus claim: "Today the

[39]In essence, the parts that were cut were symmetric rational functions, differential equations, operations on power series, anharmonic ratio, local curvature of surfaces, problems in descriptive geometry, and all of static and dynamic mechanics.

[40]Bourgoignon and Carvallo had been charged to come up with evacuation plans [AEP, I2/2].

FIGURE 7. Ceremony organized for welcoming demobilized students at the École polytechnique in 1919. © Collection École polytechnique.

liberation [of French lands including Strasbourg] is certain; ... stymied, the enemy is everywhere retreating" [Bourgoignon 1918, p. 3].[41]

On November 23, 1918, General Chapel, presiding at the meeting of the CP, emphasized the crucial role played by polytechnicians and the École polytechnique in the war. Recalling the motto of the École—"*Pour la patrie, les sciences et la gloire*" (see Figure 1)—Chapel went on:

> We will one day know what part polytechnician science played in France's gigantic scientific work, starting with the matchless cannon of 75 and culminating in the irresistible production of powerful machines for attack and defense. [This work] insured ... our strategic domination, the mastery of weapons and the decisive air domination.[42]

[41]He then recalled the speech he had given under the same circumstances in October 1917, when he underscored the event "of considerable range" that was the U.S. declaration of war on Germany. Reflecting on the reasons that had motivated the American intervention, Bourgoignon conjectured that the influence of universities on public opinion in the U.S. and the American youth's love of freedom had been decisive. On both grounds, he underscored, while recalling the link between the École polytechnique and the Military Academy at West Point, that France and its scientists and professors had played a crucial part [Bourgoignon 1917a, pp. 3–4].

[42]"*On saura un jour prochain ce qu'a été, dans l'œuvre scientifique immense de la France, la part de la science polytechnicienne, partant de l'inimitable canon de 75 pour aboutir à l'irrésistible production des plus puissants engins d'attaque et de défense ; concourant à assurer toutes les maîtrises: maîtrise stratégique, maîtrise des armes, et cette décisive maîtrise de l'air* ..." [AEP, PV of the CP, vol. 11, pp. 156–157].

3. The Postwar Period

Surviving students of the classes admitted at the École polytechnique from 1912 to 1917 had had at most one year of proper polytechnician training. In March 1919, these demobilized students are welcomed to the École (see Figure 7). Those who had had their first year of training (the classes of 1912 and 1913, as well as a fraction of the classes of 1916 and 1917) followed an accelerated six-month curriculum, while those who had received no training at all at the École (the class of 1914 and the rest of the classes of 1916 and 1917) received a course of study condensed into one year.

In the immediate postwar period, the number of students therefore was high and their level varied greatly from one individual to another. It was thus decided to double the number of courses that would be offered. Hadamard, for example, taught his first-year course in analysis first to the 1918 class from October 23, 1918, to January 22, 1919, and then once again to the other classes from July 22 to September 25, 1919. Similarly, the second-year course in analysis was taught by Humbert approximately during the same two periods (during the last of which he was, in fact, due to illness, replaced by Paul Lévy).

In 1919, in order to replenish the ranks and in the hope of reestablishing some justice between candidates who were forced to serve in the army during the war and those who directly came out of preparatory schools, there were two different sets of entrance examinations. The admission process went on normally in the spring, but a second session was held in August for candidates whose mobilization in the army had prevented them from taking part in previous admission examinations or for those whose participation in the 1914 session had abruptly been interrupted by the call to arms.[43] Altogether, some 600 students were recruited (about two-thirds through the special procedure). In view of this number, courses once again had to be doubled. In analysis, Lévy and Chapelon were each in charge of one group.

3.1. Mathematical Programs and Teaching. Despite the turmoil of the war and the attacks on its programs sustained during that time, the École polytechnique started the revision process of its curricula on January 24, 1920, as if following the usual decennial procedure. Carvallo noted: "It is a useful custom that a global revision of programs be undertaken about every ten years. The programs of the classes of *mathématiques spéciales* date from 1904 and those of the École from 1909. It will therefore come as no surprise that in 1920, after the war, a similar revision is desirable" [AEP, III3a/8]. But the revision process was, by necessity, complex since it had to take into account changes in the preparatory classes. The revision process would last until 1927 and was remarkable for its great continuity with the past.

3.1.1. *Admission Program.* The question was posed at the CP meeting of June 21, 1919, whether one should come back to the admission program of 1914, keep the reduced program of 1919, or adopt something in between [AEP, PV of the CP, vol. 11, p. 186]. The last option was chosen, and the study of curves and surfaces of the second degree, statics, and projective geometry were reintroduced into the admission program. The next year, this reduced program was maintained, but dynamics was reintroduced. In fact, the program published in 1920 (for admission in 1921) differed little from the 1904 program with the cuts made in 1914.

[43]Of the 40 people in this latter case, 34 were admitted to the École.

Only in 1924 would serious discussions be reinitiated about the admission program. A special subcommittee for the mathematical program was headed by the mechanics professor, Lecornu, and the mathematician, Lévy, wrote its report. Repeating the wish expressed by Lecornu in 1914, the subcommittee argued that mechanics should not be taught at the preparatory level, leaving room for complements in analysis and algebra. After the École's propositions were rejected by an interministerial committee, Lévy was disappointed:

> Cauchy's theorem on the comparison of series and integrals was not included [in the new programs as was asked]; neither were complex variables for power series, so that only intervals, and not radii, of convergence are discussed.... In total, we got almost nothing ... and we will not be able to count on the progam of [*Mathématiques*] *spéciales* to reduce ours.... I had the impression that in the face of university professors' unanimous wish not to overload the *spéciales* program, we were forced to back down [AEP, PV of the CP, vol. 12, p. 111].

Adopted by the CP of the École polytechnique on July 25, 1925, the new program was—one is again forced to remark—surprisingly similar to that of 1904, including mechanics, whose place had always been debated the most.[44] Additions to the 1904 program were negligible and the only important reductions were the whole section on trigonometry, applications of projective geometry, and the variation of curvature of normal sections at simple points of surfaces. More than ten years after the start of WWI, after many calls for change and even more controversies, the status quo was maintained.

3.1.2. *Internal Mathematical Teaching Program.* As far as the mathematics curriculum at the École polytechnique was concerned, stability seemed once again to be the general rule. In a memo written in January 1920, Carvallo restricted the role of mathematics in the polytechnicians' training:

> Being for the great majority of the students a means for their work and not a goal, [mathematics] must be taught, not for itself nor for its indisputable beauty, but in view of intellectual education and of its application to the experimental sciences.[45]

As we have seen, this was standard polytechnician rhetoric. But this renewed insistence on applications however portended importance changes concerning probability theory. This topic was nothing new to the École and had even been part of Bertrand's course of analysis until the time of his retirement in 1894 [Crépel 1994]. Afterwards, probability only figured in the introduction of the lessons dealing with error theory in the course on astronomy and geodesy (taught by Callandreau, Poincaré, and Bourgeois).

---

[44] About the 1904 program, Bruno Belhoste has noted: "remarkably written, it would remain in application, with very little change, for almost half a century" [Belhoste 1992, p. 10].

[45] "*Etant pour la masse des élèves une méthode, un moyen de travail et non pas un but, elles doivent être enseignées, non pour elles-mêmes et pour leur beauté qui est incontestable, mais en vue de la formation intellectuelle et de l'application aux sciences expérimentales*" [AEP, III3a/8, p. 2].

In the immediate postwar period, however, the teaching of probability was reintroduced and significantly developed within the course of analysis in view of its various applications in mechanics, physics, and the social sciences as well as artillery. In the speech quoted above, Carvallo argued that the objectives of the mathematical training he had just set out implied that probability theory needed to be added to the program. "Because of its applications to the social sciences, to the observational sciences, and especially to cannon firing and to the kinetic theory [of gases], it is necessary to introduce this important branch of mathematics in the course of analysis."[46] In his autobiography, Lévy attributed an essential role to Carvallo, who asked him to devote three lectures to probability "at the end of the winter of 1919" [Lévy 1970, p. 71]. Intended to serve as complements to Bourgeois's development of error theory in the course on astronomy and geodesy, they apparently played an important part in orienting Lévy toward the study of probability.[47]

Two years later, probability theory figured in the second-year course on analysis. Having meanwhile been appointed professor to replace Humbert, who had resigned due to illness, Lévy gave 4 lectures to a first group of students in December 1920 (his 85-page course was lithographed). As *professeur suppléant*, Chapelon devoted 5 lessons to probability in the course he gave to the second group in October 1920 (his lithographed course was 114 pages long). In November of that same year, Bourgeois also taught 2 lessons on error theory. In Table 2, we have summarized the courses on analysis taught at the École polytechnique from 1912 to 1924 and indicated lectures devoted to probability theory. Starting in 1921–1922, lessons on probability were taught during the first year, while second-year students were still following a course set most probably taught by Chapelon.[48]

When the whole revision process was over around 1926–1927, it seemed that the internal mathematics teaching program in analysis had changed very little in both orientation and content from that of 1909. Except for a few details (and of course for probability theory), the model for the new formulation of the program seemed to have been the course taught by Hadamard as early as 1912–1914.

In geometry, the reduction of stereotomy's place to the benefit of graphical computing in the 1909 program had already been reinforced in 1913 and 1916 at d'Ocagne's request. In 1918, in a report on the course of geometry that echoed wartime debates about the role of mathematics in modern engineering practice,

[46] "[Le but ainsi défini] *impose aux programmes actuels une addition, le Calcul des Probabilités. A cause de ses applications aux sciences sociales, aux sciences d'observation, en particulier au tir du canon et aux théories cinétiques, il est nécessaire d'introduire cette branche importante des mathématiques dans le cours d'Analyse*" [AEP, III3a/8, p. 2].

[47] This can be inferred from Lévy's autobiography [Lévy 1970, pp. 71ff]; see also [Barbut et al. 2004, p. 55]. Lévy's lectures, which must have taken place in the spring of 1919 for second-year students, do not figure in the Summary of Lessons in the archives [AEP, X2C7]. However, Lévy states that the content of these lectures was abstracted in "Compléments au cours d'Analyse, par M. P. Lévy, professeur suppléant," placed at the end of Humbert's lithograph course, 1st Division, 1918–1919 (pp. 495–523) [Lévy 1935, p. 82]. For Lévy's personal file, see [AEP, VI2a/art. 2/box no. 29].

[48] Several hints lead us to believe that Chapelon gave the lectures on probability in Hadamard's second-year course on analysis in 1921–1922. One finds in the archives [AEP] a 150-page lithographed course on probability for second-year students corresponding to that year. In the course register, the handwriting moreover changes after the 29th lesson where "fin de cours" is written and a few lessons on probability are described with a rounder handwritting that resembles Chapelon's.

TABLE 2. Analysis courses at the École polytechnique, 1912–1924. The number in the square brackets indicates the number lectures devoted to probability theory. This number is followed by the name of the teacher in case it differs from the usual professor.

	1st year	2nd year
1912–1913	Hadamard	Humbert
1913–1914	Humbert	Hadamard
1914–1915	no course	no course
1915–1916	no course	no course
1916–1917	Hadamard	no course
1917–1918	Humbert	Hadamard
1918–1919	Hadamard (winter)	Humbert (winter) [3 Lévy]
	Hadamard (summer)	Lévy (summer)
1919–1920	Lévy (1st group) Chapelon (2nd group)	Hadamard
1920–1921	Hadamard	Lévy (1st group) [4] Chapelon (2nd group) [5]
1921–1922	Lévy [3]	Hadamard [3 Chapelon]
1922–1923	Hadamard [3]	Lévy
1923–1924	Lévy [3]	Hadamard

d'Ocagne strongly reaffirmed: "The strict needs of [the engineer's] daily practice must not be used in order to determine the parts of the mathematical sciences that are to be put in the École's programs." Otherwise, he added, "the École polytechnique would fall from the distinguished position it has never ceased to occupy among the institutions that maintain and develop French high culture."[49] Looking at the summaries of lessons in 1916–1917, one realizes that this essentially was the seed of the program suggested in 1918, officially adopted in 1927, and taught by d'Ocagne in the meantime. For that matter, d'Ocagne specified in this 1918 project that the "spirit" of his teaching was further developed in his book [Ocagne 1917–1918], [AEP, III3a/8, p. 5].

Therefore, in analysis as well as in geometry, the same process may be observed: in 1909 new programs trying to accommodate both theoretical foundations and applications; in 1912 hiring of professors whose outlook corresponded to this orientation and whose course content would contribute to reinforce it; after the war, new programs making this evolution official.[50] This underscores the importance of hiring in the evolution of the mathematical curriculum at the École polytechnique.

[49]"*Ce ne sont donc pas les stricts besoins de la pratique courante qui doivent servir exclusivement à délimiter les parties de la science mathématique à inscrire dans les programmes de l'École; tout ce que requièrent de tels besoins doit s'y trouver, mais quelque chose aussi de plus, ou l'École polytechnique se verrait déchoir du rang distingué qu'elle n'a cessé de tenir parmi les institutions qui contribuent à entretenir et à développer la haute culture française*" [AEP, III3a/8, p. 3].

[50]Concerning the lessons on probability theory, we emphasize the roles, respectively, of Carvallo, who, as director of studies, decided to include them in Humbert's course of analysis as early

3.2. Postwar Hiring Among Mathematics Faculty. The alliance between mathematicians and military officers played an important part in the postwar hiring campaign. As we have seen, the number of students greatly increased in 1919–1921. During the war, the École polytechnique had decided to postpone all permanent hiring. Humbert fell ill in 1919 never to recover until his death in January 1921. Out of 22 positions in the mathematical sciences, there were therefore 9 new appointments, of which 4 went to new personnel not previously associated with the École (see Table 3) . The general impression one gets by examining the new situation is however again one of great stability, except perhaps in the field of analysis where *normaliens* were now in the majority.

The successful campaign of Paul Montel and the failure of Gaston Julia in 1920–1921 will however better show the way in which the alliance between mathematicians and officers relied on a very specific understanding of mathematics in relation to applications. In 1920, when Lévy was elected professor to replace Humbert, a game of musical chairs ensued. To replace Lévy, as *répétiteur adjoint*, the easy choice was Chapelon, who was already *répétiteur auxiliaire* and who, like Lévy, had already substituted for Humbert in teaching analysis. To take Chapelon's place, there were three candidates: Montel, who was already a professor at the Sorbonne; Henri Larose, a polytechnician and engineer in the Post and Telegraph Administration; and the mathematician Gaston Julia, a *normalien* like Montel.

At the three levels (subcommittee of analysis, CI, and CP), the vote largely went in Montel's favor, a clear choice for a high-level mathematician. But not everyone in the committees favored this option. The physicist Lafay, for example, deplored that "by always preferring professors, one pushed away engineers wishing to devote themselves to science" [AEP, PV of the CI, vol. 11, p. 338]. The physicist Alfred Pérot and the director of the Post and Telegraphs School Alfred Dennery similarly supported Larose's candidacy, arguing in favor of experienced engineers. But after the war, the argument was easily countered. Before both the CI and the CP, Hadamard could point to Montel's war work in ballistics: "he therefore has had to deal with practical matters" [AEP, PV of the CI, vol. 11, p. 339]. This patriotic work at the head of a computing bureau was not Montel's only incursion into the practical side of mathematics. At the École des Beaux-Arts, the École polytechnique's professor of architecture Gustave Umbdenstock noted that Montel had designed a whole new course of study on building practices. And while Montel had shown that he was capable of applied work, Hadamard emphasized that his scientific worth, like Julia's, was incomparably greater than his competitor's: "M. Larose has simply applied the lessons he received from Henri Poincaré, while MM. Montel and Julia have accomplished deep and personal work, a necessary condition for reaching the high teaching level that should be provided at the École polytechnique" [AEP, PV of the CP, vol. 11, p. 265].

One might consider Montel's success as of little significance, due to his main competitor's scientific weakness. But Julia's failure the next year showed that the case was more complex. In 1921, a position was opened to replace Boulanger, who was named director of studies after Carvallo's retirement. Larose and Julia again opposed each other. The subcommittee for analysis easily went for Julia (by a 5-to-1 margin); the CI was more divided but voted for the mathematician

---

as 1919, and of Paul Lévy, who took on this assignment before he was hired as the new professor of analysis in 1920.

TABLE 3. Faculty with permanent positions in mathematics in 1920. A * next to a name indicates a promotion; names in bold are new recruits since 1914. As before, the following abbreviations are used in the tables: PR = professor; *Rép. tit.* = *répétiteur titulaire*; *Rép. adj.* = *répétiteur adjoint*; *Rép. aux.* = *répétiteur auxiliaire*; Dr = doctorate in the mathematical sciences; *agr.* = *agrégé* (certified high school teacher); *Ing.* = *ingénieur* (engineer); mil. = military officer; Acad = member of the Academy of Sciences; X = polytechnician; N = *normalien.*

Analysis:	
PR:	Jacques Hadamard (N - Dr - Acad 1912), Paul Lévy* (X - *Ing.* - Dr - Acad 1964).
Rép. tit.:	Édouard Goursat (N - Dr - Acad 1919), Gabriel Kœnigs (N - Dr - Acad 1918).
Rép. adj.:	Auguste Boulanger (X - *agr.* - Dr), Jacques Chapelon* (X - *Ing.* - Dr).
Rép. aux.:	Ernest Vessiot (N - Dr - Acad 1943), **Paul Montel** (N - Dr - Acad 1937).
Total:	5 N and 3 X (all Dr.).
Mechanics:	
PR:	Léon Lecornu (X - *Ing.* - Dr - Acad 1910), Paul Painlevé (N - Dr - Acad 1900).
Rép. tit.:	Jacques Jouguet (X - *Ing.* - Acad 1930), Alfred Liénard* (X - *Ing.*).
Rép. adj.:	Charles Plâtrier* (X - mil. - Dr), **Jean-Baptiste Pomey** (X - *Ing.*).
Rép. aux.:	Jean Réveille (X - mil. - Dr), **Émile Dumanois** (X - *Ing.*).
Total:	1 N and 7 X.
Geometry:	
PR:	Maurice d'Ocagne (X - *Ing.* -Acad 1922).
Rép. tit.:	Joseph Fouché* (X - *agr.*).
Rép. adj.:	**Louis Barré** (X - mil. - Dr).
Total:	3 X.
Astronomy and geodesy:	
PR:	Robert Bourgeois (X - mil. - Acad 1917).
Rép. tit.:	Henri Noirel* (X - mil.).
Rép. adj.:	**Eugène Fichot** (X - *Ing.* - Acad. 1925).
Total:	3 X.

(by 13 to 12). On December 24, 1921, however, the majority of the CP elected Larose by 13 to 10! Looking at each candidate's main supporters, one observes that applied scientists and Special School directors opposed the mathematicians and the military establishment of the École. Despite being acknowledged as a first-rate mathematician, Julia failed regularly in his attempts to be hired by the École polytechnique. Only in 1924 would he get the low-ranking permanent position of *répétiteur auxiliaire*.

After the war, the experience gained during the scientific mobilization was taken into consideration for hiring professors at the École polytechnique. The case of Paul Montel is significant since his war experience seemed a decisive argument to favor him over the polytechnician engineer Larose. Much younger than Montel, Julia had however been drafted and served on the front, where he was severely wounded in the face. Immobilized for several years, he then did, among other things, the work on rational function iterations for which he is now renowned, but he did not take part in the scientific mobilization. His loss to the same candidate, Larose, and his repeated difficulty in getting a position at the École show that his "glorious wound" (an expression found in the minutes of the analysis subcommittee) was not enough to counterbalance his mathematical leanings viewed as too much toward the pure (see his own "Notice sur les travaux" [Julia 1968]).

By looking at the elections and the various declarations made on these occasions, we are led to the claim that a de facto alliance had been struck between mathematicians, and most notably analysts, and the military authorities of the École (the commander of the École and the president of the CP, who were both generals and polytechnicians) in order to hire mathematical faculty of a high scientific level, especially in analysis. On the other side, professors in other disciplines, and especially physicists, joined forces with representatives from the Special Schools, whose weight on the CP allowed them to change the ranking established by other committees.

As can be gathered from the prewar cases of hiring above, this system of alliance was not new to the postwar period.[51] But one may say that it manifested itself with a special force in the period that immediately followed WWI. As far as mathematicians were concerned, this alliance went hand in hand with the favoring of analytical work oriented toward but not subsumed under applications in mechanics and mathematical physics.

Conclusion

A comparison of the situation at the École polytechnique before and after the war can give the impression that strikingly little changed during this period of great turmoil. Mathematical programs and the place they occupied in the polytechnician's training, as well as the general outlook of the mathematical faculty and of the new recruits, all exhibited a remarkable degree of continuity. We have shown however that continuity was the result of an active process of reevaluation of the place of mathematics at the École and in engineers' training. In other words, continuity with the past was consciously chosen by the École's administration as a response to recent events and a positive valuation of tradition, not by inertia.

[51]For another manifestation of this alliance between mathematical scientists and military officers, see Martina Schiavon's work, where the scientific links between Poincaré and Bourgeois as head of the army's Geographical Services are described [Schiavon 2006, pp. 87–88].

The war period indeed witnessed, from the fall of 1916 onward, a strong offensive from several quarters to fundamentally reform the teaching of mathematics at the École polytechnique and, more generally, in the training of all engineers. Had this effort succeeded, there is no doubt that the place of mathematics at the École would have been significantly reduced. But one would be greatly mistaken in seeing this simply as a fight of "reformers" against the "conservatism" of the École polytechnique. Rather, this was a conflict between two views about engineers' training and more generally about the relationship between science and industry: one focusing on immediate practical utility and the other insisting on the importance of fundamental theoretical knowledge. By favoring the second view, the administration of the École polytechnique remained faithful to its main purpose: to train civil servants for the state rather than engineers for industry. In our analysis, the convergence of opinion between the head of the École, General Curmer, and the head of the governement, Painlevé, seemed crucial to the victory of the second view in 1917–1918.

More generally, we have noted that a de facto alliance was forged between mathematicians (including nonpolytechnicians) and the military authorities of the École polytechnique. This alliance appeared before the start of WWI, most notably when analysts were hired. The desire of the École's military leaders to maintain the polytechnical tradition of high scientific learning, most remarkably in mathematical analysis, was linked to the privileging of research areas in mathematics that were susceptible to sustaining a broad range of applications. Before the outbreak of the war, emphasis was thus placed on second-order partial differential equations and, in 1919, on probability theory, a field of mathematics to which military engineers had paid early attention. Our analysis of the case of the École polytechnique has therefore shown that prewar conditions were favorable to the swift and large participation of mathematicians to the scientific mobilization organized by Painlevé after 1915.[52] In the aftermath of WWI, finally, the experience of mobilization helped to maintain and to consolidate mathematics' privileged position not only at the École polytechnique but also in French society.

List of abbreviations

- CI: *Conseil d'instruction* of the École polytechnique.
- CNAM: *Conservatoire national des arts et métiers.*
- CP: *Conseil de perfectionnement* of the École polytechnique.
- ECAM: *École centrale des arts et manufactures.*
- ENPC: *École nationale des ponts et chaussées.*
- ENS: *École normale supérieure.*
- ICMI: International Commission for Mathematical Instruction.
- SAEP: *Société des amis de l'École polytechnique* (Society of Friends of the École polytechnique).
- SICF: *Société des ingénieurs civils de France* (French Society for Civil Engineers).

[52]On this mobilization, see [Roussel 1989] and [Prochasson & Rasmussen 1996]. On the role of Painlevé in the development of national scientific research, see [Fontanon 2005] and [Pinault 2006].

Bibliography

Archival Sources

[AEP] Archives of the École polytechnique, Palaiseau.

[Bourgoignon 1916] Jean-Antoine-Henri Bourgoignon, *Conférence par le Commandant en second*, lithograph, November 4, 1916 [AEP, III3a/7].

[Bourgoignon 1917a] Jean-Antoine-Henri Bourgoignon, *Conférence par le Lieutenant-Colonel Bourgoignon, commandant en second*, lithograph, October 6, 1917 [AEP, III3a/7]; excerpts repr. in *Extraits de la conférence...*, Bulletin de la Société des amis de l'École polytechnique, no. 15 (November 1917), pp. 112–115.

[Bourgoignon 1918] Jean-Antoine-Henri Bourgoignon, *Conférence par le Lieutenant-Colonel Bourgoignon, commandant en second*, lithograph, October 1918 [AEP, III3a/8].

[Carvallo 1916] Moïse-Emmanuel Carvallo, *Conférence d'introduction sur l'enseignement de l'École polytechnique*, lithograph, 1916 [AEP, III3a/7].

[Chapel 1914] Frédéric-Antoine Chapel, *Rapport de l'inspection générale de l'École polytechnique en 1913*, typed manuscript, February 1, 1914 [AEP, I2/2].

[Curmer 1917] Fernand-Alexandre Curmer, *Allocution du général commandant l'École*, lithograph, November 8, 1917 [AEP, I2/2].

[Joffre 1908] Joseph Joffre, *Au sujet des programmes d'enseignement* (Note au Général commandant de l'École polytechnique), July 17, 1908 [AEP, III3a/6].

[Villermet 1992] Bernard Villermet, *Une grande École dans la tourmente: l'École polytechnique pendant la Première Guerre mondiale (1914–1920)*, mémoire de maîtrise de l'Université de Savoie, Chambéry, 1992 [AEP, VIII3b (1992) vol. 1–2].

Published Works

[Appel 1904] Paul Appell, *L'enseignement supérieur des sciences*, Revue générale des sciences pures et appliquées **15** (1904), pp. 287–299.

[Aubin & Bret 2003] David Aubin & Patrice Bret, eds., *Le Sabre et l'éprouvette: invention d'une science de guerre, 1914–1939*, "14-18 Aujourd'hui," no. 6, Agnès Viénot, Paris, 2003.

[Barbut et al. 2004] Marc Barbut, Bernard Locker, & Laurent Mazliak, eds., *Paul Lévy–Maurice Fréchet, 50 ans de correspondance en 107 lettres*, Paris, Hermann, 2004.

[Belhoste 1992] Bruno Belhoste, *La fin de la tutelle de l'École polytechnique sur les "prépas"*, Pour la Science **180** (October 1992), pp. 8–10.

[Belhoste 2003] Bruno Belhoste, *La Formation d'une technocratie : l'École polytechnique et ses élèves de la Révolution au Second Empire*, Belin, Paris, 2003.

[Belhoste et al. 1994] Bruno Belhoste, Amy Dahan Dalmedico, & Antoine Picon, eds., *La Formation polytechnicienne, 1794–1994*, Dunod, Paris, 1994.

[Belhoste et al. 1995] Bruno Belhoste, Amy Dahan Dalmedico, Dominique Pestre, & Antoine Picon, eds., *La France des X : deux siècles d'histoire*, Economica, Paris, 1995.

[Belhoste et al. 1996] Bruno Belhoste, Hélène Gispert, & Nicole Hulin, eds., *Les Sciences au lycée: un siècle de réformes des mathématiques et de la physique en France et à l'étranger*, Vuibert, Paris, 1996.

[Bourgoignon 1917b] Jean-Antoine-Henri Bourgoignon, *L'École polytechnique pendant la guerre*, Bulletin de la Société des amis de l'École polytechnique **15** (November 1917), pp. 81–111.

[Cornu 1896] Alfred Cornu, *L'École polytechnique, le but de son enseignement, l'esprit qui doit inspirer ses programmes*, Revue générale des sciences pures et appliquées **7** (1896), pp. 898–902.

[Crépel 1994] Pierre Crépel, Le calcul des probabilités: de l'arithmétique sociale à l'art militaire, in [Belhoste et al. 1994, pp. 197–215].

[Dahan Dalmedico 1994] Amy Dahan Dalmedico, Rénover sans se renier: l'École polytechnique de 1945 à nos jours, in [Belhoste et al. 1994, pp. 299–332].

[Dahan Dalmedico 1995] Amy Dahan Dalmedico, Polytechnique et l'école française de mathématiques appliquées, in [Belhoste et al. 1995, pp. 283–295].

[Fontanon 2005] Claudine Fontanon, Paul Painlevé et l'aviation: aux origines de l'étatisation de la recherche scientifique, in *Paul Painlevé (1863–1933): un savant en politique*, ed. Claudine Fontanon & Robert Frank, Presses Universitaires de Rennes, Rennes, 2005, pp. 41–56.

[Fontanon & Grelon 1994] Claudine Fontanon & André Grelon, eds., *Dictionnaire biographique des professeurs du CNAM (1794–1955)*, 2 vols., INRP and CNAM, Paris, 1994.

[Gilain 1989] Christian Gilain, *Cauchy et le cours d'analyse de l'École polytechnique*, Bulletin de la Société des amis de la Bibliothèque de l'École polytechnique **5** (1989), pp. 3–145.

[Gispert 1991] Hélène Gispert, ed., *La France mathématique: La Société mathématique de France (1870–1914)*, SFHST & SMF, Paris, "Cahiers d'histoire et de philosophie des sciences," **34**, 1991; 2nd ed., 2014.

[Gispert 1994] Hélène Gispert, De Bertrand à Hadamard, quel enseignement d'analyse pour les polytechniciens?, in [Belhoste et al. 1994, pp. 181–196].

[Goldstein & Aubin forthcoming] Catherine Goldstein & David Aubin, *La Grande Guerre des mathématiciens français*, forthcoming.

[Grison 1989] Emmanuel Grison, *François Arago et l'École polytechnique*, Bulletin de la Société des amis de la Bibliothèque de l'École polytechnique **4** (1989), pp. 1–28.

[Guillet 1916] Léon Guillet, *L'enseignement technique supérieur*, Bulletin de la Société des ingénieurs civils de France **10** (October–November 1916), pp. 634–705.

[Guillet 1918] Léon Guillet, *L'Enseignement technique supérieur à l'après-guerre*, preface by Henry Le Chatelier, Payot, Paris, 1918.

[Jacomy 1984] Bruno Jacomy, *À la recherche de sa mission: la Société des ingénieurs civils*, Culture technique **12** (1984), pp. 209–219.

[Julia 1968] Gaston Julia, *Œuvres*, vol. 1, Gauthier-Villars, Paris, 1968.

[Letté 2004] Michel Letté, *Henry Le Chatelier (1850-1936) ou la science appliquée à l'industrie*, Presses Universitaires de Rennes, Rennes, 2004.

[Lévy & Pinet 1894] Albert Lévy & Gaston Pinet, *L'Argot de l'X illustré par les X*, Émile Testard, Paris, 1894.

[Lévy 1935] Paul Lévy, *Notice sur les travaux scientifiques de M. Paul Lévy*, Hermann, Paris, 1935.

[Lévy 1970] Paul Lévy, *Quelques aspects de la pensée d'un mathématicien*, Blanchard, Paris, 1970.

[Mattatia 1994] Fabrice Mattatia, *Dictionnaire d'argot de l'X*, École polytechnique & SABIX, Palaiseau, 1994.

[Ocagne 1917–1918] Maurice d'Ocagne, *Cours de géométrie pure et appliquée de l'École polytechnique*, 2 vols., Gauthier-Villars, Paris, 1917–1918.

[Picon 1994] Antoine Picon, Les années d'enlisement: l'École polytechnique de 1870 à l'entre-deux-guerres, in [Belhoste et al. 1994, pp. 143–179].

[Pinault 2006] Michel Pinault, *La Science au Parlement: les débuts d'une politique des recherches scientifiques en France*, CNRS Éditions, Paris, 2006.

[Prochasson & Rasmussen 1996] Christophe Prochasson and Anne Rasmussen, *Au nom de la patrie: les intellectuels et la Première Guerre mondiale (1910–1919)*, La Découverte, Paris, 1996.

[Roussel 1989] Yves Roussel, *L'histoire d'une politique des inventions, 1887–1918*, Cahiers pour l'histoire du CNRS **3** (1989), pp. 19–57.

[Schiavon 2003] Martina Schiavon, Des savants-officiers entre science, armée, état et industrie de précision: les géodésiens du Service géographique de l'Armée, 1887–1920, in [Aubin & Bret 2003, pp. 60–73].

[Schiavon 2006] Martina Schiavon, *Les officiers géodésiens du Service géographique de l'armée et la mesure de l'arc de méridien de Quito (1901–1906)*, Histoire & Mesure **21**, no. 2 (2006), 55–94.

[Shinn 1980] Terry Shinn, *L'École polytechnique 1794–1914*, Presses de la Fondation nationale des sciences politiques, Paris, 1980.

[Staeckel 1914] Paul Staeckel, *La préparation mathématique des ingénieurs dans les différents pays*, L'Enseignement mathématique **16** (1914), pp. 307–328.

[Thooris 2004] Marie-Christine Thooris, *Une grande École dans la Grande Guerrre: catalogue de l'exposition d'octobre 2004 à août 2005*, École polytechnique, Palaiseau, 2004.

[Tournès 2000] Dominique Tournès, *Pour une histoire du calcul graphique*, Revue d'histoire des mathématiques **6** (2000), pp. 127–161.

[Villermet 1993] Bernard Villermet, *L'École polytechnique de 1914 à 1920*, Bulletin de la Société des amis de la Bibliothèque de l'Ecole polytechnique **10** (1993), pp. 5–62.

"I'm Just a Mathematician": Why and How Mathematicians Collaborated with Military Ballisticians at Gâvre

David Aubin

On September 18, 1915, Jules Haag (Figure 1), a young professor of rational mechanics from the University of Clermont-Ferrand, wrote to his old mentor, Professor Paul Appell, in Paris. Mobilized in a nonfighting unit of the army, Haag was in charge of overseeing a workshop of the Michelin Tire Corporation that produced ammunition for the celebrated 75 mm cannon. Asked to compute ballistic trajectories for a new airplane bomb design called "*bombe Michelin*," he first tried, as he wrote, to apply "the artillerymen's classical methods."[1] Having computed logarithms for three half-days in order to construct the required curves, Haag spent the little time he could spare to try to improve the methods. Rather surprisingly—since ballistics was, after all, a sensitive matter that the Academy of Sciences had expressly indicated it wanted to study further for the benefit of the French Army—but luckily, as we will see, the short paper Haag wrote and sent to Appell was published a week later in the *Comptes rendus hebdomadaires des séances de l'Académie des sciences* (hereafter CRAS) [Haag 1915a].[2]

Characteristically for a scientist in the first months of World War I, Haag felt that his skills were not used to their fullest extent. His superiors apparently showed no interest in the results he was sending to Appell: "In their eyes," Haag complained, "*I am just a mathematician*, without practical use other that serving

David Aubin's address is Université Pierre et Marie Curie, Institut de mathématiques de Jussieu-Paris Rive Gauche, Case 247, 4 place Jussieu, F-75252 Paris, France. His email address is david.aubin@upmc.fr.

In the writing of this chapter, I used documents provided to me by Cécile Aguillaume, June Barrow-Green, Alain Carrière, Claudine Fontanon, Sébastien Gauthier, Hélène Gispert, Catherine Goldstein, and Anne-Sandrine Paumier, whom I wish to thank here for their kind help. I would also like to thank the members of the WWI study group at the Institut de mathématiques de Jussieu-Paris Rive Gauche, as well as Patrice Bret and Pierre Mounier-Kuhn. This work was supported in part by the French Agence Nationale de la Recherche.

[1] "Les méthodes classiques des artilleurs" [Academy of Sciences, Haag to Appell, September 18, 1915; pochette de séance, September 27, 1915]. All translations are the author's.

[2] Two more papers on ballistics would then be published by Haag in the same year [Haag 1915b, Haag 1915c]. Some context about the work of the Michelin Tire Corporation in WWI can be found in [Chapeaux 2006, pp. 145–146]. The first planes produced by Michelin (in collaboration with the Breguet Company) were delivered in July 1915. On Jules Haag (1882–1953), see [Broglie 1953, Mesnage 1953, Châtelet & Chazy 1955, Meyer 1989].

FIGURE 1. Jules Haag as correspondent to the Academy of Sciences. With permission of the Académie des sciences–Institut de France.

as a computing machine when the occasion arises."[3] A mere sergeant (*maréchal des logis*), Haag had been barred from the local branch of the Commission of Inventions since only officers (or civilians) could sit on it.[4] His only recourse, he explained, was to study a bit of ballistics in his spare time and wish that one would let him devote more time to it. He admitted that this might have no immediate effect in helping to "drive out the Germans," but "the questions I am asked, without being

[3]"*Je ne suis, à leurs yeux, qu'un mathématicien, sans utilité pratique autre que celle de servir, quand l'occasion se présente, de machine à calculer*" [Academy of Sciences, Haag to Appell, September 18, 1915; pochette de séance, September 27, 1915, emphasis added]. The misuse of scientific personnel at the start of World War I is a common theme among contemporary scientists and historians. For reviews, see [Kevles 1978, Hartcup 1988, Aubin & Bret 2003], and references therein.

[4]A promotion to the rank of second lieutenant (*sous-lieutenant*) in the artillery had indeed been considered for Haag and rejected. On the commission, see [Roussel 1989] as well as the contribution by David Aubin, Hélène Gispert, and Catherine Goldstein to this volume.

told more about the mysterious studies that give rise to them" made him suspect that his contribution might indeed be directly pertinent to the war effort.[5]

Less than a month later, on October 12, 1915, the professional ballistician General Prosper-Jules Charbonnier (1882–1936) (Figure 2) sent a memo to his superiors calling attention to the tremendous ballistic effort that war operations now demanded. A navy officer trained at the École polytechnique who had served in Africa and in the Far East, Charbonnier was at the time president of the so-called *Commission d'expériences d'artillerie navale de Gâvre*, which was both a proving ground and the main military body in charge of ballistic computations for the French Navy and Army. In his memo, President Charbonnier explained that his overworked personnel were now unable to face the huge quantity of experiments and computations the War Ministry asked them to carry out. Acknowledging that most artillerymen were, of course, otherwise busy on the battlefields, he noted that university professors possessed an "intellectual and professional training that would quickly make them useable by the commission for computations and even experiments." A regular reader of the *Comptes rendus*, Charbonnier put forward Haag's name as a likely candidate.[6] Two weeks later, on October 29, 1915, the mathematician left Clermont-Ferrand to reach his new assignment post at Gâvre, near the seaport of Lorient in Brittany [SHD–Terre, 6–Ye-17966, Haag's military file].

As a result of Charbonnier's memo, a dozen mathematicians, physicists, and astronomers would join the Gâvre Commission over the course of the war, including Albert Châtelet (1883–1960), Georges Valiron (1884–1955), Joseph Kampé de Fériet (1893–1982), and Arnaud Denjoy (1884–1974). The work they did at Gâvre was part of the establishment of new theoretical foundations for computing in exterior ballistics. New experimental methods for studying ballistics were also pursued, while the astronomer and future director of the Paris Observatory, Ernest Esclangon (1876–1954), carried out highly praised work on sound-ranging, that is, the localization of enemy batteries by an analysis of the sound waves they emitted.[7] Other prominent mathematicians such as Henri Lebesgue (1875–1941), Émile Borel (1871–1956), and Paul Montel (1876–1975) took part in ballistic research and enrolled several mathematics teachers in their effort.

Exterior ballistics was one of the few areas where French mathematicians truly were able, as mathematicians, to play a prominent part in World War I. This chapter examines the way in which young mathematicians such as Haag were drawn into military work at a time when the general belief was that there was no better way for them to contribute to the war effort than by fighting on the front. As the 22-year-old Pierre Abeille wrote in his last letter to his parents:

[5]"*Cela ne serait peut-être d'aucune utilité pratique pour chasser les Allemands. Cependant, les questions qu'on me pose, sans m'initier davantage aux mystérieuses recherches qui les font naître, pourraient bien laisser supposer le contraire.*" All quotes come from [Academy of Sciences, Haag to Appell, September 18, 1915; pochette de séance, September 27, 1915].

[6]Charbonnier to the minister (October 12, 1915), quoted in [Patard 1930, p. 274]. It is not known whether Charbonnier was in contact with Appell.

[7]Although it properly belonged to the field of ballistics at the time, sound-ranging will not be examined in detail here. On this topic, see especially [Schiavon 2003a, Schiavon 2003b]. See also [Jones 1921–1922, Kevles 1969, Palazzo 1999], as well as other chapters by June Barrow-Green and by the present author in this volume.

FIGURE 2. Colonel, later General, Prosper-Jules Charbonnier pondering a ballistic problem on the beach of Gâvre, undated. From [Patard 1930, p. 253].

> Shame on intellectuals who fail to understand that they have ... the sacred duty of putting their arms and chests in the same location as the arms and chests of their brothers.... To us, the privileged, the guardians of tradition, the transmitters of the Ideal, [the duty] of risking our lives and happily sacrificing ourselves for the preservation, the extension, the exaltation of all this beauty, of all this pride that we are the first to feel and to take advantage of.[8]

Historians have shown that in all warring countries the exact contours of research scientists' participation in the defense effort were far from being clearly delineated in 1914 and that many of them did not survive the first years of the war.[9] Most historians' accounts of the war work of mathematicians and scientists often, if not always, focus on the stories told by surviving scientists themselves. Most, therefore, view scientists' contributions on their own terms. But I would like to argue that to understand correctly the value of such contributions, it is necessary to widen the scope of our investigations. One needs to have a clear view of the military demands from the points of view both of general staffs and of men in the field. One also needs to pay special attention to the military technical structures that existed before the outbreak of the war to see why officers serving in these structures felt the need to enlist civilian expertise.

I want to argue that the incorporation of university mathematicians into military research structures was neither preordained nor straightforward. One may well argue that some mathematical knowledge has always been mustered in warfare.[10] But before 1914 there also seems to have been a widespread feeling that professional mathematicians—that is, the university professors teaching and carrying out research in mathematics—had little to do with it.[11]

In the United States, Princeton University mathematician Oswald Veblen, despite his enthusiasm for war service, had difficulties finding a place where he could apply his mathematical skills. But he quickly found a befitting assignment at the Aberdeen Proving Ground, which was the U.S. counterpart of the Gâvre Commission. According to the internal history of this institution, "Exterior ballistics—that part of the science dealing with the behaviour of projectiles in flight—underwent a major revolution during World War I."[12] Military historians in the U.S. easily acknowledged that scientists were major players in this revolution: Forest Ray Moulton, an astronomer from the University of Chicago, was the first head of the

[8]A public servant, Sergeant Abeille had volunteered for active duty and was killed on November 12, 1914, in Vingré, Aisne. His letter, dated September 26, 1914, is repr. in [Foch 1922, p. 13].

[9]For France, see [Roussel 1989], [Aubin & Bret 2003], and [Galvez-Behar 2008].

[10]On the history of mathematics and war, see among others [Mehrtens 1996, Booß-Bavnbek & Høyrup 2003, Steele & Dorland 2005]. On the WWI period especially, see [Siegmund-Schultze 2003].

[11]Even after the war, this feeling remained strong. A reviewer of a book on the history of ballistics [Charbonnier 1928a]—"the value of whose results I was able to appreciate for more than four years and the perfection of which I have known without joy"—expressed his surprise at encountering the names of the "greatest perhaps of the history of science: Galileo, Huygens, Newton, Euler, among others" [L.G. 1931, pp. 376–377].

[12][U.S. n.d., p. 3]. The military work of American mathematicians is considered in the contribution by Thomas Archibald, Della Dumbaugh, and Deborah Kent in this volume. On Veblen, see also the introduction to this volume. For contemporary assessments of American methods by the French, see [Schwartz 1927, Dodier & Valiron 1927].

Ballistic Branch, and even though the computation of firing tables followed standard European procedures, their implementation and improvement were the work of three mathematicians linked to Princeton (Oswald Veblen, Gilbert Ames Bliss, and Thomas Hakon Gronwall). Although historians assessed that World War I had little effect on the postwar development of mathematics in the U.S., which remained characterized by the high value placed on abstractness, the field of ballistics appeared as a paradigmatic example of a research problem until this time unsolvable by military structures and that only university scientists' special skills could solve.[13] Likewise, the history of British ballistics has been centered on the scientist, the team assembled by the Cambridge biophysicist Archibald Vivian Hill and involving the mathematicians John E. Littlewood, Edward Arthur Milne, and Ralph Howard Fowler, the last two being definitively turned away from prior involvement in pure mathematics.[14]

In France, by contrast, the situation seemed rather more complex and perhaps also more interesting. It was more complex because for at least a century most military officers had received their first academic training at the École polytechnique, before going on with their careers as military officers. As a result, they often had the mathematical sophistication required for efficiently tackling the main problems of ballistics. Indeed French military ballisticians had a long acquaintance with the use of advanced mathematical techniques, but also physical and chemical theories and experimental procedures, to tackle every aspect of the problem of artillery firing.[15] The French situation was also more complex because many more mathematicians were involved almost from the start in ballistic research, joining various military institutions that already possessed a high degree of expertise that civilians—no matter how *savant*—did not necessarily have. The circumstances under which civilian, or formerly civilian, mathematicians were drawn to the problem and the specific contributions they were able to bring to it therefore need to be assessed from the point of view of the encounter of people coming from various perspectives. This encounter was forced upon them by the special circumstances of war and the specific demands that emerged from frontline fighting experience.

In the rearlines, mathematicians and ballisticians engaged in the production of range tables and computing procedures, which were hybrid entities straddling the various worlds of the fighting artilleryman, the military specialist, and the academic mathematician. In so far as it was a hybrid product hastily put together in a time of emergency, the new ballistics that came out of the First World War only found its place with difficulty within institutional and epistemological frameworks after the Armistice. While military strategists insisted on the renewed importance of the artillery in modern warfare, resistances persisted to the complete transformation of ballistics into a branch of applied mathematics. From the early 1920s onward, civilian scientists were, however, associated to the Gâvre Commission on a permanent basis. Almost nonexistent before the war, the relationship between university mathematicians and military research bodies was made permanent.

[13]For an assessment of the lack of effect of World War I on mathematics in the U.S., see [Parshall & Rowe 1994, p. 444]. A nuanced review may be found in [Parshall 2000].

[14]See [Smith 1990, Hill n.d.]. The war work of the Cambridge mathematicians is discussed in June Barrow-Green's contribution to this volume.

[15]On the École polytechnique, see Jean-Luc Chabert and Christian Gilain's contribution to this volume.

Seeing in the research bodies hastily put in place during the war by military as well as by civil authorities a prefiguration of the state-controlled system of science funding that was fully developed during the Cold War, historians of science have mostly focused on the scientists' place in new organizations such as boards of inventions and research councils and have turned a blind eye to older military research traditions.[16] Our look at French ballistic research shows that scientists' involvement in these institutions in 1914–1918 resulted in their lasting association with military research structures that existed before the start of the First World War and endure to this day.[17]

1. Tradition, the Scientific Method, and the "Gâvrais Virtues"

Early in the twentieth century, Gâvre was a small fishing village sitting at the tip of a long and thin peninsula that separates the harbor of Lorient from the Atlantic Ocean on the south shore of Brittany.[18] On the isthmus connecting it to the mainland, a proving ground was established in 1829, when the Gâvre Commission was created by the Navy Minister. The commission was composed of military officers and engineers, coming in proportions that varied over the years from the naval, colonial, and sometimes land artilleries. Its main objective was to carry out all types of research connected to gunnery, including experimentation with cannons, rockets, and guns, with powder, and with projectiles. Their effect on various types of steel plates was also investigated. But most of all, the Gâvre Commission was famous for its expertise in ballistics, both interior and exterior, both theoretical and experimental. Since its foundation, its principal task was the compilation of numerical tables giving the range as a function of the initial shooting angle (or line of departure) for each new cannon model introduced in the navy.

When he reached the Gâvre Proving Ground, Haag was hardly stepping into virgin territory. There, he and other civilian scientists found cannons of all sizes, cranes, railways, telegraphic and telephonic lines, chronographs and photographic cameras, all types of ballistic instruments needed to measure the initial velocity of projectiles, their power of penetration in steel plates, etc. (Figure 3). But above all, they came into contact with a handful of navy officers who shared a strong scientific "ethos" attached to its own tradition and characterized by its own distinctive values. For mathematicians who were suddenly confronted with them, the "Gâvrais virtues [*les vertus gâvraises*]," as Charbonnier called them, might at first have been discomforting [Charbonnier 1906, p. 425].[19] This is what Léon Patard, the official

[16]On Britain, see [MacLeod & Andrews 1971, Pattison 1983, Hull 1999]. On the U.S., see [Kevles 1968]. On France, see [Roussel 1989]. On Germany and Italy, see [Mehrtens 1996, Epple et al. 2005]. A more detailed study of the Italian case is given in Pietro Nastasi and Rossanna Tazzioli's contribution to this volume.

[17]Let us note here that the successor institution of the Gâvre Commission, called GERBAM (*Groupement d'études et de recherches balistiques, armes et munitions*), was closed down on January 1, 2010. Only a naval training center for shooting now remains at the Gâvre polygon.

[18]After much debate, the standard usage today is to write Gâvres (with an *s* at the end). At the time of the First World War, however, military officials had decided to use Gâvre, which is much more common in contemporary documents. I will thus follow the latter usage.

[19]The idea that the Gâvre "virtues" may be understood in relation to the scientific ethos was first popularized by [Merton 1942]. More recently, John Ziman suggested that the Mertonian ethos (which he calls *cudos*, for communalism, universality, disinterested, originality, skepticism) might have coexisted with an industrial research ethos he called *place* (for proprietary, local, authoritarian, commission, expertise), which might be closer to the Gâvre ethos [Ziman 2000].

FIGURE 3. A view of the Gâvre polygon in the early 1930s. From *Le centenaire de la Commission de Gâvre*, Revue maritime et coloniale (2nd semester 1930), pp. 233–237; on p. 235.

historian of the Gâvre Commission at the time he was its president, alluded to in the following passage written in 1930:

> A success of the Naval Artillery during this war, and not the least, was the use of the good will of all these men, some of whom [were] eminent savants but little prepared for their new role. Their trust had to be won; they had to be shown that, while they were not their equals in the domain of pure science, naval artillery engineers were able to understand their ideas, discuss their theories, and answer their objections. One can only infer the tact needed to subject them to long training courses in routine computing, to stem their professional attempts at perfecting computing methods as soon as they got there, to make them comply with the strict rules of experimentation and observation, and finally to lead them progressively from the role of auxiliaries to that of collaborators [Patard 1930, p. 279].[20]

Charbonnier also insisted on the "scientific-technical" organization put in place at Gâvre, whereby collaborative work was done by "savants and officers." At Gâvre, "theoreticians remained in close and permanent contact with users" of cannons [Charbonnier 1929, pp. 27–28]. Léon Lecornu (1854–1940), a professor of mechanics at the École polytechnique, hinted at something similar when he praised the fact that "several science professors transformed by mobilization into improvised artillerymen, brought a precious collaboration to the professionals: together they contributed in a large part to the final victory" [Lecornu 1924, p. 38].

Compare this emphasis on collaboration to scientists' accounts with which historians of science are much more familiar. The general tone used in the latter makes them fall into one of two categories: while many are self-congratulatory and emphasize the unique contributions civilian scientists were able to bring to the war effort, many others are overfull with bitterness, resentment, and acrimony towards the military authorities' incompetence and their criminal misuse of the

[20]Patard (1872–1963) was president of the Gâvre Commission from January 26, 1925, to February 20, 1931.

mathematicians' or of the scientists' special abilities. An instructive example of the second attitude can be found in the pamphlet privately published by the physicist and industrialist Georges Claude: "Our savants? Ah! If only you knew what was done of them [*ce que l'on a fait d'eux*]! If only you knew their hopes, their efforts, their struggles, and in the end their powerlessness" [Claude 1919, p. 33].[21] By contrast, Patard underscored the difficulty faced by the military members of the Gâvre Commission when they endeavored to integrate civilian scientists into their working procedures. Although Haag was asked to work on the solution of a new problem (anti-aircraft gunnery) and, in the process, introduced new mathematical methods at Gâvre (error analysis), the main problem, as Patard saw it, lay in containing scientists' impatience, acquainting them with the Gâvre tradition, and making them comply with entrenched working procedures.

Some scientists seemed to have been quite aware of the fact that they were stepping on other experts' turf when they offered their help to military engineers. Aimé Cotton (1869–1951) was an established physicist and already recognized by military officials for having developed, in collaboration with Pierre Weiss (1865–1940), one of the most successful instruments used in sound-ranging. In 1916, Cotton wrote General Hubert Gossot (1853–1935), a former president of the Gâvre Commission, to offer his help on the problem of determining the effect of weather conditions on artillery fire. It is interesting to note the very cautious wording used by Cotton in his letter: "if you judge that the reflections of a physicist who is perforce incompetent on many points may be of some interest to you, I put myself at your disposal to extract from them results that would seem useable and to give them [the results] a form more easily applicable in practice (relying on the advice of men of the trade)."[22] Clearly, rather than assuming that he knew better, the physicist placed himself at the disposal of the military and not merely of his country.

Tradition at Gâvre was no empty rhetoric. It was the bedrock of its scientific credibility. One is struck not only by the palimpsest like manner in which the history of the Gâvre Commission has been written and rewritten on several occasions by some of the major ballisticians who worked there, but also by the insistence put on tradition, despite important breaks in the methods and despite rapid changes in the gunpowder and materials used by artilleries.[23] As Charbonnier wrote in 1906, Gâvre was characterized by its special virtues: "respect for the continuity of doctrines, freedom of thought, faith in experimentation, [and] patience for long computations and for careful verifications" [Charbonnier 1906, pp. 425–426]. Another ballistic engineer, Maxime Crémieux, would concur in 1930: "beyond all experimental and theoretical investigations, one senses a working method that presided over the whole: ... submission to experimental facts, scientific probity, respect of tradition, minute consideration of details, clarity of deductions, freedom in technical

[21] On Claude, see [Aubin 2003] and [Baillot 2010].

[22] "*Si vous estimez que les réflexions d'un physicien, nécessairement incompétent sur bien des points, présentent quand même quelque intérêt, je me mets à votre disposition pour en extraire les résultats qui paraîtraient pouvoir être utilisés et leur donner une forme plus facile à appliquer pratiquement (en m'aidant des conseils des gens du métier)*" [SHD–Terre, 2W292, Cotton to Gossot, February 8, 1916].

[23] Besides [Charbonnier 1906] and [Patard 1930] already cited, other insiders' histories of Gâvre are [Poyen-Bellisle 1889–1893] and [Crémieux 1930]. Secondary literature about the Gâvre Commission is however much rarer, but see [Bru 1996], [Seddik-Ameur 2003], and [Mounier-Kuhn 2010, pp. 194–195]. On ballistics in France more generally, see also [Belhoste & Lemaître 1990] and [Fontanon 2005].

opinions" [Crémieux 1930, p. 145]. To be more specific, although it has remained in the shadows of more public institutions, like observatories and laboratories, or institutions recently studied in more detail, like factory workshops, Gâvre was one of the main theaters for the development of strict experimental procedures in the nineteenth century. Indeed, it is probably no exaggeration to say that it was one of the places where modern procedures for the mathematical and quantitative testing of theories against experiments were designed and tested on a large scale.[24] In ballistics, the comparative merits of empiricism and theory were often discussed explicitly. Like the observatory, it played a leading role in the slow convergence between empirical procedures and theoretical approaches derived from Newton's first principles.[25] "While mathematicians subsequent to Cauchy were usually more interested in proofs of existence and in functional relations among solutions obtained, than in numerical results, there were already cultivated two fields which especially called for numerical methods, namely astronomy and ballistics" [Bennett et al. 1956, p. 61]. While one may contest Bennett's view of mathematicians on the grounds that the latter did manifest some interest in computation, one cannot deny the importance of astronomy and ballistics for the history of computing.

After the establishment of the Gâvre Commission, it was the professor of hydrology at the Lorient Naval School, Félix Hélie (1795–1885), who soon assumed the scientific leadership.[26] Starting in 1834, after summers devoted to experimentation and cannon testings, Hélie alone would carry out all the computations needed to interpret the experimental results and draft all the reports sent to the Navy Ministry. A staunch empiricist, he distrusted theories developed from first principles, which gave poor results when tested on the proving ground. Charbonnier thought that, in his work, Hélie showed more patience than originality [Charbonnier 1906, p. 413]. In Hélie's mind, every new range table had to be established through extensive experimental work, and the laws and formulas derived from this massive work could hope for no more than an ephemeral existence. In his *Traité de balistique expérimentale*, Hélie developed his method: summarize experimental results by as simple a mathematical formula as possible that should not be applied outside of the experimental limits used to derive it. He opened this treatise with the following words:

> The principles of rational mechanics are not sufficient to solve [all the questions relative to artillery shooting and its effects]; the forces and resistances at play can only be appreciated through observation. A treatise on ballistics must therefore be in large part composed of descriptions and discussions of experiments whose results are often the only possible demonstration for the propositions which one may consider as having been established [Hélie 1865, p. 1].

Hélie's attitude may in part be explained by the fact that, compared to the academic setting, the Gâvre Commission was never allowed to lose sight of its practical

[24]On the role of the military in scientific research in France in the period preceding the establishment of the Gâvre Commission, see [Bret 2002].

[25]On the history of the observatory in the nineteenth century, see [Aubin et al. 2010]. On the importance of computing between theory and practice, see [Warwick 1995].

[26]On Hélie, see [Delauney 1892] and compare with the history of ballistics at Metz in [Bru 1996].

mission: to produce the knowledge necessary for aiming guns accurately. So, even when later ballisticians expressed the wish to reevaluate the role of theory in their field (see below), they came up with stringent criteria.[27] According to the Gâvre ballisticians, a theory was judged satisfactory only in so far as it could be checked numerically, even if this entailed a considerable amount of work [Charbonnier 1906, p. 432]. According to Crémieux, the Gâvrais character was shaped in this period and entailed an "absolute respect" for carefully documented experimental results and repugnance toward erasing disagreement between theory and experiment. On the contrary, all discordances had to be underscored in written reports so that one knew exactly where methods were in need of improvement [Crémieux 1930, p. 149]. This was, Charbonnier insisted, the application of Francis Bacon's method, which led to practical prescriptions: (1) never to fire a useless shot; (2) never to fire a necessary shot without having first computed all experimental results expected from it; and (3) "perform all experiments necessary in the toughest circumstances, such that disagreement between theory and experiment has the greatest chance of manifesting itself."[28]

For theoretical considerations to be put in practice, he added, three precautions were required:

(1) compute all necessary tables in full, without disdainfully leaving this vulgar care to technicians (*praticiens*);
(2) provide full and detailed numerical examples;
(3) prepare with great care computation skeletons, which can be lithographed and whose columns are merely left, so to speak, to be filled out numerically [Charbonnier 1906, p. 432, footnote].

In short, Charbonnier wrote, the Gâvrais character could be summarized as "a very practical outlook ... which holds a theory as satisfactory only in so far as it has been numerically checked, compared to all the known experiments, and when necessary numerical tables have been computed often at the price of considerable and off-putting labor an idea of which only those who have themselves executed a similar task can have" [Charbonnier 1906, p. 432].

In 1915, procedures followed by the Gâvre Commission therefore were particularly stringent. They were solutions adopted over the course of almost a hundred years of confrontation with intricate problems and practical demands from fighting artillerymen on ships around the world. If these procedures ensured the international high regard in which results coming out of Gâvre were held, "doctrinal" thinking, the requirements of efficiency, and respect for hierarchy also rigidified procedures. Once adopted after extensive series of tests, computing procedures were rarely changed and then only after much debate. To understand what was at stake in theoretical and practical ballistics at the time when Haag and his colleagues reached the Gâvre peninsula, let us now briefly review the status of exterior ballistics at that time.

---

[27] Note that this was not only felt at Gâvre. See the way in which an engineer in the artillery branch of the Creusot Factories argues in favor of theory and for the insufficiency of experiments alone in [Morel 1904, p. 7].

[28] [Charbonnier 1906, p. 417]. One should here be reminded that in the course of the nineteenth century it became increasingly expensive to fire a cannon shot.

2. Exterior ballistics before 1915

Ballistics is a complex science in which theoretical, experimental, and computational uncertainties always clash with one another. Moreover, technological innovation in cannon and projectile design sometimes increase the complexity of the problem. The first complication to consider came from the fact that the motion of a projectile actually involves two very different sets of problems: interior ballistics, which deals with what happens inside the cannon muzzle and which can be tackled mostly by mobilizing the tools of thermodynamics and chemistry, and exterior ballistics which is mainly a mechanical and mathematical problem. Both problems, moreover, require some input from fluid mechanics. It is a remarkable fact that up to WWI, and beyond, military ballisticians would address both interior and exterior ballistics, often with an equal degree of competency.

Restricting our attention to exterior ballistics, while the laws of dynamics easily allowed the determination of the differential equations governing the trajectory of a projectile through the air, various effects combined to make its general solution very hard to find. At the beginning of the twentieth century, Charbonnier drew inspiration from astronomical methods to introduce a useful distinction that considerably simplified the presentation of exterior ballistics [Charbonnier 1907].[29] "Secondary" problems, as he called them, were considered perturbations to the "principal" problem. They were supposed to take into account the effects of the wind, of variations in atmospheric temperature and pressure, of the wear of the piece, of the projectile's spin around its axis, of the earth's rotation, etc. The principal ballistic problem therefore amounted to solving an ordinary differential equation, called the "hodograph," whose simplest form was

$$dv_x = \frac{cv}{g} F(v) d\tau, \tag{2.1}$$

where v is the magnitude and $v_x = v \cos \tau$ is the horizontal component of the velocity v of the projectile in the (x, y)-plane, expressed as a function of the uniformly decreasing angle τ with respect to the horizontal at each point of the trajectory (Figure 4); g is the acceleration due to gravity at the surface of the earth (taken in first approximation to be constant); c, the so-called "ballistic coefficient" (varying according to the size and shape of the projectile); and $F(v) = f(v)/v^2$, the law of resistance of the air to the motion of the projectile supposed in first-order approximation to be a function of v only.[30]

The principal ballistic problem is therefore mathematically equivalent to finding the solution (or at least a fairly accurate approximation of certain quantities derivable from the solution) of an ordinary differential equation involving some unknown functions. Issues connected to the resolution of the principal ballistic problem were of three sorts: (1) how to compute the ballistic coefficient c; (2) what was the exact form of the air resistance function $F(v)$; and (3) in case these quantities gave rise

[29] Many publications give the state of the art in the first decade of the twentieth century; see, e.g., [Gilman 1905] and [Cranz & Vallier 1913].

[30] Because of its great simplicity, I adapt here Charbonnier's own presentation in [Charbonnier 1929]. Note however that he also included as another factor the air density $H(y)$ that prior to WWI was generally taken to be a constant in the principal ballistic problem (corresponding to low altitude trajectories). When vertical differences in air density were taken into account (as a secondary problem), an exponential law $H(y) = e^{-hy}$ was usually assumed. A very similar, equivalent expression is found in [Charbonnier 1906, p. 449].

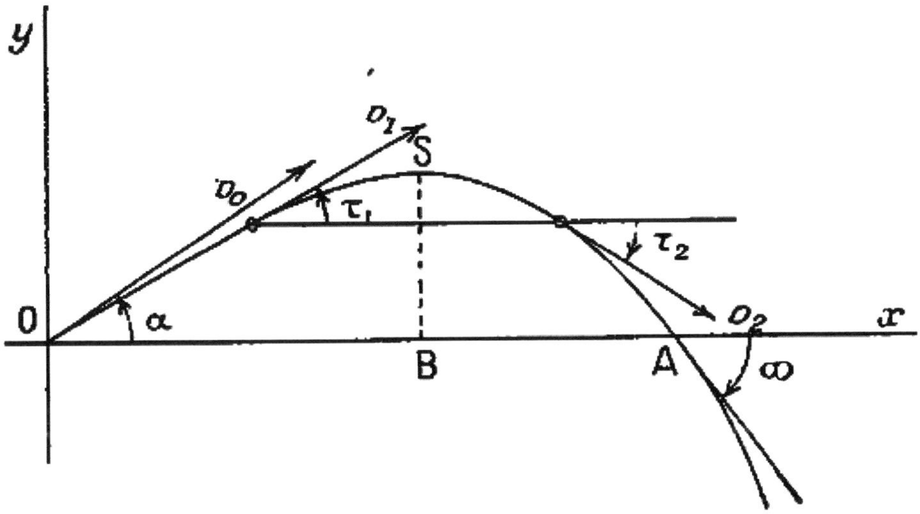

FIGURE 4. Schematic representation of a projectile's trajectory, displaying variables relevant to the hodograph. On this view, taken from [Adhémar 1934, p. 7], two steps in the piecewise integration method, at angles τ_1 and τ_2, have been pictured. At a generic point on the trajectory, the magnitude of the velocity was simply noted by v and the angle with the horizon by τ. Courtesy of MIR Library, Paris.

to differential equations that were insoluble by formal means, what was the best method for finding a good approximation of the solution? Although at first sight the former two questions would seem to be soluble by the tools of fluid mechanics and the latter to be a purely mathematical problem, all questions were intertwined, a fact that can be illustrated by recalling that in the first experiments carried out by the Gâvre Commission, before the ballistic pendulum was introduced there, initial bullet velocities were essentially determined by measuring the range of their trajectories![31]

2.1. Ballistic Coefficient and Air Resistance. At the beginning of WWI, the Gâvre Commission had, under Charbonnier's leadership, adopted a theoretical framework and strict computational procedures for dealing with exterior ballistics. If the practical needs of the artillery were thereby met by the Gâvre savants, any unforeseen innovation modifying shooting parameters would immediately throw ballisticians into uncharted territory. For most purposes, reliable results seemed to be obtained using the following formula for the ballistic coefficient:

$$c = \frac{i\Delta_0 a^2}{p}$$

[31]Invented by Benjamin Robins in the eighteenth century, the ballistic pendulum was greatly improved by the ballisticians of the Metz Artillery School, Guillaume Piobert, Arthur Morin, and Isidore Didion, in the 1840s [Didion 1860, pp. 321–326], [Morin 1855, pp. 228–233]. It was adopted by the Gâvre Commission in the same decade. On the Metz artillery school, see [Belhoste & Picon 1996].

where Δ_0 was the air density on the ground; a and p, respectively, the caliber and the weight of the projectile; and i its "form factor", often considered proportional to $\sin\gamma$, with γ the penetrating angle at the tip of the projectile. But all attempts at securing a theoretical foundation for this formula remained elusive.

As far as the resistance law was concerned, ballisticians first used the Newton resistance law proportional to the square of the velocity (v^2). When rifled barrels were introduced, initial speeds greatly increased and the resistance law suggested by Alexandre-Hippolyte Piton-Bressant (1820–1847), proportional to v^4, was adopted for a while.[32] In the second half of the century, smokeless powder again increased initial projectile speed. New extensive series of experiments were carried out in the 1860s by Nikolai Maievski near Saint Petersburg in Russia and by Francis Bashforth with the help of his electric chronographs in Woolwich, England. In the next decades, new evidence from experiments performed from 1879 onward by the Krupp Company on the Meppen shooting range in Germany and by Colonel Hojel in Holland in 1884 showed that the resistance law decreased at higher velocity. As a result the functional form of the law became tremendously complicated. In Italy, Francesco Siacci (1839–1907) suggested the following expression [Cranz & Vallier 1913, p. 16]:

$$f(v) = 0.2002v - 48.05 + \sqrt{(0.1648v - 47.95)^2 + 9.6} + \frac{0.442v(v-300)}{371 + \left(\frac{v}{200}\right)^{10}}.$$

In 1896, an empirical resistance law was adopted at Gâvre: the famous "*fonction de Gâvre.*" Thousands of numerical results, derived from firing tests with initial speeds from 400 to 1,200 meters per second with all calibers and all types of projectiles, were used to determine this function, which was presented as a table, as a graph, and, finally, as an analytic expression introduced by *chef d'escadron* Demogue [Crémieux 1930, p. 152].[33] According to Haag,[34] the function used during WWI had the following form, where the exponential term was introduced by the naval engineer Maurice Garnier:

$$F(v) = v^2 \left[0.255 + \frac{\sqrt[4]{1 + 0.0392\left(\frac{v-300}{500}\right)^8}}{27{,}226 + 494\left(\frac{v-330}{50}\right)^2} \arctan\frac{v-330}{50}\right] \exp\left(\frac{v-600}{10^6}\right).$$

[32]Many other laws $f(v)$ were suggested and used in the nineteenth century: bv^3 by Bashforth; $av^2(1+bv)$ by Saint-Robert; $av^2(1+bv^2)$ by Didion; $a+bv$ by Chapel, etc. See [Charbonnier 1906, p. 444]. A short and simple mathematically oriented introduction to the question of air resistance is to be found in [Long & Weiss 1999].

[33]The resistance law was first expressed as a numerical table and presented in report nos. 1414 and 1429 of the Gâvre Commission (March 14 and July 21, 1896), giving $\log f(v)$ or $\log[f(v)/v^2]$ for values of v from 0 to 1,200 meters per second; see Table I1 in [Gâvre 1916, pp. 3–6]; "Tableaux nécessaires au calcul des trajectoires par arcs successifs" [SHD–Terre, 2W292], and [Bingen 1928, pp. 20–22]. One finds a graphic representation of the laws and the various experimental trials that led to it in [Cranz 1903, p. 197], reproduced in Figure 13, Chapter 1.

[34][Haag 1921, p. 3]. Note that there is a mistake in this function. To reproduce the numerical data in published tables, the coefficient of the arctangent, and this one alone, should be multiplied roughly by a factor of 3500. Both expressions of the function above moreover differ by a factor of 1000.

2.2. Siacci's Direct-Fire Approximation. Shunning simple power laws for the air resistance often meant that resulting differential equations became intractable by formal methods.[35] Two types of approximation methods were used by the Gâvre Commission: (1) step-by-step integration methods (in French, *calcul des trajectoires par arc*) and (2) approximations derived using the "direct-fire" assumption (in French, "*tir de plein fouet*"). We review how they were successively adopted by the Gâvre Commission for the computation of firing tables.

Step-by-step integration methods, which French ballisticians called the "arc method," dated back to Leonhard Euler (1753).[36] In this approach, the equation was solved on a small interval only assuming that the air resistance was quadratic over the interval. Interest in these methods was given a boost when the second edition of Hélie's treatise was published in 1884.[37] After the Franco-Prussian War and the death of Hélie in 1887, the "struggle engaged between traditionalism and progress was vigorously undertaken" [Charbonnier 1906, p. 416] and ballisticians at Gâvre adopted a more theoretical approach. The second edition of the treatise received significant contributions from the polytechnician mathematical physicist Pierre-Henry Hugoniot (1851–1887), whose inclination for mathematical theories was much greater than Hélie's.[38] While the work of Bashforth and George Greenhill in Britain, of Siacci in Italy, and of Carl Cranz (1858–1945) in Germany was renewing the field of ballistics, Hélie and Hugoniot showed that old step-by-step methods could be as accurate as desired, provided the intervals chosen were numerous and small enough [Hélie & Hugoniot 1884, vol. 2, p. 289].

After Hugoniot's untimely death in 1887, Hubert Gossot, also a graduate of the École polytechnique (1874) who had joined the Naval Artillery Corps, took over the Gâvre Commission.[39] In 1887 and 1888 he used the step-by-step method to compute firing tables for new 34 cm and 90 mm caliber cannons. At long last, the Gâvre Commission had mastered a method for computing firing tables that was safe, accurate, and dependable. But it was also time consuming and, many felt, inelegant: it was "a computing process, not a theory" [Patard 1930, p. 243].

> The Commission thus consented to submit itself deliberately and for many long years to the boredom of very long, very fastidious and very inelegant computations using the step-by-step method

[35]During the war, several mathematicians studied conditions to be imposed on $F(v)$ in order for the hodograph to remain integrable. This is point 2 in the report on ballistics drafted after the war [Hadamard 1920], discussed in David Aubin, Hélène Gispert, and Catherine Goldstein's contribution to this volume. Jules Drach suggested a method using Galois theory in 1914. Denjoy wrote a report about this method while he was at the Gâvre Commission in July 1917. For more on this, see [Drach 1914, Drach 1920]. See also the note written by Paul Montel in [Charbonnier 1927, vol. 1, pp. 498–515]. Several notes and letters concerning Denjoy's role are to be found in [Academy of Sciences, "Ballistics" file of the "*Commission nommée pour l'étude des questions la défense nationale*"].

[36]On the early history of exterior ballistics, see [Charbonnier 1929], [Hall 1952], and [Steele 1994].

[37]Note that while the second edition of the treatise was published by two authors [Hélie & Hugoniot 1884], the first one had only one [Hélie 1865].

[38]On the life and work of Hugoniot, see [Chéret 1990]. A clear offshoot of his ballistic studies, Hugoniot's seminal paper on shock compression was published in two parts in the *Journal de l'École polytechnique* in 1887 and 1889.

[39]On Gossot, see [Challéat 1933–1935, vol. 2, p. 314] and [SHD–Terre, 11–Yd–47]. Gossot was later central director of Naval Artillery (1905–1909) and *Inspecteur des études et expériences techniques de l'artillerie* from July 1915 to 1917.

> ... because it understood what were the consequences of its liberation from empirical methods and of the return of ballistics back to its natural source: possibility a priori of computing any trajectory from any cannon; extreme reduction of ballistic shots needed to establish a firing table; ... exact determination of unobserved elements of the trajectory ([Charbonnier 1906, p. 423]; quoted in [Patard 1930, p. 243]).

A second group of methods was concurrently developed for approximately solving the hodograph. The most successful was introduced in 1880 by Siacci, whose treatise was quickly translated into French. The principle of the method was to replace the velocity v by a pseudo-velocity u defined as [Siacci 1892, p. 47]

$$v \cos \tau = u \cos \alpha,$$

where α, called the angle of projection (or line of departure), corresponded to the actual initial shooting angle.[40] The pseudo-velocity was equal to the velocity only at the origin and in the descending branch of the trajectory when $\tau = -\alpha$, but in the case of direct fire when the angle τ was close to zero over the whole trajectory, the difference between velocity and pseudo-velocity also remained small. Writing the hodograph equation in terms of the pseudo-velocity u, Siacci found a general equation for the trajectory [Siacci 1892, p. 49]:

$$y = x \tan \alpha - \frac{gx^2}{2v_0^2 \cos^2 \alpha} G\left(\frac{x}{c}\right), \tag{2.2}$$

where G was a function that depended on the air resistance law $F(u)$ that was adopted, but that could be computed, provided four secondary functions $J(u)$, $S(u)$, $D(u)$, and $A(u)$ defined as follows were computed and tabulated for every value of u:

$$J(u) = -g \int \frac{du}{uF(u)}, \qquad D(u) = -\int \frac{udu}{F(u)},$$
$$S(u) = -\int \frac{du}{F(u)}, \qquad A(u) = -\int \frac{uJ(u)du}{F(u)}.$$

Using the tabulated values of these functions, most elements of a trajectory could be computed rather quickly, typically in less than twenty to thirty simple operations. For example, given the value of the pseudo-velocity u at a point on the trajectory and the initial velocity v_0, the x-coordinate of the projectile was simply given by the formula $x = C[D(v_0) - D(u)]$, for a certain constant C. Siacci also relied on the assumption that a certain variable β was constant and equal to 1. Under this assumption, the results computed using Siacci's method remained relatively reliable in the case of direct fire, that is, provided $\alpha < 20°$ [Charbonnier 1907].

To compute deviations due to secondary problems such as the rotation of the projectile or wind, a few other secondary functions were required, all of which were computed and tabulated. Written by the ballistician Anne, Gâvre reports from December 23, 1912, and July 28, 1913, contained the required tables. Having joined the Gâvre Commission in 1907, Anne died in October 1921, "worn out by the exhausting labor he was submitted to during the war" [Patard 1930, p. 293]. This extensive computing effort was completed and printed in January 1916 [Gâvre 1916]. By that time, however, war had made them obsolete!

[40]Note that Siacci used θ instead of τ and φ instead of α.

FIGURE 5. The celebrated 75 mm cannon of the French field artillery, here adapted for anti-aircraft shooting. Note the high angle of shooting. From La Nature **44–2** (1916), p. 135. Courtesy of the MPIWG, Berlin.

3. The Mathematical War Viewed from the Front

"One of the surprises of the present war," the ballistician Emmanuel Vallier (1849–1921) from the Academy of Sciences wrote in May 1915, "certainly is the great development of indirect fire" [Vallier 1915, p. 297]. Few indeed had foreseen the tremendous change of fortune that artillery—and as a result ballistics—would undergo during WWI. To understand the evolution of problems and solutions considered worthwhile at Gâvre, one needs to take into account fighting men's reactions to what the commission had to offer. While it struck the "everyman at war" that mathematics played a crucial part in his predicament, the way this was translated into practice is hardly straightforward.[41] Indeed, the very usefulness of the mathematical apparatus for ballistics was drastically questioned. Due to the rapid evolution in the tactical and strategic use of artillery, commanding officers immediately found that mathematical support for directing fire was imprecise, confusing, or simply lacking. At the hostilities' outbreak, there even seemed to have been a widespread sentiment among artillerymen that, as a practical science, mathematical ballistics had failed them and a "conflict arose between artillerymen and their [firing table] suppliers" [Boissonnet 1920, p. 36].

[41]This alludes to British Private Edgar Norman Gladden's feeling that this was a "war of guns and mathematics" [Gladden 1930, p. 121]. See the introduction to this volume by David Aubin and Catherine Goldstein.

At a strategic level, it was quickly realized that artillery was to play a major role in this war. Less than a week after the declaration of war, Captain Lombal observed that the standard 75 mm cannon adopted by the French Army in 1897 (Figure 5) was unexpectedly deadly. With just 16 shots fired on August 7, 1914, he estimated he had taken down 600 to 700 German cavalrymen. Some computed that this amounted to one dead per kilo of explosive—and marveled at such a high return![42]

Among the first "lessons" drawn from the emergence of scientific warfare was the conviction that heavy artillery now played a much bigger role than expected (see, e.g., [Bos 1923] and [Rouquerol 1920]). Following General Hippolyte Langlois's doctrine, the French Army had hitherto emphasized the auxiliary role of artillery with respect to infantry in open warfare. A regulation of 1913 that is often quoted stated: "artillery does not prepare the attacks, it supports them."[43] Despite the success of "the little Frenchman" as the light rapid-firing 75 mm cannon was sometimes called [Sainean 1916, p. 145], in 1914 the French troops' morale greatly suffered from the German domination in terms of heavy artillery. Shell shock was made worse by the evidence that the long range of German guns kept them out of reach of their French counterparts, which were overwhelmed, too few in numbers, and generally outdated. The President of the Republic, Raymond Poincaré, lamented the lack of cannons and ammunition. Marshall Ferdinand Foch himself underscored the inefficiency of the 75 mm cannons for indirect fire. In this domain as well as in many others, unpreparedness seemed appalling not only to military officers but also to some astute civilian observers.[44] What made this realization even more painful was the fact that it had been discussed at length in artillery circles just before the war.[45] In haste, naval artillery and siege cannons were brought to the frontline and positioned on terrain they had not been designed to occupy. As is well known, new material was ordered en masse. In May, the socialist member of Parliament Albert Thomas was put in charge of armament production at the ministerial level, and the former director of artillery General Louis Baquet was sent back to the front: in his eyes, a politician was "better armed than a general to sustain political assaults."[46] In the summer of 1915, the French Army started to receive new pieces of large caliber.

While the strategic importance of artillery, and especially heavy artillery, increased significantly, the tactical use of cannons underwent important changes as

[42] [Gascouin 1920, pp. 78–81]. On the 75 mm cannon, literature is abundant; see especially an early praise [Houllevigue 1914] and more informed studies in [Challéat 1933–1935, vol. 2, pp. 338–364] and [Rouquerol 1919, pp. 58–77].

[43] "*L'artillerie ne prépare pas les attaques, elle les appuie*" (quoted in [Gascouin 1920, p. 56]). For the French combat doctrine before WWI, see [Percin 1914, p. 165] and [Foch 1903, p. 314]. This doctrine was by and large based on [Langlois 1892] and [Langlois 1906]. On Langlois, see [P.N. 1907], [Glück 1919], and [Ripperger 1995]. For contrary opinions expressed just before the outbreak of WWI, see [Herr 1913] and [Rouquerol 1914].

[44] For example [Lebon 1915, p. 220] and [Reinach 1916, pp. 253–260]. See also the testimonies of the commanders, such as [Poincaré 1928, vol. 5, pp. 333–334] or [Foch 1931, vol. 1, p. 19]. On the insufficiencies of the 75 mm cannon, see [Percin 1914, p. 264].

[45] See [Ripperger 1995], as well as [Bédier 1919, pp. 147–157].

[46] [Baquet 1921, p. 15]. For an overview of the evolution of the materials in the artillery during the war, see [D'Aubigny 1921].

well. Anti-aircraft and antizeppelin firing, mountain engagements, and the general need for shooting from entrenched positions behind a protective monticule—which was called indirect fire—required new procedures. In August 1915, General Frédéric-Georges Herr was put in charge of the defense of Verdun and suffered the full blow of the German offensive in February 1916. In 1912, Herr had observed the Balkan wars and predicted many of the features of the war he was now fighting [Herr 1913]. By 1915 it had become clear to him that the main problem facing artillery was no longer a simple question of supply and materials: "a gigantic intellectual effort was required." In particular, shooting had become the focus of attention. "It had become necessary to [be able to] shoot at invisible targets, during the night, and, in all circumstances, shots had to be of the utmost accuracy ... so as not to hit friendly troops" [Herr 1923, p. 39]. Now, to shoot accurately had become a complex technoscientific problem, which involved the identification of targets using sound-ranging or aerial reconnaissance, their localization on large-scale maps that needed to be produced in large quantities, the development of reliable telephonic communication between observers and gunners, better knowledge of meteorological data at various altitudes (wind, pressure, and temperature), and precise firing tables. "Meteorology, acoustics, optics, cartography, what branch of science was not drafted into artillery's service?" [Bédier 1919, p. 180]. As a testimony of the importance of artillery in the war, it is worth noting that, in all these areas, significant technological and scientific advances were made from 1914 to 1918.[47]

The new guns required new firing tables. The new uses demanded that existing tables be extended to higher shooting angles. It had made sense to compute direct fire tables for guns to be used at sea, just as there was no harm in relying on time-consuming procedures to adjust the fire of siege cannons. This was no longer the case. Information was needed about the whole trajectory of projectiles, and not just their range. Hence the change in terminology, whereby what used to be called "range tables" was now known as "firing tables." But ballistic military structures were totally unprepared to supply them. When war broke out in August 1914, Gâvre was no exception to this general trend. All but five officers deserted the proving ground and joined fighting units. For lack of personnel, most technical activities were abruptly interrupted, and the five military engineers remained desperately idled at Gâvre while their repeated requests to be reassigned to the front were denied one after the other.[48]

To make matters worse, while the navy had paid small attention to ballistics, the army had paid almost no attention to it. Before 1914, low regard, and indeed disdain, for ballistics was entrenched among artillerymen.[49] Despite a rich tradition of ballistic studies, in the first decades of the century the Officer Training School for Artillery and Engineering in Fontainebleau (the *École d'application d'artillerie et du génie* formerly located in Metz, mentioned earlier) allowed time for barely four lessons in the officers' training curriculum devoted to ballistics and the principles

[47]For more on observation, see [Morgan 1959–1960]. On map-making, see [Laves 1919], [Winterbotham 1919], and [Heffernan 1996]. On sound-ranging, see footnote 7 above. On meteorology, see [Launay 1919, Launay 1922].

[48][Patard 1930, p. 270, footnote] and [Crémieux 1930, p. 158].

[49]See, e.g., [Baquet 1921, p. 28, footnote] and [Anonymous 1921].

of firing tables.[50] A conscript once complained that during instruction, "ballistics [was] nothing other than a soporific lesson in terminology" [Malloué 1911, p. 136]. As a result, the philosopher Alain [Émile Chartier] who served in the artillery during WWI was hardly impressed by the mathematical skills of the polytechnicians he met on the front: "Our artillerymen seemed poor mathematicians to me."[51] More shockingly perhaps, among the staggering number of casualties there were not only young mathematicians who might later have turned out to be precious resources for ballistic work, but also fully trained ballisticians whose expertise was acutely missed at Gâvre. Born in 1871, Commandant Henri Batailler had been in charge of the ballistics course at the Fontainebleau school, and he had already published several articles on the topic in the *Revue d'artillerie*. But on June 9, 1915, Commandant Batailler was killed on the Marne front.[52]

"Up until 1916, the artilleryman and the ballistician lived on knowledge acquired in peace time" [Boissonnet 1920, p. 41]. In fact, artillerymen in fighting units more often than not were forced to resort to their own means for improving the accuracy of their fire. On September 9, 1914, Jules-Émile Henches wrote from the Marne front that "each day" he became more and more convinced that in the fighting "science is necessary, but its application must be [made] the simplest possible" [Henches 1918, p. 8].[53] "War," wrote the physicist and astronomer Charles Nordmann, who actively served both in the artillery and on the Commission of Inventions, "had ceased to be an art to become an experimental science like physics" [Nordmann 1917, p. 9].[54] The front itself was likened to a gigantic experiment in ballistics that should be exploited to increase shooting accuracy: "Because of the rich harvest in experimental shooting that everyone is able daily to reap on the front, firing tables produced before the war were found to hold insufficient, and even erroneous, information" [Herr 1923, p. 40].

In fact, badly equipped experimental polygons sometimes lagged behind the front, notably in the study of atmospheric densities [Boissonnet 1920]. Using accurate ballistic coefficients (provided by ballisticians) and atmospheric studies made by the meteorologists enrolled by the army, fighting artillerymen were able to find inaccuracies in the Gâvre air resistance function, which had until then remained within the margin of error. Frontline expertise was developed, with which ballisticians from the rearlines were hard pressed to compete. Reports had been written by officers on the front questioning either the accuracy of the air resistance function $F(v)$ or the firing tables themselves. Some officers were worried "not only about the firing tables at their disposal, but also about those that could still be delivered to them, if they are computed on the basis of the present inaccuracies."[55] A specialist

[50][Challéat 1933–1935, vol. 2, p. 64, footnote 1; see also p. 259]; Challéat's course at Fontainebleau most probably condensed the lessons of his predecessor *chef d'escadron* Pierre-Émile Henry, who was among those who introduced Siacci's methods in France [Henry 1894]. On ballistics at Metz and Fontainebleau, see [Bru 1996].

[51]"*Nos artilleurs m'ont paru assez peu géomètres*" [Alain 1937, p. 115].

[52][Sebert 1915]; see also [Garnier 1920, p. 8].

[53]Also interesting in regard to the scientific work of artillerymen on the front are the following testimonies: [Pastre 1918, Lintier 1916, Cardot 1987].

[54]On Nordmann, see David Aubin's contribution in [Goldstein & Aubin forthcoming].

[55]"*Quelques officiers ... manifestent quelque inquiétude, non seulement sur les tables de tir dont ils disposent, mais sur celles qu'on peut encore leur délivrer, si elles sont calculés par les errements actuels*" [Garnier 1918a, p. 1]. This book, which gathers various secret notes written

in anti-aircraft shooting since September 1914, Henches wrote from his command post that he called "Aviatik-City" on June 18, 1915:

> There have been, concerning those airplane shootings, papers overfilled with mathematics done by people who could never see that the instinct of a "hunting animal" is necessary here. They reach, for that matter, by very complicated, inapplicable processes, results that are close to, but not as valuable as, those I have been using for the last six months [Henches 1918, p. 59 & p. 85].[56]

On the front, artillery material was put to much more extensive testing than it ever was on proving grounds, and, in the process, firing tables computed in the rear-lines appeared defective. In a letter from General Curières de Castelnau written weeks before he was replaced by Philippe Pétain as the head of the Second Army, it was stated that regulatory tables for the 75 mm cannon differed from nomograms (or abacs) given to motorized artillery sections. "These discordances not only concern trajectories corresponding to high angles but even trajectories that are regularly used" on the field [SHD–Terre, 2W309, Général commandant la IIe armée à M. le colonel Leleu, chef de la section technique de l'artillerie, place Saint-Thomas d'Aquin, May 19, 1915]. As a result, commanding officers had their own tables of corrections, or networks of trajectories computed [Boissonnet 1920, p. 39]. Others designed special slide rulers or mechanical aiming devices [Garnier 1922, p. 111]. Some artillery units lost confidence in the tables computed by theoretical means and corrected them on their own by experimental means. "Thus instead of having a single range table carefully established in the rearlines, there were many built with the help of a very large number of shots carried out in lousy experimental conditions."[57]

The electrical engineer Hippolyte Parodi, who would come to play an important role in computing new ballistic tables for the French Army using graphical methods, underscored that he had first become aware of the initial insufficiency of firing tables while he was fighting on the front:

> When I was called to head the *Service de balistique et de préparation des tables de tir*, I had long been aware, according to the shots I had taken or controlled on the front, that the near totality of firing tables in use in the army were clearly false and that they had been established ... through "*archaic*," *inexact* and *simplified* methods.[58]

---

during the war by Garnier, can be found in various slightly different versions. I quote from one of the differing copies kept at the Bibliothèque nationale de France, in Paris, call number 8-V-42188.

[56] On the lack of practical knowledge of heavy artillery officers, see also [Henches 1918, pp. 149–150].

[57] "*On l'a bien vu au cours de la guerre : il est arrivé souvent, en effet, qu'une batterie, avertie par des insuccès de l'inexactitude de sa table de tir établie par des procédés théoriques, l'a rectifiée expérimentalement sur le front. Ainsi, au lieu d'une seule table de tir bien établie à l'arrière, il y en avait plusieurs construites a l'aide de très nombreux tirs, faits dans de mauvaises conditions expérimentales et présentant souvent de gros inconvénients militaires.*" [Academy of Sciences, Lebesgue to Hadamard, October 23, 1919; Commission de la défense nationale, dossier "Balistique"].

[58] "*Quand j'ai été appelé à la direction du Service de balistique et de préparation des tables de tir, il y avait longtemps que je savais, d'après les tirs que j'avais dirigés ou contrôlés au front, que la presque totalité des tables de tir en usage aux Armées étaient nettement*

The tables, Parodi went on, were not only wrong; they were also inconsistent and self-contradictory. Some projectile might, for example, have, for the same line of departure, a greater range for a smaller initial velocity.

> These errors and these incoherencies, which thankfully were uncovered only by a handful of artillery officers, were susceptible of arousing doubts in the minds of the combatants and to withdraw all confidence in the technical documents that were distributed to them.... Yet, who knows whether by making some battery commanders excessively prudent this distrust would not have allowed, at the beginning of the war, for the saving of precious human lives?[59]

Parodi's hopes notwithstanding, officers did notice deficiencies in firing tables. "Artillerymen demanded experimental firing tables, established by cannon fire [*à coups de cannon*]," not by theory which they distrusted [Boissonnet 1920, p. 40]. In the face of such criticism, professional ballisticians repeated that "exterior ballistics was, before the war, brought by the Gâvre Commission to a degree of perfection which fully satisfied all practical requirements" [Garnier 1918b, p. i].[60] This meant that there existed accurate firing tables corresponding to initial speed up to 850 meters per second and initial angle up to 20°. But fighting officers sometimes not only questioned the accuracy of the firing tables they had at their disposal but also the very possibility of computing them with enough precision [Garnier 1918a]. At the *Section technique de l'artillerie*, Parodi concluded that many cannon shots were necessary to establish a firing table: "One should not forget that firing tables are only worth what the experiments used as a basis for computations are worth, and that the mathematical apparatus [*appareil mathématique*] in which they have been enclosed is incapable in itself of increasing their precision. Many cannon shots must be fired [*Il faut consentir à tirer beaucoup de coups de cannon*]" [Academy of Sciences, Parodi to Hadamard, October 30, 1919]. Even the mathematician, Henri Lebesgue (1875–1941), who had collaborated with Parodi in the Mathematics Section placed under the Undersecretary of Inventions concurred: "What is this computing sickness, when experiments are (apparently), and in all cases can be, carried out. There are certainly places in France where real shooting by 75 mm cannons is done ... and there is the front."[61]

---

fausses et qu'elles avaient été établies ... par des méthodes 'archaïques,' inexactes ou simplifiées" [Academy of Sciences, Parodi to Hadamard, October 30, 1919; Commission de la défense nationale, dossier "Balistique"; emphasized by Parodi]. On Parodi, see [Darrieus 1968] and [Merger 1994].

[59] "*Ces erreurs et ces incohérences qui, heureusement n'ont été décelées que par un très petit nombre d'officiers d'artillerie, étaient de nature à porter le doute dans l'esprit du combattant et à lui enlever toute confiance dans les documents techniques qui lui ont été remis... Qui sait pourtant si cette défiance, en rendant systématiquement prudents certains commandants de batteries, n'aurait pas permis, au début de la guerre, d'économiser de précieuses vies humaines ?*" [Academy of Sciences, Parodi to Hadamard, October 30, 1919; Commission de la défense nationale, dossier "Balistique"].

[60] See also [Peira 1955–1957, vol. 1, p. 36].

[61] [Lebesgue 1991, letter CCXIII, p. 319]; this letter is tentatively dated early 1915 by Pierre Dugac. On the Lebesgue-Borel correspondence during WWI, see Hélène Gispert's contribution to [Goldstein & Aubin forthcoming].

The ballisticians at Gâvre therefore felt the need to emphasize that the "mathematical toolbox [*outillage mathématique*]" they used was no "smokescreen [*trompe-l'œil*]" [Garnier 1918a, p. 8]. On the contrary, ballisticians emphasized, mathematics was indispensable for correctly evaluating perturbations depending on the particular circumstances of shooting. As the war unfolded, hopes indeed increased as to the possibility of firing without preparation. Was it possible to open fire on a target, having allowed for all modifications due to the special circumstances by means of computations, without any prior warning? This question had a solution, and all belligerents were looking for it from the moment the war had started.

4. Scientific Work at Gâvre

How the requirements of fighting artillerymen impacted the Gâvre Commission and the specific role played by mathematicians over there remain to be examined. As was hinted at above, Gâvre was not the only institution involved in the French ballistic effort of 1915–1918. Under Parodi's energetic leadership, the Calais Commission played a major part in the computing effort together with the Board of Inventions (the *Direction des inventions intéressant la défense nationale*, in which the Paris mathematicians Lebesgue and Montel were also involved). To fulfill his responsibilities, Parodi suggested that high school teachers and university professors be mobilized for the task. More than 400 answered positively and about 300 effectively worked on the project. The team computed trajectories for every initial angle α a multiple of 5° from 0° to 90°, for initial velocities varying from 0 to 1,000 meters per second, and for various values of ballistic coefficients.[62] Intense computational work was moreover carried out by the so-called "Commission ALVF" (*artillerie lourde sur voie ferrée*, that is, heavy artillery on rail), the committee in charge of organizing railway artillery headed by Lieutenant-Colonel Girardville, which produced extensive tables giving range as a function of initial angle, initial velocity, and ballistic coefficient. Contrary to all previous practices, these tables, called the "(α, V_0, c) tables" (here with a capital "V") were abstract constructs corresponding to no specific cannon or projectile.[63] At Gâvre, the naval engineer Georges Sugot also introduced a new method for speeding up computations, called the fictitious speed method (*méthode des vitesses fictives*) [Sugot 1918].

As was pointed out, all this work, whose immediate usefulness was questioned, was only as good as the physical assumptions on which it was based. Knowledge of air resistance laws and the atmospheric density variation with respect to altitude might be improved with unknown effects on the computed tables. From both a

[62]Details above come from [Ottenheimer 1924, pp. 51–52]. One should however take these numbers with care. According to a secret report from December 1916, the computing board put together by Lebesgue and Montel had only recruited between 30 and 40 voluntary collaborators including actuaries and astronomers [Fonds Painlevé, 313/AP/62]. Numbers cited by Ottenheimer are indeed high considering the available pool of mathematics professors and teachers at the time. Before the war, there were only 65 university professors in mathematics in the whole of France. To approximate the number of high school teachers, one may consider that the professional union (*Association des professeurs de l'enseignement secondaire public*) counted 501 members in 1913, of which 102 were in Paris [Barbazo & Pombourcq 2010].

[63]Established in 1916 by the War Ministry, the Commission ALVF had computed several general tables for angles of 22°, 28°, 33°, and 44°, with initial velocities between 300 to 900 meters per second and various ballistic coefficients. In 1919, the tables used the Gâvre air resistance law and a better estimate of the variation of atmospheric density with altitude [ALVF 1921].

mathematical and a practical point of view, it was the step-by-step method developed collaboratively by the mathematician Haag and the military ballistician Maurice Garnier that was the most innovative as well as the most lasting effort in ballistics during WWI.

On June 16, 1917, Haag was assigned to the testing center of Vitry-le-François where he instructed training officers. On November 29, he was promoted to the rank of lieutenant. In his commendation, Charbonnier wrote:

> Ever since he arrived at Gâvre, M. Haag has studied the improvement and practical application of the new ballistic methods required by present shooting conditions, and especially by the problems of aerial shooting. ... Monsieur Haag significantly contributed to this work and, among the important questions to whose solution he contributed I must name the following:
> a) the improvement of step-by-step computing processes;
> b) the invention and development of a computation method for differential coefficients;
> c) new and original applications of probability theory to the determination of the detonation ellipses of projectiles with fuses.[64]

Charbonnier's commendation clearly distinguished between principal and secondary problems in ballistics, as well as problems linked with probability theory. After the war, Haag summarized some of his work on probability theory in the fourth volume of Borel's *Treatise on Probability* [Haag 1926]. As we shall see below, Haag's contribution to secondary ballistics problems played a crucial part in increasing the accuracy of scientific shooting procedures. But from a mathematical point of view, it was his contribution to the principal problem of ballistics—to solve the hodograph—that is the most significant. In the following, we shall examine the mathematicians' work at Gâvre from two perspectives: (1) we will look in more detail at the specific ways in which Haag was able to work on and improve the step-by-step method that, as we have seen, can be dated back to Euler; and (2) we will review collaborative experimental and computing procedures followed at Gâvre and the role played by mathematicians in them.

4.1. The Theory of Errors. Already in the note he sent to the CRAS in 1915, Haag made an interesting innovation in exterior ballistics. Since, in practice, initial conditions were never known exactly, every gunner was perfectly aware of the need for determining the effects of small variations in the main three parameters: initial velocity, line of departure, and ballistic coefficient. This was usually done by simple interpolation in the tables. Alternatively, Siacci had analyzed, in the case

[64] Jules Haag's military file [SHD–Terre, 6–Ye–17966]: "*Depuis son arrivée à Gâvre, M. Haag a été employé à l'étude du perfectionnement et à la mise en application pratique des méthodes balistiques nouvelles qu'exigeaient les conditions de tir actuelles et en particulier les problèmes de tirs aériens, qui ont pris une importance si grande, aussi bien pour le Département de la Guerre que celui de la Marine.*

Monsieur Haag a contribué dans une large mesure à ces travaux et, parmi les questions importantes à la solution desquelles il a collaboré, je dois citer les suivantes :

a) *Amélioration des procédés de calculs des trajectoires par arcs successifs.*

b) *Élaboration et mise au point d'une méthode de calcul des coefficients différentiels.*

c) *Applications nouvelles et originales du calcul des probabilités à la détermination des ellipses d'éclatement des projectiles fusants.*"

FIGURE 6. Preparing for aerial shooting at the training center for practical anti-aircraft shooting in Arnouville, 1917. From left to right in the foreground, Jules Haag (taking a photo), Garnier, Pagézy (the head of the training center), Rouch (head of the Army Meteorological Service), and an unidentified officer from the training center. From [Patard 1930, p. 288].

of direct fire, the effect of substituting $v_0 + \Delta v_0$, $\alpha + \delta\alpha$, and $c + \Delta c$ in the equation of the trajectory (equation (2.2)). No method existed, however, for providing such estimates in the Eulerian step-by-step method to which one was forced to resort in the case of general trajectories, in particular those involved in airplane bombings with which Haag was at first concerned [Haag 1915a]. With his formal training in mathematical analysis, Haag was easily able to evaluate the size of the error without having to integrate the equation.[65] In a formal analogy with celestial mechanics, he simply tracked down, at any given order of approximation, the errors thus produced. This is the work that had caught Charbonnier's eye.

When he reached Gâvre at the end of 1915, Haag undertook the systematic study of errors in ballistic theory. The first to work on this problem, he published his results after the war [Haag 1921]. In this account, Eulerian integration methods produced two kinds of error. The *ballistic* error came from the assumption that air resistance was quadratic over the small integration step. The *geometric* error was due to approximations in the method of integration. Assuming that the arc

[65] Note that Haag's work in ballistics is *not* directly related to his previous research in differential geometry. Supervised by Gaston Darboux, his doctoral thesis dealt with Lamé surfaces [Haag 1910].

FIGURE 7. Osée Marcus in the 1920s. Courtesy of Simone Marcus.

was infinitely small, Haag therefore produced a complete analysis of both types of error. Hopefully, this method would provide an estimate of the maximal arc lengths that gave the precision needed with minimal computation time, allowing the considerable acceleration of the laborious process of computation. At Gâvre, piecewise integration procedures used the angle τ of the projectile velocity with respect to the horizontal as the independent variable, whereas American and British ballisticians instead chose time. Following an ad hoc rule, arcs were selected so that the angle decreased by less than 5° and the velocity loss did not exceed 50 meters per second over the length of the arc. Using his method, Haag could show that this rule was no guarantee against imprecision or inefficiency: "parts of the trajectory are computed with a precision that is much too high, while others are insufficiently precise" [Haag 1921, p. 21]. A new, more complicated rule was derived for determining the arcs for which the relative error remained smaller than 1/500.

Together with the naval engineer Maurice Garnier, Haag applied this rule and designed a new computing procedure. Computing skeletons can be found in Haag's article [Haag 1921, pp. 18–19, 27, & 30]. Convergence radii of the expansion series were studied carefully to achieve the given precision. This at last provided a basis for making the "rational choice of the amplitude" of integration steps [Garnier 1922, p. 126]. This computing method was called the GH method (for Garnier-Haag) and was presented in a special report in January 1917.

But this scientific method was not as efficient as one might have wanted. Later that year, the mathematics teacher (*agrégé de mathématiques*) Osée Marcus (Figure 7), who was employed as a computer, pointed out that computing procedures could be quickened by dispensing with the consideration of radii of convergence and by relying instead on simple Taylor approximations.[66] Incorporating other minor

[66] A Jewish immigrant from the Baltics, Marcus appears not to have been drafted in a fighting unit. For the rest of his life, he remained on the French Navy's payroll, carrying out computations for them in his home in Neuilly-sur-Seine. I wish to thank Marcus's niece Simone Marcus for this information.

points made by Captain Lévy (from the Mining Corps)[67] and from Georges Valiron,[68] the GHM (Garnier-Haag-Marcus) method was adopted in September 1917 and remained in use until after World War II in France.[69]

Discussions about the approximation introduced by Marcus highlight a divergence of viewpoints between theory and practice:

> In the case of practical computations involving numerical integration, it is more often harmful than useful, more complicated than it is advantageous, to perform the rigorous quadratures provided by analysis. Thus are we sometimes led to reject certain analytical formulae—no matter how elegant and seductive they may seem to a mathematician's eyes—to perform quadratures by simple approximation methods, apparently coarser, but in truth quicker and more precise [Garnier 1918b, p. vi].

As one can see, theses debates lay at the root of the later development of applied mathematics and numerical analysis.

4.2. Practical Work and Collaborative Procedures. At Gâvre, Haag's work took place in a special unit devoted to anti-aircraft gunnery. As discussed above, anti-aircraft gunnery posed a difficult challenge to older ballistic methods. It was the engineer Gustave Lyon (Figure 8, on the right) who pushed for its creation. A polytechnician (X 1877) working as an acoustical engineer, Lyon had volunteered in 1914 and was soon in charge of the protection of the port of Cherbourg against airborne assaults. The *Mission du tir aérien de Gâvre* (MTAG) was set up under Lyon and Garnier's authority in April 1916. Soon, this "scientifico-technical organization" [Charbonnier 1928b, p. 580] counted a dozen mathematicians and physicists working under military guidance (Table 1). Charbonnier described the procedure followed by the MTAG in a memo written in December 1917. This led to its formal establishment by the Navy Ministry, on January 30, 1918, under the name of *Mission balistique du tir aérien* (MBTA), with a section based in Paris for theoretical investigations and another in Lorient in charge of experimental studies.[70] It is interesting to examine the procedure outlined by Charbonnier to get a precise understanding of the type of work mathematicians were doing at Gâvre.

First, for all types of guns and projectiles, the ballistic coefficient was evaluated with one or two shots. A network of 19 trajectories was completely computed using the labor-intensive step-by-step method and was computed twice to insure that no mistake was made. The second step of the procedure was new with respect to

[67]This is probably an allusion to the probabilist Paul Lévy, who was in the Mining Corps and an anti-aircraft officer during WWI. Although he served as an instructor at Arnouville-lès-Gonesses (Figure 6) in 1916, Lévy was mobilized in a fighting unit for most of the war and did not mention his work at Gâvre in his autobiography [Lévy 1970, pp. 54–55]. He is, however, mentioned in [Charbonnier 1928b, p. 580].

[68]Later a professor of analysis at the Sorbonne, Valiron was then a mathematics teacher at the *lycée* of Lyons (in the *classes préparatoires*). On the French system of higher education, see the contribution by Jean-Luc Chabert and Christian Gilain to this volume.

[69]In the GH method, the final velocity of the arc was first computed by approximating the arc as a parabola. The variations of the elements of motion Ds and $D\sigma$ were first computed by exact quadratures. Dx and Dy were approximately computed with Ds and Dt with $D\sigma$. In the GHM method, variations were computed directly using Taylor series and averages [Garnier 1918b].

[70]Charbonnier, Rapport No. 7, Note adressée à l'Ingénieur général, directeur central de l'Artillerie navale, December 17, 1917 [SHD–Terre, 10W73].

FIGURE 8. From left to right, Maurice Garnier, Louis Fort, and Gustave Lyon with measuring instruments, undated. From [Patard 1930, p. 275].

previous customs. Extensive series of experiments were performed to measure the trajectory and duration of flight using a simplified theodolite designed by Lyon and a chronophotographic instrument. This led to an immense amount of work. For 60 to 80 shots, 1,000 to 1,200 observations were taken that had to be analyzed. Atmospheric conditions were measured using weather balloons. Using all this material, the third step in the procedure consisted of preparing a corrected table using the computing method developed by Haag and Garnier. For each series of 9 shots, the piecewise integration had to be performed on about 10 arcs. Finally, the MTAG produced networks of curves with shell trajectories and isochrone lines. "Refraining from imposing a method or an apparatus, the MTAG strictly limited itself to its ballistic role: to give combatants (army and navy) the networks they wanted."[71]

[71]Charbonnier, Rapport No. 7, p. 6 [SHD–Terre, 10W73].

TABLE 1. Mobilized scientific personnel at the Gâvre Commission during WWI. From [Patard 1930, pp. 277–278]. In addition, Ernest Esclangon, from the Bordeaux Observatory, and Gabriel Foëx, from Zürich, were present as nonmobilized scientific personnel.

Name	Rank and title	Functions
Châtelet	2nd-class officer in health services; assistant professor at University of Lille; Doctor of Sciences	First assigned to the MTAG until Sept. 1917, then assistant to Naval Engineer Anne: helps with the preparation and execution of shootings; drafts reports and computations
Denjoy	Soldier in infantry; Doctor of Sciences; professor at University of Montpellier	General service
Ferdinand Dreyfus	Soldier in infantry; Bachelor in Science; actuary at Ministry of Work	Assistant to Naval Artillery Engineer Sugot: same functions as Châtelet
Fort	Warrant officer in the infantry; mathematics teacher at naval school and *lycée* of Neuilly	Preparation, execution, and interpretation of aerial shootings
Goullins	Lieutenant in reserve metropolitan artillery; Naval Artillery Engineer	Helps with, then directs, shots; main drafter of reports (flares and incendiary shells) after he becomes member of the Commission in Sept. 1917. Assist. to Sugot on gunpowders
Haag	Second lieutenant in artillery; professor at University of Clermont-Ferrand	Improvement and application of new methods for computing trajectories
Kampé de Fériet	Auxiliary in infantry; Doctor of Sciences; Assistant Astronomer at Paris Observatory	Assistant to Sugot: helps with shooting; drafts reports and computations
Marcus	Ordnance soldier; mathematics teacher	Assistant to Garnier for experiments and computing
Pélissier	Auxiliary in infantry, then second lieutenant in artillery; student at École normale supérieure	Assistant to Anne; same functions as Châtelet
Sauvigny	Temporary second lieutenant in artillery; mathematics teacher at *lycée* of Nancy	Assistant to Anne; same functions as Châtelet
Valiron	Soldier in the infantry; special mathematics teacher at *lycée* of Lyons	Assistant to Garnier for experiments and computing

This led them however to produce extensive networks: for the 75 mm cannon, for example, it contained more than 40 sheets for each type of projectile.

Notes written for a series of experiments undertaken in the spring of 1916 can help to convey an even clearer sense of the mathematicians' activities at Gâvre. In one set, Garnier wrote down very specific instructions: "In view of coordinating efforts in the best possible way and to achieve [our objectives] as fast as possible

... I indicate in the following the detailed division of labor."[72] The workload was divided into field and office work. In the field, Garnier, another officer, and a soldier operated the battery; mathematicians (Haag and Châtelet), officers, workers, and apprentices manned three observation stations, while others tended to the registering instrument (Figure 8). In order to communicate between observation stations, mathematicians were asked to study Morse code.[73]

The office work was also strictly divided among participants. Haag, for example, assisted by an apprentice named Guillaouic, was supposed to compute the fundamental trajectories and the differential coefficients that allowed for correct trajectories for given experimental conditions and, on this basis, draw the networks. Together with a military officer, Louis Fort, a mathematics teacher at the *lycée* of Neuilly, was assigned the task of preparing the shots and analyzing shooting conditions. On the chronophotographic plates, Fort was in charge of measuring the Cartesian coordinates of every explosion of the fuses that were recorded. Assisted by an apprentice, Albert Châtelet, a *maître de conférences* at the University of Lille, was supposed to perform all the operations needed to draw isochrone lines on the networks.[74]

On June 12, 1916, experiments had been performed and computational assignments were ready to be carried out. In a very explicit report, Garnier listed all the operations that everyone had to do in sequence.[75] But two weeks later, orders were modified due to changes in personnel and in priorities. All personnel not otherwise busy were to contribute to the computation of the networks under Haag's supervision. A 16-step procedure explained the work in an even more detailed manner. All apprentices received precise assignments; sample tables were drawn for computers to fill in; the number of copies to be made and the destination of each copy was specified; every computation and drawing to be done was described in detail.[76]

While military work at Gâvre may have seemed enviable compared to the lot of soldiers on the front, it was not without danger. In a talk delivered in Lille in 1924, Châtelet recalled an incident that had happened to him. In one of the shootings, observers behind the cannon saw huge red flares and abundant black smoke coming out of the mouthpiece that was expelling schrapnel for about one minute. The shell had exploded in the barrel and the cannoneers had nowhere to hide: "I can assure you, even if this is not in the written report, that observers ... felt that one minute can be very long."[77]

---

[72]"*En vue de coordonner le mieux possible les efforts, pour aboutir dans les délais les plus rapides, ... j'indique ci-après la répartition détaillée du travail.*" Comptes rendus de la Commission de Gâvre, note no. 6, May 29, 1916 [SHD–Terre, 10W73]. This series of 13 numbered notes from May and June 1916 follow from the establishement of the MTAG in April 1916. All are signed by Garnier and can be found in [SHD–Terre, 10W73].

[73]Comptes rendus de la Commission de Gâvre, note no. 6, May 29, 1916, & note no. 9, June 2, 1916 [SHD–Terre, 10W73].

[74]In fact, Châtelet had been called to Lille on August 5, 1914, and only took up his position there in 1919. See [Condette 2009] and Sébastien Gauthier's contribution to [Goldstein & Aubin forthcoming].

[75]Comptes rendus de la Commission de Gâvre, note no. 12, June 12, 1916 [SHD–Terre, 10W73].

[76]Comptes rendus de la Commission de Gâvre, note no. 16, June 26, 1916 [SHD–Terre, 10W73].

[77]"*je puis vous assurer, quoique ce ne soit pas sans le procès verbal, que les observateurs ... trouvèrent qu'une minute, ça peut être très long.*" The typed text of Châtelet's speech with

From the minute description of the work done at Gâvre, one gets a rich impression of what it meant to use the mathematicians' skills in the First World War. That impression is far from the romantic vision of genius solving a problem that had frustrated ignorant militarymen for ages. On the contrary, ballisticians at Gâvre had enough mathematical sophistication to be able to see the added value mathematicians were susceptible to bringing to their trade. It also seems clear that mathematicians were not "well prepared for their new role" [Patard 1930, p. 279] and that, at first, their experience at Gâvre may have been rather humbling. But in time, all could see how tight collaboration and division of labor were necessary for producing important results such as firing tables. Only then could collaborative procedures drafted by ballisticians make room for mathematicians' special abilities and sometimes assign them to positions of leadership.

4.3. Application to the Battlefield. Firing tables were not the end of the story. Artillerymen needed to be trained in order to make good use of them. In this, mathematicians had, once again, a crucial part to play. Officers testified that a "*carnet de Haag*" [Boissonnet 1920, p. 39] circulated underground, one year before it was printed and distributed officially (see Figure 9). In this course designed for trainee officers, Haag wrote that the exact solution of the ballistic problem was a "chimera." But, he added, "one can, as in all experimental problems, look for an approximate solution. I will try to show you what this solution is and *what degree of confidence we may grant to it*" (emphasis added).[78]

In this course, Haag argued for the usefulness of the mathematical approach to firing, which was presented as a mathematical problem. "Given a target, a cannon, [and] ammunition, it is asked to send one or several shells to the target, or at least to its immediate neighborhood."[79] The problem had two solutions: the first was to fire several shots and observe the effect; the second was to prepare the shooting in such a way that the first shot fell, if not on the target, then at least in its vicinity. If the first method was simpler and applicable without special training, if it did the trick for the daily operations of 75 mm cannons, it was too expensive, too time consuming as far as heavy artillery was concerned.

The various mathematical operations required in the artillery and the different procedures developed on the field were evaluated in terms of efficiency: computing in the command post, the use of double-entry tables for deviations, the manipulation of slide rules, the considerations of graphs, and mechanical instruments. Figure 10 shows a nomogram prepared by the army in order to adjust the fire given two observers at different spots. Seven hundred and fifty copies of this nomogram were produced by November 1917. But were they used? Haag's course clearly shows that the hierarchy seemed intent on increasing the mathematical level of artillerymen serving on the front.

Testimonies from men in the field show that this intention was not illusory. Take the case of the young Jean-Alexandre Cardot. He was merely 16 years old

---

handwritten annotations can be found among his papers [Fonds Châtelet, 81 J 117]. I wish to thank Sébastien Gauthier for making this document available to me.

[78]"*Le problème est ardu. La solution exacte est une chimère. Mais on peut, comme pour tous les problèmes d'ordre expérimental, en chercher une solution approchée. Je vais essayer de vous montrer en quoi consiste cette solution et quel est le degré de confiance qu'on peut lui accorder*" [Haag 1917, p. 3].

[79]"*Étant donné un objectif, un cannon, des munitions, il s'agit d'envoyer un ou plusieurs obus sur l'objectif, ou au moins dans un voisinage immédiat*" [Haag 1917, p. 3].

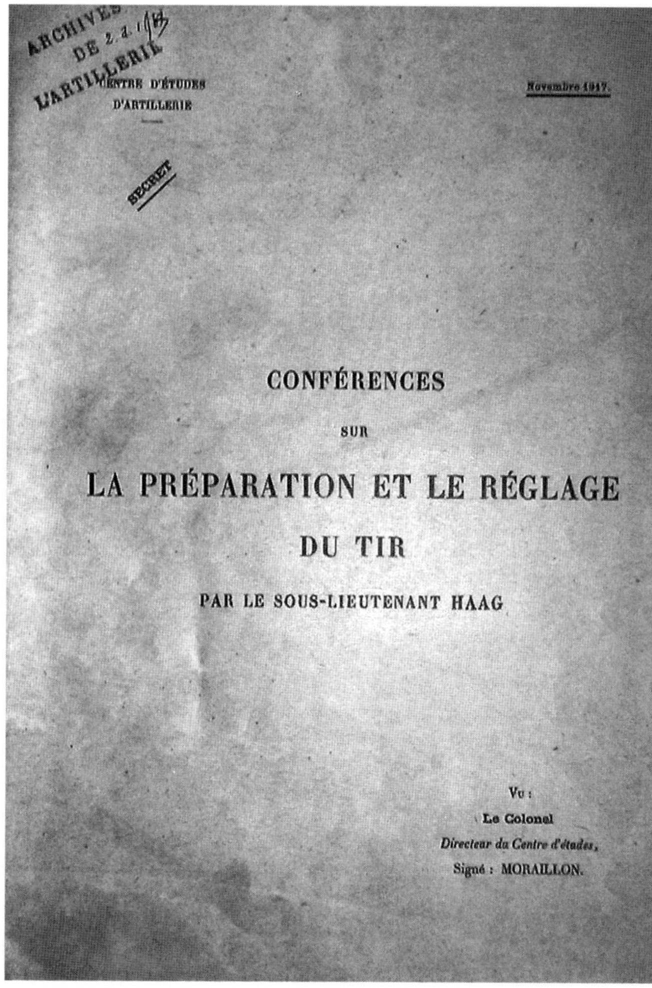
ARCHIVES DE L'ARTILLERIE

CENTRE D'ÉTUDES
D'ARTILLERIE

Novembre 1917.

SECRET

CONFÉRENCES

SUR

LA PRÉPARATION ET LE RÉGLAGE

DU TIR

PAR LE SOUS-LIEUTENANT HAAG

Vu :
Le Colonel
Directeur du Centre d'études,
Signé : MORAILLON.

FIGURE 9. Le *carnet de Haag*: Haag's course to artillery officers in training [Haag 1917]. © Service historique de la Défense Vincennes, SHD–Terre, 2-W-292.

when war broke out and he pursued his mathematical education in the troubled circumstances of war. In 1917, he may have been among Haag's audience at the artillery school at Fontainebleau. He gave a vivid account of his first campaign in a battery of 75 mm cannons in Lorraine in 1918. In Cardot's description, firing had indeed become a scientific exercise. Every target was located on maps by their coordinates. Weather reports transmitted daily the force and direction of the wind. Communications between observers and cannoneers were ensured via the telephone. One day he went out with his lieutenant to test new shells that could reach a target at 11 kilometers. Cardot described firing manouvers in detail. He explained how he was asked to use simple mathematical instruments and computations. The lieutenant told him: "You will direct the fire.... Go. Here's the firing table" [Cardot 1987, p. 65]. Although the lieutenant tried to trick him by handing him

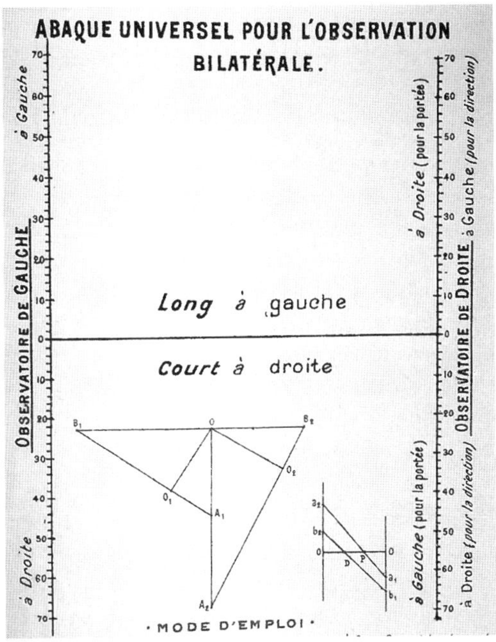

FIGURE 10. Nomogram used to adjust artillery fire developed by Haag during WWI. Repr. from [Haag 1917]. © Service historique de la Défense Vincennes, SHD–Terre, 2-W-292.

the wrong table, Cardot was wiser and the German train in the distance was finally shot down.

At the level of the general staff, scientific firing was now taken for granted. At the end of 1917, according to General Herr, "the time ha[d] come when the French artillery at long last found, if not a complete and definitive [answer], at least one that was precise enough to be applicable henceforth." The solution to the problem of firing by surprise was "the scientific preparation of shooting [*la préparation scientifique du tir*]" [Herr 1923, p. 93]. In October 1917, the French artillery was able to fire at night or in the fog: "a fearful innovation" [Herr 1923, p. 94]. It now possessed:

> a scientific shooting method [that allowed] for shooting under any weather condition, at any time, on every terrain, in all circumstances.... It [was] able to open fire almost instantaneously on any seen or unseen point merely identified by its coordinates on a map [Herr 1923, p. 95].

By the end of the war, the French artillery was finally using "scientific ballistics" [Challéat 1933–1935, vol. 2, p. 314]. Those were "the French methods of firing" that American artillerymen were taught in 1917 [Grotelueschen 2001, p. 20]. Although less praised than tanks and less bedeviled than poison gas, the new firing methods played no small part in the outcome of the war, and the cannon was called the "artisan of Victory" [Rouquerol 1920]. "It was the massive surprise action of our artillery which, from July 18, 1918, onwards, insured the success of our great offensives until Germany's capitulation" [Campana 1923, p. 122]. Ballisticians, it would seem, had successfuly fulfilled their mission. As we have argued, this success owed much to their ability to enroll the effective collaboration of some mathematicians.

5. Mathematicians' Attitudes towards Ballistics

After the war, ballisticians wished to assert the new scientific status of their trade. Charbonnier praised the progress on ballistics achieved by the "alliance, made by the war, of ballisticians and savants" (Charbonnier's preface in [Garnier 1918b, p. viii]). While it used to be the exclusive domain of engineers and technicians, ballistics had, he wanted to say, truly become a first-rank science. "Because of its intrinsic interest, as pure science as well as for national interest, ballistics deserves to become, as it once was, a topic of research for pure scientists."[80] Mathematicians, however, appeared rather more circumspect about the mathematical value of wartime ballistic research. Asked by Jacques Hadamard to report on the ballistic computations in which he had taken part, Lebesgue wrote on October 23, 1919:

> Such a report would perhaps be more pertinent if it were addressed to a Committee awarding the *prix de vertu* [attributed at that time for the most courageous act on the part of a poor Frenchman], since none of those who worked on the construction of systems of ballistic trajectories claims to have done scientific work by performing numerical computations using well-known procedures; all merely tried to be useful.[81]

Lebesgue went on to explain that the team he headed (together with Montel) computed more than a thousand trajectories, each of them independently computed by two persons. This work, Lebesgue concluded, was useful because it allowed the quick construction of range tables at a time when few shooting tests could be carried out due to the overload of work encountered by proving grounds and military research institutions. "But the extent of the effort should not lead one to think that the tool that was constructed is worth so much that it could be used for a long time to come."[82] Recalling the two main uncertainties in ballistic theory (the air resistance law $F(v)$ and the air density law $\Delta(y)$), Lebesgue stated that no computation could compete with pure empiricism. Past experience during the war had shown that the latter method was simpler, cheaper, and more efficient from a military point of view.

Lebesgue's poor opinion of the value of computational ballistics was formed early on. Asked by Borel to get involved in both ballistics and sound-ranging studies, Lebesgue reluctantly agreed. "If this is useful to you, I am willing to do the computations. But then you have to tell me explicitly and in detail the operations I must carry out. I am willing to be a computing machine, but nothing

---

[80] See Charbonnier's preface in [Garnier 1918b, p. viii]. Charbonnier reached out again to mathematicians at the International Congress of Mathematicians in Toronto in 1924 [Charbonnier 1928b].

[81] "*Un tel rapport serait peut-être plus légitime s'il était adressé à la Commission des prix de vertu, car aucun de ceux qui ont travaillé à la construction du réseau ne prétend avoir fait œuvre scientifique en effectuant des calculs numériques par des procédés bien connus ; tous se sont efforcés simplement d'être utiles*" [Academy of Sciences, Lebesgue to Hadamard, October 23, 1919].

[82] "*Mais il ne faudrait pas que la grandeur de l'effort accompli fasse croire que l'outil construit a une valeur telle qu'on pourra l'employer encore longtemps.*" [Academy of Sciences, Lebesgue to Hadamard, October 23, 1919].

more." He went on: "I will do the computations you ask in the manner of a stupid clerk; but nothing more."[83] As a result of this episode, relations soured between Borel and Lebesgue whose *amour-propre* seemed to have been hurt by Borel's managerial style. But I think that Lebesgue's reluctance, which was due to many factors that were deeply personal to him, was also the result of his very high ethical ground which forbade him to take credit he did not believe his war work had earned him. Lebesgue insisted that one should not put too much importance on the war contribution of the "*Sorbonnoids*" at the *Bureau des calculs* of the Board of Inventions.[84] His reticence toward computational ballistics was rooted in his lack of satisfaction with the experimental basis of theoretical ballistics. Distancing himself from many mathematicians who had taken part in the war effort, he underscored that "mathematics cannot create the world." In the absence of a solid experimental basis, Lebesgue thought, the time was not ripe for mathematical ballistics.[85]

During the war and after, Lebesgue's opinion about the value of the mathematical war effort varied little. He thought that in ballistics as well as in sound-ranging, the physical problems overshadowed the mathematical effort. The only contribution mathematicians could make—and effectively made, according to him—was to organize computing methods and their execution by technical assistants. However ingenious the methods developed during the war (although Lebesgue hardly believed they were), this contribution would only last as long as deficient experimental bases would—not very long, Lebesgue thought.

Hadamard shared Lebesgue's opinion and was not overly impressed by the mathematics and science coming out of war work. As with most commentators, he believed that the great scientific war was mostly one of the application of known results rather than one of striking innovation. The ballistics papers he reviewed, he wrote, "for the most part bring, not scientific improvements concerning the principles, but rather modifications of a purely technical nature aimed at applying these principles more or less easily in specific practical consequences. These are topics with which the academy wishes to remain involved, but that nonetheless are on the sidelines of its proper function" [Hadamard 1920, p. 437].

Conclusion

Our study has shown that, although it was not always trivial, the mathematics of ballistics was, for the most part, tedious. Mathematicians at Gâvre did work on topics that had some wider implications, but, for the most part, they were

[83] "*Si cela vous est utile, je veux bien faire des calculs. Mais alors dîtes-moi expressément, et dans le détail, les opérations que je dois faire. Je veux bien être une machine à calcul, mais rien de plus.... Je ferai donc les calculs que vous me demandez à la façon d'un bon employé idiot ; rien de plus*" [Lebesgue 1991, letter CCXIII, n.d., pp. 318–319]. In this correspondence with Borel, letters explicitly dealing with war work are often undated; editors believe they can date them from early 1915.

[84] Lebesgue to Borel, letter n.d. [April 1917], in [Lebesgue 1991, p. 339]. On the fate of the friendship between Borel and Lebesgue during WWI, see Hélène Gispert's contribution to [Goldstein & Aubin forthcoming].

[85] "*les mathématiques ne peuvent pas créer le monde ; qu'elles ne peuvent suppléer l'expérience et l'observation mais, tout au plus, les résumer ; que le moment n'est donc pas encore venu de faire de la balistique mathématique et que l'effort a faire actuellement est d'ordre expérimental*" [Academy of Sciences, Lebesgue to Hadamard, October 23, 1919].

involved in menial tasks of experimentation, computation, and education. Still, their contribution to the war effort was significant, useful, and perhaps crucial to the final victory—and it was recognized as such. Ballisticians could legitimately pride themselves on having been able to use mathematicians' special abilities to their own ends.

Many prominent mathematicians were involved in ballistics during the war. To Lebesgue, Hadamard, Montel, and Drach already mentioned, we may add Arnaud Denjoy, who worked on the solubility of the hodograph at Gâvre in July 1917, Ernest Vessiot, later to succeed Borel as director of the École normale supérieure, and René Garnier, who taught at the Sorbonne after the war. But the result was paradoxical. They were happy to close the parenthesis and leave behind the work that had seemed most useful from a ballistics standpoint, as it had seemed to them rather trivial from a mathematical point of view. But of Drach's and Montel's work, most admired by Hadamard and which led to further work in the abstract theory of differential equations after the war, Charbonnier bluntly wrote: "this conquest, which honors the mathematicians, does not seem susceptible to providing ballisticians with new resources in view of applications" [Charbonnier 1928b, p. 574].

Yet, one wonders whether the ballistics experience of some French mathematicians and physicists did not have a deeper influence on the work they later did. Most mathematicians who were active at Gâvre indeed kept their connection to the military research institution. In 1921, civilians joined the commission for the first time since Hélie. Several mathematicians were nominated including Haag, Châtelet, Valiron, and Joseph Kampé de Fériet by then at the University of Lille.

Some pure mathematicians never returned to their earlier concerns. Haag later became the director of the chronometric school in Besançon and produced some of the most important work in France on the theory of dynamical systems. Joining the Lille faculty, Kampé de Fériet became a specialist in fluid mechanics and a world leader in turbulence studies after World War II. Also at Lille, Châtelet was so invested in the institutional rebuilding of his university that his mathematical research in number theory and algebra begun before the war was relegated to a minor place [Goldstein 2009]. The style of mathematical research this generation of mathematicians perpetuated was characterized by very formal approaches to problems directly inspired by applications.[86] Although they were pushed aside by the turbulent Bourbaki generation, these mathematicians planted the seeds of applied mathematics research in France.

Abbreviations

- ALVF: *Artillerie lourde sur voie ferrée* (heavy artillery on rail).
- CRAS: *Comptes rendus hebdomadaires des séances de l'Académie des sciences.*
- GHM: Garnier-Haag-Marcus (ballistic computing method).
- MTAG: *Mission du tir aérien de Gâvre* (Gâvre's mission for air firing).
- MBTA: *Mission balistique du tir aérien* (Air Firing Ballistic Mission).

[86]The work of Henri Villat in fluid mechanics, which rose to high prominence in interwar France, fitted squarely into these approaches. On this, see [Aubin, forthcoming].

Bibliography

Archival Sources

[Academy of Sciences] Archives of the Academy of Sciences, Paris.

[Garnier 1918a] Maurice Garnier, *Note technique n° 2 : définition et détermination du coefficient balistique, tables de tir (tableau principal et tableaux de correction)*, dated March 27, 1918, [SHD–Terre, 10W73].

[Garnier 1920] Maurice Garnier, *Balistique extérieure : conférences au Centre d'études tactiques d'artillerie de Metz*, 3rd ed., April 1920, [SHD–Terre, 2W293].

[Gâvre 1916] Commission de Gâvre, *Tables balistiques de la Commission de Gâvre 1. Tir de plein fouet*, Imprimerie des constructions navales, Lorient, 1916, [SHD–Terre, 2W292].

[Fonds Châtelet] Fonds Albert Châtelet, Archives départementales du Pas-de-Calais, Arras, 81 J.

[Fonds Painlevé] Fonds Paul Painlevé, Archives Nationales, Paris, 313/AP.

[Haag 1917] Jules Haag, *Conférences sur la préparation et le réglage du tir*, Centre d'études d'artillerie, November 1917, [SHD–Terre, 2W292].

[Hill n.d.] Archibald V. Hill, *Memories and Reflections*, 3 vols., Churchill College Archives, 920 HIL.

[SHD–Terre] Service historique de la défense, armée de Terre, Vincennes.

Published Works

[Adhémar 1934] Robert d'Adhémar, *Balistique extérieure*, Gauthier-Villars, Paris, 1934.

[Alain 1937] Alain [Émile Chartier], *Souvenirs de guerre*, Paul Hartmann, Paris, 1937; repr. "Les classiques de sciences sociales," UQAC, Chicoutimi, 2003.

[ALVF 1921] Ministère de la Guerre – Commission ALVF [Artillerie lourde sur voie ferrée], *Tables de balistiques extérieure*, 3 vols., n. p., 1921. Copy available at Bibliothèque nationale de France.

[Anonymous 1921] Anonymous [Général *** (sic)], *Devant Saint-Mihel (1914)*, Archives de la Grande Guerre **8** (1921), pp. 445–480.

[Aubin 2003] David Aubin, La Guerre du froid, la défaite de l'inventeur: bombes à oxygène liquide et production d'hélium, in [Aubin & Bret 2003], pp. 105–116.

[Aubin 2005] David Aubin, George David Birkhoff's Dynamical Systems (1927), in *Landmark Writings in Western Mathematics, 1640–1940*, ed. Ivor Grattan-Guinness, Elsevier, Amsterdam, 2005, pp. 871–881.

[Aubin, forthcoming] David Aubin, *'Audacity or Precision': The Paradoxes of Henri Villat's Fluid Mechanics in Interwar France*, paper presented at a "Workshop on the History of Fluid Mechanics", Rauischholzhausen, October 15–18, 2006; preprint available at http://hal.upmc.fr/hal-00639877.

[Aubin & Bret 2003] David Aubin & Patrice Bret, eds., *Le Sabre et l'éprouvette: invention d'une science de guerre, 1914–1939*, "14–18 Aujourd'hui," no. 6, Agnès Viénot, Paris, 2003.

[Aubin et al. 2010] David Aubin, Charlotte Bigg, & H. Otto Sibum, eds., *The Heavens on Earth: Observatories and Astronomy in Nineteenth-Century Science and Culture*, Duke University Press, Durham, 2010.

[Baquet 1921] Louis Baquet, *Souvenirs d'un directeur de l'Artillerie: les canons, les munitions, novembre 1914–mai 1915*, Henri Charles-Lavauzelle, Paris, 1921.

[Baillot 2010] Rémi Baillot, *Georges Claude, Le Génie fourvoyé : créateur de l'Air Liquide, du tube au néon, de l'énergie thermique des mers*, EDP Sciences, Paris, 2010.

[Barbazo & Pombourcq 2010] Éric Barbazo & Pascale Pombourcq, *Cent ans d'APMEP*, Association des professeurs de mathématiques de l'enseignement public, Paris, 2010.

[Bédier 1919] Jospeh Bédier, *L'Effort français : quelques aspects de la guerre*, Renaissance du Livre, Paris, 1919.

[Belhoste & Lemaître 1990] Bruno Belhoste & Louis Lemaître, J. V. Poncelet, les ingénieurs militaires et les roues et turbines hydrauliques, in *Le moteur hydraulique en France au XIXe siècle : concepteurs, inventeurs et constructeurs*, Cahiers d'histoire et de philosophie des sciences **29**, SFHST & Belin, Paris, 1990, pp. 33–89.

[Belhoste & Picon 1996] Bruno Belhoste & Antoine Picon, eds., *L'École d'application de l'artillerie et du génie de Metz, 1802–1870: enseignement et recherches*, Musée des plans-reliefs, Paris, 1996.

[Bennett et al. 1956] Albert A. Bennett, William E. Milne & Harry Bateman, *Numerical Integration of Differential Equations*, Dover, New York, 1956; repr. from a report of the Committee of the Division of Physical Sciences of the National Research Council, Bulletin of the National Research Council **92**, 1931.

[Bingen 1928] Jean-Nicolas Bingen, *Tables de calcul pour l'emploi de la méthode de Gâvre pour le calcul des trajectoires par arcs*, Institut cartographique militaire, Bruxelles, 1928.

[Boissonnet 1920] J. Boissonnet, *Les tables de tir pendant la guerre: combattants et balisticiens*, Revue d'artillerie **86** (1920), pp. 35–56 and 146–164.

[Booß-Bavnbek & Høyrup 2003] Bernhelm Booß-Bavnbek & Jens Høyrup, eds., *Mathematics and War*, Birkhäuser, Basel, 2003.

[Bos 1923] Charles Bos, *Les Leçons qu'il faut tirer de la guerre de 1914 à 1918*, Archives de la Grande Guerre **16** (1920), pp. 288–316.

[Bret 2002] Patrice Bret, *L'État, l'armée, la science: l'invention de la recherche publique en France, 1763–1830*, Presses universitaires de Rennes, Rennes, 2002.

[Broglie 1953] Maurice de Broglie, *Jules Haag*, CRAS **236** (1953), pp. 761–763.

[Bru 1996] Bernard Bru, *Problème de l'efficacité du tir à l'école d'artillerie de Metz: aspects théoriques et expérimentaux*, Mathématiques et sciences humaines **136** (1996), pp. 29–42.

[Campana 1923] J. Campana, *Les Progrès de l'artillerie: l'artillerie française pendant la guerre de 1914–1918*, Lavauzelle, Paris, 1923.

[Cardot 1987] Jean-Alexandre Cardot, *Artilleurs de campagne, 1918*, La Pensée universelle, Paris, 1987.

[Challéat 1933–1935] Jules Challéat, *L'Artillerie de Terre en France pendant un siècle : histoire technique (1816–1919)*, 2 vols., Charles Lavauzelles, Paris, suppl. to the *Mémorial de l'artillerie française*, 1933–1935.

[Chapeaux 2006] Antoine Chapeaux, *Michelin et l'aviation, 1896–1945 : patriotisme industriel et innovation*, Lavauzelle, Paris, 2006.

[Charbonnier 1906] Prosper-Jules Charbonnier, *Historique de la balistique extérieure à la commission de Gâvre*, Revue maritime et coloniale **168** (1906), pp. 411–476.

[Charbonnier 1907] Prosper-Jules Charbonnier, *Balistique extérieure rationnelle*, O. Douin, Paris, 1907.

[Charbonnier 1927] Prosper-Jules Charbonnier, *Traité de balistique extérieure*, 2 vols., Gauthier-Villars, Paris, 1927.

[Charbonnier 1928a] Prosper-Jules Charbonnier, *Essais sur l'histoire de la balistique*, Société d'éditions géographiques, maritimes et coloniales, Paris, 1928.

[Charbonnier 1928b] Prosper-Jules Charbonnier, Sur l'état actuel de la balistique extérieure théorique, in *Proceedings of the International Mathematical Congress Held in Toronto, August 11–16*, 2 vols., ed. John Charles Fields, University of Toronto Press, Toronto, 1928, vol. 2, pp. 571–594.

[Charbonnier 1929] Prosper-Jules Charbonnier, *Notice sur les travaux scientifiques*, Gauthier-Villars, Paris, 1929.

[Châtelet & Chazy 1955] Albert Châtelet & Jean Chazy, *Jules Haag*, Annuaire de l'association amicale des anciens élèves de l'École normale supérieure, 1955, pp. 34–37.

[Chéret 1990] Roger Chéret, The Life and Work of Pierre Henri Hugoniot, in *Shock Compression of Condensed Matter 1989: Proceedings of the American Physical Society Topical Conference Held in Albuquerque, New Mexico, August 14–17, 1989*, ed. Steven C. Schmidt, James N. Johnson, & Lee W. Davison, North-Holland, Amsterdam, 1990, pp. 12–19; repr. James N. Johnson & Roger Chéret, eds., *Classic Papers in Shock Compression Science*, Springer, Berlin, 1998, pp. 148–160.

[Claude 1919] Georges Claude, *Au creuset de la guerre: politiciens et Polytechniciens*, author's ed., Boulogne-sur-Seine, 1919.

[Condette 2009] Jean-François Condette, *Albert Châtelet: La République par l'école (1883–1960)*, Artois Presse Université, Arras, 2009.

[Cranz 1903] Carl Kranz, Ballistik, in *Encyklopädie der mathematischen Wissenschaften mit Einschluss ihrer Anwendungen*, IV-18 Teubner, Leipzig, 1903, pp. 185–279.

[Cranz & Vallier 1913] Carl Cranz & Emmanuel Vallier, Ballistique extérieure, in *Encyclopédie des sciences mathématiques pures et appliquées*, ed. Jules Molk & Paul Appell, tome IV, 6e volume, 1er fascicule, Gauthier-Villars, Paris & H. G. Teubner, Leipzig, 1913, pp. 1–105.

[Crémieux 1930] Maxime Crémieux, *La Commission de Gâvre*, Revue maritime et coloniale, 2e sem. (1930), pp. 143–161.

[Darrieus 1968] Georges Darrieus, *Notice nécrologique sur Hippolyte Parodi (1874–1968), Membre de la Division des Applications des sciences à l'industrie*, CRAS **267** (1968), pp. 81–86.

[D'Aubigny 1921] Alain Albert Leret d'Aubigny, *Rapport sur les armements*, Archives de la Grande Guerre **8** (1921), pp. 493–536.

[Delauney 1892] Julien Delauney, *Notice sur le professeur Hélie*, Mémorial de l'artillerie de la marine **20** (1892), pp. 788–794.

[Didion 1860] Isidore Didion, *Traité de balistique*, 2nd ed. J. Demaine, 1860; orig. publ. in 1848.

[Dodier & Valiron 1927] Henri Dodier & Georges Valiron, *Calcul des trajectoires curvilignes par arcs et des altérations d'après les méthodes américaines*, Mémorial de l'artillerie française **6** (1927), pp. 425–471.

[Drach 1914] Jules Drach, *Sur les équations différentielles du premier ordre et du premier degré*, CRAS **158** (1914), pp. 926–929.

[Drach 1920] Jules Drach, *L'équation différentielle de la balistique extérieure et son intégration par quadratures*, Annales scientifiques de l'École normale supérieure **37** (1920), pp. 1–94.

[Epple et al. 2005] Moritz Epple, Andreas Karachalios, & Volker R. Remmert, *Aerodynamics and Mathematics in National Socialist Germany and Frascist Italy: A Comparison of Research Institutes*, Osiris **20** (2005), pp. 131–158.

[Foch 1903] Ferdinand Foch, *Des principes de la guerre : conférences faites à l'École supérieure de guerre*, Berger-Levrault, Paris, 1903.

[Foch 1922] Ferdinand Foch, ed., *La Dernière Lettre écrite par des soldats français tombés au champ d'honneur 1914–1918, choisies par des pères qui pleurent un enfant mort pour la France et par d'anciens combattants*, Union des Pères et Mères dont les fils sont morts pour la Patrie, Ligue des Chefs de Section et des Soldats combattants & E. Flammarion, Paris, 1922.

[Foch 1931] Ferdinand Foch, *Mémoires pour servir à l'histoire de la guerre 1914–1918*, 2 vols., Plon, Paris, 1931.

[Fontanon 2005] Claudine Fontanon, L'obus Chilowski et la soufflerie balistique de Paul Langevin. Une recherche militaire oubliée de la mobilisation scientifique (1915–1919), in *Deux siècles d'histoire de l'armement. De Gribeauval à la force de frappe*, ed. Dominique Pestre, CNRS Éditions, Paris, pp. 81–110.

[Galvez-Behar 2008] Gabriel Galvez-Behar, *La République des inventeurs: propriété et organisation de l'innovation en France (1791–1922)*, Coll. "Carnot," Presses universitaires de Rennes, 2008.

[Garnier 1918b] Maurice Garnier, *Balistique extérieure: calcul des trajectoires par arcs successifs* Gauthier-Villars, Paris, 1918.

[Garnier 1922] Maurice Garnier, *Conférences sur la balistique extérieure*, Mémorial de l'artillerie française **1** (1922), pp. 109–185.

[Gascouin 1920] Firmin-Émile Gascouin, *L'Évolution de l'artillerie pendant la guerre*, Flammarion, Paris, 1920.

[Gilman 1905] Franck Gilman, *The Ballistic Problem*, Annals of Mathematics **6** (1905), pp. 79–89.

[Gladden 1930] Edgar Norman Gladden, *At Messines Ridge in 1917*, in *Everyman at War*, ed. Charles B. Purdom, J. M. Dent, London, 1930, pp. 114–121.

[Glück 1919] Gabriel-Leopoldt Glück, *L'artillerie lourde pendant la guerre*, Mercure de France **131** (1919), pp. 444–456.

[Goldstein 2009] Catherine Golstein, *La théorie des nombres en France dans l'entre-deux-guerres : de quelques effets de la première guerre mondiale*, Revue d'histoire des sciences **62-1** (2009), pp. 143–175.

[Goldstein & Aubin forthcoming] Catherine Goldstein & David Aubin, *La Grande Guerre des mathématiciens français*, forthcoming.

[Grotelueschen 2001] Mark E. Grotelueschen, *Doctrine Under Trial: American Artillery Employment in World War I*, Greenwood Press, Westport, Conn., 2001.

[Haag 1910] Jules Haag, *Familles de Lamé de surfaces égales : généralisation et applications*, Annales scientifiques de l''École normale supérieures **27** (1910), pp. 257–337.

[Haag 1915a] Jules Haag, *Sur un système de formules différentielles concernant les éléments de tir d'un projectile soumis à une résistance quadratique de l'air*, CRAS **161** (1915), pp. 379–381.

[Haag 1915b] Jules Haag, *Sur la méthode d'Otto*, CRAS **161** (1915), pp. 524–526.

[Haag 1915c] Jules Haag, *Sur le calcul du temps*, CRAS **161** (1915), pp. 633–634.

[Haag 1921] Jules Haag, *Sur le calcul des trajectoires et de leurs altérations*, Journal de l'École polytechnique **21** (1921), pp. 1–52.

[Haag 1926] Jules Haag, *Applications au tir*, Gauthier-Villars, Paris, 1926 [= vol. 4 of Émile Borel's *Traité des probabilités et de ses applications*].

[Hadamard 1920] Jacques Hadamard, *Rapport sur les travaux examinés et retenus par la Commission de Balistique pendant la durée de la guerre*, CRAS **170** (1920), pp. 436–445.

[Hall 1952] A. Rupert Hall, *Ballistics in the Seventeenth Century: A Study of the Relation of Science and War with Reference Principally to England*, Cambridge University Press, Cambridge, 1952.

[Hartcup 1988] Guy Hartcup, *The War of Invention: Scientific Development, 1914–18*, Brassey's Defence Publ., London, 1988.

[Heffernan 1996] Michael Heffernan, *Geography, Cartography, and Military Intelligence: The Royal Geographical Society and the First World War*, Transactions of the Institute of British Geographers **21** (1996), pp. 504–533.

[Hélie 1865] Félix Hélie, *Traité de balistique expérimentale: Exposé général des principales expériences d'artillerie executes à Gâvre de 1830 à 1864*, J. Dumaine, Paris, 1865.

[Hélie & Hugoniot 1884] Félix Hélie & Pierre-Henry Hugoniot, *Traité de balistique expérimentale*, 2nd ed., 2 vols., Gauthier-Villars, Paris, 1884.

[Henches 1918] Jules-Émile Henches, *À l'école de la guerre : lettres d'un artilleur (août 1914 – octobre 1916)*, Hachette, Paris, 1918.

[Henry 1894] Pierre-Émile Henry, *Balistique extérieure: cours d'artillerie de l'École d'application de l'artillerie et du génie*, lithograph, 1894.

[Herr 1913] Frédéric-Georges Herr, *La Guerre des Balkans. Quelques enseignements sur l'emploi de l'artillerie*, Berger-Levrault, Paris, 1913.

[Herr 1923] Frédéric-Georges Herr, *L'Artillerie, ce qu'elle a été, ce qu'elle est, ce qu'elle doit être*, Berger-Levrault, Paris, 1923.

[Houllevigue 1914] Louis Houllevigue, *Le Canon français de 75*, Revue de Paris (1914), pp. 200–217.

[Hull 1999] Andrew Hull, *War of Words: The Public Science of the British Scientific Community and the Origins of the Department of Scientific and Industrial Research, 1914–16*, British Journal for the History of Science **32** (1999), pp. 461–481.

[Jones 1921–1922] E. Lancaster Jones, *Sound-ranging*, Proceedings of the Musical Association **48** (1921–1922), pp. 77–89.

[Kevles 1968] Daniel J. Kevles, *George Ellery Hale, the First World War, and the Advancement of Science in America*, Isis **59** (1968), pp. 427–437.

[Kevles 1969] Daniel J. Kevles, *Flash and Sound in the A[merican] E[xpeditionary] F[orce]: The History of a Technical Service*, Military Affairs **33** (1969), pp. 374–384.

[Kevles 1978] Daniel J. Kevles, *The Physicists: The History of a Scientific Community in Modern America*, Alfred A. Knopf, London, 1978.

[L.G. 1931] L. G., Review of [Charbonnier 1928a], Isis **15** (1931), pp. 376–380.

[Langlois 1892] Hippolyte Langlois, *L'Artillerie de campagne en liaison avec les autres armes*, 3 vols., L. Baudouin, Paris, 1892.

[Langlois 1906] Hippolyte Langlois, *Question de défense nationale*, Berger-Levrault, Paris, 1906.

[Launay 1919] Roland Bréauté [Robert Launay], *Un universitaire aux Armées*, Brossard, Paris, 1919.

[Launay 1922] Robert Launay, *La météorologie pendant la Grande Guerre*, Archives de la Grande Guerre **12** (1922), pp. 695–704.

[Laves 1919] Kurt Laves, *How the Map-Problem was Solved during the War*, American Mathematical Monthly **26** (1919), pp. 181–187.

[Lebesgue 1991] Henri Lebesgue, *Lettres d'Henri Lebesgue à Émile Borel*, Cahiers du séminaire d'histoire des mathématiques **12** (1991), pp. 1-506.

[Lebon 1915] Gustave Lebon, *Enseignements psychologiques de la guerre européenne*, Flammarion, Paris, 1915.

[Lecornu 1924] Léon Lecornu, *Applications diverses à la mécanique*, Bulletin de la Société mathématique de France **52** (1924), pp. 36–40.

[Lévy 1970] Paul Lévy, *Quelques aspects de la pensée d'un mathématicien*, Blanchard, Paris, 1970.

[Lintier 1916] Paul Lintier, *Avec une batterie de 75 : ma pièce. Souvenirs d'un canonnier 1914*, Plon-Nourrit, Paris, 1916; repr. L'Oiseau de Minerve, 1998.

[Long & Weiss 1999] Lyle N. Long & Howard Weiss, *The velocity dependence of aerodynamic drag: a primer for mathematicians*, American Mathematical Monthly **106**, No. 2 (February 1999), pp. 127–135.

[MacLeod & Andrews 1971] Roy MacLeod & Elizabeth Kay Andrews, *Scientific Advice for the War at Sea, 1915–1917: The Board of Invention and Research*, Journal of Contemporary History **6** (1971), pp. 3–40.

[Malloué 1911] Armand Malloué, *Comment on prend la ligne de mire*, Culture physique **8** (1911), pp. 136–138.

[Mehrtens 1996] Herbert Mehrtens, Mathematics and War: Germany, 1900–1945, in *National Military Establishments and the Advancement of Science and Technology*, ed. Paul Forman & José M. Sánchez-Ron, Kluwer, Amsterdam, 1996, pp. 87–134.

[Merger 1994] Michèle Merger, Parodi, Hippolyte (1874–1955), professeur d'électricité appliquée, s.v. in *Les Professeurs du CNAM: Dictionnaire biographique 1794–1955*, ed. Claudine Fontanon & André Grelon, INRP & CNAM, Paris, 1994, pp. 338–348.

[Merton 1942] Robert Merton, *Science and Technology in a Democratic Order*, Journal of Legal and Political Sociology **1** (1942), pp. 115–126; repr. The Ethos of Science, in R. Merton, *On Social Structure and Science*, University of Chicago Press, Chicago, 1996, pp. 267–276.

[Mesnage 1953] Pierre Mesnage, *La vie et l'œuvre de Jules Haag*, Annales françaises de chronométrie, 2nd ser. **7** (1953), pp. 141–150.

[Meyer 1989] C. Meyer, Haag (Jules), s. v. in *Dictionnaire de biographie française*, Letouzey & Ané, Paris, vol. 17, 1989, pp. 444–445.

[Morel 1904] Marie-Auguste Morel, *La Balistique graphique et son application dans le calcul des tables de tir*, Gauthier-Villars, Paris, 1904.

[Morgan 1959–1960] Prentice G. Morgan, *The Forward Observer*, Military Affairs **23** (1959–1960), pp. 209–212.

[Morin 1855] Arthur Morin, *Notions fondamentales de mécanique et données d'expérience*, 2nd ed., Hachette, Paris, 1855.

[Mounier-Kuhn 2010] Pierre Mounier-Kuhn, *L'informatique en France de la seconde guerre mondiale au Plan Calcul : l'émergence d'une science*, Presses universitaires de Paris-Sorbonne, Paris, 2010.

[Nordmann 1917] Charles Nordmann, *À coups de canons: notes d'un combattant*, Perrin, Paris, 1917.

[Ottenheimer 1924] Jean Ottenheimer, *Balistique extérieure*, Armand Colin, Paris, 1924.

[P.N. 1907] Commandant P.N., *Notre artillerie et les sanctions de la dernière guerre*, Revue militaire générale **5** (1907), pp. 534–561.

[Palazzo 1999] Albert P. Palazzo, *The British Army's Counter-Battery Staff Office and Control of the Enemy in World War I*, Journal of Military History **63** (1999), pp. 55–74.

[Parshall 2000] Karen H. Parshall, *Perspectives on American Mathematics*, Bulletin of the American Mathematical Society **37** (2000), pp. 381–405.

[Parshall & Rowe 1994] Karen H. Parshall & David E. Rowe, *The Emergence of the American Mathematical Research Community, 1876-1900: J. J. Sylvester, Felix Klein, and E. H. Moore*, American Mathematical Society & London Mathematical Society, Providence & London, 1994.

[Pastre 1918] Jules-Louis Gaston Pastre, *Trois ans de front : Belgique, Aisne et Champagne, Verdun, Argonne, Lorraine. Notes et impressions d'un artilleur*, Berger-Levrault, Paris & Nancy, 1918; repr. Presses universitaires de Nancy & Secrétariat d'État chargé des anciens combattants et des victimes de guerre, Nancy, 1990.

[Patard 1930] Léon Patard, *Historique de la Commission d'expériences de Gâvre (1829–1930)*, suppl. Mémorial de l'artillerie française, Impr. nationale, Paris, 1930.

[Pattison 1983] Michael Pattison, *Scientists, Inventors and the Military in Britain, 1915–19: the Munitions Inventions Department*, Social Studies of Science **13** (1983), pp. 521–568.

[Peira 1955–1957] Pierre Marius Peira, *Historique de la conduite du tir dans la marine, 1900–1940*, 3 vols., Mémorial de l'artillerie française, Paris, 1955–1957.

[Percin 1914] Alexandre Percin, *Le Combat*, F. Alcan, Paris, 1914.

[Poincaré 1928] Raymond Poincaré, *Au service de la France : neuf années de souvenirs*, 10 vols., Plon, Paris, 1928.

[Poyen-Bellisle 1889–1893] Henri de Poyen-Bellisle, *Notice historique sur la Commission d'expérience de Gâvre*, Revue maritime et coloniale **101** (1889), pp. 312–329; **102** (1889), pp. 32–51 & 377–391; **116** (1893), pp. 487–519.

[Reinach 1916] Joseph Reinach, *La Guerre sur le front occidental : étude stratégique 1914–1915*, Charpentier, Paris, 1916.

[Ripperger 1995] Robert M. Ripperger, *The Development of the French Artillery for the Offensive, 1890–1914*, Journal of Military History **59** (1995), pp. 599–618.

[Rouquerol 1914] Gabriel Rouquerol, *Méconnaissance du rôle et la puissance de l'artillerie dans l'organisation actuelle de l'arme et dans l'armée*, Revue militaire générale **16** (July 1914), pp. 5–22.

[Rouquerol 1919] Gabriel Rouquerol, *Après la victoire : notes et critiques*, Berger-Levrault, Nancy, Paris, & Strasbourg, 1919.

[Rouquerol 1920] Gabriel Rouquerol, *Le Canon : artisan de la victoire*, Berger-Levrault, Nancy, Paris, & Strasbourg, 1920.

[Roussel 1989] Yves Roussel, *Histoire d'une politique des inventions, 1887–1918*, Cahiers pour l'histoire du CNRS **3** (1989), pp. 19–57.

[Sainean 1916] Lazare Sainean, *L'Argot des tranchées : d'après les lettres des poilus et les journaux du front*, E. de Boccard, Paris, 1919; repr. Slatkine, Geneva, 1973.

[Schiavon 2003a] Martina Schiavon, *Itinéraires de la précision : géodésiens, artilleurs, savants et fabricants d'instruments en France, 1870–1930 (environ)*, doctoral thesis, EHESS, Paris, 2003.

[Schiavon 2003b] Schiavon, Martina, Des savants-officiers entre science, armée, État et industrie de précision : les géodésiens du Service géographique de l'Armée, 1887–1920, in [Aubin & Bret 2003], pp. 60–73.

[Schwartz 1927] Philip Schwartz, *La balistique appliquée aux Etats-Unis: étude critique des progrès réalisés en balistique appliquée par le département de l'artillerie de l'armée des Etats-Unis depuis 1917*, Mémorial de l'artillerie française **6** (1927), p. 589.

[Sebert 1915] Hippolyte Sebert, *Rapport sur les travaux de M. le commandant Batailler*, CRAS **161** (1915), pp. 895–897.

[Seddik-Ameur 2003] Nacira Hadjadji Seddik-Ameur, *Les tests de normalité de Lhoste*, Mathématiques et sciences humaines **162** (2003), pp. 19–43.

[Siacci 1892] Francesco Siacci, *Balistique extérieure*, trans. P. Laurent, Berger-Levrault, Paris & Nancy, 1892.

[Siegmund-Schultze 2003] Reinhard Siegmund-Schultze, Military Work in Mathematics 1914–1945: An Attempt at an International Perspective, in [Booß-Bavnbek & Høyrup 2003], pp. 23–82.

[Smith 1990] Meg Weston Smith, *E. A. Milne and the Creation of Air Defence: Some Letters from an Unprincipled Brigand, 1916–1919*, Notes and Records of the Royal Society of London **44** (1990), pp. 241–255.

[Steele 1994] Brett D. Steele, *Muskets and Pendulums: Benjamin Robins, Leonhard Euler, and the Ballistic Revolution*, Technology and Culture **35** (1994), pp. 348–382.

[Steele & Dorland 2005] Brett D. Steele & Tamara Dorland, eds., *The Heirs of Archimedes: Science and the Art of War through the Age of the Enlightenment*, The MIT Press, Cambridge, Mass., 2005.

[Sugot 1918] Georges Sugot, *Cours de balistique: École d'application de l'artillerie navale*, Imprimerie nationale, Paris, 1918.

[U.S. n.d.] U.S. Government, Aberdeen Proving Ground, *Ballisticians in War and Peace*, vol. 1: *A History of the United States Army Ballistics Research Laboratories 1914–1956*, n.d.

[Vallier 1915] Emmanuel Vallier, *Le tir indirect en 1915*, Revue générale des sciences pures et appliquées **16** (1915), pp. 297–299.

[Warwick 1995] Andrew Warwick, The Laboratory of Theory, or What's Exact About the Exact Sciences?, in *The Values of Precision*, ed. M. Norton Wise, Princeton University Press, Princeton, 1995, pp. 311–351.

[Winterbotham 1919] Harold St. John Lloyd Winterbotham, *Geographical Work with the Army in France*, Geographical Journal **54** (1919), pp. 12-23.

[Ziman 2000] John Ziman, *Real Science: What it is and what it means*, Cambridge Univ. Press, Cambridge, 2000.

Why Aerodynamics Failed to Take off in Nancy: An Unexpected Casualty of World War I

Laurent Rollet and Philippe Nabonnand

Even in small university centers, World War I could have significant effects on teaching and research in the mathematical sciences. As personnel and students were massively displaced due to mobilization, as local industrial concerns, national strategic priorities, and international geopolitics changed abruptly, as financial conditions fluctuated, conditions for establishing and developing the mathematical sciences likewise varied. To make this clear, the case of the University of Nancy is especially instructive. At the turn of the nineteenth century, the University of Nancy occupied a special place in the French system of higher education. Close to the new border with Germany and situated in one of France's industrial regions, the city of Nancy was home to a recent university—founded in 1854—whose scientific project hinged on the development of the applied sciences from a patriotic perspective. In 1872 the historian Edgar Quinet gave a vivid description of this nationalistic agenda:

> Make of Nancy another Metz, another Strasbourg, from moral and intellectual points of view. You may [achieve] this by attracting the new generations from Alsace to the great scientific teaching center we shall establish on our new border ([Quinet 1872, p. 251], rep. in [Dusart & Moulin 1998, p. 10]).

Convinced that defeat in the Franco-Prussian War of 1870–1871 was partly due to the failure of French science to support industrial development, many universities decided to foster research in the applied sciences and to establish technical institutes for the training of technicians and engineers needed by industry. In Nancy this general trend took a special turn due to its symbolic character as a frontier

---

Laurent Rollet's address is Laboratoire d'histoire des sciences et de philosophie – Archives Henri Poincaré, UMR 7117, Université de Lorraine, CNRS & Maison des sciences de l'homme Lorraine, USR 3261, Nancy, France. His email address is laurent.rollet@univ-lorraine.fr.

Philippe Nabonnand's address is Laboratoire d'histoire des sciences et de philosophie – Archives Henri Poincaré, UMR 7117, Université de Lorraine, CNRS & Maison des sciences de l'homme Lorraine, USR 3261, Nancy, France. His email address is philippe.nabonnand@univ-lorraine.fr.

The authors wish to thank Prosper Doh for his careful rereading and Jean-René Cussenot for his very helpful historical database devoted to the professors of the Faculty of Sciences of Nancy: http://cussenot-fst-nancy.ahp-numerique.fr/cussenot2/.

town. As opposed to episcopal Metz, the old ducal capital of Lorraine indeed remained a French territory and therefore took on a new role in the region as well as at the national level. During that period, the population of Nancy skyrocketed from about 50,000 to some 120,000.[1] Industrialization partly accounts for the strong demographic growth, as mining, iron, and steel industries were developed in Lorraine. But Nancy's new geopolitical situation was also responsible for this growth. After the war, governments of the Third Republic conferred a strategic role on Nancy through the installation of garrisons.[2] Moreover, refusing to adopt German nationality, a large number of the inhabitants of Alsace and Lorraine decided to settle in Nancy (more than 13,000 in 1877).

This demographic shift impacted the intellectual and scientific life of the city. The University of Nancy now welcomed a large number of students coming from Alsace and Moselle. As the University of Strasbourg was handed over to the Germans, many of its professors chose to move to Nancy, thereby reinforcing its institutions of higher education.[3] As a consequence, Nancy's Preparatory Medical School (*École préparatoire de médecine*) was transformed into a full-fledged Faculty of Medicine. Important intellectual and cultural repercussions were felt outside the university, as the defeat of 1870 also led to the emergence of an intellectual and artistic golden age symbolized by the avant-garde Nancy School (*École de Nancy*) famous for its promotion of Art Nouveau in France.

The last decades of the nineteenth century were also a time of expansion for mathematics in Nancy. The study of the very small community of mathematicians (in 1914 there were barely 4 mathematical chairs in Nancy) often boils down to analyzing the capacity of its actors to elaborate ambitious projects and get support from the university, government ministries, and the general public. In a period of rapid industrialization, the development of technical research and training institutions indeed created a crucial need for mathematical instruction. This in turn was the bedrock for tightening collaborations between mathematicians and physicists, as well as between the Faculty of Sciences, local authorities, industrial firms, and the army.

The first part of this chapter surveys the "golden age" of the Nancy Faculty of Sciences before World War I, focusing especially on the pioneering development of technical institutes and the role played by mathematicians in this process. The second part focuses on the elaboration of an ambitious project—emerging through the tight collaboration between mathematicians and physicists—to establish there a center for aerodynamics and meteorology of national importance. The third part deals with the involvement of the mathematical community in the war on both the military and the home fronts. In conclusion, we survey the damaging consequences the war had on Nancy, where victory in 1918 at times led to effects opposite to those of the defeat of 1870 and, paradoxically, the demise of its aeronautical dreams at a time when aviation would become a national priority.

---

[1] In censuses taken in 1861 and 1872, the population of Nancy was, respectively, 49,305 and 52,978. Figures from 1911 and 1921 are, respectively, 119,949 and 113,226. Data from `http://cassini.ehess.fr`.

[2] In 1913, the military represented almost 9% of the whole population. In 1866, this proportion was far less important (around 1.2%). See [Sicard-Lenattier 2002].

[3] On the history of the University of Strasbourg, see [Crawford & Olff-Nathan 2005].

1. The Nancy Faculty of Sciences before World War I

In the decades before the outbreak of WWI, the lessons from the crushing defeat of 1870, together with the rapid process of industrialization, led to the establishment of several technical institutes for the training of engineers in Nancy. Their creation was financed by the Ministry of Education, by local political institutions (the General Council (*Conseil général*) of the Meurthe-et-Moselle department, Nancy's city council, etc.), and by industrial firms. The period from 1887 to 1914 saw the founding of the Chemical Institute (1887), a brewery school (1893), the Electrotechnic Institute (1900), the Agricultural and Colonial Botanic Institute and the ephemeral Aerodynamic and Meteorological Institute (*Institut d'aérodynamique et de météorologie*, hereafter IAM), about which there will be more below (see Section 2).[4] The most influential actors in this story were the Alsatian chemist Albin Haller (1849–1925), who was the first director of the school of chemistry, and the physicist Ernest Bichat (1845–1905). As president of the General Council of the Meurthe-et-Moselle department, Bichat was an influential political and economic figure in Lorraine. A staunch promoter of closing the gap between science and industry, Bichat played a crucial role in the foundation of the Chemical and the Electrotechnical Institutes. From 1888 until his death, he was dean of the Faculty of Sciences and a member of the French National Education Board (*Conseil supérieur de l'instruction publique*). His successor to the deanship, the mathematician Gaston Floquet (1847–1920), pursued his work by promoting the development of aeronautics and meteorology. The Belgian manufacturer Ernest Solvay (1838–1922) also played a leading role in this process by giving generous subsidies to the technical institutes between 1890 and 1914. Thanks to funds from the municipal council and the General Council, the Faculty of Sciences was able to develop new courses in applied physics and applied mechanics. This strengthened the relationship with local industry. In 1899, industrial physics courses were taught by the physicist François Perreau and by the mathematician Henri Vogt (1864–1927), called to the chair of applied mathematics in 1899.

As an auxiliary discipline, mathematics was a crucial component of the curricula in the technical institutes. But the level of the mathematical teaching required for engineering was, in Nancy as elsewhere, hotly debated.[5] The problem faced by Nancy mathematicians was to find the correct balance between pure and applied mathematics in engineers' education, knowing that the students' former education in mathematics was often very poor.[6] Concerning the aims of mathematical training in engineering education, Vogt, for example, declared in 1911: "One must not weaken the teaching of mathematics but direct it toward the goal one wishes to attain: if [mathematical training] is reduced in scope, it must be deepened and geared toward applications; it must be pushed up to the derivation of numerical values and their approximations" [Nabonnand 2006, p. 138].

[4] On the history of institutions of higher education in Nancy, see [Nye 1986, pp. 33–78] and [Grelon & Birck 1998, Birck & Grelon 2006, Rollet & Choffel-Mailfert 2007].

[5] On this point, see the contribution by Jean-Luc Chabert and Christian Gilain to this volume.

[6] To attend the technical institutes, the *baccalauréat*, the final *lycée* examination, was not required and students did not always attend the mathematically intensive preparatory classes (called *classes de mathématiques spéciales préparatoires aux Grandes Écoles*). For a presentation of the French system of higher education, see the contribution by Jean-Luc Chabert and Christian Gilain to this volume.

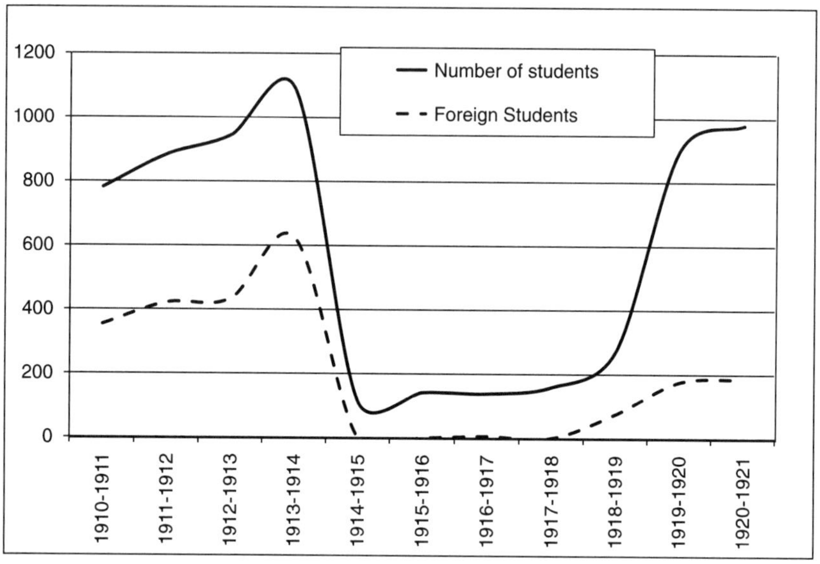

FIGURE 1. French and foreign students at the Faculty of Sciences of Nancy (1910–1921). Source of data: [*Séances*, passim].

In 1896, university reforms (which gave official statutes to the University of Nancy) led to the creation of the certificates (*certificats de licence*), that is, specialized one-year college degrees that could be collected to earn a bachelor's degree (*licence*). Mathematics students were offered certificates in differential and integral calculus, rational mechanics, higher geometry, analysis and astronomy, etc. But some students lacked the prerequisite training to succeed in some of these certificates. This was especially true of foreign students who were then attracted in large numbers to Nancy (see Figure 1). Similar problems were faced by many French universities, in Paris as well as in the provinces. For these reasons, a new mathematical certificate was created in 1908. Analogous to the program of the preparatory classes (*classes de mathématiques spéciales*), the new general mathematics certificate (*certificat de mathématiques générales*) became the core of engineers' mathematical training. During their courses, they also needed to study in view of getting certificates in other scientific disciplines (such as general and applied physics, rational mechanics, etc.). But this was not enough for all students. In Nancy, the large influx of foreign students without solid initial training in mathematics and in the sciences led to the creation of new courses in 1908 organized into a one-year prerequisite preparation for the technical institutes. The Faculty of Sciences was keenly aware that foreign students constituted one of the keys to the success of these institutes and accordingly devoted time and money to cater to and retain this new student population.

Was the beginning of the twentieth century a golden age for the University of Nancy? Before 1914, the University of Nancy was one of the most active provincial universities, especially in the scientific and medical fields. Prior to the infamous N-Rays affair, René Blondlot (1849–1930) enjoyed an enviable international reputation for electromagnetic experiments performed in the 1890s [Nye 1986, pp. 53–78].

From 1907, Edmond Rothé was heavily involved in wireless telegraphy (this technique allowed the Paris Observatory to communicate the exact time to the Nancy Faculty of Sciences in 1910, [Saint-Martin 2008, pp. 278–283]). He also shared with Floquet a great interest in aerodynamics and meteorology.[7] In 1912, the chemist Victor Grignard (1871–1935) was awarded the Nobel Prize for his work on organometallic reactions (the so-called Grignard reaction). New buildings and new laboratories were erected: in 1897, Haller obtained a very important subsidy from the Société industrielle de l'Est for the creation of a laboratory of electrochemistry and physical chemistry. The Solvay firm also donated 100,000 francs for this project and, three years later, an equivalent subsidy for the creation of a laboratory of electricity and electrical engineering. In 1906, the Mathematics and Physics Institute planned by Bichat was built close to the technical institutes. In 1913, Solvay donated 500,000 francs to extend the electrical engineering laboratory, to establish a chair of electrical engineering—attributed to Alexandre Mauduit (1874–1956)—and to transform the former position of associate professor (*maître de conférences*) in electrical engineering into an analogous position in physics, on the condition that the holder also teach mathematics. This position was held by the physicist Albert Grumbach [Birck 2006, Choffel-Mailfert 2007]. At the same time, Louis Maurice Roy was recruited as chief assistant in mathematics.

Consequently, material conditions in Nancy were quite good for scientific research and higher instruction. Nevertheless, this profusion of means hid the artificiality of this development. As in other provincial universities, the number of students remained rather low. The best students from local preparatory classes usually opted for the Parisian *grandes écoles*: mostly the École polytechnique and the École normale supérieure. As a result, the technical institutes widened their recruitment pools and accepted students from the primary education system then existing in parallel with the elite *lycée* system (*école primaire supérieure* and *école professionnelle de l'Est*). To solve their demographic and financial problem, the Faculty of Sciences and its institutes, following the model of the German technical universities (*Technische Hochschulen*), also opened up to foreign students beginning in 1902. The University of Nancy was one of the first to develop an active policy of attracting foreign students with the help of the Society of Friends of the University of Nancy (*Société des amis de l'université*), which engaged in active lobbying outside France. The first consequence of this drive was to increase the demographic weight of foreign students in Nancy. In 1914, for instance, out of 1,089 students enrolled at the Faculty of Sciences, more than 600 came from abroad, most of whom enrolled in the schools of chemistry and electricity.[8] The second consequence was the high rate of failure in the examinations for scientific certificates (see Figure 2).

Despite these problems, official rhetoric before 1914 was very optimistic. It uniformly praised the harmonious relationship between academic research, industrial development, and patriotism and put Nancy at the forefront of dynamic modernity. In 1912, the chemist Paul Petit (1862–1936), then assistant to Dean Floquet, expressed this very clearly:

---

[7] As we shall see, he would be the first (and only) director of the Aerodynamic and Meteorological Institute, officially founded in 1913.

[8] Most foreign students came from Russia. See [Gouzévitch & Gouzévitch 2006] and [Birck 2006, pp. 37–41].

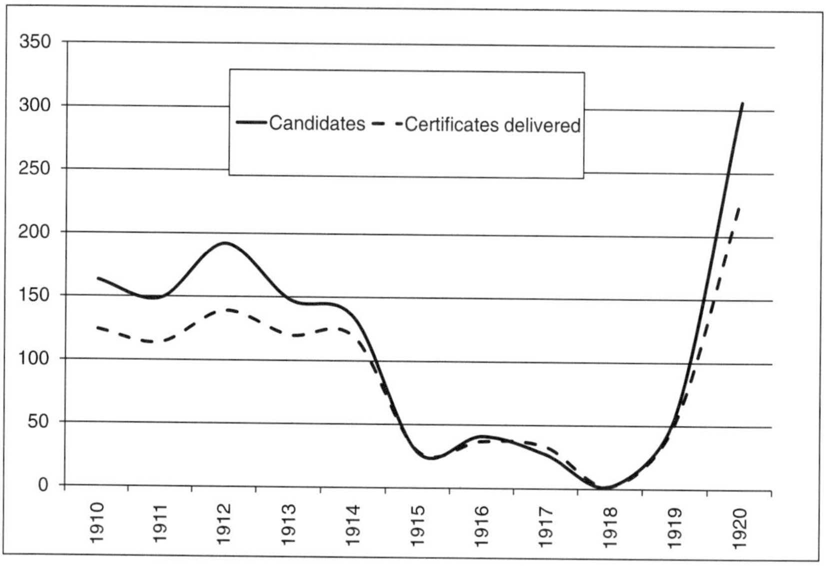

FIGURE 2. Evolution of the number of certificates delivered at the Nancy Faculty of Sciences (1900–1920) for all disciplines. Source of data: [*Séances*, passim]. See also [Nabonnand 2006].

> To our institute's toolkit, [which was] already very extensive, apparatuses for testing the strength of materials have been added. ... As one can see, the Mechanical Institute preserves its well-justified reputation as a model institute, while being simultaneously mindful of assisting regional industry with all the services that are within its province. In Germany, entrepreneurs routinely make use of the testing of the strength of materials and our institute will now be able to answer the requests that have been addressed to it (quoted from [*Séances*, 1912, p. 29]).

In other words, with its constellation of technical institutes, Nancy was successfully emulating the Germans and fostering its close ties with local industry. With the establishment of an institute for aeronautics and meteorology, this trend of development now involved mathematicians.

2. The Founding of the Aerodynamic and Meteorological Institute

As we saw, the central role played by technical institutes in Nancy favored a strong cooperation between "applied" and "pure" scientific disciplines. In the 1910s, the establishment of the Aerodynamic and Meteorological Institute (IAM), a center devoted to aerodynamic and meteorological research and teaching, was an exciting project where mathematicians, physicists, and engineers engaged in intense collaboration. For this project, the university also relied on close partnerships with the army and with civic associations.

On July 12, 1914, the Society of Friends of Aviation (*Société des amis de l'aviation*, hereafter SAA) officially donated to the French Army two double aviation hangars and a workshop hangar for the maintenance of airplanes, at the military aerodrome of Villers-les-Nancy. This airfield had been built in 1913 and was, at the beginning, the home base of a squadron of 6 Farman airplanes. The society's donation, amounting to 105,000 francs, was made possible thanks to several grants and subscriptions that had been obtained by the association since 1912: the cities of Nancy and Saint-Dié (Vosges) contributed importantly, so did the General Student Association of Nancy (*Société générale des étudiants de Nancy*) and newspapers such as l'*Est républicain* and l'*Étoile de l'Est*. These solemn festivities were part of the celebrations of the French National Day, the Fourteenth of July, involving official and public torchlight processions, republican banquets, etc.

The hangar ceremony was followed by another donation by the SAA. A five-hectare plot of land next to the military buildings was given to the University of Nancy for the establishment of the future IAM. Official inauguration speeches were delivered by the Minister of Education Jean-Victor Augagneur and by both of the professors from Nancy who were involved at the start of the project, Rothé and Floquet.

In fact, this ceremony was the consecration of their long-held commitment in favor of aviation. On March 20, 1912, the Republican Union of Nancy invited the mathematician Paul Painlevé for a public lecture on aviation. At that time, Painlevé was simultaneously a member of the Academy of Sciences, a member of Parliament representing Paris, the president of the National Aeronautic Commission (*Commission nationale d'aéronautique*), and the vice president of the Air National League (*Ligue nationale aérienne*).[9] Painlevé came with some already famous pioneers of aviation, like captain Albert Étévé (1880–1976), the director of the first military aviation school in Versailles.[10] During his lecture, Painlevé mentioned several requirements for the development of aviation in the provinces, such as the military's need for more airplanes, air fields, and hangars.

After Painlevé's lecture, a hangar committee was created by Floquet and Rothé with the double purpose of raising funds and buying a large plot of land near Nancy. A few days later (April 7–8), at a meeting chaired by Floquet and held at the air field of Jarville (a village close to Nancy), the hangar committee merged with the Eastern Air League (*Ligue aérienne de l'Est*). The result of this merger was the SAA, whose aims were to promote aviation—and especially military aviation—in Nancy through the establishment and the installation of air fields, the construction of hangars, the organization of aviation contests and festivals, and the founding of a university institute for aerodynamics.

Rothé's and Floquet's interest in aerodynamics and in the promotion of aviation was not new.[11] Since his nomination in 1905, Rothé taught aerology and aerodynamics in the third part of his physics course [Rothé 1917] and devoted most of his research to these disciplines. Founder and director of the Eastern Air League and then of the SAA, Floquet, for his part, had spared no effort toward the development of aeronautics in Nancy. In 1909, Floquet himself had suggested Nancy's candidacy to host the fourth International Congress of Aeronautics. The event took place there from September 18–23 of that year and was the occasion

[9]Presided over by René Quinton, the league included several members of the Academy of Sciences, such as Paul Appell, Jean Bouquet de la Grye, Pierre Léauté, etc. On Painlevé's involvement in aviation, see [Fontanon 2003].

[10]An aeronautical engineer, Étévé developed various important devices such as an anemometer (anemometer Étévé) and a machine gun turret (*système Étévé*).

[11]For more details on Floquet's career, see [Rollet & Nabonnand 2011].

of several air shows (including one by the dirigible *La Ville de Nancy* that was a fiasco; see [Chimello 2005, pp. 83–93]). In 1909–1910, Floquet actively campaigned for the choice of Nancy as a staging point of the Eastern Tour (*Circuit de l'Est*), an important air race with the following stages: Paris to Troyes to Nancy to Charlevilles-Mézières to Douai to Amiens to Paris. The race was organized by the daily newspaper *Le Matin* and the winner received by a prize of 100,000 francs. This event took place in Nancy during a sports week (August 7–14) and was linked to an aviation festival ending with a cavalcade and the crowning of an "aviation queen."

The board of the SAA was chaired by Floquet, while Rothé was technical secretary. It brought together military officers, such as Colonel Lecomte, and the local economic, social, and political elite, including the glass artist Antonin Daum, François Villain (the president of the very influential Société industrielle de l'Est), and Henry Brun (who was adviser on foreign trade). Rothé was the first to suggest that an aerodynamic school be set up in Nancy, insisting that it would be in close interaction with military and civilian organizations. Rothé clearly expressed this aim in the *Est républicain* on July 22, 1914:

> A very special character was conferred on the founding of the Nancy Center by its promotors' intent to create something useful to the military by providing Nancy with first-rank squadrons and hangars, as well as to civilian society and general instruction by helping the Faculty of Sciences to establish a new institute with courses, research laboratories and observation platforms. To this end, the society [SAA] donated a plot of land close to the military airfield where the buildings of the new institute will be erected. Officers from the air squadron will be freely admitted there to perform the observations and experiments they will judge useful to do [Rothé 1914].

In the conclusion of his article, Rothé emphasized the undertaking's patriotic dimension: "Considering the progress achieved by Germany [and] the threats posed by German aviation, [the SAA] will endeavor to attract new contests [and] new subscriptions" [Rothé 1914]. Flight was pursued by emphasizing patriotic motives, as can be seen from a 1910 letterhead of the Air National League that stated: "Join the league, enroll your parents, enroll your children,... the German Naval League has a million members" [Nancy Arch., file "Course aérienne de l'Est," (k) I2–11].

What was the position of the army concerning the development of aeronautics in Nancy? Military authorities were, in general, happy to benefit from the popular fascination with aeronautics and a national fund supported the development of French military aeronautics. In 1912, Colonel Édouard Hirschauer (1857–1943), who was one of the fathers of military aviation and, later, a major figure at the battle of Craonne in 1917, wrote a report about the strategic organization of French air squadrons in which he stressed the active participation of several cities including Nancy:

> The administration in several cities and *départements* helped the national fund for military aviation by contributing airfields and hangars.... The city of Arras thus erected a hangar on a military training field; the city of Bar-le-Duc bought a plot of land which it equipped with hangars; in Nancy, the local committee is

FIGURE 3. Victor Prouvé, "Beyond the Clouds, for Peace." Menu of the banquet of the Société des amis de l'aviation on July 12, 1914. Archives municipales de Nancy, 17 Fi 30.

also suggesting the acquisition of a plot of land and hangars, etc. To dispose of a very remarkable aerial tool, these initiatives only need to be encouraged and channeled so as to constitute a network of airfields and hangars along the main routes throughout France.[12]

[12] "*Beaucoup de Municipalités ou de Départements ont apporté leur concours à la souscription nationale pour l'Aviation Militaire sous forme d'offres de terrains d'atterrissage et de hangars. D'autre part, le Comité de souscription paraît tout à fait désireux de consacrer une*

In 1912, Hirschauer's priority seemed strategically to distinguish between training centers (with the finest equipment) and depots with their annexes. The training centers he wished to develop included Châlons, Avor, Pau, and Versailles–Saint-Cyr, but not Nancy. A year later, however, Hirschauer now listed Nancy in another report concerning the allocation of military funds for the development of airfields in France. But at nearly 86,000 francs, the cost he advocated for the Nancy development was nothing compared with the estimated costs of the aviation centers for Dijon, Versailles, or Lyon (for each more than 500,000 francs) [SHD–Air, AA 13163, doc. AA14].

The IAM, first conceived by Rothé, became a reality in November 1912 with Rothé himself acting as director. In June 1913, the institute started to offer a one-year course and a degree in the field of aerodynamics (*diplôme d'études supérieures aérodynamiques*). One of the aims of this new diploma was to train airplane designers and aviators, as chief education officer Charles Adam explained in an official speech:

> The teaching of the Institute concerns all students interested in aeronautics in order to be hired by building firms; it is useful to the officers who wish to be assigned to aeronautical units and to the young people who want to fulfill their military duties there; it indeed fulfills the theoretical requirements for a military pilot licence [*Séances*, 1913, pp. 143–144].

In 1912, the IAM already counted 37 students, among whom several were in possession of an engineering diploma from the Electrical Engineering and Applied Mechanics Institute and 8 were noncommissioned officers or officers on active duty. In 1913–1914, the War Ministry sent three lieutenants (Frelut, Guillemin, and Rigaut) to follow the instruction of the institute. In the absence of its own building at the Villers-les-Nancy aerodrome, the IAM provided courses that took place in the Mathematics and Physics Institute. Floquet was in charge of general lectures on aeronautics. Rothé taught aerodynamics, meteorology, and wireless telegraphy. The mathematics professor Édouard Husson (1872–1959) lectured on airplane balance and on the different types of flight. Course curriculum also included lectures on engines and screw-propellers (by the professor of applied mechanics Ernest Hahn), on strength of materials (by Antoine Daniel Dumas, an engineer who had graduated from the Eidgenössische Technische Hochschule Zürich (ETH) in Zürich), on electrical engines (by Alexandre Mauduit), and on meteorology (by Auguste Marsal, a

---

partie des fonds de la souscription au développement de cette organisation. C'est ainsi que déjà la ville d'Arras a élevé un hangar sur le terrain de manœuvres, que la ville de Bar-le-Duc a acheté un terrain et y a fait édifier des hangars, qu'à Nancy le Comité local se propose également d'avoir son terrain, et ses hangars, etc. Il suffit d'encourager ces mouvements, de les canaliser afin d'avoir ces terrains et ces hangars jalonnant les principaux parcours à travers la France pour posséder un outil aérien tout à fait remarquable" [SHD–Air, AA 13163, doc. AA5].

physics teacher at the *lycée* of Nancy). Pol Simon[13] and Maurice Guéritot[14] offered some physics and mathematics classes. The training at the IAM was completed by several demonstrations on the Villers-les-Nancy airfield where airplanes and kites were flown and working engines were examined. On the airfield, Captain Marie-Joseph Loubigniac (1877–1937) was praised by Hirschauer for his organization of the aviation center of Épinal; he later moved to Nancy, where he gave inspirational talks to the students [*Séances*, 1913–1915].

From the start, Rothé believed that the teaching of aerodynamics had to be coupled with the teaching of meteorology. According to him, both sciences had strong historical and conceptual relationships. Inasmuch as the aim of IAM was to train aviators and plane designers and builders, this association seemed all the more pertinent. In the introduction to the third volume of his *Cours de physique*, which was devoted to aerology and aerodynamics, Rothé recalled that in his lectures in Nancy he used to say:

> Today it is a duty for physicists to direct their attention to the physics of the atmosphere and to contribute to aviation by pursuing new research—research that will only be successful through the tight collaboration between the aviator and the physicist [Rothé 1917, p. vii].

To solve problems raised by the development of flight, Rothé pleaded for a methodological and experimental shift from pure mechanics to meteorology. "We need to look for solutions to the problems aviators will pose to us. In particular, the atmosphere needs to be mapped out, just as the seas were" [Rothé 1917, p. vii]. His point of view was far from the romantic visions that prevailed at the beginning of aviation. He emphasized the fact that most of the heavier-than-air pioneers were

---

[13]Pol Simon (1880–1922) was a pure product of the Nancy Faculty of Sciences. After his high school degree in science (*baccalauréat ès sciences*), Simon was appointed chief assistant in mathematics (*chef de travaux en mathématiques*) in 1905. This was a newly created position which was meant to solve the problem raised by the significant increase in foreign students. In addition to his teaching at the Faculty of Sciences, Simon taught mathematics at the Chemical Institute and at the Electrotechnical Institute. At the beginning of the war, he served as a lieutenant-instructor in Autun at the 69th Infantry Regiment where he organized a cycle of conferences for wounded soldiers. Some of these conferences were published in 1916 in a collective book entitled *Civilisation et "Kultur"*, which was a direct answer to the call of German intellectuals to civilized nations (known as the Manifesto of the 93; see the introduction by David Aubin and Catherine Goldstein to this volume). He was sent to the front in October 1915 and ended the war with the rank of captain and a summons to the Order of the Army. His memories of the front in 1917 were published in a book entitled *Ligne des lignes, impressions d'un combattant (Octobre 1915–Juillet 1917)*. After the war he went back to teach mathematics in Nancy, in particular at the Metallurgical and Mining Institute of Nancy, a newly founded engineering school (1919) [Rollet forthcoming].

[14]Maurice Guéritot (1885–1923) was an assistant in physics in Nancy and he taught aerodynamics at the IAM before the war. Together with Rothé, he published two articles in the *Comptes rendus de l'Académie des sciences*, one dealing with wireless telegraphy [Rothé & Guéritot 1913] and the other dealing with the determination of the metacentric curves of the airplane [Rothé & Guéritot 1914]. During the war, Guéritot spent the first year at the front, assigned to the 8th Engineering Corps under the orders of Colonel Gustave Ferrié (then inspector of the military telegraph). He was then transferred to the staff of the 1st Army Corps where he worked on the adaptation of wireless telegraphy to planes. From October 1918 to March 1919, he was sent to Toulon to test remote control devices for airplanes and boats. After this period in the army, he went back to civilian life and in March 1919 was hired as an engineer by the Gaumont Company in Paris, where he stayed until his death in April 1923 [Bolmont forthcoming].

engineers and that the first aeronautic feats relied on prior studies concerning birds, gliders, or kites. Rothé therefore insisted on the experimental nature of his aerodynamic teachings, and only the mechanical notions necessary for the understanding of the experimental method were introduced in his book.

But Rothé was clearly conversant with more sophisticated works on air resistance by Gustave Eiffel, Nikolai Zhukovsky, Tullio Levi-Civita, and Henri Villat, and he himself proposed a new theory of air flow around surfaces. He discussed the fundamental formula for the theory of air resistance:

$$R = KSV^2,$$

where R was the resistance, S the cross-section, and V the speed of the air flow. The constant K depended on atmospheric conditions, such as temperature, pressure, and hygrometry. The introduction of this constant in the fundamental law of air resistance, Rothé argued, showed the crucial interdependence of aerodynamics, meteorology, and flight theory, since characteristic features of an airplane had to be measured under standard atmospheric conditions:

> The workings of the motor-propeller system depend on these factors [pressure, temperature, and hygrometry] to such an extent that airplane trials ought not to be done under any condition; they should be reduced to ideal conditions [or] to a so-called standard atmosphere [Rothé 1917, p. x].

A few weeks before the outbreak of WWI, Floquet and Rothé were very confident about the future of the IAM. The project had received unanimous support from the university and the city of Nancy, from major regional economic leaders, as well as from the Ministries of War and of Public Instruction. On June 1, 1914, the chief education officer, Charles Adam, and the dean of the Faculty of Sciences, Floquet, approved an estimate by university architect Émile André. The construction of buildings for the IAM would cost 18,500 francs, an amount to be covered by the 20,000 francs Rothé had secured from the town council. The war put an end to the project, and the building never saw the light of day.

3. The Nancy Mathematicians during the War

In the fall of 1914, the declaration of war brought the university almost to a standstill. In 1915, the Faculty of Sciences counted that it had lost the services of 12 of its 22 professors.[15] Almost all of the mathematics teachers were likewise mobilized. Out of the seven professors and instructors of mathematics in Nancy (leaving aside a few physicists such as Albert Grumbach, who was sometimes involved in the teaching of mathematics), only two seem to have remained active. In 1914, there were four chairs of mathematics in Nancy: the chair of analysis was held by Floquet, the chair of applied mathematics by Vogt, and the chairs of rational mechanics and of differential and integral calculus both by Husson.[16] In addition to

[15] The situation was comparable in the other faculties: 7 professors out of 17 in the Faculty of Law, 5 out of 18 in the Faculty of Medicine, 7 out of 14 in the Faculty of Arts, and 9 out of 10 in the *École supérieure de pharmacie* were no longer available for teaching [*Séances*, 1916].

[16] Husson had just been elected to the chair of rational mechanics when he was mobilized. As a consequence, he kept the formal responsibility for his old chair for the duration of the war, during which the chair of rational mechanics was unoccupied.

the three professors, four part-time lecturers were in charge of courses in the technical institutes: a professor at the *École professionnelle de l'Est*, Ernest Bertrand (1873–1949), was responsible for practical training at the Institute of Electrical Engineering (from 1906 to 1920); a teacher at the *lycée* of Nancy (to be named the Lycée Henri Poincaré in 1913), Lucien Chanzy, was in charge of practical training, of mathematical courses and graphical works at the Chemical Institute from 1903 to 1908 and then at the Electrotechnic Institute until 1920; Simon, a graduate from the Faculty of Sciences, replaced Chanzy at the Chemical Institute; and, as already mentioned, the engineer Louis Roy had been hired in 1913 as chief assistant (*chef de travaux pratiques et chargé de cours de mathématiques*).[17]

While the overall picture is clear, it is rather difficult to get precise information about the wartime activities of the mathematics teachers of Nancy. According to the official chart of the university [*Séances*] and to administrative decrees of nominations, Bertrand was no longer on the faculty after November 1914. Roy probably had to leave the faculty at the end of 1915 or at the beginning of 1916. Were they sent to the army? This hypothesis is indeed plausible, but one could also suppose that their part-time lectureships were suppressed due to a lack of students. In the absence of precise information about the two instructors, it is reasonable to assume that they were more than 40 years old and therefore not automatically mobilizable. Despite his poor health, Simon enlisted in the army at the beginning of the conflict and served as a lieutenant and instructor in the infantry. But he also saw the front since, after the war, he was mentioned in dispatches for his courage in battle. As for Chanzy, his presence in Nancy seems to be affirmed as late as 1916.[18] At the beginning of the war, Husson was on active duty as sergeant in the territorial infantry. Assigned to Toul near Nancy, he most probably worked part-time at the university until November 1916.[19] Husson was later named technical attaché to an aviation center and finally joined the Aviation School in Pau as technical instructor. As for the physicists, the professor of electrical engineering, Mauduit, was also mobilized in Toul, while Edmond Rothé joined the radiographic service of the military hospital in Nancy (and went on with his teaching at the onset of the war). Some professors of the Faculty of Sciences were also involved in the invention and development of technical devices for the army: Camille Gutton built a grenade detector and Rothé built a field radiographic car and an anemometer; Guéritot, who was in charge of practical training in electricity, developed with Rothé a mechanism for the use of wireless telegraphy in airplanes; Blondlot, who was then professor emeritus, worked on periscopes....

In practical terms, this meant that Vogt and Floquet, who were from the older generation, therefore remained alone in charge of all the courses during the war. In 1914–1915, Vogt gave more than eight lectures per week and Floquet took charge

[17] A graduate from the Higher School of Electricity (*École Supérieure d'électricité*), Roy specialized in rational and applied mechanics after the war and pursued a brilliant career at the University of Toulouse.

[18] But the official charts of the university contain many errors. For instance, Simon's name is listed until 1917 although he was in the army beginning in 1914.

[19] His name is mentioned in the *Séances de rentrée de l'université de Nancy* [*Séances*], but his courses seem to have been taught by Floquet.

TABLE 1. Number of certificates delivered in mathematics and physics at the Nancy Faculty of Sciences (1900–1920): A (differential and integral calculus); B (general mathematics); C (higher analysis); D (higher algebra); E (higher geometry); F (astronomy); G (rational mechanics); H (applied mechanics); I (general physics); J (applied physics). Source of the data: [*Séances*].

Year	A	B	C	D	E	F	G	H	I	J
1900	3		0	0	3	4			9	10
1901	16		0	1	0	1			2	1
1902	16		2	0	1	3			8	3
1903	16		0	0	0	6			10	5
1904	22		1	2	0	4			7	11
1905	14		4	2	0	2			16	8
1906	20		0	1	3	0			9	8
1907	29		4	0	1	9			20	8
1908	0	29	4	0	0	4		6	15	14
1909	8	10	6	0	0	1		7	12	12
1910	3	18	4	0	0	0		5	11	19
1911	3	21	4	0	0	2	4	8	13	7
1912	5	22	4	0	0	5	7	7	20	10
1913	11	14	7	0	2	7	9	4	11	13
1914	4	14	5	0	3	3	9	7	11	0
1915	5	10	3	0	0	3	1	0	2	0
1916	2	7	2	0	0	0	6	2	2	1
1917	3	9	0	1	0	3	0	5	2	0
1918	1	0	0	0	0	0	0	0	1	0
1919	4	19	0	0	0	0	8	7	0	0
1920	6	96	0	0	3	0	10	19	10	6

of differential and integral calculus.[20] In 1915–1916, Floquet taught rational mechanics and advanced analysis and Vogt taught general mathematics. During the next academic year, Floquet taught differential and integral calculus as well as astronomy and Vogt kept the charge of general mathematics. Vogt also gave courses in applied mathematics and provided (perhaps with Chanzy) practical instruction in mathematics.

Besides losing the services of its professors, the university also witnessed a sharp drop in the number of its students. In 1913–1914, there were 1,089 students registered at the Faculty of Sciences, with a large proportion of foreign students. A year later, there were only 114 registered students, and not a single foreign student was left (see Figure 1).[21] And this was not without repercussions on the number of diplomas and certificates delivered (see Figure 2 and Table 1).

[20]At the beginning of the 1914–1915 academic year, all the courses where some students registered were open. For lack of students, the courses on rational mechanics, astronomy, aerodynamics, and brewery were cancelled.

[21]In 1914–1915, the Faculty of Law had 66 students, the Faculty of Medicine 27, the Faculty of Arts 18, and the *École supérieure de pharmacie* 25 [*Séances*, 1916].

Despite the sharp decrease in the number of teachers and students, university officials issued self-congratulatory statements about the way they had adapted to the situation until the end of 1915. Speeches praised professors who served in the army and those who, because of their age, had stayed in Nancy, on the home front. Some saw the war as an opportunity for the university, as shown by the following quotation from Arthur Meyer, chief technician of the physics laboratory, who declared at the start of the academic year in November 1915:

> Far from us the thought of lamenting this temporary decrease: those who have left today fulfill the noblest of duties; while the deans have been unanimously satisfied with the work and assiduity of those who were allowed to stay. Results were excellent. Often, the usual lecture has almost become a personal tutorial to the students' great benefit; practical training sometimes resembles research experiments that the master carries out with his laboratory assistants.... Sometimes the lecture ... is momentarily interrupted by the sound of the alarm and the firing that signals the incursion of airplanes; once the incident has passed, the session goes on; and one could not insist too much on the quiet firmness with which our students have adapted to the extraordinary conditions of university life in Nancy.
>
> This quick sketch of our nonmobilized colleagues' activities would not be complete if the annual report did not insist on the part they are taking, besides their [normal] occupations, to work more directly concerning national defense, on special committees, in assistance organizations, in ... hospitals, in the technical services of the institutes on questions of chemistry, radiography, aerostation, meteorology,... [*Séances*, 1916].

Fortified by its past achievements, hopeful for its future, the university maintained its faith in its mission and contributed to the establishment of a moral atmosphere "purified" from past troubles.[22] Until the end of 1915, the material conditions at the Faculty of Sciences apparently remained satisfactory.

When a long-range 380 mm quick-loading cannon was set up in Hampont (Moselle) in January 1916, however, the situation changed considerably. This cannon, a Krupp type SK-L/45 Max (called "*le gros Max*" by Nancy's inhabitants) targeted Nancy and Lunéville. Nancy moreover increasingly suffered severe air shelling by airplanes and dirigibles. Between 1914 and 1918, 101 bombardments resulted in 181 dead and 310 injured in Nancy [Anonymous 1918]. These attacks notwithstanding, official speeches retained an optimistic tone. In his report, biologist Lucien Cuénot explained that Nancy had effectively received several hundred bombs but consequences for the Faculty of Sciences had been limited: while some buildings had suffered important degradations (broken glass at the Chemical Institute, at the Physics Institute, and at the School of Brewery), the *baccalauréat* examinations had to be held in the cellars of the Faculty of Sciences. Fortunately, no loss occurred among the students and the professors.

What Lucien Cuénot could not foresee was that much destruction at the university would occur over the two years that followed his rather upbeat speech. As

[22]This, at least, was the wish expressed in November 1914 by the law professor François Geny [*Séances*, 1915].

FIGURE 4. The university library after the fire. Photo Dufey, repr. from *Nancy sous les obus, images de guerre*, Albert Barbier imprimeur, Nancy, 1918. Bibliothèque municipale de Nancy.

early as February 11, 1917, the Minister of Education decreed that the University of Nancy be closed due to the bombings. In 1917, the laboratory of electrical engineering was bombed and suffered nearly complete destruction. On October 31, 1918, two incendiary bombs hit the university library destroying 56,000 out of its 113,000 books (see Figure 4).

The last few months of conflict were therefore to have catastrophic consequences for all the faculties of the University of Nancy. After the Armistice the physical destruction put a heavy financial burden on the faculties and the engineering school, which received state funding that remained identical to that of 1914. This funding had in fact remained stable for more than thirty year, but inflation from the war and buildings that had to be rebuilt now put considerable stress on university officials who had to manage it. Financial strain was compounded by a severe drop in personnel since many were lost to the war (see Figure 5). Student enrollment had moreover steeply decreased and their fees (especially those paid by foreign students) were insufficient to cover expenditures for reconstruction.

At the end of the war, the city urged the university to reopen as quickly as possible and to follow Lille's example, where war destruction due to German occupation was far more important. On December 14, 1918, the city council decreed that the university was to be reopened within 15 days. Furthermore, it enjoined the Faculty of Arts to restart French courses for foreign students and asked the government for a quick demobilization of the professors. This demobilization was very disappointing for Nancy because of its slowness and also because of the return

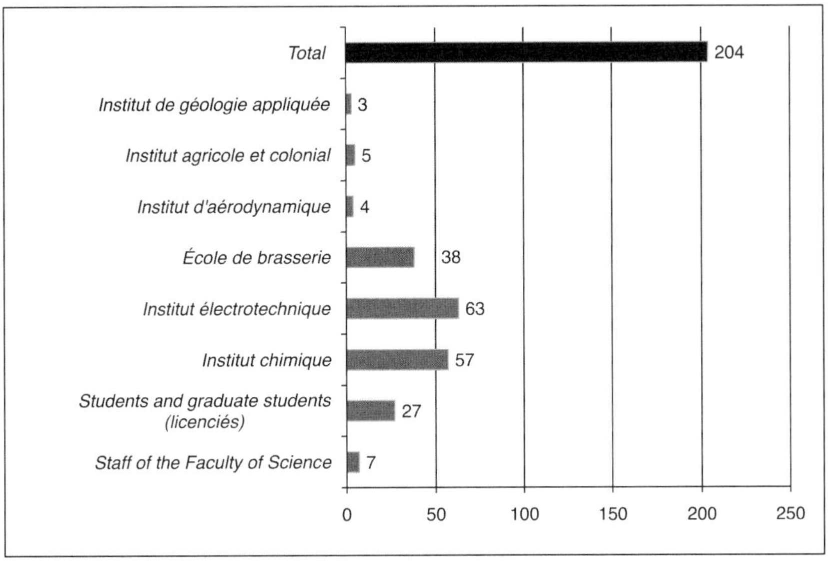

FIGURE 5. The human losses due to war for the Nancy Faculty of Sciences.

of Alsace-Moselle to the French fold. The reopening of the University of Strasbourg offered very interesting opportunities for many professors from Nancy. As a consequence, Paul Müller (physical chemistry), Louis Hackspill (industrial chemistry), and Edmond Rothé chose to pursue their careers in Strasbourg.

The war surely had underscored the importance of aerodynamics and meteorology for the modern, warring state. But with Rothé gone to Strasbourg, this hardly translated into renewed excitement for the Aerodynamic and Meteorological Institute. The building planned before the war had not been built, and the existence of the institute, merely virtual, was made manifest only by the few courses it offered. Now intent on developing its own network of airfelds and meteorological stations, the air force deserted the Villers aerodrome. The SAA was moribund. Although it could still boast 214 members, 109 of them voted in favor of dissolution on July 9, 1920. Floquet was appointed as liquidator and donated the 10,000 francs left in its accounts to the Faculty of Sciences [ADMM, 4M86]. Floquet died on October 7, 1920, and with him, the IAM vanished.

Conclusion

The period from 1914 to 1918 marked a turning point for the University of Nancy. The *Belle Époque* which had been characterized by strong intellectual effervescence and modernity claims was followed by a difficult period. After 1918, the Faculty of Sciences faced serious obstacles, such as the departure of many professors for more attractive positions in Strasbourg. These departures weakened Nancy's scientific institutions which, during the interwar years, had to cope with a sharp demographic drop among students and severe financial crises.

An important consequence of the war was the emergence of new strategies centered on the development of engineering training more specifically geared toward local or regional needs. Before 1914, Bichat, Haller, and Floquet fervently stressed the patriotic dimensions of the applied sciences, and aeronautics especially. These developments then seemed to be part of an avant-garde national and international strategy. After the war, ambitions took a more modest turn. Following the recommendations of the ministry enjoining universities to reinforce their links with the local industry, a new Metallurgy and Mining Institute (*Institut métallurgique et minier*) was created at the Faculty of Sciences in 1919. The new dean of the Faculty of Sciences Paul Petit stressed the fact that the new institute was meant to fulfill local needs, that it was partly financed by the Nancy Board of Trade and Industry, and that the national government only contributed the initial motivation but none of the funding for it.[23]

The historical sketch presented here highlights the crucial role played by several individuals. Floquet's death and Rothé's departure for Strasbourg definitively closed the aeronautics parenthesis in Nancy. In 1918 only older and more solid academic institutions were able to survive; this was the case for most of the engineering training institutes created between 1880 and 1910 but not for the more recently created IAM. In 1929, when a national drive in favor of aerodynamical research in French universities was sponsored by the recently established Ministry of Air, the Nancy Faculty of Sciences asked for a chair of fluid mechanics to be established in the name of the moribund IAM, but that obviously came too late: Nancy was not selected.

Abbreviations

- ETH: Eidgenössische Technische Hochschule Zürich.
- IAM: Institut d'aérodymanique et de météorologie.
- SAA: Société des amis de l'aviation.

Bibliography

Archival Sources

[ADMM] Archives départementales de Meurthe-et-Moselle, Nancy.
[Nancy Arch.] Archives municipales de Nancy.
[SHD–Air] Service historique de la défense, Armée de l'air, Vincennes.

Published Works

[Anonymous 1918] Anonymous, *Les bombardements de Nancy, du 4 septembre 1914 au 31 octobre 1918*, Nancy.

[Appell 1925] Paul Appell, *Henri Poincaré*, Plon, Paris, 1925.

[Appell et al. 1913] Paul Appell et al., *Hommage à Henri Poincaré*, Association des anciens élèves des lycées de Nancy, Metz, Strasbourg et Colmar, Imprimerie Moderne A. Colin, Nancy, 1913.

[Birck 2006] Françoise Birck, De l'institut électrotechnique de Nancy à l'École nationale supérieure d'électricité et de mécanique (1900–1960), in [Birck & Grelon 2006, pp. 23–88].

[Birck 2013] Françoise Birck, *L'École des mines de Nancy (ENSMN), 1919–2012 : entre université, grand corps d'État et industrie*, Presses universitaires de Lorraine, Nancy, 2013.

[23]Petit stated: "*Jusqu'ici l'État n'est intervenu en rien, ni pour la première installation, ni pour le fonctionnement de l'Institut métallurgique et minier ; nous espérons que cette création semblant répondre tout à fait aux vues exprimées par M. le Ministre de l'Instruction publique, celui-ci voudra bien accorder largement son appui financier au nouvel Institut, comme aussi aux autres Instituts, constitués dans la même pensée d'adaptation aux besoins de la région*" [*Séances*, 1919]. For a detailed history of the Nancy Mining Institute, see [Birck 2013].

[Birck & Grelon 2006] Françoise Birck & André Grelon, eds., *Un siècle de formation d'ingénieurs électriciens, ancrage local et dynamique européenne : L'exemple de Nancy (actes du colloque organisé en 2001 à Nancy à l'occasion du centenaire de l'ENSEM)*, Éditions de la Maison des Sciences de l'Homme, Paris, 2006.

[Bolmont forthcoming] Étienne Bolmont, Maurice Guéritot (1885–1923), in *Dictionnaire biographique des enseignants de la Faculté des sciences de Nancy*, forthcoming.

[Chimello 2005] Sylvain Chimello, *L'épopée des dirigeables en Lorraine 1900–1918*, Éditions Serpenoise, Metz, 2005.

[Choffel-Mailfert 2007] Marie-Jeanne Choffel-Mailfert, Architecture et organisation spatiale du patrimoine universitaire nancéien (1770–1930), in [Rollet & Choffel-Mailfert 2007, pp. 17–119].

[CPIA 1909] Commission permanente internationale d'aéronautique (CPIA), *IVe congrès international d'aéronautique, Nancy, 18–23 septembre 1909 : Procès-verbaux, rapports et mémoires*, Dunod & Pinat, Paris, 1909.

[Crawford & Olff-Nathan 2005] Elisabeth Crawford & Josiane Olff-Nathan, eds., *La Science sous infuence: l'université de Strasbourg, enjeu des conflits franco-allemands, 1872–1945*, La Nuée Bleue, Strasbourg, 2005.

[Dusart & Moulin 1998] Alain Dusart & François Moulin, *Art nouveau, l'épopée lorraine*, Éditions de l'Est, Strasbourg, 1998.

[Fontanon 2003] Claudine Fontanon, Paul Painlevé et l'aviation : aux origines de l'étatisation de la recherche scientifique, in *Paul Painlevé (1863–1933): un savant en politique*, ed. Claudine Fontanon & Robert Frank, Presses Universitaires de Rennes, Rennes, pp. 41–56.

[Gouzévitch & Gouzévitch 2006] Irina Gouzévitch & Dmitri Gouzévitch, Les étudiants d'Europe de l'Est à l'Institut électrotechnique de Nancy (1900–1939), in [Birck & Grelon 2006, pp. 271–319].

[Grelon & Birck 1998] André Grelon & Françoise Birck, eds., *Des ingénieurs pour la Lorraine 19ème-20ème siècles*, Éditions Serpenoise, Metz, 1998; repr. Presses Universitaires de Nancy, Nancy, 2007.

[Nabonnand 2006] Philippe Nabonnand, La création du certificat de mathématiques générales à Nancy, in [Birck & Grelon 2006, pp. 123–154].

[Nye 1986] Mary Jo Nye, *Science in the Provinces: Scientific Communities and Provincial Leadership in France, 1860–1930*, University of California Press, Berkeley, 1986.

[Quinet 1872] Edgar Quinet, *La République, condition de la régénération de la France*, E. Dentu, Paris, 1872.

[Rollet & Choffel-Mailfert 2007] Laurent Rollet & Marie-Jeanne Choffel-Mailfert, eds., *Aux origines d'un pôle scientifique: faculté des sciences et écoles d'ingénieurs à Nancy du Second Empire aux années 1960*, Presses Universitaires de Nancy, Nancy, 2007.

[Rollet forthcoming] Laurent Rollet, Pol Simon (1880–1922), in *Dictionnaire biographique des enseignants de la Faculté des sciences de Nancy*, forthcoming.

[Rollet & Nabonnand 2011] Laurent Rollet & Philippe Nabonnand, La carrière du mathématicien Gaston Floquet à Nancy : le choix de l'Est, in *Espaces de l'enseignement scientifique et technique*, ed. Virginie Fonteneau & Renaud d'Enfert, Hermann, Paris, 2011, pp. 137–151.

[Rothé 1914] Edmond Rothé, *L'historique des hangars d'aviation de Nancy qui seront inaugurés dimanche*, Est républicain (Nancy) (July 11, 1914).

[Rothé 1917] Edmond Rothé, *Cours de physique professé à la Faculté des Sciences de Nancy, 3ème partie, aérologie et aérodynamique*, Gauthier-Villars, Paris, 1917.

[Rothé & Guéritot 1913] Edmond Rothé & Maurice Guéritot, *Sur une méthode permettant d'effectuer des essais réduits en télégraphie sans fil*, Comptes rendus hebdomadaires des séances de l'Académie des sciences **157** (1913), p. 370.

[Rothé & Guéritot 1914] Edmond Rothé & Maurice Guéritot, *Sur une méthode expérimentale de détermination des courbes métacentriques de l'aéroplane*, Comptes rendus hebdomadaires des séances de l'Académie des sciences **159** (1914), p. 361.

[Saint-Martin 2008] Arnaud Saint-Martin, *L'Office et le télescope : une sociologie historique de l'astronomie française, 1900–1940*, doctoral thesis, University Paris–Sorbonne, 2008.

[*Séances*] *Séances de rentrée de l'université de Nancy*, Imprimeries de l'Est, Nancy, 1900–1921.

[Sicard-Lenattier 2002] Hélène Sicard-Lenattier, *Les Alsaciens Lorrains à Nancy, 1870–1914*, éd. Gérard Louis, Haroué, 2002.

Index

Selected Published Titles in This Series

42 **David Aubin and Catherine Goldstein, Editors,** The War of Guns and Mathematics, 2014

41 **Christopher Hollings,** Mathematics across the Iron Curtain, 2014

40 **Roman Duda,** Pearls from a Lost City, 2014

39 **Richard Dedekind and Heinrich Weber,** Theory of Algebraic Functions of One Variable, 2012

38 **Daniel S. Alexander, Felice Iavernaro, and Alessandro Rosa,** Early Days in Complex Dynamics, 2011

37 **Henri Poincaré,** Papers on Topology, 2010

36 **Éric Charpentier, Étienne Ghys, and Annick Lesne, Editors,** The Scientific Legacy of Poincaré, 2010

35 **William J. Adams,** The Life and Times of the Central Limit Theorem, Second Edition, 2009

34 **Judy Green and Jeanne LaDuke,** Pioneering Women in American Mathematics, 2009

33 **Eckart Menzler-Trott,** Logic's Lost Genius, 2007

32 **Jeremy J. Gray and Karen Hunger Parshall, Editors,** Episodes in the History of Modern Algebra (1800–1950), 2007

31 **Judith R. Goodstein,** The Volterra Chronicles, 2007

30 **Michael Rosen, Editor,** Exposition by Emil Artin: A Selection, 2006

29 **J. L. Berggren and R. S. D. Thomas,** Euclid's *Phaenomena*, 2006

28 **Simon Altmann and Eduardo L. Ortiz, Editors,** Mathematics and Social Utopias in France, 2005

27 **Miklós Rédei, Editor,** John von Neumann: Selected Letters, 2005

26 **B. N. Delone,** The St. Petersburg School of Number Theory, 2005

25 **J. M. Plotkin, Editor,** Hausdorff on Ordered Sets, 2005

24 **Hans Niels Jahnke, Editor,** A History of Analysis, 2003

23 **Karen Hunger Parshall and Adrian C. Rice, Editors,** Mathematics Unbound: The Evolution of an International Mathematical Research Community, 1800–1945, 2002

22 **Bruce C. Berndt and Robert A. Rankin, Editors,** Ramanujan: Essays and Surveys, 2001

21 **Armand Borel,** Essays in the History of Lie Groups and Algebraic Groups, 2001

20 Kolmogorov in Perspective, 2000

19 **Hermann Grassmann,** Extension Theory, 2000

18 **Joe Albree, David C. Arney, and V. Frederick Rickey,** A Station Favorable to the Pursuits of Science: Primary Materials in the History of Mathematics at the United States Military Academy, 2000

17 **Jacques Hadamard,** Non-Euclidean Geometry in the Theory of Automorphic Functions, 2000

16 **P. G. L. Dirichlet and R. Dedekind,** Lectures on Number Theory, 1999

15 **Charles W. Curtis,** Pioneers of Representation Theory: Frobenius, Burnside, Schur, and Brauer, 1999

14 **Vladimir Maz'ya and Tatyana Shaposhnikova,** Jacques Hadamard, A Universal Mathematician, 1998

13 **Lars Gårding,** Mathematics and Mathematicians, 1998

12 **Walter Rudin,** The Way I Remember It, 1997

11 **June Barrow-Green,** Poincaré and the Three Body Problem, 1997

10 **John Stillwell,** Sources of hyperbolic geometry, 1996